T0142309

Studies in Computational Intelligence

Volume 974

The series "Studies in Computational Intelligence" (SCI) publishes new developments and advances in the various areas of computational intelligence—quickly and with a high quality. The intent is to cover the theory, applications, and design methods of computational intelligence, as embedded in the fields of engineering, computer science, physics and life sciences, as well as the methodologies behind them. The series contains monographs, lecture notes and edited volumes in computational intelligence spanning the areas of neural networks, connectionist systems, genetic algorithms, evolutionary computation, artificial intelligence, cellular automata, self-organizing systems, soft computing, fuzzy systems, and hybrid intelligent systems. Of particular value to both the contributors and the readership are the short publication timeframe and the world-wide distribution, which enable both wide and rapid dissemination of research output.

Indexed by SCOPUS, DBLP, WTI Frankfurt eG, zbMATH, SCImago.

All books published in the series are submitted for consideration in Web of Science.

More information about this series at http://www.springer.com/series/7092

Abdalmuttaleb M. A. Musleh Al-Sartawi
Editor

The Big Data-Driven Digital Economy: Artificial and Computational Intelligence

Editor
Abdalmuttaleb M. A. Musleh Al-Sartawi
Accounting Finance and Banking
Ahlia University
Manama, Bahrain

ISSN 1860-949X ISSN 1860-9503 (electronic)
Studies in Computational Intelligence
ISBN 978-3-030-73059-8 ISBN 978-3-030-73057-4 (eBook)
https://doi.org/10.1007/978-3-030-73057-4

This Springer imprint is published by the registered company Springer Nature Switzerland AG
The registered company address is: Gewerbestrasse 11, 6330 Cham, Switzerland

Preface

This book presents chapters that discuss contemporary issues related to the digital economy, mainly in relation to the challenges and opportunities by artificial intelligence and computational intelligence in big data-driven digital economies. The main theme of the book arises from the dynamic relationship between the advancement in information and communication technologies and the development of real-world issues that has been continuously accelerating the progress of the digital economy. As such, data has become an increasingly valuable driver transitioning societies to post-Fordist data-driven economies characterized by rapid innovation and the high value placed on education. Today, big data is an economic resource not unlike oil; those who are able to collect, retain, transform and use data are in power. Therefore, intelligence gathered by the process of analysing and learning from data improves our knowledge through artificial intelligence (AI) and computational intelligence (CI). CI systems form a set of human information processing methodologies which have been developed to address complex real-world big data-driven problems which can no longer be solved and optimized using traditional techniques. It a major role in developing successful intelligent systems, including games and cognitive developmental systems. Recently, there more attention has been given to issues such as convolutional neural networks as it has become the fundamental basis of AI, where nearly most successful AI systems are based on CI.

In lieu of the United Nation's efforts to spread the benefits of the digital economy globally, this book discusses and debates the implications of both AI and CI in the era of the digital economy and provides a holistic perspective on a wide range of topics including intelligence education, economics, finance, sustainability, ethics, governance, cybersecurity and knowledge management. Moreover, the chapters investigate various computational intelligence concepts which are currently shaping the digital economy such as artificial intelligence and blockchain. These complex and multifaceted technologies have implications on businesses as well as engineering in the wave of the Fourth Industrial Revolution, whereby the combination of the two provides the depth and breadth of the digital economy. In addition, due to the high premium placed on education, the book looks into AI and CI approaches and algorithms for developing the digital learning and intelligent pedological systems.

Due to the holistic nature of the topics, this book is of interest to economists, financial managers, engineers, ICT specialists, digital managers, data managers, policymakers, regulators, researchers, academics and students. What differentiates this book from others is that it brings together two important areas, intelligence systems and big data in the digital economy, with special attention given to the opportunities, challenges and threats for education, business growth and economic progression of nations. It hereby focuses on how societies can take advantage and manage data, as well as the limitations they face due to the complexity of resources in the form of digital data and intelligence. Unlike other books, the chapters offer an in-depth insight into approaches which will support economic development strategies and add to the efforts made by the UN towards achieving their goal.

Organization of the Book:

The Call for Papers asked for submissions in one main category, i.e. full research papers, and accordingly attracted many submissions from around the world. Submissions were reviewed in a double-blind process by academics in the relevant fields.

The book is organized into thirty-four chapters and covers various topics as follows. Chapter 1 evaluates how big data advancements could positively impact the supply chain inefficiencies. It also discusses what challenges organizations have to face for implementing big data framework in their organization structure. The chapter takes into consideration the issues of global supply chain vis-à-vis the Indian Supply Chain Industry.

Chapter 2 aims to critically present a comprehensive literature to review the mobile computing for enhanced living environments and mobile health applications published from 2016 to 2020. The main contribution of this paper is to present a comprehensive literature review on mobile computing technology applicability to support older adults and to promote the overall public health and well-being.

Additionally, Chapter 3 discusses an approach based on the genetic folding (GF) algorithm to discriminate between positive and negative Arabic reviews in social media. Sentiment analysis is a set of techniques for collecting subjective opinions from a document.

On the other hand, Chapter 4 examines the e-learning system, Moodle, adoption by the students in three Gulf universities "Saudi Arabia, United Arab Emirates and Kingdom of Bahrain" and examines the students' perceptions towards this system from the framework of two theories: the theory of reasoned action (TRA) and the theory of planned behaviour (TPB).

Chapter 5 gives us another perspective on a company's earnings per share in the digital economy in a developing country. Accordingly, the chapter examines whether relational capital (RC) affects the relationship between intellectual capital (IC) and earnings per share (EPS), for 50 industrial companies on the Amman Stock Exchange (ASE).

Along similar lines, Chapter 6 discusses earnings management practices and then rereviews the previous studies that discussed the role of machine learning in uncovering earnings management practices. The study also suggests future research directions.

Chapter 7 was able to critique literature and blend two schools of thoughts: artificial intelligence and IT governance to comprehend the role of one over the other. Seldomly, these two research domains have been addressed in the literature: through an important area of discussion. Also, implications and future research agendas are also depicted in this paper. Such a research topic is unique and significant since this opens the doors to a new phenomenon in financial performance.

Moreover, Chapter 8 presents a literature review paper which proposes a conceptual framework for decision-making style within virtual platforms for improving service quality. Thus, making one proposition viable for future quantitative empirical evidencing, bearing theoretical and managerial implications.

Similarly, Chapter 9 aims to express a review of current literature revealing a want to know the impact of participants' social capital for enhancing their virtually shared knowledge, in turn for improving their readiness to learn as they participate in an online (virtual) community. Furthermore, there is further elaboration on the theoretical and practical implications.

In line with the educational topics in the digital economy, Chapter 10 critically reviews some of the recently published articles tackling the concept and the importance of game-based learning to come up with some practical recommendation that can be used in developing the mechanism of applying the concept and how we can use it during the COVID-19. The study recommends that to develop teaching methods and students' skills as well by investing more in AI techniques which will create more usable games in teaching and learnings during and after COVID-19.

Furthermore, in the digital economy blockchain is a public ledger that is secured and trusted and automatically records and verifies huge number of digital transactions. While paper 11 highlights the importance of blockchain technology and provides an overview on how blockchain can be implemented in the healthcare sector, paper 12 examines how the digitization of financial services can play a major role in increasing inclusive growth and entrepreneurship with a special emphasis on the Indian and Egyptian cases.

Likewise, financial sectors use artificial intelligence (AI) nowadays to manage transactions, invest in capitals and manage housing to achieve well-being. Chapter 13 gives an overview of artificial intelligence during the coronavirus, also known as the COVID-19 pandemic.

On the other hand, Chapter 14 explores the advanced technology of augmented reality (AR) as a digital tool of Industry 4.0. As such, it investigates the potential use of AR for the on-board training of seafarers, as a new technological tool aiming to increase their safety and operational effectiveness.

Moreover, with regard to the gaming industry in the digital economy, Chapter 15 identifies the extent of emotion impact on the consumer electronic game procurement decision. Qualitative research method has been followed in this research via case study strategy. Interviews were conducted with a group of respondents who are used to play electronic games in Omani malls.

Chapter 16 throws light on another study conducted in Oman, where it analyses the role of Higher Education Institutions (HEIs) on the E-Innovation of Oman. It was found that there are several conditions for the expansion of the contribution of universities to regional development of E-Innovation system. These conditions are predominantly related to a broad set of factors that relate to characteristics of HEIs, characteristics of the regional firms, aspects of the collaborative relationship and characteristics of environmental context in which HEIs and firms are embedded.

In the light of the advancements in technology, Chapter 17 aims to predict the foreign currency exchange rate over twenty-two different currencies based on the US dollar. This chapter proposes three machine learning algorithms, such as ridge regression, lasso regression, decision tree and a deep learning algorithm named bidirectional long short-term memory (Bi-LSTM) to predict the foreign currency exchange rate.

With regard to corporate governance issues in the digital economy, Chapter 18 provides a new perspective to social media disclosure. The chapter addresses several research questions, mainly the association between boards of directors' composition: board diversity and board independence, and the level of social media financial disclosure in the UAE.

Chapter 19 explores organizational strategies for the development of digital skills. It analyses the manager's DS perception of a multinational firm present in 18 countries and operating in a manufacturing sector.

Chapter 20 brings to the attention of readers the importance of sustainable development. It discusses the concept of citizen science which refers to a scientific project, managed by researchers, where volunteers are involved in studying and acquiring data related to a natural phenomenon. The chapter contributes to the literature discussing three topics related to citizen science and proposing recommendations to address these actions, based on a literature review. The study has certain implications for practitioners and researchers. In the same vein, Chapter 21 explores the outcomes of smart tourism on-site applications for sustainable tourism.

Similarly, Chapter 21 presents the conceptual framework of financial inclusion with linking to individuals, SMEs and banks as these entities considering as demand and supply sides of financial inclusion. The chapter addresses the definition and main factors of financial inclusion, and also the importance and influence of financial inclusion in economic development and sustainability.

Chapter 23 continues the exploration of the theme of sustainability in relation to smart cities. The chapter determines the various vital functions concerning mobility, water and energy supply, transport networks and waste management are reviewed to respond to new ecological issues and the comfort requirements of city dwellers.

Chapter 24, however, illustrates the issues and opportunities for accounting in the digital economy, whereby it proposes how to use the artificial intelligence practically to automate removing the audit weaknesses. This, in turn, leads to earnings manipulation detection, reduces detection risk and control risk and enhances audit quality by minimizing the risk accounting information and earnings manipulation detection.

Chapter 25 identifies the role of digital transformation in increasing the efficiency of the performance of banks listed on the Palestine Stock Exchange, identifying the digital transformation in banks and the role of digital transformation in increasing the efficiency of bank performance to enhance competitive advantage and achieve a stable financial situation. Meanwhile, Chapter 26 investigates the impact of financial technology on the performance of Bahraini banks.

Moreover, Chapter 27 is a conceptual paper which investigates the effect of behavioural factors on job performance and the mediating effect of employee engagement between behavioural factors and job performance in Jordanian commercial banks.

Chapter 28 combines the two important topics of education and social media. It investigates the effect of WhatsApp on the academic performance of mass communication education students.

Chapter 29 analyses the expected effect of using big data technology on the performance of human resources management in Jordanian universities.

For business growth, it is essential for companies to manage knowledge and integrate external information in individual innovation capacity. As such, paper 30 provides an evaluation of the capacity of individual innovation in the literature of information systems. This chapter reviews the individual innovation framework to reduce any uncertainties by enhancing awareness of individual innovation.

Chapter 31 explores broadband marketing strategy mainly buzz marketing to build customers' relationship management (CRM). The chapter analyses the marketing strategy of virgin, the case study, vis-à-vis university student's needs and requirements.

In the digital economy, the advent of the Internet has revolutionized the procurement's sourcing processes in firms among multiple sectors and industries. It has stimulated cost-effective ways besides various innovative applications that enterprises could leverage from it. Chapter 32 investigates the main factors which are impacting the supply chain performance. These factors are E-procurement, supplier relationship and supplier integration.

Chapter 33 explores virtual reality (VR) in education, particularly in relation to disciplines such as Science, Technology, Engineering and Mathematics (STEM) fields. This chapter critically compares educational VR methods and discusses some challenges of the use of VR in education.

Finally, Chapter 34 investigates the role of artificial intelligence and digital economy tools in developing electronic human resource management systems. This chapter discusses the capabilities and the features of the electronic human resource management systems and the difficulties which can be addressed by adopting artificial intelligence tools.

Abdalmuttaleb M. A. Musleh Al-Sartawi

Acknowledgements

This book discusses a very important issue that links the importance of artificial and computational intelligence with big data in the transformation of economies towards digitalization. This is in line with the sustainable developed goals of nations.

This book would not have been written if not for our inspiration, **Prof. Abdulla Yusuf Al Hawaj**, the Founding President of Ahlia University, the Kingdom of Bahrain. He is the biggest advocate of economic diversification in the digital economy and sustainable technology in education in the Arabian Gulf.

First, I would like to thank all contributing authors for their strong chapter submissions on the increasingly critical topics discussed throughout the book. Their dedication to join the conversation on educational sustainable development is appreciated.

Second, the Editor would like to extend his gratitude to the reviewers for reviewing the chapters. Most of the authors also served as reviewers, and their constructive suggestions are essential to the success of the book.

Finally, I would like to thank the Editor of the book series **Prof. Dr. Janusz Kacprzyk** and the Springer editorial team including Editorial Director **Dr. Thomas Ditzinger** and Senior Editorial Assistant **Mr. Holger Schaepe** for giving me the opportunity to be a member of their global family and for the support attained for developing this book.

Abdalmuttaleb M. A. Musleh Al-Sartawi

Contents

Leveraging Big Data to Accelerate Supply Chain Management in Covid-19

Ruchika Gupta, Priyank Srivastava, Shubham Sharma, and Melfi Alrasheedi

Abstract Pre Covid 19 (CoV), supply chains over the globe were getting progressively mind boggling, on account of globalization and the ever-changing elements of supply and demand. During Covid-19, as a precautionary measure, governments all across world announced full lockdown. This lockdown during CoV, disrupted well established supply chains throughout the world, as movement of people was restricted, factories were shut down, transportation facility came to stand still. With unlock of manufacturing activities in many countries, the organizations are tackling the intensity of technology innovation, or all the more decisively: big data analytics, to acquire troublesome changes at all degrees of Supply Chain. In this context, big data allows vast amounts of organized and unstructured data from multiple sources to be seen rapidly. This will help improve perceptibility for the supply chain, and provide additional bits of information to the entire supply chain. Looking at the potential benefits of big data, the aim of this chapter is to discuss the numerous inefficiencies in the supply chain and their negative effect on the competitiveness of companies. The chapter then evaluates how Big Data advancements could positively impact the supply chain inefficiencies. It also discusses what challenges organizations have to face for implementing big data framework in their organization structure. In the last, it discusses the future prospects for Big Data in supply chain and how Big Data will impact the future of the industry workforce. The chapter takes into consideration the issues of global supply chain vis-à-vis the Indian Supply Chain Industry.

R. Gupta (✉)
Amity Bussiness School, Amity University, Gr. Noida, India
e-mail: rgupta@gn.amity.edu

P. Srivastava · S. Sharma
Department of ME, Amity University, Noida, India
e-mail: psrivastava5@amity.edu

S. Sharma
e-mail: ssharma32@amity.edu

M. Alrasheedi
School of Business Study, King Faisal University, Hofuf, Saudi Arabia
e-mail: malrasheedy@kfu.edu.sa

A. M. A. Musleh Al-Sartawi (ed.), *The Big Data-Driven Digital Economy: Artificial and Computational Intelligence*, Studies in Computational Intelligence 974,
https://doi.org/10.1007/978-3-030-73057-4_1

Keywords Covid-19 · Supply chain management · Big data · Analytics · Big data analytics · Logistics · SCM · Supply chain inefficiencies · Global supply chain · Supply chain optimization · Predictive analytics · Inventory management

1 Introduction

The human history has been plagued with numerous incidents which had tremendous impact on the future transitions of world. These transitions were felt in socio-economics structure of society worldwide. If we take account of past 100 years, there have been catastrophic events of World Wars I and II, Pandemics, Natural Disasters and Man-made disasters. All these events were somehow, proved turning point in the world we live. Every catastrophic event is risk for human civilization. But every "Risk is an Opportunity". If we look in the perspective of pandemics, it is low frequency and high impact (LFHI) event that poses greater challenges to well established supply chains [1–4]. The Covid-19 ripple effect in supply chain is now well established. The decrease in demands of the products, inbound and outbound supply chain issues, low production issues, negative impacts on the international trade and financial markets are some of the impacts of CoV. The chapter has been written in the context of the discrepancies and black spots in supply chain. The CoV situation, now can be seen as a major factor in bringing disruption in supply chain. This situation has made the stakeholders across the supply chain to resort to different internet tool (IT). Looking in the prospects of black spots and framing the suggestion in line with CoV, is the highlight of this chapter.

The market is being increasingly sophisticated and tailored in modern society. In the market, this implies that, in order to satisfy the demands of individuals and companies, it is necessary to adopt a more effective and informed manufacturing approach, optimizing the productivity of all processes involved, dramatically reducing costs and variable output times. In brief, output optimisation. Supply chains are no longer simply mechanisms for keeping track of goods along the line, but they have become a way of gaining a competitive advantage and even building up a brand of their own.

The main problems currently facing supply chains are lack of transparency along the line, as well as the difficulties of tracking the products that go through it. In fact, inefficiencies in a supply chain are costly; it is estimated to bring about a yearly expense of $2 billion in the United Kingdom alone, while in India it is also costing $65 billion annually [5, 6]. Administrators of Supply chain expect to build proficiency, save cost and market pace but battle a dynamic environment, powered by numerous players (manufacturers, suppliers, distributors, etc.), networks (online, ofline, omni) and factors.

Likewise, most information generated during supply chain process cascades beyond the level of only one undertaking or segment which makes assessment through all the more testing. As the supply chain management (SCM) scene is turning out to be progressively mind boggling and conventional frameworks are ending up being deficient, inventory network directors are going to implement big data analytics.

It is clear in the review directed by APQC [7], wherein the top needs among supply chain associations in 2018 are improving quality of services, concentrating on management of performance, and putting resources into data analytics.

Data Analytics or Big Data Analytics represents a blend of apparatuses, preparing frameworks, & calculations which could decipher bits of knowledge from information. Customarily, SCM has depended on ERP and other divergent stockpiling frameworks for information. In any case, the supply chain analytics has moved the needle from only robotization to ground breaking information mix and better dynamic. Utilizing steady information—which is a mix of both filtered through and unstructured plans—and the intensity of the 3Vs (variety, volume, and velocity), store networks examination has associated with joint exertion of supplier frameworks and all the way deal in the most veritable sense.

Therefore, this chapter explores the various supply chain inefficiencies and their negative effect on business profitability. The chapter then evaluates how Big Data advancements could positively impact the supply chain inefficiencies. It also discusses what challenges organizations have to face. In the last, it discusses the future prospects for Big Data in supply chain and how Big Data will impact the future of the industry workforce. The chapter takes into consideration the issues of global supply chain vis-à-vis the Indian Supply Chain Industry.

2 Supply Chain Inefficiencies: Present Situation

The supply chain has truly been similar to a black box for endeavours, with clients not knowing where and what condition their merchandise are in. Makers are losing a great deal of time, cash and stock because of unpredicted cargo development. India alone spends about $160 billion on road logistics, twice [what is gone through by] nations with a proficient transportation framework [6].

In our worldwide economy, supply chains regularly spread across nations. Thus, organizations may not comprehend where parts of their item are being made and how they are finding a workable pace. Supply chains are multi-layered, which can make issues when something turns out badly. In the event that a specific piece of an item is being harmed for instance, it might be amazingly hard to tell where and when the harm is occurring. This absence of perceivability is a typical obstruction for organizations in each industry. In the global supply chain resilience report 2017 by Business Continuity Institute [8], 69% of respondents said they don't have full perceivability of their supply chains. Aside from this, there various elements liable for the wasteful aspects in the worldwide supply chains. A few of these are:-

- Different Methods of Transportation: Organizations have developments of raw material to their office or finished products going out from their office. Consequently, their products are dealt with by different methods of transportation, stockrooms and geologies. Because of various hand-offs, and organizations managing the products, there are for the most part no standard working methods, a solitary window or dashboard to screen this throughput through the shopper point.

- Middle person (Representative) System Impact: Because of exceptionally divided and unstructured nature, the logistics division has an immense convergence of conventional merchants/booking specialists intermediating the exchanges. Therefore, there is a tremendous cover of haziness in valuing and the general costs that customers pay for the exchanges. Customer shippers could take backing of different advanced cargo stages to get straightforward and fair-minded evaluating and a reliable accomplice to encourage their logistics.
- Absence of Responsibility: In many organizations, the obligation of ensuring the supply chain runs easily falls on the Chief Procurement Officer (CPO). While this is fine, issues emerge when the CPO doesn't consider anybody in the supply chain responsible. Numerous if not most organizations neglect to place anybody or any gathering of individuals set up that can be dependable to disregard activities. At the point when issues emerge, they just lead to disarray and fault as opposed to a composed arrangement of distinguishing and taking care of the issue. Also, when representatives realize they are responsible they will perform better.
- Poor co-ordination: Deviations from creation/assembling or vessel plans at port or deferral in picking and pressing or stacking or slant during month-closes, postponed documentation and subsequently co-appointment between sea liners/cargo forwarders. This influences the worth chain, creation plans and furthermore the expenses.
- Lack of Innovation: Supply chain is an evolving world, and organizations need to adjust to new innovations to turn out to be increasingly effective. Late advancements for instance permit distribution centers to store items naturally through an arrangement of robotized transport lines, scanners, and so on. This spares distribution centers room and time and forestalls wasteful aspects that accompany human blunder. Notwithstanding the way that numerous distribution centers have utilized a specific strategy for quite a long time, change and advancement is fundamental in the quick paced universe of business.
- High Administration and Organization Cost: One of the key shade of cost to any calculated and supply chain tasks is the high density of labor assigned to exercises relating to the exchanges throughput. Most conventional coordinations players are gigantically labor subordinate and have the thriving to be affected by the non-straightforward in the exchanges or dealings. The market is gigantically divided and dark and has high counter gathering issues. Vintage strategic players have a decent feeling of how the plague of debasement happens on their business. Subsequently most have turned to having labor picketed at every single twist in the business bringing about gigantic organization and overhead expenses.
- Right Asset to Load Matching: One of the most well-known events of cost heightening is the Benefit (truck/trailer/compartment) coordinating to the heap. The vast majority of the occasions the explanation could be mis-understanding the prerequisites mutually by the shipper and in that the arrangements offered from the calculated accomplice and inability to prescribe the privilege streamlined

resource dependent on volumetric effectiveness to the shipper or freight proprietors. Absence of involvement with the shipper's end in co-ordinating for the correct prerequisite likewise adds to the wastefulness.

The above given inefficiencies also have a huge impact on the logistics performance of the countries which is apparent in the World Bank's Logistics Performance Index [9]. This index tests and develops countries logistics friendliness profiles focused on qualitative and quantitative factors such as networks, consistency & competency, monitoring & tracing, and so on. According to this index, India's Logistic Performance Index has increased considerably from 3.07 in 2007 to 3.42 in 2016.

India's exhibition, while rising, has a generally poor rating contrasted with other significant economies. Insufficient store network framework, complex tax collection laws, significant levels of mediators, item expansion, poor following and absence of production network visibility are a couple of supply chain issues looked by the retail business in India.

In any case, right now the Indian Supply chain experiences a couple of key difficulties:

- The fragmented warehousing industry in India experiences low degrees of innovation and absence of talented work. ~ the unorganized sector hold almost 90% of the Indian warehousing space [8].
- Transportation industry in India is mostly split with fleet owners of small scale. 74 per cent control of vehicles is with limited vehicle operators with 1–5 cars [8].

3 Probable Cost of Supply Chain Inefficiencies

Many supply chain organizations don't know about the colossal wholes of cash they are losing each day because of operational wasteful aspects and manual arranging forms. These imperfect procedures and practices lead to covered up, heightening costs that include and negatively affect primary concerns.

In spite of the fact that organizations need to wow clients with excellent bundling, refined item plan and basic arrangement guidelines, yet while these components are significant, the client venture really begins well before the client gets an item—and an organization's supply chain is pivotal in deciding if that experience is certain or negative.

A wasteful and inadequately working supply chain can contrarily affect each part of an association. Most customers who experience issues with quality or conveyance don't quickly think 'this is a supply chain issue.' Rather, they ascribe it to the wastefulness of the organization, the sloppiness and the issues they face accordingly. This is the reason supply chains must work easily and productively.

While supply chain wasteful aspects may appear to be a business issue, it is, indeed, generally a client issue. To clients, it once in a while matters what the reason or coordinations of a postponement—they realize that they paid for expedited service and they didn't get their bundle in time. Additionally, the client's desires in the present serious commercial center progressively determined by web based business are high. Individuals need to recognize what they are requesting and when and how it will show up, and they have little persistence for absent or late shipments. They have a lot of alternatives, so one postponed conveyance can make them drop their request and proceed onward to another organization.

A point by point investigation of more than 100 shipments from a cross area of UK enterprises by Zencargo, an advanced cargo forwarder proposes that UK organizations that exchange universally are squandering more than three hours on normal for each individual shipment. In aggregate, in excess of 100 million hours of time are squandered every year in acquirement, supplier management and cargo organization capacities for a complete yearly expense of £1.5 billion ($1.98 billion) [10].

The Indian retail industry also has been in high-development mode, and experts expected that it will develop more than $879 billion by 2018. Be that as it may, the wasteful supply chain structures have sadly threatened the growth. The latest report, led by the Confederation of Indian Industry (CII) and Amarthi Counselling, says India annually loses $65 billion because of structures in the inefficient supply chain. The country, which ranked 47th on logistics, falls behind other countries, such as Japan, the US, Germany, and China, and seemingly notable difficulties in the supply chain. The cost of the supply chain in India is about 12% to 13% of the total national production (Gross Domestic Product) compared to 7% of the GDP in the developed nations [11, 12].

Supply chain activities (schedule/source/create/communicate) typically direct a greater portion of an organization's income as supplies, products purchased, and personnel/work costs. For example, at any random time an typical assembly company might have up to 45 per cent or more of its working capital tied up in the supply chain. A lot of cash in play can have a recognizable and negative impact on the primary concern with wasteful aspects with even a little level of effect.

For illustration, we should be taking a gander at a manufacturing company worth $250 million. With at least $110 million tied up in the supply chain, squandering territories will result in tens of thousands of these estimated blunders, unrecovered transporter/transportation prices, missing stock, late payment penalties and unavoidable early payment limits can run into tens of dollars every month. This is lost and for the most part unrecoverable capital that could be put to all the more likely use.

For mid-organizations that typically have tighter edge controls, the lack of accessibility of this extra working capital not only inhibits market and item development, it restricts the ability to adapt and scale to the business's more drawn-period intensity. At the other hand, put another way, companies are in a much more grounded position to respond to changing economic conditions and reserve growth activities by finding a way to reduce the inefficient aspects of the supply chain and turn them into working capital.

4 Leveraging Big Data to Optimize Supply Chains

IoT and Big Data are uncovering supply chain inefficiencies by disposing of vulnerable sides from logistics processes. One of the numerous approaches to use IoT for the supply chain is IoT empowered freight checking. Today the visibility into the development of merchandise is as yet negligible and generally done by means of procedures intensely reliant on calls, administrative work and email correspondence. The conventional supply chain perceivability idea is achievement based, bringing about logistics blind spots for the shippers. Information about the status of goods is possibly recorded when merchandise arrive at specific checkpoints or achievement. In the middle of the checkpoints, organizations come up short on the visibility to know where their stock is found, regardless of whether transporters are giving the correct assistance levels, and where the bottlenecks and inefficiencies are in the supply chain (Fig. 1).

Much the same as Excel drastically changed the manner in which supply chain checking and revealing was executed, Big Data platforms are permitting experts to concentrate on greater picture things rather than tedious errands. Supply chain management—as it generally has existed—has just ascended in prole and significance with the ascent of Big Data.

64% of supply chain managers find Big Data Analytics to be a problematic and important breakthrough, setting the establishment to adjust their association's executives for long haul [8]. Whereas, 97% of supply chain managers' report having an understanding of how big data analytics can profit from their supply chain. Be that as it may, in at least one supply chain capability only 17% report having only actualized analytics [10].

Big data allows enormous quantities of organized and unstructured data from various sources to be seen rapidly. This can help the executives increase perceptibility for the supply chain and give additional bits of knowledge into the entire supply chain. Using big data, the supply chain connections will, for example, boost the reaction to

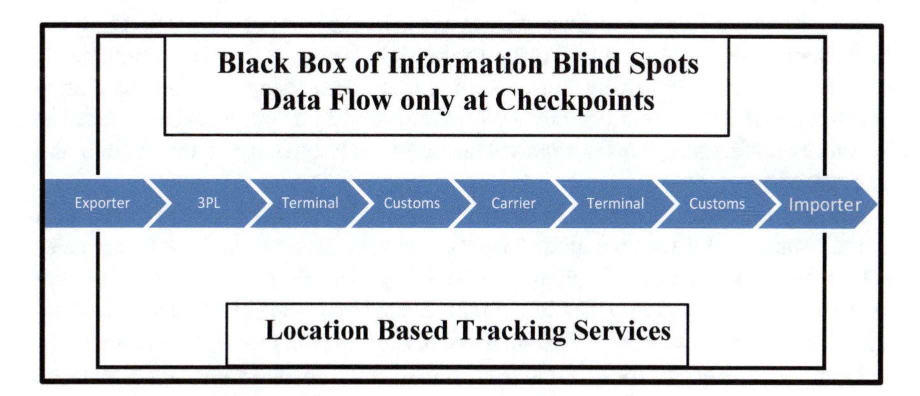

Fig. 1 Supply chain blind spots. *Source* [13]

Table 1 Stages in supply chain process and benefits offered by big data

Stage	Benefit
Planning	Forecast the product demand more accurately
Sourcing and development	Evaluate the contractor performance in real time and identify hidden costs
Execution	Maximization of resources and production output
Delivery	Improve performance significantly in terms of efficiency, speed and accuracy
Return	Reduced return costs and provide greater visibility to the process

Source [15]

unforeseeable interest or supply chain threat and reduce the issues associated with the current problem. Organizing the development of a big data analytics method would allow association to solve various supply chain problems [14]. Let's look at how big data in every point of the supply chain will make a difference (Table 1):

- **Planning**

Demand forecasting is one of the key strides in concocting an effective supply chain procedure. Robotized demand forecasting engages organizations to react to the vacillations in the commercial centre and keep up ideal stock levels consistently. In spite of the fact that conventional gauging programming kept on being utilized by some little and-medium-sized endeavours, they are tedious and include a ton of manual work. With the commercial centre getting increasingly unstable and rivalry progressively wild, it has gotten basic for organizations to react to request supply elements continuously and get the correct item conveyed at the perfect place and ideal time. Mechanized demand forecasting has, in this way, expected an essential job in the supply chain industry.

Big Data analytics also support predicting the customer needs and preferences. More than 90% of disappointed customers will not be working with a brand that has neglected to live up to their wishes [16]. In the client's age, delivering the right item at the right time and place to the right person is crucial to picking up (or holding) customer loyalty and reliability. Keen partnerships can use big data to provide a full 360-degree perspective on the customer to more readily anticipate consumer preferences, consider individual inclinations and have an insightful brand understanding.

During this stage, incorporated data over the whole network organize joined with the utilization of measurable models permit request (for example stock levels, sales estimates) to be figure all the more precisely. E.g., we can guarantee that there are no out - of-stock circumstances in the retail sector by having stock and recharging systems to express. Not only have such simulations take in past and current details, but they also talk of macroeconomic factors, market trends, and knowledge regarding rivals.

Through prudent utilization of these stages, organizations can keep up ideal stock, oversee circulation arranges adequately and improve distribution centre cost effectiveness. They can design out their deals and dispersion methodology well ahead of time to maintain a strategic distance from any very late glitches.

- **Sourcing and Development**

Progression in innovation has permitted organizations to group data on various providers. Utilizing proper big data devices, organizations can use this data to increase important data about the authentic record of any provider. Organizations can evaluate these providers against a few key exhibition measurements, for example, client audits, gainfulness, area, nature of administration, level of consistence and select the most suitable provider for any product and area.

An organization can outline supply chain and incorporate it with ongoing data on climate changes, strikes and traffic to follow interruptions in the supply chain and set themselves up well ahead of time for any deviation from typical conveyance designs.

Costs of raw materials differ with the changing commercial centre elements. Variables like regular vacillations, new item dispatches and changes in promoting technique can fundamentally affect the cost of any ware. Big data assists organizations with looking at the synchronous effect of every one of these factors on ware evaluating and sets them up at any conceivable cost climb.

On average, procurement costs approximately 43% of the overall expenses of an organization [17]. Given the enormous productivity potential in this field, supply chain analytics are utilized to evaluate contractual worker execution and consistence progressively as opposed to in quarterly or yearly cycles when it might be past the point where it is possible to intercede with and decrease costs. Regardless of whether the vendors are analyzed, quantitative methodologies will render the cost structure increasingly straightforward by helping pioneers distinguish hidden expenses.

Cost proficiency, cost reduction, and investment optimization must follow as the top company needs in the executives supply chain. Inserting big data analytics into activities prompts a 2.6× increase of 10% or more notable supply chain competence [10].

- Execution

Big Data can support rearranging the many moving parts during execution to refresh the available resources (space, PCs, assets, people, and so on.) as well as to increment the yield. For example, in the assembly sector, IoT sensors can provide ongoing hardware details that can be optimized along the way to enhance production capacity and resource execution. Statistical applications such as fault prediction or schedule management also makes use of analytics. For eg., Intel uses predictive analytics to save $656 million per annum [4].

61% organizations considered to be pioneers in the supply chain found that supply chain risk management was essential. Similarly, those equivalent heads perceive the necessity for abilities that make their supply chains more notable perceptibility and continuity [10]. Big data may help survey the likelihood of a problem and its latent capacity impact, and support systems to identify the threat to the supply chain.

Consolidating the review of chronic data, danger analysis, and arrangement of the situation will empower the board approach to early warning of a threat.

Data originating from IoT gadgets fitted along creation lines gives a colossal chance to streamlining item quality and amount. Right now modern IoT, makers utilize ground-breaking big data devices to pinpoint wasteful aspects and openings in each section of the creation procedure. This causes them track failing to meet expectations parts and procedures.

Data exuding from sensors in hardware and cameras can be joined with money related and operational data for an increasingly all-encompassing image of the creation procedure. With authoritative consent, industrial facilities can screen specialist efficiency by utilizing IoT empowered worker identifications which trade data with sensors introduced underway line units.

Predictive analytics in big data has huge applications in item testing. To keep up item quality, makers for the most part need to perform several tests on a solitary item. Utilizing prescient analytics, organizations can decrease the quantity of tests required for an item and perform just those tests which are completely fundamental.

Predictive analytics likewise discovers applications in item customization. Till as of late, makers concentrated exclusively for enormous scope creation to spare expenses. Customization was left in the hands of players serving the specialty showcase. Changing innovation has made it feasible for the fabricates to anticipate the market for modified products and produce them in-house without overspending assets.

Since makers can now intently and precisely track the whole creation cycle, it is workable for them to distinguish focuses in the creation line where they can helpfully and gainfully embed another custom procedure. They can likewise defer the creation procedure and permit a merchant to do customization before the assembling procedure is finished. Producers would thus be able to execute huge scope customization and take specially made demands gainfully.

Organizations depend vigorously on their machines and spotlight on upgrading their presentation. Brilliant sensors introduced in machines catch and transmit execution data continuously and help pinpoint abandons assuming any. By utilizing scientific devices, makers can investigate huge amounts of such data to anticipate and plan preventive maintenance for all the machines. Staying up to date with mileage of hardware causes them to maintain a strategic distance from resource breakdowns and forestall any unscheduled personal time. Preventive maintenance additionally drags out the life expectancy of machines by forestalling irreversible disappointments.

- **Delivery**

It's everything about pace (getting the thing out on time), accuracy (ensuring the groups meet the correct objective) and unwavering quality (finding the ideal course/joining movements) at this point. Real-time conveyance information superimposed on outer subtleties, for example, traffic and environment planning, will lead to significant changes in logistics and supply chain management.

90% of organizations claim weakness and pace are essential or vital to their company [17]. The ability to reach customer satisfaction goals efficiently and deftly are measured as the second most significant driver of the upper hand in all projects. Installing big data analytics in activities will influence the response time of associations to supply chain issues (41%) and can cause an increase of 4.25× as full-conveyance times [10].

Fleet Managers utilize big data answers for advance delivery routes. These arrangements coordinate data from a few sources progressively, for example, GPS, meteorological forecasts, traffic data, street maintenance data, work force timetable and vehicle maintenance plan into a framework that encourages the vehicle to take the most ideal course and reroute at whatever point required.

For instance, UPS, the American SCM mammoth has utilized large information to further their potential benefit in a major manner. By analyzing their information, UPS understood that their trucks turning left was costing them a ton. The vehicle transforming into approaching traffic squandered fuel, caused delays in conveyance and expanded dangers. UPS trucks presently turn left just when completely vital, not over 10% of the time. This single change has helped them decrease their armada by 1100 and convey 350,000 additional bundles each year.

Analytics can likewise help monstrously in scope quantification. Organizations can contrast the interest at various areas and the accessibility of vehicles and faculty and dispense resources where they are required the most. This forestalls over blockage of vehicles in a specific area and encourages them coordinate interest with supply.

Traceability is regularly straightforwardly connected to inventory network chance. Traceability and natural problems for 30 per cent of companies are the key things to look for [14]. Traceability and feedback are generally concentrated in details. Huge information can possibly give improved discernibility execution; it can likewise decrease huge number of hours engaged with getting to, coordinating, and overseeing item databases that catch items that ought to be reviewed or retrofitted. Vehicles outfitted with visual and acoustic sensors empower constant following of the stock from the time it is dispatched to the time it arrives at the client. Upgraded Traceability acquires more straightforwardness in tasks.

Organizations can follow the exhibition of their drivers and make greater responsibility for them. This improved access to ongoing information on dispatch vehicles can assist organizations with improving help and assemble long haul relationship with clients.

- **Return**

One of the most significant advantages of large information for internet business organizations is that it can spare item bring costs back. By and large, the expense of item returns is 1.5 occasions that of the genuine delivery. Today, object returns for certain component classes [18] are anticipated to be 30%, which is a significant obstruction for endeavours to continue their profitability. Instances of invert logistics costs incorporate re-putting away costs, transportation costs in restoring the thing to the retailer/extra space, shipment overheads in conveying another thing to the purchaser and choosing costs in assessing the thing got back. Big Data Analytics will help to reduce these costs and provide insight by consolidating production and distribution details and inbound and outbound movements to generate consistent returns.

By incorporating enormous information examination, organizations can evaluate the conceivable outcomes of items being returned. These apparatuses can distinguish the items that are well on the way to be returned, and can permit organizations to take the fundamental measures to diminish misfortunes and expenses.

The most widely recognized things that are returned, or traded, are pieces of clothing, shoes, and style extras, to give some examples. The regular elements prompting item returns incorporate defective items, mistaken sizes, neglected models, and so on. Organizations can discover which urban communities have the most item returns, or which client as often as possible trades products, by incorporating large information innovations. They can likewise receive a proactive methodology, calling clients to look for criticism on an item. This can assist organizations with eliminating costs in transportation and logistics.

With big data analytics, merchants can rapidly recover exchange data, in spite of the fact that the first deals channel may not be known around then. That ability causes it conceivable to better to execute the turnaround logistics related with returns.

At the point when venders can process returns utilizing the first buy data, they can make a consistent encounter for clients and wipe out compromise mistakes toward the back. Likewise, utilizing the first buy data naturally accommodates the first price tag, including limits or coupons, with the sum to be discounted. Furthermore, it assists with forestalling false returns, and the local following adjusts stock data with certifiable accessibility while likewise empowering product to move into the invert logistics supply chain quickly.

Adjusting data goes far toward decreasing the expenses of profits preparing, while likewise conceivably improving the client experience. However, handling returns despite everything stays a significant expense recommendation for the retailer. Quite a bit of that cost can be credited to the high-contact nature of profits preparing, where people need to deal with the product, interface with clients, review the product and at last decide if the product can be restocked, come back to the seller, reused, fixed, gave or sold.

Moreover, Big data analytics is giving business strategists and supply chain supervisors with new data they can use towards different capacities-

- Proactive maintenance and activity: With regards to the proficient activity of warehousing offices and their support, Big Data instruments and A.I. fueled devices, combined with associated sensors and contraptions, assume a significant job. On account of the constant access to data about different triggers and issues, for example, prompt upkeep needs of vehicles or the finish of administration cycle for parts, organizations currently can take progressively proactive choices and measures bringing about more proficiency and sure-footedness. The development of the Modern Web of Things (IIoT), bragging another web network across contraptions and sensors, is helping supply chain administrators to diminish personal time and make the procedure increasingly streamlined. In addition, proactive support and streamlined tasks are giving a superior cost bit of leeway to the organizations in regard of supply chain and transportation.

- Overseeing dangers and possibilities outside human ability to control: In supply chain the executives, the outer dangers and possibilities can have a critical effect, and this is the place the most recent Big Data analytics and A.I. fueled apparatuses can assume a big job. For instance, the ascent of unrefined petroleum costs in the universal market can play devastation with the expense of transport and logistics. In such cases, if the supply chain administrators have any earlier piece of information or clue about the chance, they can take proactive measures to manage the up and coming possibilities. In a similar way, there can be an abrupt upsurge sought after on account of a frenzy about deficiencies because of war or expectations about a flood or other regular catastrophe. Disregarding having your standard alternate course of action set up, the organization may not be squeezed to manage such abrupt changes constantly. This is the place data-driven bits of knowledge and forecasts can assume a noteworthy job.
- Stock sensors for distribution centre administration: For certain businesses where merchandise have a lesser time span of usability, stock and stockroom the board is of preeminent significance. For instance, for nourishment supplies, stock sensors can send convenient alarms about its condition, days left before expiry, and give continuous cautions about the need to restocking the nourishment in the event of need. Convenient alarms with associated sensors will help decrease warehousing and stock administration cost for nourishment things. This will likewise assist with anticipating deficiencies regarding stock administration dependent on past requests. For example, if during occasions, there is greater probability of a spike in the requests of chicken, the associated sensors will send alarms about this.

5 Key Challenges in Adopting Big Data Analytics for Supply Chain

As indicated by a Deloitte overview [19], 37% of supervisors have simply "occupied with discussions" to actualize analytics. 27% of the individuals who have actualized big data analytics have it operationalized and inserted into basic supply chain forms. In any case, 57% of respondents utilize big data as logistics arrangements on a specially appointed premise.

As indicated by a study by Soft web Arrangements [20], retailers who utilize prescient analytics have accomplished a 73% expansion in deals, when contrasted with the individuals who didn't utilize it. Hence, retailers are using Big Data arrangements by means of client analytics to duplicate benefit and beat contenders by customizing their in-store contributions and online item. Be that as it may, there are not many hindrances for supply chain the executives while executing real time analytics.

The trial comes with management of the colossal details of Big Data. The through the data, the more complicated it's to handle. End-to-end operation of this facility is problematic because of the unavailability of the state-of - the-art trained operators. Apart from that, the application of Big Data Analytics in Supply Chain Management poses several other challenges:

- **Visibility**: business logistic specialists may have enormous volumes of data yet come up short on the mastery to see and break down it as the realistic interface isn't easy to use.
- **Use**: individuals may approach data however don't have a clue how to provide them guidance in wording their ideal results, key and everyday directions. As it were, administrators have monstrous measures of data accessible, however don't have the foggiest idea how to manage it.
- **Volume:** the measure of high amounts of data is as yet something that needs improvement in little economies and locales like Latin America, the volume of examples are accessible in a urban region or a metropolitan district, however are scarcer in conjecture traffic in country territories. With outsider organizations, it's simpler to make cooperative energies and economies of scale with big data. Something else, the person in-house utilization of big data isn't savvy.
- **Protection & security concerns:** Data sharing over a Supply Chain Network is a main consideration in gathering data from different sources, examining it and giving bits of knowledge. Albeit, local or worldwide Supply Chain Networks may confront troubles in sharing data over its various sources because of different Privacy, Security laws worried about sharing of data. Absence of shared data in such cases can influence the exactness of the bits of knowledge that Big Data Analytics may create.
- **Tedious:** Factors like the volume of Big Data, Complexity of Supply Chain and translation objectives for the datasets alongside outside elements, for example, absence of access to data contribute in making the investigation procedure tedious.
- **Inadequate assets:** For better outcomes, the accessibility of ongoing data is pivotal. Supply Chain being a stage that produces complex cross-utilitarian data for interlinked elements, assortment and capacity of cross functional data ought to be streamlined.
- **Absence of aptitudes:** The multifaceted nature of Big Data created from Supply Chain source requires a mix of good area information examination aptitudes and the capacity to decipher the convenience of data. As indicated by overviews, such a mix alongside experience is hard to discover.
- **Absence of procedures:** Incapability of a firm to use the data influences the power of the bits of knowledge created in the wake of examining the datasets. The methods used to investigate, figure, gauge and picture should be changed or updated in understanding to the multifaceted nature or volume of data.

6 Future Prospects

The Supply Chain Big Data Analytics Market was esteemed at USD 3.03 billion of every 2019 and is relied upon to reach USD 7.91 billion by 2025, at a CAGR of 17.31% over the figure time frame 2020–2025. US uses on transportation alone ($688 billion) are bigger than the GDPs' of everything except 16 nations (World Bank Gross domestic product data) [8]. With headways in data innovation, firms are currently ready to access, store, and procedure a monstrous measure of data. Associations are breaking down data sets and recognizing key bits of knowledge to apply to their activities, making it clear that Big Data has a significant task to carry out in any industry. From nourishment and drink conveyance to innovative, organizations are consolidating analytics.

Big Data Analytics in supply chain management and logistics has gathered expanding consideration because of its unpredictability and the unmistakable job of SCM in improving the general business execution. As indicated by a study directed by Accenture [10], more than 33% of the respondents revealed being occupied with genuine discussions to send analytics in SCM, while three out of ten as of now have showed a drive to execute analytics.

The worldwide supply chain and logistic industry is relied upon to outperform $15 trillion by 2023 as indicated by an examination from Transparency Market Research [20]. The US is thoroughly hoping to reinforce its assembling industry, by upgrading its efficiency by laying accentuation on improving exercises over the supply chain, inside the mechanical area in the nation. The web based business industry in the US is multiplying, attributable to which the necessity for proficient supply chain the executives is on the ascent. As per the US Trade Office, the web based business industry in the nation rose by over 40% in 2017 [20]. Accordingly, Big Data is relied upon to rise altogether, in this way, positively affecting the supply chain analytics in the nation.

Prominently, as per Reddish College's Harbert School of Business, in mid 2018, the retailers in the US were relied upon to encourage their interest in the supply chain the executives, particularly in innovation overhaul, inferable from extension and fast development in the internet business industry. Also, new companies are attempting to wander into the retail space in the locale that are raising assets to support their operational productivity through Big Data analytics and other rising advances. For example, A.S. Watson gathering (ASW) reported an organization with Rubikloud, a Toronto-based startup, essentially to put resources into growing Big Data capacities. The previous organization contributed about USD 70 million to upgrade the operational proficiency and client experience through the incorporation of perception and AI abilities. Accordingly, it is anticipated to push the supply chain big data analytics advertise development in the nation.

Despite, India being a developing country, the Indian business is quickly making through the collaboration of progression, segment of new authority associations, establishment improvement and administrative advances. According to PwC [21], there are 4 major elements molding the India's supply chain room namely- Regulatory Enablers, Infrastructure Investments, Emergence of Organized 3 PL Players and IOT & Other Technology Automations.

As a result of all these factors, we may well observe an expansion in the quantity of self-sufficient instruments utilized in business—calculations, projects, and robots that require zero client contribution to play out a vocation, yet in addition to settle on choices about that activity.

Big data is progressively getting key to having a productive supply chain and a decrease of expenses. Indeed, it's presently standard practice to assemble and investigate large amount of data to help support income. Specialists foresee the pattern will proceed to grow, and the cost-reserve funds alone in effectively re-organizing supply chains are possibly enough for noteworthy extra benefit as well as for proficient, streamlined operations pushing ahead.

Big Data is transforming supply chain chiefs into "mind readers," permitting them to foresee and respond to purchaser practices in new manners. This implies, in the future supply chain workforce, the authoritative assignments of the past will be decreased extraordinarily and a more elevated level of human analysis will be required.

In view of a multi-customer study [22], computerized switches could drive much more prominent gains in the coming decade. In light of current creation volume figures, this change could bring about a decrease in a producer's workforce of up to 25% full-time counterparts (FTE), driven by man-made consciousness or back-office computerization or big data analytics or by utilizing innovation to cultivate joint effort (and increment redistributing in the region of Research and development, for instance). Furthermore, it will cause a huge recalibration in the kind of undertakings the laborers perform.

Digitization of exercises brings about diminishing the portion of routine value-based exercises, empowering representatives to concentrate on esteem including undertakings. In any case, to utilize this capacity, the representatives should grow new abilities and become all the more carefully familiar. While preparing in process-mechanization apparatuses is moderately simple, preparing in big data analytics is progressively mind boggling: Other than the learning on the best way to utilize analytics instruments, representatives need to see how the devices work to utilize them. Just if these prerequisites are met will staff have the option to profit by such innovations. While existing staff can be prepared on the utilization of those advanced apparatuses to certain degree, there will be new position profiles for which no present representative has the abilities or can be prepared effectively, for example, data analysts.

Organizations not just need to consider their inward workforce, yet in addition of an outer workforce technique to discover, pull in, enroll, and persuade advanced experts, ability that regularly is attracted to computerized organizations, for example, Google or nimbler new companies. While supply chain organizations must expand their allure and grow progressively adaptable approaches to draw in with these abilities, they may consider helping out colleges, explore establishments, and different organizations to catch the computerized ability.

Supply Chain and logistic organizations should reevaluate their workforce arranging and enlistment procedures to be progressively unique and light-footed. Rather than focusing on existing profession structures and enrolling channels, HR chiefs need to distinguish enlistment needs for the future, finding the ideal individuals with the correct ranges of abilities, regardless of whether work at present exists for them or not.

As the innovation turns out to be progressively unpredictable and coordinates with other innovation, the industry will proceed to advance and require particular specialists for specialty work jobs and duties. Coders will be required to upskill to Big Data analytics and chiefs will be asked to upskill to business insight and situate themselves with anything to do with data for business and profession development.

As indicated by gauges, it is uncovered that the worldwide piece of the overall industry of big data and its united innovation is relied upon to reach roughly $57 billion continuously 2020. The US is the harbinger in the age, utilization and organization of data and this innovation and is trailed by India. Analysts likewise uncovered that India would develop to turn into a $8 billion industry by 2020, creating a requirement for more than 50,000 talented Big Data specialists and data researchers [22, 23].

Supply chain supervisors don't have to stress over their employments leaving, however measurements show that the requirement for supply chain experts will just develop. Their employments will simply move on account of Big Data examination, where these administrators will take a gander at the immense measure of data gave to them and have the option to identify significant things. This more significant level of expertise will likewise require propelled instruction, so supply chain directors of things to come should hope to join associations that give chances to progressing training.

7 Conclusion and Suggestions

This chapter looked at the challenges and prospects of adopting Big Data to speed up supply chain management. The scale of Big Data is known to be the key reason behind its use in the Supply Chain. In the wake of considering the Big Data era wellsprings in supply chain procedures and drills, essential bits of information about Big Data Analytics' capabilities were revealed. It has been shown that the combination of mind-boggling data from supply chain experiments and the magnitude of Big Data as far as Variety, Volume, Velocity, value and Veracity are concerned with ground applications that are likely to unravel the most common problems the supply chain has looked at even in the years to come. Considering the selection of Big Data Analytics, a moderately new marvel, it was discovered that the pace of creating infrastructure to sustain the expanding data needs to increment. It has been discovered that the inaccessibility of experts with sufficient expertise ranges will disrupt the potential of Big Data Analytics in the Supply Chain. As the Supply Chain Networks ' multifaceted nature builds far and wide, the Supply Chain Industry should work alongside the Data Analytics Industry to grow new, successful models and systems.

In addition, the Covid-19 crisis further highlights that supply chains need to adopt emerging technology to withstand potential shocks. In the current world, where the

COVID-19 pandemic has created too much confusion in global supply chains, it is quite clear to assume that exposure and coordination have never become more essential. Big Data provides the secret to detecting challenges around the whole supply chain, helping to prepare and respond to evolving situations instantaneously. Big data and analytics inside the supply chain will help push companies to the next stage, supplying the power for predictive software to track, anticipate and prepare challenges before they eventually arise. This provides the opportunity to handle disturbances months ahead of time, rather than react after they have arisen.

India despite everything has far to go for improving the proficiency and viability of its logistics and supply chain processes. With the customer industries requesting better, more secure and all the more mechanically propelled logistics models, the Logistics service providers ought to guarantee both industry explicit compliances and quicker reception of Digital innovation as well. In the Indian Logistics Industry, there are various investment opportunities for providers from both engineering and logistics administration fields that could make huge incentives. This involve–creating quantitative value chains by transferring information from all the centre points of the chain to data; and chipping away to reach top-level warehouse management systems at specific exercises.

"Logistics/ supply chain as a whole will not change with incremental innovation, it will change through disruption! What is good-to-have today will become a necessity tomorrow."

References

1. Ivanov D, Dolgui A, Sokolov B, Ivanova M (2017) Literature review on disruption recovery in the supply chain. Int J Prod Res 55(20):6158–6174
2. Kinra A, Ivanov D, Das A, Dolgui A (2019) Ripple effect quantification by supply risk exposure assessment. Int J Prod Res 58(18):5559–5578
3. Hosseini S, Ivanov D, Dolgui A (2019) Review of quantitative methods for supply chain resilience analysis. Transp Res Part E 125:285–307
4. Ivanov D, Tsipoulanidis A, Schönberger J (2019) Global supply chain and operations management: a decision-oriented introduction into the creation of value, 2nd edn. Springer Nature, Cham
5. https://www.supplychaindive.com/news/digital-freight-forwarding-supply-chain-inefficiency/527945/
6. Gupta M, Sikarwar TS (2020) Modelling credit risk management and bank's profitability. Int J Electron Bank 2(2):170–183
7. APQC 2018 Supply Chain Management Priorities and Challenges. https://www.apqc.org/resource-library/resource-listing/2018-priorities-supply-chain-management-overview
8. Global Supply Chain Resilience Report (2017) Business Continuity Institute. https://www.thebci.org/uploads/assets/uploaded/c50072bf-df5c-4c98-a5e1876aafb15bd0.pdf
9. World Bank's Logistics Performance Index (LPI) (2018). https://lpi.worldbank.org/international/global
10. Big Data Analytics in Supply Chain: Hype or Here to Stay? Accenture Global Operations Megatrends Study (2018). https://www.accenture.com/t20160106t194441__w__/fi-en/_acnmedia/accenture/conversion-assets/dotcom/documents/global/pdf/digital_1/accenture-global-operations-megatrends-study-big-data-analytics-v2.pdf

11. Hobbs B (2019) Supply chain Inefficiencies can Crush Your Customer Experience and Cost you Millions. Forbes business Development Council Post
12. One SCM (2014) The Cost of Waiting: Delaying Supply Chain Improvements Costs Both Margin and Market Share. https://www.onescm.com/holding-supply-chain-improvements-costs-margin-expansion/
13. https://www.supplychain247.com/article/internet_of_things_big_data_are_accelerating_supply_chain
14. Mcdaniel S (2019) Big Data for Supply Chain Management. https://www.talend.com/resources/big-data-supply-chain/
15. https://www.talend.com/resources/big-data-supply-chain/
16. https://customerthink.com/
17. https://www.bain.com/insights/unearthing-hidden-treasure-of-procurement
18. Memdani L (2020) Demonetisation: a move towards cashless economy in India. Int J Electron Bank 2(3):205–211
19. Deloitte (2016) Point of View: Supply Chain Analytics. https://www2.deloitte.com/content/dam/Deloitte/nl/Documents/deloitte-analytics/deloitte-nl-supply-chain-analytics.pdf
20. https://www.prnewswire.com/news-releases/global-logisticsmarket-to-reach-us155-trillion-by-2023-research-report-published-by-transparency-market-research-597595561.html
21. PWC Report (2019) India Manufacturing Barometer 2019: Building Export Competitiveness. https://www.pwc.in/assets/pdfs/research-insights/2019/india-manufacturing-barometer-2019.pdf
22. Oliver Wyman (2019) Retail and Consumer Journal. Volume 7. https://www.oliverwyman.com/content/dam/oliver-wyman/global/en/images/insights/retail-consumer-products/2019/Retail_Journal_Vol_7.pdf
23. McKinley N (2019) How big data and IoT are bringing massive change to transport and the supply chain industry. Trade Ready: Blog for International Trade Experts. https://www.tradeready.ca/2019/topics/supply-chain-management/how-big-data-and-iot-are-bringing-massive-change-to-transport-and-the-supply-chain-industry/

Mobile Computing Technologies for Enhanced Living Environments: A Literature Review

Salome Oniani, Salome Mukhashavria, Gonçalo Marques, Vera Shalikiani, and Ia Mosashvili

Abstract Mobile computing is a growing area in scientific research. There are numerous mobile health applications available. Moreover, this number increases daily, especially in the case of mHealth apps for diseases prediction and patient's remote monitoring. The objective of this paper is to review the mobile computing for enhanced living environments and mobile health applications published from 2016 to 2020. The main contribution of this paper is to present a comprehensive literature review on mobile computing technologies applicability to support older adults and to promote the overall public health and well-being. Mobile computing enables the creation of multiple enhanced applications that increase people's quality of life. Nevertheless, there are several limitations to create novel mobile health studies such as investigation mHealth app numbers, participator number and using only one app store platform. The presented literature review aims to support academics and professionals by presenting the current state of the art on mobile health and mobile computing technologies.

Keywords Mobile computing · MHealth · Enhanced living environments

S. Oniani
Faculty of Business, Technology and Education, Ilia State University, Tbilisi, Georgia
e-mail: salome.oniani@iliauni.edu.ge

S. Mukhashavria · V. Shalikiani
Faculty of Informatics and Control Systems, Georgian Technical University, Tbilisi, Georgia
e-mail: mukhashavria_sa@gtu.ge

G. Marques (✉)
Polytechnic of Coimbra, ESTGOH, Rua General Santos Costa, 3400-124 Oliveira do Hospital, Portugal
e-mail: goncalosantosmarques@gmail.com

I. Mosashvili
Faculty of Mathematics and Computer Science, Kutaisi International University, Kutaisi, Georgia
e-mail: ia.mosashvili@kiu.edu.ge

A. M. A. Musleh Al-Sartawi (ed.), *The Big Data-Driven Digital Economy: Artificial and Computational Intelligence*, Studies in Computational Intelligence 974,
https://doi.org/10.1007/978-3-030-73057-4_2

1 Introduction

Currently, mobile computing is a new and demanding trend in multiple fields. Notably, using mobile applications in the Internet of Things (IoT) and Ambient Assisted Living (AAL) [1, 2]. There are multimedia mobile health (mHealth) applications that are based on Big Data technologies.

Using wearable technology creates several opportunities to improve people's lifestyle and control people's health conditions. According to [3], there was 16,500 health and fitness application in 2015, and today, their number is over 100,000, which are designed for disease management and support regarding diabetes control, hypertension, and psychological problems. Also, numerous mobile applications focus on older adults to improves physical activity, mental performance, and health condition.

Mobile health applications can be classified according to their functionality. These applications are self-assessment apps which use data from clinical tests to increase physical activity [3] or personalized treatment apps which examine real-time data from smartphones and smartwatches.

The author of [4] propose two main mobile health applications: (1) a hybrid system to create a 3D stereoscopic virtual image of hyperbaric oxygen chamber and (2) voice interactive games to improve dependence of patients during the therapy.

According to [5], 36% of US adult has used mHealth applications, and 60% use fitness applications. Besides, mobile frameworks are developed [6, 7] for improving data equation, security, and privacy of users' medical data.

As for countries where mobile health application established there is the Unified Theory of Acceptance and Use of Technology (UTAUT) model which tested and examined in three countries: Canada, USA, and Bangladesh [8]. The model states the technologies and methodologies for mHealth developers that must be talked into account in the software development process considering the country's cultural traits. It is discovered that self-concept, social influence, and expectancy are primary factors to adopt healthcare system in citizens. The results in Saudi Arabia show [9] that approximately 92% of mental health patients used mobile application for daily physical activity, monitoring and controlling of health condition.

According to patients report, using mHealth application improve patients' mental health state, PubMed, CINAHL EMBASE and Psych Info are electronic databases that are oriented on risk-benefit assessment, security issues, design and technical concerns [10, 11].

Mobile health applications can improve the result of treatment in communicable diseases. In Africa, Asia, and Latin America, 67% of users have an application for maternal and infectious diseases [11]. Furthermore, mobile computing technologies are also used to access data from the environment. Several researchers have designed and developed multiple sensor systems for environmental monitoring and assessment, which include mobile computing technologies for data consulting and notifications [12–15]. Mobile devices are an integral element of people's daily routine. Therefore, mobile technologies used as an effective and efficient method for data access and primarily applied to the healthcare domain.

Three main differences influence the adoption of mobile applications: (1) compatibility and perceived cost; (2) availability and personalization, and (3) self-efficacy and facilitating conditions [16]. Telemedicine supports new methods to integrate mobile health in the medical field. Moreover, telemedicine is crucial in the treatment of diabetes and some long process illnesses. The main goal of telemedicine is the creation of a modern and adequate treatment environmental in developing countries [17]

The main contribution of this paper is to present a comprehensive literature review on mobile computing technologies applicability to support older adults and to promote the overall public health and well-being. Furthermore, the main objectives of this document are to introduce the related research domains and to provide a comprehensive report of the impact, challenges and opportunities for future developments. Also, this paper aims to state the prevailing trends, limitations and knowledge gaps regarding the application of mobile computing technologies for enhanced living environments.

In Sect. 2, this paper discusses new ideas about mobile computing in enhanced living environments and the relevance of mobile applications on the healthcare and AAL domains. The mobile health applications are presented in Sect. 3. Results and discussion that present the methodology of the mobile healthcare studies, research outcomes and limitations are presented in Sect. 4. Finally, the conclusions are presented in Sect. 5.

2 Mobile Computing for Enhanced Living Environments

Mobile computing for mHealth and healthcare systems is an opportunity to design and create enhanced living environments. These systems should take into account the subjective risk assessment, the behaviour of treatment, and social pressure [18]. According to [19], in 2019, US population have access to 30,132 medical applications in the Apple iTunes store and Android Google Play Store, which consider diseases and habits like smoking, atrial fibrillation, diabetes, dyslipidemia, and hypertension. From these applications, 85% has been designed for a single purpose and 15% for multiple purposes. Also, there are 74 apps for stroke diseases. A Model-Oriented Web Approach (MoWebA) that focuses on the data layer is proposed by [20]. MoWebA collects offline data and stores it before an internet connection. It is compatible with Android and Windows Phone platform. In Indonesia, there is an m-health application that uses the System Usability Scale (SUS) and Post-Study System Usability Questionnaire (PSSUQ) frameworks [21]. The application had 48 participants to test application safety and usability, which showed that application usability is better if it is a local language.

A monitoring system for Alzheimer disease is presented in [22]. This study involves low-fidelity and high-fidelity methodology. Also, the application collects data about Parkinson's disease, Lewy Body dementia, and Vascular Dementia.

A mobile application that provides a patient form and includes three main parts: CF (Cystic Fibrosis) info, Medical history, and Clinical appointments is proposed by [23]. This application was created for CF adults and tested by five adults. In total, around 325,000 mobile health applications are available worldwide [24]. Mobile applications involve behaviour change techniques (BCTs) [25], which suggests suitable medical applications to users. The BCT studies users reports and characteristics to create user-friendly environmental that means each user has its interface on the application.

A mobile app to measure users emotion condition is proposed by the authors of [26]. In total, 130 patients have been involved in experimental works. The results showed that "patients' emotional bonding with the apps mediated the impacts of autonomy and relatedness needs satisfaction on their well-being in terms of enhanced IT-enabled self-esteem and reduced post-surgery physical symptoms [26]". The myAirCoach is an application for asthma patients to use inhalator effectively [27]. Thirty participants participated in the myAirCoach research, which extended for 3–6 months. The results showed that the control coefficient of the Asthma Control Questionnaire difference was 0.70 [27].

In [28], the authors evaluate the factors that impact the adoption of mHealth apps in developing countries. The Unified Theory of Acceptance and Use of Technology (UTAUT) methodology has been used, and the study includes 296 participants. The results show the gender effect on mHealth services adoption in partial cases.

A study concerning the healthcare provider relationships where thirty-seven research papers were analyzed is presented in [29]. The results present a promising impact on the use of mobile applications. Moreover, the mobile application helps doctors to communicate with patients on time, which makes easier patients health management. Mobile Healthcare Social Network (MHSN) is for the elderly [30] population and has Authorized Private Set Intersection (APSI). MHSN can protect medical information while doctors are receiving raw data with multiple symptoms from the patient.

In India, mHealth application is popular in females (50,25%) of age 18–25 (48,44%). There are applications for heart diseases, cycles, cancer, stroke, and cholesterol [31].

3 Mobile Health Applications

The user experience of mobile health applications has been evaluated by using Twitter sentiment analysis [32]. The study focusses on fitness, diabetes, meditation, and cancer thematic apps. Its' methodology implies the following: It measures emotional level by sentimental analysis of tweets. A Twitter application that uses system dynamics, mobile health adoption, and social media analytics methodologies. It has three main functions creation of dataset, tweets sentiment analysis, and creation mHealth adoption causal loop diagram.

Health Canada is regulation for mobile application, especially for mobile healthcare apps [33]. According to [34], there are 180 mobile health application in Bangladesh, mostly healthcare information, baby care and fitness apps. As for mHealth application in national languages, there is 96 healthcare applications in Spanish [35]. The most popular apps focus on meditation, mindfulness, and fitness applications.

Currently, there are numerous opportunities for innovation in medicine [36]. Numerous researchers are working in the creation of an entirely new era in medicine. Moreover, numerous healthcare professionals support the development of online medicine [37], and the number of this technology will have reached 80% of the population of the world by 2026. Innovation is the key to the implementation of technologies and applications that will automatically collect and process information about human health. Equipping hospitals with a quick electronic medical record (EMR) [38] (the program initiated by the USA) opens up new possibilities for information gathering and integration. The application of diet tracking applications confirms the potential impact of mobile computing for enhanced life quality [39].

Sustainable technologies and applications will be used by most people in their everyday life [40], and will consequently be an essential part of our lives. In modern medicine, people have started to use mobile phone applications actively. Recent advancements in computer and software technologies can be used to build the best healthcare system in the world, the branches of which will be able to make a correct diagnosis and treat patients effectively [41].

The future of mobile health applications is promising. People may only spend ten minutes a year at a consultation with a doctor, and doctors are unable to inform patients about essential information during this short space of time [42]. However, it is also possible that a human body has undergone some significant changes before the consult. The mobile app continuously monitors even lean processes going on in the body and stores them as data [43].

Glucose monitoring has already been tested today. Soon, the right diagnostic information about human health will be given with mobile applications. This information will be useful for medical communities [44]. In this case, the mobile application will automatically provide this data with the information. Diagnostic data often incomprehensible for people will be clear for doctors, and they will respond quickly accordingly.

The number of smartphones and laptop computers are increasing on every day, which, thanks to developing new technologies, offer plenty of opportunities for engineers, designers, entrepreneurs and researchers. The access to the Cloud enables people to get information quickly, process and store it in real-time [45]. Discussions on healthcare are becoming a trending topic. The creation of interactive games for the Internet and a mobile phone has reached an incredible dynamism. Thus, there has been prepared the groundwork for a new starting point in the field of medicine - this is a new era in healthcare known as "interactive medicine."

In 2011, the World Health Organization (WHO) predicted the increase of eHealth's mobile medical technologies and their potential to change the types of healthcare systems around the world. By 2016 there were 260,000 mobile medical apps worldwide. It is possible to measure basic parameters of lung function and a heart rate, do blood tests, or while using microfluidic fluids of the body, and it can be used as an ophthalmoscope or an otoscope using smartphone applications [46].

Mobile applications have a severe influence on patients' health. Therefore, the Food and Drug Administration (FDA) has decided to certify them. The FDA monitors and regulates health apps. The applications used to diagnose a critical condition, regulate drug delivery systems, or monitor critical factors, such as blood oxygen levels, is subject to the supervision and observation of the FDA [47]. Currently, the FDA reviews only 20 applications per year.

The Ministry of Health of Georgia has created a vaccine schedule mobile application for absentminded parents. The free mobile and tablet app for the immunization schedules (working on iOS and Android platforms) was developed by the National Center for Disease Control and Public Health (NCDC) [48] and with the financial support of UNICEF for parents.

A mobile application called "112 Georgia" allows the users can call an emergency centre or talk with an operator. This mobile application also has an SOS button, which enables people to call an ambulance unnoticeably. Moreover, the application is available for Android and iOS users as well. WHO is actively reviewing the latest advances and suggests countries involved to engage in international medical research.

4 Results and Discussion

Figure 1 shows the location of the analyzed literature. According to research results,

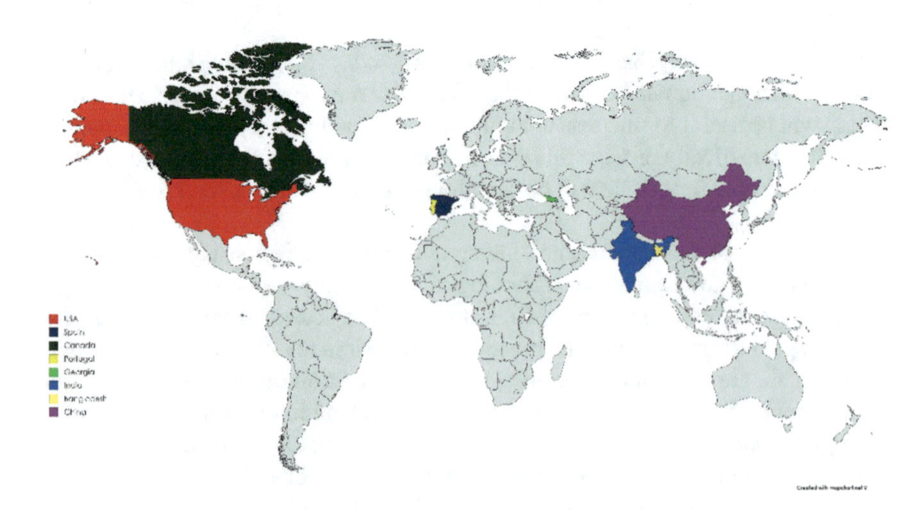

Fig. 1 Locations of the analyzed research papers (Developed by chapters authors)

the scopes of the paper cover the following countries: US, Spain, Canada, Portugal, Georgia, India, and Bangladesh.

In total, 43% of the analyzed papers have been published in 2020 and 2019. Moreover, 6% of the papers are from 2016 and 8% from 2018 (Fig. 2).

Table 1 presents the methodology and functions of the considered mobile health applications. The analyzed applications are mobile health applications evaluation, 112 Georgia, Bangla Health Guide, and Predicting dementia using EMR data (Table 1).

According to Table 1, the following mHealth application functions are highlighted: the creation of dataset, silent SOS, stress reduction, and prediction. Also, telemedicine methods are useful to measure human feelings, and patient data can be used to provide a prediction of disease conditions. Moreover, the collaboration of the mHealth applications in several domain fields, such as diet, fitness, aerobic, services brings a high number of functions to a customer.

Table 2 present the most relevant studies and details its application, research methodology and results.

According to Table 2, using social networks is an effective method to measure users' mental conditions. Also, the keywords are an appropriate method to find appropriate applications on different mobile platforms. Besides, cross-cultural adaptation and performance evaluation will be required. Table 3 details the research outcomes and limitations of the analyzed studies.

According to Table 3, mHealth applications are used worldwide with different purposes, outcomes and limitations. Mobile computing technologies are applied for multiple processes such as emotion recognition, healthcare process tracking, and patient monitoring. Furthermore, there are different methodologies applied in the development and of statistical analysis of mobile health. However, ML is a critical method to support the creation of self-monitoring and prediction healthcare applications. Fuzzy and linear regression is the most used method for data analysis. The

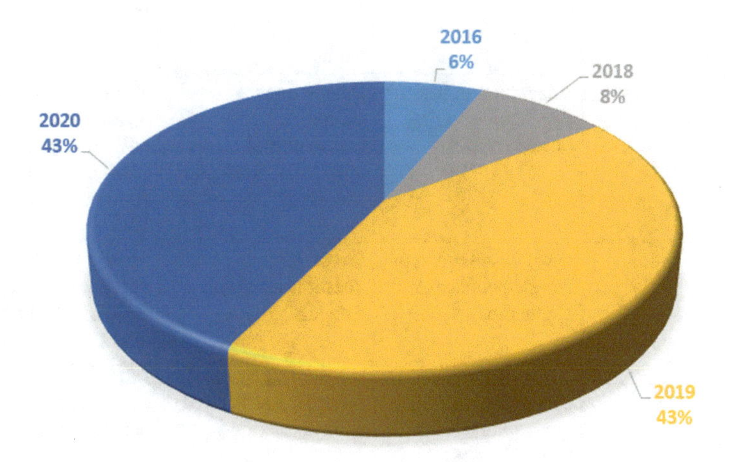

Fig. 2 Distribution of the analysed literature according to the year of publication (Developed by chapter authors)

Table 1 Relevant studies methodology and functionality

mHealth app	Methodology	Functions	Publication
Mobile apps evaluation	System dynamics, mobile health adoption, and social media analytics	Creation of dataset, tweets sentiment analysis, creation mHealth adoption causal loop diagram, and measurement of sentiment scores for mHealth apps	[32]
112 Georgia	The mobile application enables users to call emergency services	Call 112 (Emergency and Operative Response Center in Georgia), Chat with the call-taker, Silent SOS	[48]
Bangladesh health guide	Collection of about 500 useful health tips	Diet, Stress Reduction, Aerobics, Healthy Snack, Fitness Safety	[34]
Disease prediction	ML and Human-interpretable data processing	Predicting Dementia	[38]

Table 2 mHealth application studies with research methodology and results

Reference	Application	Research methodology	Results
Pai et al. [32]	Measure the stress level	Acquisition data form social network "Twitter"	Adaptation a new system into a social network
Karim et al. [34]	Health apps status in Bangladesh	Search applications on Google Play, App Store, Windows Phone App, and Blackberry App World with appropriate vital words	The most popular apps are the applications which provide healthcare information (36%)
Martin Payo et al. [35]	Adaptation of Mobile Application Rating Scale in Spain	Design three processes: cross-cultural adaptation, translation, and metric evaluation	There was 540 mobile health application, and only 41 apps were suitable for Mobile Application Rating Scale
Pai et al. [31]	Determined usage of mobile healthcare application	Structured questionnaires and data analyzed	Mobile health application is popular female (50,25%) and adults within 18–25 age (48,44%)
Miled et al. [38]	Diagnosis, prescriptions, and medical notes	Analyzing structured and unstructured	Prognosticate dementia in a year

Table 3 Summary of the analyzed literature considering the research outcomes and limitations

Methodology	Region	Research outcomes	Limitations
Twitter data analysis [32]	Worldwide	User's emotions will be measured	Demand one thousand tweets each from four different mHealth apps such as meditation, fitness, cancer, and diabetes
Risk-based model [33]	Canada	Regulation of software as a medical device	Investigation of 100,000 mHealth apps from Apple and Android app stores
Brainstorming and descriptive statistics [34]	Bangladesh	Clustered mHealth apps into Health Care Information, Body Fitness, Mother, Doctors' Information, Pregnancy and Child, Diseases Specific, Nutrition, and Health/Medical Institute	9 searching keywords which might not cover all mHealth apps
Mobile Application Rating Scale [35]	Spain	Apps' internal consistency, temporal stability, and inter-rater reliability	The research used only health and fitness apps from iTunes and Google Play
Viral load and CD4 counts [36]	USA	Define training and clinical needs for mobile care services	Describe a future project
Data Preprocessing and Model Development [38]	USA	Rx model got 70.39% and 65.63% accuracy appropriate one year and three years	The system was used in 3456, and 1738 patients appropriate within one year and three years
Linear regression model [39]	Florida	Self-monitoring rates and elucidate potential times	Participants: Adults with mean body mass index 34.7 ± 5.6)
Fuzzy geometric mean-fuzzy analytic hierarchy process [42]	Taiwan	Relaxation of the related medical laws and unobtrusiveness	Considers only the studies published over the last ten years on Google Scholar

presented literature review will help researchers and professionals about the current state of the art regarding mobile computing technologies for enhanced living environments. However, the presented study has limitations since the authors considering only papers between 2016 and 2020.

5 Conclusion

The paper presents a review of mobile computing for enhanced living environments and mobile health applications between 2016 and 2020. The results show promising applications of mobile computing for enhanced health and well-being. There are some limitations to create enhanced mobile health applications. More research initiatives are needed in the mHealth domain, including a cross-domain characterization by the involvement of computer science and healthcare researchers. Moreover, the existence of multiple marketplaces presents a relevant challenge since mobile applications need to be created for multiple operating system platforms.

References

1. Marques G (2019) Ambient assisted living and internet of things. In: Cardoso PJS, Monteiro J, Semião J, Rodrigues JMF (eds) Harnessing the internet of everything (IoE) for accelerated innovation opportunities. IGI Global, Hershey, pp 100–115
2. Dohr A, Modre-Opsrian R, Drobics M, Hayn D, Schreier G (2010) The internet of things for ambient assisted living. In: 2010 seventh international conference on information technology: new generations, Las Vegas, NV, USA, pp 804–809. https://doi.org/10.1109/ITNG.2010.104
3. Helbostad JL et al (2017) Mobile health applications to promote active and healthy ageing. Sensors 17(3):622. https://doi.org/10.3390/s17030622
4. Lv Z, Chirivella J, Gagliardo P (2016) Bigdata oriented multimedia mobile health applications. J Med Syst 40(5):1–10. https://doi.org/10.1007/s10916-016-0475-8
5. Bhuyan SS et al (2016) Use of mobile health applications for health-seeking behavior among US adults. J Med Syst 40(6):153. https://doi.org/10.1007/s10916-016-0492-7
6. Oniani S, Pires IM, Garcia NM, Mosashvili I, Pombo N (2019) A review of frameworks on continuous data acquisition for e-Health and m-Health. In: Proceedings of the 5th EAI international conference on smart objects and technologies for social good, Valencia, Spain, September 2019, pp 231–234. https://doi.org/10.1145/3342428.3342702
7. Hussain M et al (2018) Conceptual framework for the security of mobile health applications on android platform. Telematics Inform 35(5):1335–1354. https://doi.org/10.1016/j.tele.2018.03.005
8. A generalised adoption model for services: A cross-country comparison of mobile health (m-health) - ScienceDirect. https://www.sciencedirect.com/science/article/abs/pii/S0740624X15000751. Accessed 23 Feb 2020
9. Atallah N, Khalifa M, El Metwally A, Househ M (2018) The prevalence and usage of mobile health applications among mental health patients in Saudi Arabia. Comput Methods Programs Biomed 156:163–168. https://doi.org/10.1016/j.cmpb.2017.12.002
10. m-Health adoption by healthcare professionals: a systematic review | Journal of the American Medical Informatics Association | Oxford Academic. https://academic.oup.com/jamia/article/23/1/212/2379923. Accessed 23 Feb 2020
11. JMIR - Barriers to the Use of Mobile Health in Improving Health Outcomes in Developing Countries: Systematic Review | Kruse | Journal of Medical Internet Research. https://www.jmir.org/2019/10/e13263. Accessed 23 Feb 2020
12. Memdani L (2020) Demonetisation: a move towards cashless economy in India. Int J Electron Bank 2(3):205–211
13. Dhingra S, Madda RB, Gandomi AH, Patan R, Daneshmand M (2019) Internet of things mobile–air pollution monitoring system (IoT-Mobair). IEEE Internet Things J. 6(3):5577–5584. https://doi.org/10.1109/JIOT.2019.2903821

14. Marques G, Ferreira CR, Pitarma R (2019) Indoor air quality assessment using a CO2 monitoring system based on internet of things. J Med Syst 43(3):1–10. https://doi.org/10.1007/s10916-019-1184-x
15. Marques G, Miranda N, Kumar Bhoi A, Garcia-Zapirain B, Hamrioui S, de la Torre Díez I (2020) Internet of things and enhanced living environments: measuring and mapping air quality using cyber-physical systems and mobile computing technologies. Sensors 20(3):720. https://doi.org/10.3390/s20030720
16. What factors influence the mobile health service adoption? A meta-analysis and the moderating role of age - ScienceDirect. https://www.sciencedirect.com/science/article/pii/S0268401216308854. Accessed 23 Feb 2020
17. Sinyolo S (2020) Technology adoption and household food security among rural households in South Africa: the role of improved maize varieties. Technol Soc 60:101214. https://doi.org/10.1016/j.techsoc.2019.101214
18. Lee U et al (2019) Intelligent positive computing with mobile, wearable, and IoT devices: literature review and research directions. Ad Hoc Netw 83:8–24. https://doi.org/10.1016/j.adhoc.2018.08.021
19. Piran P et al (2019) Medical mobile applications for stroke survivors and caregivers. J Stroke Cerebrovasc Dis 28(11):104318. https://doi.org/10.1016/j.jstrokecerebrovasdis.2019.104318
20. Núñez M, Bonhaure D, González M, Cernuzzi L (2020) A model-driven approach for the development of native mobile applications focusing on the data layer. J Syst Softw 161:110489. https://doi.org/10.1016/j.jss.2019.110489
21. Pinem AA, Yeskafauzan A, Handayani PW, Azzahro F, Hidayanto AN, Ayuningtyas D (2020) Designing a health referral mobile application for high-mobility end users in Indonesia. Heliyon 6(1):e03174. https://doi.org/10.1016/j.heliyon.2020.e03174
22. Chávez A, Borrego G, Gutierrez-Garcia JO, Rodríguez L-F (2019) Design and evaluation of a mobile application for monitoring patients with Alzheimer's disease: a day center case study. Int J Med Informatics 131:103972. https://doi.org/10.1016/j.ijmedinf.2019.103972
23. Vagg T, Shortt C, Fleming C, McCarthy M, Tabirca S, Plant BJ (2019) Designing heterogeneous-mHealth apps for cystic fibrosis adults. In: Cystic fibrosis - heterogeneity and personalized treatment. IntechOpen, pp 1–20
24. Perry K, Shearer E, Sylvers P, Carlile J, Felker B (2019) mHealth 101: an introductory guide for mobile apps in clinical practice. J Technol Behav Sci 4(2):162–169. https://doi.org/10.1007/s41347-019-00108-8
25. Mao X, Zhao X, Liu Y (2020) mHealth App recommendation based on the prediction of suitable behavior change techniques. Decis Support Syst 132:113248. https://doi.org/10.1016/j.dss.2020.113248
26. Li J, Zhang C, Li X, Zhang C (2020) Patients' emotional bonding with MHealth apps: an attachment perspective on patients' use of MHealth applications. Int J Inf Manag 51:102054. https://doi.org/10.1016/j.ijinfomgt.2019.102054
27. Khusial RJ et al (2020) Effectiveness of myAirCoach: a mHealth self-management system in asthma. J Allergy Clin Immunol Pract 8(6):1972–1979. https://doi.org/10.1016/j.jaip.2020.02.018
28. Alam MZ, Hoque MdR, Hu W, Barua Z (2020) Factors influencing the adoption of mHealth services in a developing country: a patient-centric study. Int J Inf Manag 50:128–143. https://doi.org/10.1016/j.ijinfomgt.2019.04.016
29. Qudah B, Luetsch K (2019) The influence of mobile health applications on patient - healthcare provider relationships: a systematic, narrative review. Patient Educ Couns 102(6):1080–1089. https://doi.org/10.1016/j.pec.2019.01.021
30. Wen Y, Zhang F, Wang H, Gong Z, Miao Y, Deng Y (2020) A new secret handshake scheme with multi-symptom intersection for mobile healthcare social networks. Inf Sci 520:142–154. https://doi.org/10.1016/j.ins.2020.02.007
31. Pai RR, Alathur S (2019) Assessing awareness and use of mobile phone technology for health and wellness: Insights from India. Health Policy Technol 8(3):221–227. https://doi.org/10.1016/j.hlpt.2019.05.011

32. Pai RR, Alathur S (2018) Assessing mobile health applications with twitter analytics. Int J Med Informatics 113:72–84. https://doi.org/10.1016/j.ijmedinf.2018.02.016
33. Zawati MH, Lang M (2019) Mind the app: considerations for the future of mobile health in Canada. JMIR mHealth uHealth 7(11):e15301. https://doi.org/10.2196/15301
34. Karim M et al (2016) Mobile health applications in Bangladesh: a state-of-the-art, September 2016, pp 1–5. https://doi.org/10.1109/CEEICT.2016.7873148
35. Martin Payo R, Fernandez Álvarez MM, Blanco Díaz M, Cuesta Izquierdo M, Stoyanov SR, Llaneza Suárez E (2019) Spanish adaptation and validation of the mobile application rating scale questionnaire. Int J Med Informatics 129:95–99. https://doi.org/10.1016/j.ijmedinf.2019.06.005
36. Wulsin L, Pinkhasov A, Cunningham C, Miller L, Smith A, Oros S (2019) Innovations for integrated care: the association of medicine and psychiatry recognizes new models. Gen Hosp Psychiatry 61:90–95. https://doi.org/10.1016/j.genhosppsych.2019.04.007
37. Namin AT, Vahdat V, DiGennaro C, Amid R, Jalali MS (2020) Adoption of new medical technologies: the effects of insurance coverage vs continuing medical education. Health Policy Technol 9(1):39–41. https://doi.org/10.1016/j.hlpt.2020.01.003
38. Miled ZB et al (2020) Predicting dementia with routine care EMR data. Artif Intell Med 102:101771. https://doi.org/10.1016/j.artmed.2019.101771
39. Turner-McGrievy GM et al (2019) Defining adherence to mobile dietary self-monitoring and assessing tracking over time: tracking at least two eating occasions per day is best marker of adherence within two different mobile health randomized weight loss interventions. J Acad Nutr Diet 119(9):1516–1524. https://doi.org/10.1016/j.jand.2019.03.012
40. Senthil Kumar A, Camacho S, Searby ND, Teuben J, Balogh W (2020) Coordinated capacity development to maximize the contributions of space science, technology, and its applications in support of implementing global sustainable development agendas—a conceptual framework. Space Policy 51:101346. https://doi.org/10.1016/j.spacepol.2019.101346
41. Soni P, Pal AK, Islam SH (2019) An improved three-factor authentication scheme for patient monitoring using WSN in remote health-care system. Comput Methods Programs Biomed 182:105054. https://doi.org/10.1016/j.cmpb.2019.105054
42. Chen T (2020) Assessing factors critical to smart technology applications to mobile health care – the FGM-FAHP approach. Health Policy Technol. https://doi.org/10.1016/j.hlpt.2020.02.005
43. Yoo S et al (2020) Developing a mobile epilepsy management application integrated with an electronic health record for effective seizure management. Int J Med Inf 134:104051. https://doi.org/10.1016/j.ijmedinf.2019.104051
44. Amalba A, Abantanga FA, Scherpbier AJJA, van Mook WNKA (2019) The role of community-based education and service (COBES) in undergraduate medical education in reducing the mal-distribution of medical doctors in rural areas in Africa: a systematic review. Health Prof Educ. https://doi.org/10.1016/j.hpe.2019.09.003
45. Li S, Yu C-H, Wang Y, Babu Y (2019) Exploring adverse drug reactions of diabetes medicine using social media analytics and interactive visualizations. Int J Inf Manag 48:228–237. https://doi.org/10.1016/j.ijinfomgt.2018.12.007
46. Imison C, Castle-Clarke S, Watson R, Edwards N (2016) Delivering the benefits of digital health care, p 108
47. Dombeck CB, Hinkley T, Fordyce CB, Blanchard K, Roe MT, Corneli A (2020) Continued investigator engagement: Reasons principal investigators conduct multiple FDA-regulated drug trials. Contemp Clin Trials Commun 17:100502. https://doi.org/10.1016/j.conctc.2019.100502
48. Teitelman AM, Kim SK, Waas R, DeSenna A, Duncan R (2018) Development of the NowIKnow mobile application to promote completion of HPV vaccine series among young adult women. J Obstet Gynecol Neonatal Nurs 47(6):844–852. https://doi.org/10.1016/j.jogn.2018.06.001
49. Omarini A (2018) The Digital Transformation in Banking and The Role of Fin Techs in the New Financial Intermediation Scenario. Online at https://mpra.ub.uni-muenchen.de/85228/. MPRA Paper No. 85228, UTC, Bocconi University- Department of Finance- Via Roentegen, Milano, Italy, pp 1:12

Sentiment Analysis for Modern Standard Dialect Using Genetic Folding Algorithm

Mohammad A. Mezher

Abstract This paper presents an approach based on the Genetic Folding (GF) algorithm to discriminate between positive and negative Arabic reviews in social media. Sentiment analysis is a set of techniques for collecting subjective opinions from a document. Sentiment analysis in Arabic is difficult due to rich language morphology. In addition, it is further complicated if the assignment is applied to very informal and noisy Twitter data. This paper aims to extract and evaluate Twitter data written in the Saudi Arabic dialect. The Twitter dataset is labeled for subjectivity and sentiment analysis at the sentence level. The model uses TF-IDF representations and applies three different kernels for support vector machines to compare with the GF algorithm, intending to extract users' sentiment from written text. Experimental results showed that the GF algorithm leads to substantially better performance and showed that the mean square error was minimal, and the GF algorithm was the most accurate in predicting topic polarity. Best results were achieved using the generated GF kernel with an overall accuracy of 100% over 96% of the best other kernels.

Keywords Evolutionary algorithms · Genetic Folding · Arabic analysis · Arabic sentiment · Classification · Saudi dialect

1 Introduction

Sentiment Analysis (SA) has gained a lot of interest from the literature among the problems of Natural Language Processing (NLP) [1–3]. SA immediately accepts the opinions of the written or spoken language. SA has difficulty with Arabic because of its rich linguistic morphology. In addition, the job is complicated when extended to highly informal and noisy Twitter data. SA has identified challenges for Arabic in [4]. Dialectal Arabic (DA) was one of these difficulties. The Arabic language has a condition called diglossia where the written form of the language varies dramatically from the language encountered daily [5]. The official language is called Modern

M. A. Mezher (✉)
Fahd Bin Sultan University, Tabuk, Saudi Arabia
e-mail: mmezher@fbsu.edu.sa

A. M. A. Musleh Al-Sartawi (ed.), *The Big Data-Driven Digital Economy: Artificial and Computational Intelligence*, Studies in Computational Intelligence 974,
https://doi.org/10.1007/978-3-030-73057-4_3

Standard Arabic (MSA), and in several Arabic countries, multiple Arabic dialects are created in the spoken language. Social media vocabulary is itself highly dialectal [6].

SA methods have become necessary to support knowledge discovery as the digital text platforms of social media, networks, and organizations are constantly growing. SA program and document sources are different. Though many research groups do not agree on a definition [7], SA can be defined as a series of processes for evaluating unstructured information and understanding emotions previously unknown [8].

A common option for solving these tasks is to calculate words or embed a vector-based representation of all words in an expression. Incorporations are then used to predict the existence or absence of emotion in the text using supervised learning algorithms.

The rest of this paper is organized as follows: Sect. 2 presents in more detail the related works. Section 3 explains the subtasks of the GF algorithm. Section 4 describes the Arabic datasets and how they are pre-processed as well as the experimental setup, results, and discussions. Section 5 is the conclusion and points to some possible directions for future works.

2 Literature Review

SA, also known as opinion mining, has become one of the most important areas of NLP science. It has become a central factor for decision-makers and market executives to grasp emotions and viewpoints. Arabic sentiment analysis is one of the more complicated sentiment analysis tools of social media due to the informal, noisy content and the rich morphology of the Arabic language. Several works have been proposed for Arabic sentiment analysis. The approach [9] was based on a Discriminative Multinerential Naïve Bayes (DMNB) method with an N-grams tokenizer. Since more resources and tools have recently become available for SA, Arabic is now an evolving language.

The rich morphology and abundance of dialectal in Twitter make it both fascinating and challenging. Early Arabic studies focused on SA in health sciences, but recently there has been a lot more work on social media, especially Twitter, where Twitter has a huge number of Arabic users who mostly post and write their tweets using the Arabic language. The [10] paper introduces an Arabic language dataset about opinions on health services. Machine Learning algorithms were utilized in their experiments of SA on their health dataset.

Some work studied the utility of machine translation for sentiment analysis of Arabic texts [11]. The development of a standard Arabic Twitter dataset for sentiment, and particularly with respect to topics, will encourage further research in this regard. User demographic information in Twitter has been studied and analysed using network analysis. In Arabic, there are multiple dialects in different regions of the Arab world. Users commonly communicate in social media using their local dialect rather than the formal MSA. This introduces a core NLP problem for Arabic, namely, dialect identification. Arabic Dialect Identification (ADI) enables a finer-grained demographic identification [12].

Arabic SA has been examined in many NLP tasks [13, 14], including state-of-the-art language identification and sentiment analysis of the modern standards and the dialects of Arabic languages. This paper [15] focuses on the use of Arabic Sentiment Classification's Rough Set theory approach. Rough Set theory is a mathematical method for classifying unknown, incomplete, or ambiguous details. It can be used to significantly reduce data dimensionality without much loss of information content.

Arabic is one of the most complex languages of the world, semantically and syntactically. The text summary is a big obstacle to data mining. Noisy results, duplication, decreased readability, and incoherent sentences are part of the problem. It is suggested to use an unsupervised approach that incorporates the space model, the Continuous Bag Of Words (CBOW), clustering, and statistically dependent approaches [16].

The need for research into Arabic SA is accentuated by driving developments in various Arab regions. The [17], presents the pros and cons of the various methods used for Arabic SA and illustrates the difficulties of it including deep learning [18, 19]. In [20], they constructed Slang Sentimental Words and Idioms Lexicon (SSWIL) for SA using SVM classifier for Gaussian kernel to label the Facebook [20, 21] commentary of Arabic news.

3 GF Algorithm

A GF [22, 23] is an evolutionary algorithm [24, 25] influenced by nature's "RNA folds" that can be used to solve problems of optimization. A GF produces a chromosome population consisting of several chromosomes. The population size is a parameter that is defined by the user. The optimization problem is solved by every chromosome. To determine how successful the optimization problem is resolved, each chromosome is evaluated.

The evaluation is achieved by calculating each chromosome's fitness. A chromosome with higher fitness is better for a maximization problem than a chromosome with lower fitness. This study uses GFs to optimize the correct (positive or negative) classification of different text dialects, which contain ambiguous words in the text. Although certain algorithms use large word dictionaries with associated feeling values [25], the GFs propose how to learn the form and associated feelings applicable to each of the terms.

3.1 GF Algorithm

Algorithm 1 shows the GF pseudocode. The initial chromosome population is randomly generated in Step 2, and each chromosome is evaluated in Step 3 to see whether the initial population can solve the optimization problem. In step 5, the algorithm begins a generational loop until the full generation is completed or a solution to the problem of optimization is sought. The maximum number of generations is a parameter defined by the user.

```
Algorithm 1: Genetic Folding algorithm
Input: parameters (params) shown in Table I
1 Begin
2   Generate (a pool of floating-numbers population)
3   Evaluate (the initial population)
4   i = 0
5   While i ≤ Generations do
6       i = i + 1
7       Select (parents)
8       Perform (GF genetic operators, params)
9       Replace (new offspring) in step 8
10  Evaluate (current population)
11 Draw (best GF chromosome)
12 Return (Best GF)
```

3.2 *GF for SA: Model and Code*

While formatting a file, a type of scaling and SVM for a pre-defined kernel and GF for generating a kernel and Tfidvectorizer was performed in the proposed model. In Fig. 1, to verify which type of scaling the input text is and to take advantage of the concepts of SVM, a very wide range of kernels were used. For experimenting with Saudi dialect tweets, the dataset was broken down into a train set and a test set.

The set of hyperparameters used along with their corresponding values are:

```
# GF parameters used in a run
params['type'] = 'binary' # problem type
params['kernel'] = 'rbf, linear, polynomial, gf'
params['mutProb'] = 0.7 # mutation probability
params['crossProb'] = 0.5 # crossover probability
params['maxGen'] = 10 # max generation
params['popSize'] = 50 # population size
params['crossVal'] = 5 # cross validation slots
params[GFLength] =20 #chromosome length
#GF operators and operands
params['opList'] = [' + _s', '-_s', '*_s', ' + _v', '-_v', 'x', 'y']
# change the text file format of the experiment data
trainX, testX, trainY, testY = text_to_data(data.txt)
```

Fig. 1 GF Algorithm for SA Model (Developed by the chapter author)

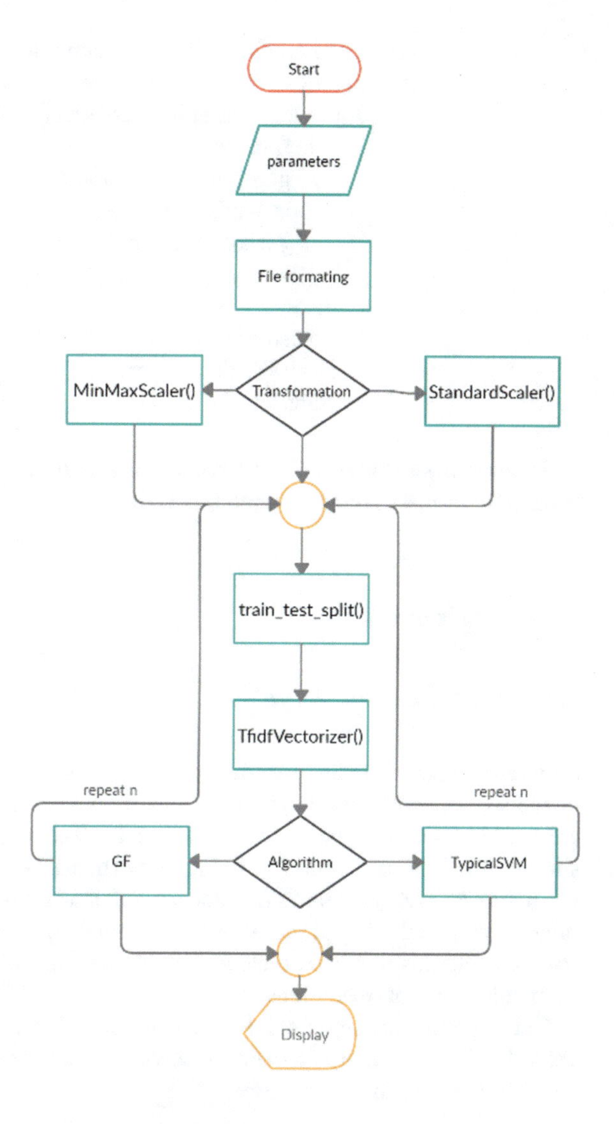


```
# Update the param values for the dataset inserted
params['trainX'] = trainX # train data attribute
params['testX'] = testX # test data attribute
params['trainY'] = trainY # label of train data
params['testY'] = testY # label of test data
```

Generally, experiments parameters were performed over a set amount of generation and population to achieve minimal error or maximum accuracy values.

```
#run GF a number of times to compute the average
for i in range(5): # number of times
    for index, kernel in enumerate(kernels):
        if kernel == 'gf':
            # generate full population
            pop = inipop(param)
            # get the best population
            mse = genpop(pop, params, i)
        else:
            mse = typicalsvm(params)
        totalMSE[kernel].append(mse)
        print('\n')
```

In order to help the Arabic NLP community in further research activities, the code is publicly available on the Github link [26].

4 Experiments

4.1 Experimental Setup

Some experimental findings are discussed in the next subsection to show the efficacy of the strategies proposed. The functions used are standard, and MinMax scalers are used to overcome the vectorization results. The so-call MinMax scaling (often simply called "normalization"—a common explanation for ambiguity) is an alternative approach to Z-score normalization (or standardization). Data are scaled to a fixed range in this method, typically 0 to 1. In comparison to standardization, the advantage of using this restricted scale is that the influence of outliers can be minimized by smaller standard deviations.

The experimental setup will be evaluated with GF and the parameters defined in Table 1. We have performed many cross-validation tests and extraction functions for most applications in the GF kernel.

4.2 Dataset

The Saudi dataset [14] which was conducted in the experiment contains more than 4k Saudi dialects. The Saudi dataset was obtained from the Twitter social network and published in Kaggle. Arabic data was provided by Saudi dialect tweets and manually labeled by the creator of the dataset as a split (positive 1, negative 0) for each tweet. This set of data focuses on the effects of political transformations and social events on Saudi Arabia's 2030 vision on Arab societies and polarizing trends and policies.

Table 1 GF Experimental parameters

Parameter types	GF
Scaling	Standard, MinMax
Mutation rate	0.7
Crossover rate	0.5
Population size	50
Generations	10
random state	2019
Chromosome length	20
Folding level	6

Table 2 Experimental dataset

Label type	# instances	# tokens
Positive	2436	52,553
Negative	1816	45,100

The Twitter-based data collection phase spanned two months, and tweets were collected in Saudi dialects for these polarizing events along with other controversial topics. These collected tweets were involved in incidents such as encouraging women to drive, to enter football stadiums, etc. Dataset statistics are shown in Table 2.

4.3 Results

Table 3 illustrates the experimental results of applying the GF algorithm to multiple scalars—Standard Scalar and MinMax scalar—compared to predefined kernel results.

Results in Table 3 show that GF improved results by 37 points in accuracy scoring, 100% compared with the highest accuracy scored with predefined kernels, making GF the new state-of-the-art for StandardScalar() on different cross-validation models. Testing GF with different cross-validation showed similar results in the StandardScalar() model.

The reason this occurred is that the GF begins producing randomly, which is then compared to the newly generated entities. An example of this can be seen in Fig. 2 with plus_s label as a root, providing the number of folding levels at 6. In testing the accuracy of the generated GF kernel, it proved efficient as we received generations higher than the fifth generations as shown in Fig. 2.

In Fig. 3 the best chromosome string was generated using the GF algorithm represented as an array of strings as follows:

['Plus_s', 'Multi_s', 'Plus_v', 'Plus_v', 'Minus_v', 'Minus_s', 'Minus_v', 'Minus_s', 'Multi_s', 'Minus_v', 'Minus_v', 'y', 'x', 'y', 'Minus_v'].

Table 3 Results of tree types of cross-validation

Cross-validation	Accuracy results of StandardScalar()			
	Polynomial	RBF	Linear	GF
25%	61.7	63.3	61.7	**100**
15%	59.7	61.0	59.7	**100**
45%	58.9	58.9	58.9	**100**
Cross-validation	Loss-value results of StandardScalar ()			
	Polynomial	RBF	Linear	GF
25%	0.38	0.37	0.38	0.38
15%	0.40	0.39	0.40	0.37
45%	0.41	0.41	0.41	**0.33**
Cross-validation	Accuracy results of MinMaxScalar()			
	Polynomial	RBF	Linear	GF
25%	61.7	63.3	61.7	79.1
15%	59.7	61.0	59.7	70.1
45%	58.9	58.9	58.9	**99.1**
Cross-validation	Loss-value results of MinMaxScalar ()			
	Polynomial	RBF	Linear	GF
25%	0.38	0.37	0.38	0.38
15%	0.40	0.38	0.40	**0.37**
45%	0.41	0.41	0.41	0.39

Where the best GF chromosome numbers in a folding point model are represented as follows:

['1.2', '3.4', '5.6', '7.8', '0.4', '9.10', '0.6', '11.12', '13.14', '0.9', '0.10', '0.11', '0.12', '0.13', '0.14'].

Both the chromosome styles, in the array of string and the array of folding-point style, are shown as a tree in Fig. 3.

Figure 4 shows that the Mean Square Errors (MSE) were tested with all kernel types, predefined, and GF kernels. The results show that the MSE of all specified kernels was overcome by the GF. Even though GF achieved the highest MSE values, the polynomial kernel was the lowest kernel to refer to that dataset.

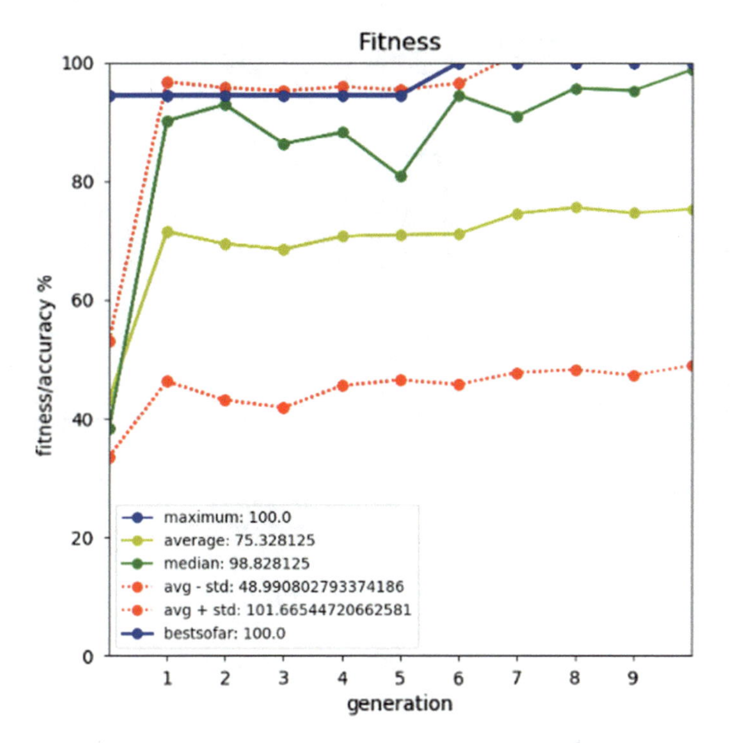

Fig. 2 Example of GF accuracy using StandardScalar (Developed by the chapter author)

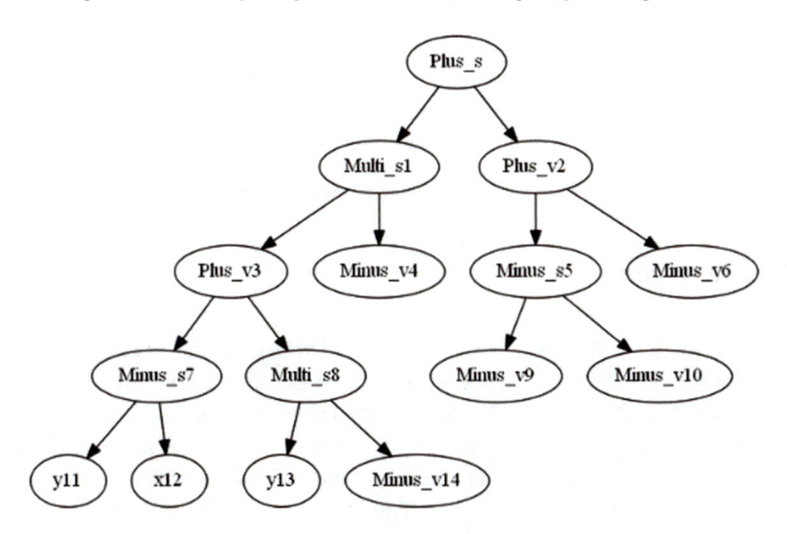

Fig. 3 Example of GF output chromosome drawn using StandardScalar (developed by the chapter author)

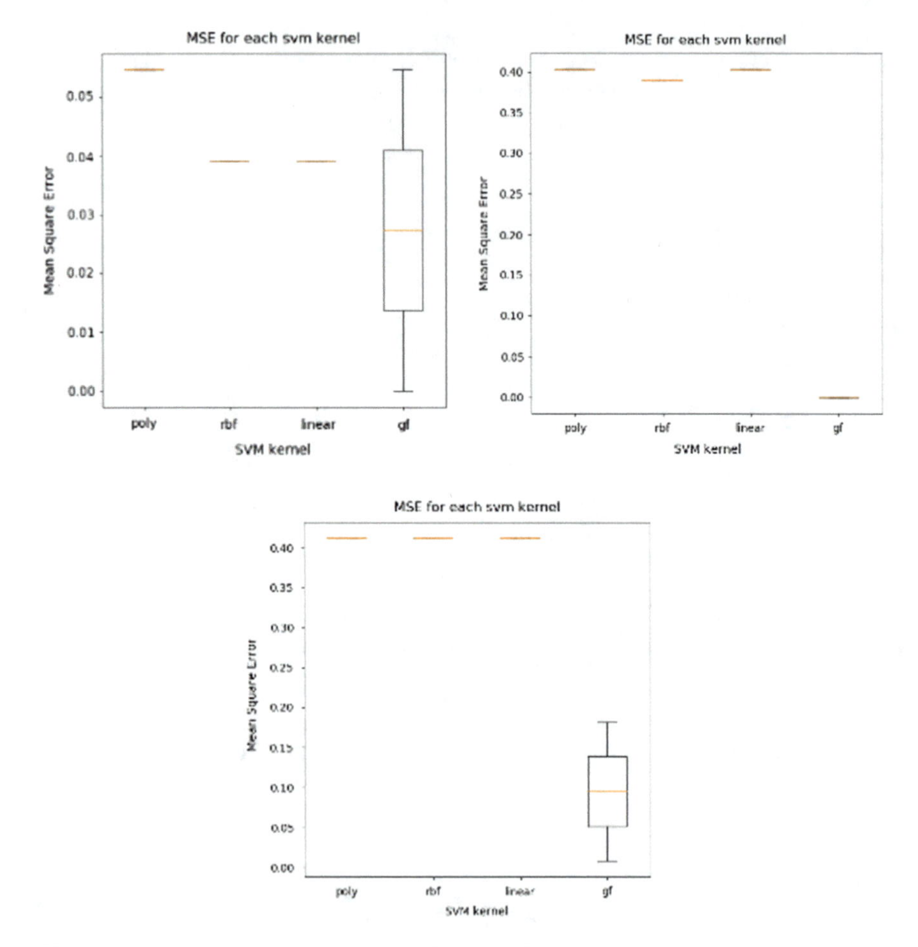

Fig. 4 Example of an image with an acceptable resolution (Developed by the chapter author)

4.4 Discussion

Figure 4 shows GF achieved state-of-the-art performance on SA, namely in the task of opinion mining. This adds truth to the assumption that evolutionary algorithms on a dialectic language only surpass the performance of predefined kernels. It is also noted that the pre-processing applied to the training data took into consideration the complexities of the Arabic dialect. Therefore, effective language vocabulary is improved by excluding excessive redundant dialects with certain common hashtags and helping to improve modeling by reducing the complexity of expression. These factors helped to reach the best results in 3 different Cross-validation and 4 different kernels.

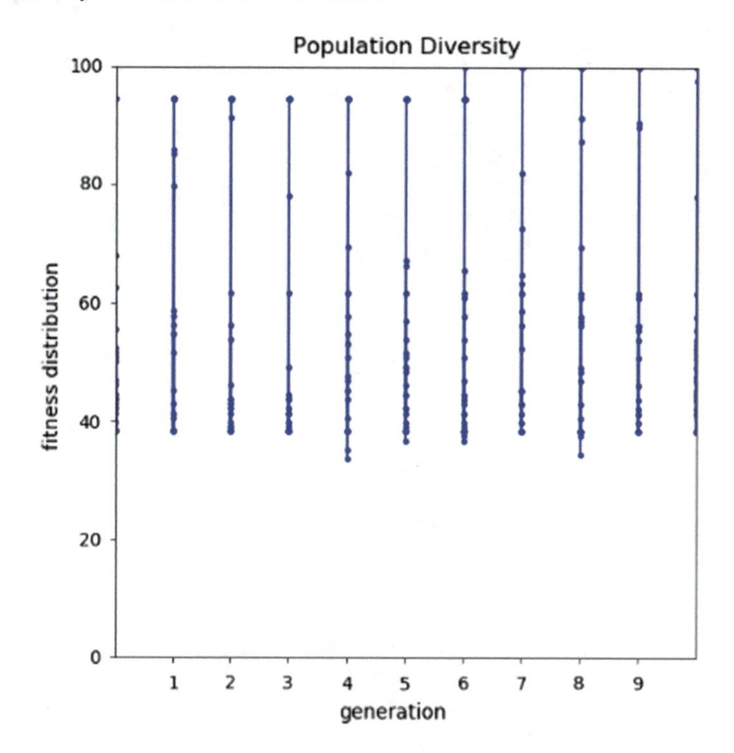

Fig. 5 Example of fitness values in a population diversity using StandardScalar (Developed by the chapter author)

Obtained results indicate that the advantage we got in the GF algorithm is better understood in an Arabic dialect model rather than that of a general Arabic language model trained on predefined kernels such as Polynomial and RBF kernels. Also, the diversity which GF can generate during the population decades can guarantee that the fitness values achieved the best results as shown in Fig. 4 (Fig. 5).

The following digraph tree shows how GF toolbox illustrates the result of the best-folded chromosome. The first group of GF digraph tree shows the terminals variables in which the last group presents the non-terminal group. For example, following relationships are captured by the GF toolbox to represent the final GF tree:

```
digraph tree {
    # terminal group
    "Plus_s";
    "Multi_s1";
    "Minus_s3";
    "Plus_s7";
    "Minus_v13";
    "y14";
    "x8";
    "Plus_v4";
    "x9";
    "Minus_v10";
    "Plus_v2";
    "Minus_s5";
    "Minus_v11";
    "x12";
    "x6";
    #Non-terminal group
    "Plus_s" -> "Multi_s1";
    "Plus_s" -> "Plus_v2";
    "Multi_s1" -> "Minus_s3";
    "Multi_s1" -> "Plus_v4";
    "Minus_s3" -> "Plus_s7";
    "Minus_s3" -> "x8";
    "Plus_s7" -> "Minus_v13";
    "Plus_s7" -> "y14";
    "Plus_v4" -> "x9";
    "Plus_v4" -> "Minus_v10";
    "Plus_v2" -> "Minus_s5";
    "Plus_v2" -> "x6";
    "Minus_s5" -> "Minus_v11";
    "Minus_s5" -> "x12";
}
```

5 Conclusions

The paper presented the first sentiment analysis for the Saudi dialect dataset using the Genetic Folding algorithm. Sentiment analysis is a relevant problem in NLP. In previous work, GF has been used to solve classification and regression problems. This paper is effective in predicting sentiment with a finer level of detail as a binary classification problem for sentiment analysis using the GF algorithm.

However, the hypothesis was that the GF algorithm compared to classical kernels leads to the flexibility to capture more subtle differences in datasets. Therefore, in this paper, I have proposed a modified GF algorithm as a way to evolve a new structure of kernels. By addressing the creation of kernels as a multi-objective problem, we have been able to generate kernels that simultaneously optimize two of the accuracy metrics and also, are optimized for computational complexity.

One way to understand how language differs across dialects is to study the speech patterns used for different purposes, such as communication with computers or as a form of branding or advertising. This project is part of the genetic folding toolbox. To help the Arabic NLP community in further research activities, the code is publicly available on the Github link [26].

References

1. Deriu J, Lucchi A, De Luca V, Severyn A, Müller S, Cieliebak M, Hofmann T, Jaggi M (2017) Leveraging large amounts of weakly supervised data for multi-language sentiment classification. In: Proceedings of the 26th international conference on world wide web, WWW 2017. Republic and Canton of Geneva, Switzerland, pp 1045–1052
2. Socher R, Pennington J, Huang EH, Ng AY, Manning CD (2011) Semisupervised recursive autoencoders for predicting sentiment distributions. In: Proceedings of the conference on empirical methods in natural language processing, EMNLP 2011, Stroudsburg, PA, USA, pp 151–161
3. Beck D (2017) Modelling representation noise in emotion analysis using Gaussian processes. In: Proceedings of the eighth international joint conference on natural language processing, Taipei, Taiwan, vol 2, pp 140–145. https://doi.org/10.1109/ICCAIS48893.2020.9096850
4. Al-Twairesh N, Al-Khalifa H, Al-Salman A (2016) AraSenTi: large-scale TwitterSpecific Arabic sentiment lexicons. In: Proceedings of the 54th annual meeting on association for computational linguistics, Berlin, Germany. Association for Computational Linguistics
5. Habash N, Eskander R, Hawwari A (2012) A morphological analyzer for Egyptian Arabic. In: Proceedings of the twelfth meeting of the special interest group on computational morphology and phonology. Association for Computational Linguistics, pp 1–9
6. Darwish K, Magdy W (2014) Arabic information retrieval. Found Trends Inf Retrieval 7(4):239–342
7. Proksch SO et al (2019) Multilingual sentiment analysis: a new approach to measuring conflict in legislative speeches. Legislative Stud Q 44(1):97–131
8. Abo MEM et al (2019) SSA-SDA: subjectivity and sentiment analysis of Sudanese dialect Arabic. In: 2019 international conference on computer and information sciences (ICCIS). IEEE
9. Alayba AM, Palade V, England M, Iqbal R (2017) Arabic language sentiment analysis on health services. In: 2017 1st international workshop on arabic script analysis and recognition (ASAR), Nancy, pp 114–118. https://doi.org/10.1109/ASAR.2017.8067771
10. Zakraoui J, Saleh M, Al-Maadeed S, AlJa'am JM (2020) Evaluation of Arabic to English machine translation systems. In: 2020 11th international conference on information and communication systems (ICICS), Irbid, Jordan, pp 185–190. https://doi.org/10.1109/ICICS49469.2020.239518
11. AlYami R, AlZaidy R (2020) Arabic dialect identification in social media. In: 2020 3rd international conference on computer applications & information security (ICCAIS), Riyadh, Saudi Arabia, pp 1–2. https://doi.org/10.1109/ICCAIS48893.2020.9096847

12. Oueslati O, Cambria E, HajHmida MB, Ounelli H (2020) A review of sentiment analysis research in Arabic language. Future Gener Comput Syst 112:408–430
13. Abo MEM, Raj RG, Qazi A (2019) A review on Arabic sentiment analysis: state-of-the-art, taxonomy and open research challenges. IEEE Access 7:162008–162024. https://doi.org/10.1109/ACCESS.2019.2951530
14. Al-Radaideh QA, Twaiq LM (2014) Rough set theory for Arabic sentiment classification. In: International conference on future internet of things and cloud, Barcelona, pp 559–564. https://doi.org/10.1109/FiCloud
15. Abdulateef S, Khan NA, Chen B, Shang X (2020) Multidocument Arabic text summarization based on clustering and Word2Vec to reduce redundancy. Information 11:59
16. Alyami S (2018) Arabic Sentiment Analysis Dataset SS2030 Dataset. Sentiment Analysis of Social Events in Arabic Saudi Dialect. https://www.kaggle.com/snalyami3/arabic-sentiment-analysis-dataset-ss2030-dataset
17. El-Masri M, Altrabsheh N, Mansour H (2017) Successes and challenges of Arabic sentiment analysis research: a literature review. Soc Netw Anal Min 7:54. https://doi.org/10.1007/s13278-017-0474-x
18. Singhal P, Bhattacharyya P (2016) Sentiment analysis and deep learning: a survey. https://www.cfilt.iitb.ac.in/resources/surveys/sentiment-deeplearning-2016-prerna.pdf
19. Ain QT, Ali M, Riaz A, Noureen A, Kamran M, Hayat B, Rehman A (2017) Sentiment analysis using deep learning techniques: a review. Int J Adv Comput Sci Appl 8(6):424
20. Soliman TH, Elmasry M, Hedar A, Doss M (2014) Sentiment analysis of Arabic slang comments on facebook. Int J Comput Technol 12(5):3470
21. Ortigosa A, Martin JM, Carro RM (2014) Sentiment analysis in Facebook and its application to e-learning. Comput Hum Behav 31:527
22. Mezher M, Abbod M (2010) Genetic folding: a new class of evolutionary algorithms, pp 279–284
23. Mezher M (2019) GFLIB: an open source library for genetic folding solving optimization problems. Artif Intell Adv 11–17
24. Holland JH (1975) Adaptation in natural and artificial systems. University of Michigan Press, Ann Arbor
25. Koza JR (1992) Genetic programming - on the programming of computers by means of natural selection. Complex adaptive systems
26. https://github.com/mohabedalgani/GFSA

The Usage of E- Learning Among Mass Communication Students

Mahmoud Gamal Sayed Abd Elrahman, Abdulsadek Hassan, Hanan Gunied, Kadhim Moans Aziz Al Saedi, and Fahema Abdulla Mohammed

Abstract The study examined the Moodle system adoption by the students in three Gulf universities "Saudi Arabia, United Arab Emirates and Kingdom of Bahrain" and examines the students perceptions towards this system from the framework of two theories: The Theory of Reasoned Action (TRA) and the Theory of Planned behavior (TPB). This study is a descriptive survey research. The convenience sampling technique was adopted in the selection of 231 respondents from the sampled universities. The results revealed that the students used the Moodle system for many purposes such as: doing assignments, to check upcoming events, downloading the course materials and to participate in discussion with my instructors. The results also revealed that the most impactful perceived ease of use factors: easy usage, flexibility and interaction. Finally, the study is suggested to investigate academic concerns and needs to develop the usage of Moodle in Arabian and Gulf universities.

Keywords Technology · Education · GCC · Moodle system · Information and communication technology (ICT)

1 Introduction

The fast growth of Information and communication technology (ICT) has influenced all aspects of contemporary life, including education that have affected by this technology. E-Learning has become widely one of the educational systems beside the traditional structure education. E-Learning is a teaching process that takes place

M. G. S. Abd Elrahman (✉)
Faculty of Mass Communication- Radio and Television Department, Beni-Suef University, Beni Suef, Egypt

A. Hassan · K. M. A. Al Saedi · F. A. Mohammed
Ahlia University, Manama, Bahrain
e-mail: kalsaedi@ahlia.edu.bh

H. Gunied
Faculty of Mass Communication, Department of Public Relations, Cairo University, Giza, Egypt

A. M. A. Musleh Al-Sartawi (ed.), *The Big Data-Driven Digital Economy: Artificial and Computational Intelligence*, Studies in Computational Intelligence 974,
https://doi.org/10.1007/978-3-030-73057-4_4

in the Internet environment making this process more interactive and collaborative between the instructor and his students. In this process, the knowledge is stored and updated regularly by the faculty members who follow the evolution of this process and monitor student performance in the tasks assign to them.

The word "Moodle" is an acronym for "Modular object- oriented dynamic learning environment", the Moodle system begun as a research of PHD by Australian Martin Dougiamas in 1999 to help educator to study at effective collaborative on – line learning community [1].

The universities–all around the world–have adopted E- Learning system as a tool of curriculum teaching staff development and student's performance by effective use of technologies that facilities the educational process and move it forward. This process enables staff members to provide their students with all materials belong the different courses not located in the same place and they can evaluate the learning process. The universities in the Middle East- especially in Gulf area- have introduced the Moodle system in E-learning to improve the quality of learning and teaching. The universities have encouraged their students to use this system. However, the traditional method of teaching and learning is still preferred. Mass Communication education requires a variety of materials; educational, physical, graphical and actual information. The traditional Mass Communication education includes the use of texts, images, lectures and books. This old style of Mass Communication education is reinforced through blending E-learning strategies that are essential in the field of Mass Communication education especially in practical courses because of the development in software programs and Radio and Television production techniques. In the Gulf region, the regulators are starting to move toward the knowledge-based economy and digital economy [2, 3]. Which required to spend more in developing the intellectual capital by investing more in developing the teaching and learning techniques and by encouraging the companies to be more social responsibilities [4, 5]. Furthermore, all the interested bodies in the information are become responsible for developing their ways when interacting with the society through utilizing the benefits of social media to disclose about their data and to help the educational institutions for adopting the cut edge technology which can be used in preparing the new generation of employee [6]. Accordingly, the universities have recently adopted the Moodle system as the traditional teaching process was changed after using Information and multimedia technology as a new method of teaching, at the same time, the learning environment was also changed by using electronic media. In spite of this, there are many barriers still face E- learning successful application where this system is still in its infancy phrase. Consequently, this study investigates Moodle system adoption by the students in three gulf universities "Saudi Arabia, United Arab Emirates and Kingdom of Bahrain" and examines the students perceptions towards this system from the framework of two theories: The Theory of Reasoned Action (TRA) and the Theory of Planned behavior (TPB).

2 Literature Review

Numerous studies have been directed towards the adoption of Moodle in leaning and teaching in educational institutions. These studies showed that the Moodle systems allows better co-operation between instructor and students and makes teaching and learning more easily through saving teaching materials on the Moodle system and download materials [1, 7]. The Moodle system provides interactivity also between instructor and students by using interactive tools to facilitate the good communication such as: Chats, assignments submission, Forums, online news and announcements, files downloads grading/marks, and online quiz [8]. Moodle produces some benefits for students like: Availability, less technical difficulties, continuous progress, flexibility, and most accepted by top universities all over the world [9]. The Moodle also provides other benefits such as: getting courses and topics, collaborating with peers in doing homework, facilitating the learning method, facilitate the self-assignments and online assessment is more objective than traditional method "Face-to-Face" learning [10], introducing the course content, course specification and course feedback [11]. The most used information materials in the Moodle are texts and images [12]. In other side, some instructors viewed that the online method of learning must combine with traditional learning [13]. Other studies showed that in higher education institutions, there is an essential need to use Information Communication Technologies (ICT) to cope with new trends and challenges in education and commit to E-learning education. The perception of the ease of use of Moodle is significantly influenced by technology complexity and trialability [14]. While Behavioral intentions has a significant influence of Moodle usage [15]. From the side of students, the studies indicated that the students must be core focus in the adoption of technology and the instructors must take part in developing teaching and learning by using Moodle system [16], the knowledge sharing and quality have a significant role among students on E-learning acceptance, at the same time, innovativeness and trust have no significant role among students on E-learning acceptance [17]. The innovativeness is influenced by the technology acceptance of teachers and perceived ease of use, perceived usefulness and subjective norms determine the behavioral intent [18].The adoption of E-learning has also a good influence in academic performance so the students must have good skills to perform the academic assignments and must be active in learning process through effective engagement that support cognitive and non-cognitive for academic performance in the universities [19]. In general, the students indicated Moodle system effectiveness and efficiency made them more satisfied with the Moodle system available at universities [20].

Objectives: To establish the relationship between variables to determine the current usage and attitude towards Moodle system among mass Communication students in Gulf universities. To determine the factors of (TRA) and (TPB) influence the Moodle system among mass Communication students in Gulf universities.

To explore the influence of other factors: Technical support, Communication and Perceived Satisfaction that affect the use of Moodle system among mass Communication students in Gulf universities.

Questions: What is the frequency of Moodle system usage among mass Communication students in Gulf universities? What are the purposes of Moodle system usage among mass Communication students in Gulf universities?

What are the factors of (TRA) and (TPB) that influence the Moodle system among mass Communication students in Gulf universities? What is the influence of other factors: Technical support, Communication and Perceived Satisfaction that affect the use of Moodle system among mass Communication students in Gulf universities?

3 Sample

Using a convenience sampling technique, the participants consist of 231 undergraduate students who are available for and willing to participate in the study (16).

Of the 231 respondents who answered gender question, slightly more than half of the respondents were males constituting 57.14% (N = 132), while women constituting 42.86% (N = 99) of the total respondents. With regard the type of university, the majority of respondents were from private universities that contributes 60.61% (N = 140) of total respondents while governmental universities only contribute 39.39% (N = 91). Regarding the Moodle experience among Mass Communication students, the majority of students came from one year and more 52.38% (N = 121), followed by from 6 months to 1 y 38.10% (N = 88) and 9.52%(N = 22) are experienced Less than 6 months. Lastly, the results revealed that the majority of respondents came from age group Less than 20 ys old 46.75% (N = 108), followed by age group from 20 to 22 yrs 42.42% (N = 98) and 10.82% (N = 25) are aged 22 yrs and more.

4 Data Analysis

To analyze the collected data, descriptive statistics were used to calculate the frequencies, percentage, mean standard deviation and Pearson using SPSS 23 version.

Table (1) displays Perceived ease of Moodle use of Mass Communication students in Gulf Universities. Table (1) shows the most impactful perceived ease of use factors, the respective mean score (M) and standard deviation (SD), and the positioning of the elements based on students' view: Learning to use Moodle system is easy (M = 2.39, SD = 0.594), it is easy to use Moodle system(M = 2.18, SD = 0.452), I found Moodle system is flexible to deal with(M = 1.99, SD = 0.555), I feel time passes quickly during Moodle system usage(M = 1.95, SD = 0.680), the interaction with Moodle system is understandable and clear (M = 1.86, SD = 0.694).

Table (2) displays perceived usefulness of Moodle use for Mass Communication students in Gulf Universities. Table (2) shows the most impactful perceived usefulness factors, the respective mean score (M) and standard deviation (SD), and the positioning of the elements based on students' view: Using Moodle system enables

Table 1 Mass Communication students' attitudes towards perceived ease of Moodle use in Gulf Universities

Statement	Strongly agree		Agreed		Disagree		Mean	SD
	F	%	F	%	F	%		
Learning to use Moodle system is easy	114	49.35	104	45.02	13	5.63	2.39	0.594
It is easy to use Moodle system	176	76.19	49	21.21	6	2.60	2.18	0.452
It is easy for me to be more skillful at using Moodle system	83	35.93	112	48.48	36	15.58	1.67	0.731
The interaction with Moodle system is understandable and clear	116	50.22	73	31.60	42	18.18	1.86	0.694
I found Moodle system is flexible to deal with	160	69.26	36	15.58	35	15.15	1.99	0.555
I feel time passes quickly during Moodle system usage	104	45.02	89	38.53	38	16.45	1.95	0.680

me to accomplish my assignments more rapidly (M = 2.42, SD = 0.685), using Moodle system allows instructors to submit quizzes with model answers (M = 2.41, SD = 0.717), Moodle system allows me to get information faster (M = 2.26, SD = 0.668), Moodle system improves my scientific performance (M = 2.21, SD = 0.713) and Moodle system provides course syllabus and specification (M = 2.01, SD = 0.688).

Table (3) displays Behavioral Control of Mass Communication students in Gulf Universities. Table (3) shows the most impactful Behavioral Control factors, the respective mean score (M) and standard deviation (SD), and the positioning of the elements based on students' view: Using Moodle system would be entirely within my control (M = 2.04, SD = 0.664), I have ability, resources and knowledge to use Moodle system (M = 1.99, SD = 0.771) and I am using Moodle system without help (M = 1.90, SD = 0.795).The mean scores of selected factors are shown respectively.

Table (4) displays Subjective Norms of Moodle use among Mass Communication students in Gulf universities. Table (4) shows the most impactful subjective norms factors, the respective mean score (M) and standard deviation (SD), and the positioning of the elements based on students' view: I think that the university would support the Moodle system usage (M = 2.06, SD = 0.791), my colleagues think

Table 2 Mass Communication students' attitudes towards perceived usefulness in Gulf Universities

Statement	Strongly agree		Agreed		Disagree		Mean	SD
	F	%	F	%	F	%		
Moodle system improves my scientific performance	103	44.59	89	38.53	39	16.88	2.21	0.713
Moodle system allows me to get information faster	112	48.48	90	38.96	29	12.55	2.26	0.668
Using Moodle system enables me to accomplish my assignments more rapidly	83	35.93	122	52.81	26	11.26	2.42	0.685
Using Moodle system enhances my learning effectiveness	107	46.32	67	29.00	57	24.68	2.00	0.704
Using Moodle system allows instructors to submit quizzes with model answers	72	31.17	128	55.41	31	13.42	2.41	0.717

Table 3 Mass Communication students' attitudes towards Behavioral Control in Gulf Universities

Statement	Strongly agree		Agreed		Disagree		Mean	SD
	F	%	F	%	F	%		
I am using Moodle system without help	85	36.80	83	35.93	63	27.27	1.90	0.795
Using Moodle system would be entirely within my control	129	55.84	56	24.24	46	19.91	2.04	0.664
I have ability, resources and knowledge to use Moodle system	94	40.69	69	29.87	68	29.44	1.99	0.771

Table 4 Mass Communication students' attitudes towards Subjective Norms in Gulf Universities

Statement	Strongly agree		Agreed		Disagree		Mean	SD
	F	%	F	%	F	%		
My colleagues think that I should use Moodle system	64	27.71	129	55.84	56	24.24	2.03	0.653
My colleagues think that I must use Moodle system	69	29.87	94	40.69	68	29.44	1.97	0.771
I recommend my colleagues to strongly use Moodle system	97	41.99	93	40.26	41	17.75	1.76	0.735
My instructors think that I should participate in Moodle system activities	101	43.72	85	36.80	45	19.48	2.00	0.751
I think that the university would support the Moodle system usage	103	44.59	105	45.45	23	9.96	2.06	0.791

that I should use Moodle system (M = 2.03, SD = 0.653), my instructors think that I should participate in Moodle system activities (M = 2.00, SD = 0.751) and my colleagues think that I must use Moodle system (M = 1.97, SD = 0.771).

Table (5) displays technical support of Moodle among Mass Communication students in Gulf Universities. Table (5) shows the most impactful technical support factors, the respective mean score (M) and standard deviation (SD), and the positioning of the elements based on students' view: When I face any difficulty in Moodle system usage, I get rapidly assistance (M = 2.61, SD = 0.614), the university offers technicians to provide assistance (M = 2.53, SD = 0.656)and when I face any problem in Moodle system usage, I know where to get assistance (M = 2.17, SD = 0.651). The mean scores of selected factors are shown respectively.

Table (6) displays Social influence of Moodle for Mass Communication students in Gulf Universities. Table (6) shows the most impactful Social influence factors, the respective mean score (M) and standard deviation (SD), and the positioning of the elements based on students' view: My instructors want me to use Moodle system frequently (M = 2.06, SD = 0.765), my instructors support me to use Moodle system (M = 1.98, SD = 0.645), my instructors expect to carry on Moodle system usage (M = 1.84, SD = 0.699) and my colleagues want me to use Moodle system frequently (M = 1.65, SD = 0.494).The mean scores of selected factors are shown respectively.

Table 5 Mass Communication students' attitudes towards technical support of Moodle in Gulf Universities

Statement	Strongly agree		Agreed		Disagree		Mean	SD
	F	%	F	%	F	%		
The university offers technicians to provide assistance	131	56.71	92	39.83	8	3.46	2.53	0.565
When I face any problem in Moodle system usage, I know where to get assistance	127	54.98	71	30.74	33	14.29	2.17	0.651
When I face any difficulty in Moodle system usage, I get rapidly assistance	158	68.40	57	24.68	16	6.93	2.61	0.614

Table 6 Mass Communication students' attitudes towards Social influence of Moodle in Gulf Universities

Statement	Strongly agree		Agreed		Disagree		Mean	SD
	F	%	F	%	F	%		
My instructors expect to carry on Moodle system usage	77	33.33	113	48.92	41	17.75	1.84	0.699
My instructors want me to use Moodle system frequently	60	25.97	95	41.13	76	32.90	2.06	0.765
My instructors support me to use Moodle system	46	19.91	135	58.44	30	12.99	1.98	0.645
My colleagues want me to use Moodle system frequently	82	35.50	147	63.64	2	0.87	1.65	0.494

5 Conclusion

Moodle has become ubiquitous and has a part of university life. consequently, students are spending fundamental part of their time on Moodle system in all universities all around the world. University students are considered the largest users of Moodle system. Despite a ubiquitous of this system, there is a dearth of studies in Gulf universities on how Moodle system affects the students' performance and academic achievement by changing the way students interact.

The study addressed this issue by examining the nature of Moodle usage and all factors affected it the frequency of this usage and addressed the purposes why students use Moodle system. The researcher used survey method to collect the data from Gulf universities that applied this system.

The study analyzed the main factors of the theory of Reasoned Action (TRA) and the theory of Planned behavior (TPB) applying to a sample of 231 students from Gulf universities who used the Moodle system in these universities.

The results contribute to understand – to some extent- the usage of Moodle system in Gulf universities in some fields such as:

- The purposes of using Moodle system: The results revealed that the students used the Moodle system for many purposes such as: doing assignments, to check upcoming events, downloading the course materials and to participate in discussion with my instructors. The study agreed with [8] who concluded that Moodle system provides interactivity also between instructor and students by using interactive tools to facilitate the good communication such as: Chats, assignments submission.
- Perceived ease of Moodle use: The results revealed that the most impactful perceived ease of use factors: easy usage, flexibility and interaction. The study agreed with [9] who concluded that Moodle systems allows availability, less technical difficulties, continuous progress, flexibility, and most accepted by top universities all over the world.
- Perceived usefulness: The results revealed that the most impactful perceived usefulness factors: allow instructors to submit quizzes with model answers, allow students to get information faster, improve the scientific performance of students and enhance learning effectiveness. The results contradict the study of [19], whose study revealed that the innovativeness is influenced by the technology acceptance of teachers and perceived ease of use, perceived usefulness and subjective norms determine the behavioral intentions, not the Moodle only.
- Behavioral Control: The results revealed that the most impactful behavioral control factors: Using Moodle system would be entirely within the control of students, ability, resources and knowledge to use and using without help. The results contradict the study of [17] who confirmed the students must have good skills to perform the academic assignments.
- Subjective Norms: The results revealed that the most impactful Subjective Norms factors: the university supports the Moodle system usage; the colleagues facilitate the usage among themselves and the instructors think that their students use

Moodle system. the instructors must take part in developing teaching and learning by using Moodle system. The results agree with the study of) [16] who confirmed the instructors must take part in developing teaching and learning by using Moodle system.

- Moodle system adoption: The results revealed that the most Moodle system adoption factors: The instructors support the use of Moodle system, the students plan to use Moodle system in the future, the students feel pressure from my instructors to adopt Moodle system, they have basic knowledge of Moodle system tools to use it in their courses. The results contradict the study of [16] who confirmed that the students must be core focus in the adoption of technology and the instructors must take part in developing teaching and learning by using Moodle system.
- Technical support: The results revealed that the most Moodle system adoption factors: When the students face any difficulty in Moodle system usage, they get rapidly assistance, the university offers technicians to provide assistance and when they face any problem in Moodle system usage, they know where to get assistance. The results agree the study of [9] who confirmed that Moodle system has less technical difficulties.
- Social influence: The results revealed that the most Social influence factors: The instructors want the students to use Moodle system frequently, they support them to use Moodle system and their colleagues want them to use Moodle system frequently. The results agree the study of [10] who confirmed that Moodle contributes in collaborating with peers in doing homework.
- Behaviorial intention: The results revealed that the most behaviorial intention factors: the students intend to use Moodle system constantly in the future, If the Moodle system is easy, they will use it as much as possible and If the Moodle system is easy, they will download/upload course materials. The results agree the study of [11] who confirmed that the Moodle facilitate in introducing the course content, course specification and course feedback.

References

1. Alhothli NI (2015) Investigating the impact of using moodle as an e-learning tool for students in an english language institute. Un-Published Master Thesis, The University of Montana, the Graduate School at Scholar Works, p 21
2. Al-Sartawi A (2020) Social media disclosure of intellectual capital and firm value. Int J Learn Intellect Capital 17(4):312–323
3. Al-Sartawi A (2020) Does it pay to be socially responsible? Empirical evidence from the GCC countries. Int J Law Manage 62(5):381–394
4. Al-Sartawi A (2020) Information technology governance and cybersecurity at the board level. Int J Crit Infrastruct 16(2):150–161
5. Al-Sartawi A (2019) Assessing the relationship between information transparency through social media disclosure and firm value. Manage Acc Rev 18(2):1–20
6. Al-Sartawi A (2018) Ownership structure and intellectual capital: evidence from the GCC countries. Int J Learn Intellect Capital 15(3):277–291

7. Dharmendra C, Kumar C, Abhishek B, Soni CA (2011). Effective e-learning through moodle. Int J Adv Technol Eng Res (IJATER) 1(1):34–38
8. Costa C, Alvelos H, Teixeira L (2012). The use of Moodle e-learning platform: a study in a Portuguese University. In CENTERIS - Conference on ENTER prise Information Systems, pp 334–343
9. Prashant B, Londhe BR (2014). From teaching, learning to assessment: Moodle experience at b'school in India. In Symbiosis Institute of Management Studies Annual Research Conference (SIMSARC13), Procedia Economics and Finance 11, pp 857–865
10. Oproiu GC (2015). A study about using e-learning platform (moodle) in university teaching process. In The 6th international conference education world 2014 "education facing contemporary world issues", 7th - 9th November 2014, Procedia - social and behavioral sciences 180, pp 426–432
11. Abdulrasool FE, Turnbull SI (2020) Exploring security, risk, and compliance driven IT governance model for universities: applied research based on the COBIT framework. Int J Electron Bank 2(3):237–265
12. Zainuddin N, Idrus R, Jamal AFM (2016). Moodle as an ODL teaching tool: a perspective of students and academics. Electron J e-Learning 14(4):282−290
13. Rymanova I, Baryshnikov N, Grishaeva A (2015). E-course based on the LMS moodle for english language teaching: development and implementation of results. In XV international conference "linguistic and cultural studies: traditions and innovations", LKTI 2015, 9–11 November 2015, Tomsk, Russia, pp 236–240
14. Teo T, Zhou M, Fan AC, Huang F (2019). Factors that influence university students' intention to use moodle: a study in macau, educational technology. Res Dev New York 67(3):749–766
15. Aliyu OA, Arasanmi C, Ekundayo S (2019). Do demographic characteristics moderate the acceptance and use of the Moodle learning system among business students? Int J Educ Dev Inf Commun Technol (IJEDICT) 15(1):179−192
16. van de Heyde V, Siebrits A (2019) The ecosystem of e-learning model for Authors: higher education. South African J Sci 115(5/6):78–83
17. Kim HJ, Hong AJ, Song HD (2019) The roles of academic engagement and digital readiness in students' achievements in university e-learning environments. Int J Educ Technol High Educ 16(21):1−18
18. Musleh Al-Sartawi AMA (2020) E-Learning improves accounting education: case of the higher education sector of bahrain. In Themistocleous M, Papadaki M, Kamal MM (eds) Information Systems. EMCIS 2020. Lecture Notes in Business Information Processing, vol 402. Springer, Cham
19. Suradi Z, Baqwir JAM, Yusoff NH (2018) Factors affecting the use of moodle system among students in Dhofar university. In Proceedings of 130th The IRES International Conference, Taipei, Taiwan, 26th -27th July, p 1
20. Govender I, Khumalo S (2014) Reasoned action analysis theory as a vehicle to explore female students' intention to major in information systems. J Commun 5(1):39

The Influence of Relational Capital on the Relationship Between Intellectual Capital and Earnings Per Share in the Digital Economy in the Jordanian Industrial Sector

Kamelia Moh'd Khier Al Momani, Nurasyikin Jamaludin, Wan Zalani Wan Zanani Wan Abdullah, and Abdul-Naser Ibra-him Nour

Abstract This study aims to examine whether relational capital (RC) affects the relationship between intellectual capital (IC) and earnings per share (EPS), for 50 industrial companies on the Amman Stock Exchange (ASE) during the period 2008–2017. The value-added intellectual coefficient (VAIC™) was used to measure IC. The study firstly investigates if the EPS is affected by VAIC™ after that with its' components. Secondly, it investigates whether RC affects the relationship between VAIC™ then with its' components, with EPS. the results of this study show that VAIC™ has a significant positive influence on EPS. On its component side, human capital efficiency (HCE) and structural capital efficiency (SCE) have a significant positive influence on EPS, but capital employed efficiency (CEE) did not have any influence on the EPS. Based on the panel data and hierarchical regression analysis of the moderating effect of RC, the results reveal that RC weakly moderates the relationship between CEE and EPS. Lastly, this study does not find any significant influence of RC on the relationship between the VAIC™, HCE, SCE, and EPS. The results suggest that industrial companies in Jordan and in the Middle East must pay attention to the IC elements, especially the HC, which is the main pillar of VAIC™ and make new researches about RC. In addition, the current study is useful to policymakers and managers in their decisions regarding the development of the industry that leads to improved and increased corporate returns.

Keywords Relational capital · Intellectual capital · Earnings per sher

---

K. M. K. Al Momani (✉) · N. Jamaludin · W. Z. W. Z. W. Abdullah
Universiti Malaysia Terengganu, Kuala Terengganu, Malaysia

N. Jamaludin
e-mail: asyikin@umt.edu.my

W. Z. W. Z. W. Abdullah
e-mail: zanani@umt.edu.my

A.-N. I. Nour
Al-Najah National University, Nablus, Palestine
e-mail: a.nour@najah.edu

A. M. A. Musleh Al-Sartawi (ed.), *The Big Data-Driven Digital Economy: Artificial and Computational Intelligence*, Studies in Computational Intelligence 974,
https://doi.org/10.1007/978-3-030-73057-4_5

1 Introduction

Since the 1970s, global economic growth has changed due to the rapid development of communications, computers, biological engineering, and digital engineering, which led to the emergence of the digital economy. Where the digital economy focuses on employees rather than machines, as they are the main drivers of wealth, moreover, they are the actual users of technology, and responsible for achieving sustainable economic growth [1–3].

The digital economy is a concept that refers to the widespread use of information and communication technology in all social and economic endeavors, which expands opportunities, stimulates economic growth, and improves the delivery of public services [4]. The digital economy plays an important role in building "smart societies" that enhance the capabilities of all actors, public authorities, government, companies, and citizens, especially youth and women, to make better and informed decisions and reduce inequality. The digital economy revolution or the fourth industrial revolution IR4.0 is just as important as the previous industrial revolutions that coincided with the introduction of steam power, combustion engines, and electricity [5].

In recent years, the economy reliance on artificial intelligence (AI) which aims to make machines thinking like humans [6], and intellectual capital (IC) which depend on the intangible assets more than tangible assets, became an alternative to land, monetary, and physical capital [7]. Moreover, IC became considered more important than financial and physical capital, especially in the digital economy that based on technological development. Furthermore, the previous studies indicate that companies' competitive advantage and performance are affected by their IC [8].

IC defined as all of the intangible or knowledge assets in the company [9]. The IC has three main components are human capital (HC) which depends on the employees who have a set of cognitive and organizational capabilities, that enable them to produce new ideas, structural capital (SC) relying on organizations such as database, computers, software, patents and trademarks, and relational capital (RC) based on coordinating the relation between the company and the market channels such as customers and suppliers [10–12].

IC cannot be measured and its value cannot be determined because it is a set of intangible assets that are not purchased and thus cannot be controlled by the companies; such as the skills and experiences of employees [13]. So, Pulic in 1998 developed a new model, value added intellectual coefficient (VAIC™) to measure the total value creation efficiency of the companies that comes of three key resources: Capital Employed Efficiency (CEE), Human Capital Efficiency (HCE) and Structural Capital Efficiency (SCE) [14].

The present study will utilize the VAIC™ model, which gained widespread use in many countries because it is easy to measure, and at the same time, allows the comparison to be made between companies [15]. The data used in the model are based on audited financial statements, which is considered reliable [16].

This study came for two main reasons. First, to study the effect of VAIC™ on earnings per share (EPS) in Jordanian industrial companies that lack such studies. Secondly, the VAIC™ model suffers from the lack of inclusion of RC. In this regard, RC defined as a wealth of knowledge from marketing channels [17]. RC refers to the relationship between the companies and their customers, suppliers, and industry association [18]. According to [19], the importance of RC to companies, is to gain a unique competitive advantage, through obtaining the right feedback from external channels such as customers and suppliers. Besides, RC is the most important component for any company because it is the main source of income for the companies in sustaining its business activities [20].

This study will be a few of its kind in Jordan and the Middle East, according to the researcher's knowledge, to study the relationship between VAIC™ and EPS in the industrial sector, using RC as a moderating variable, by using the data of the industrial companies listed on the Amman Financial Market from 2008 to 2017.

Therefore, this study aims to answer the following questions:

1) What is the relationship between VAIC™ and EPS of the industrial sector in ASE?
2) What is the relationship between VAIC™ components and EPS of the industrial sector in ASE?
3) What is the relationship between RC and EPS of the industrial sector in ASE?
4) Does the RC effectively moderate the relationship between VAIC™ and EPS of the industrial sector in ASE?
5) Does the RC effectively moderate the relationship between VAIC™ components and EPS of the industrial sector in ASE?

The remainder of the study is regulated as follows. section two describes the literature review and hypotheses. While section three explains the methodology of the study. Section four presents and discusses the results. The study concludes with the main conclusions and future studies.

2 Literature Review

As seen above, AI and IC have become the most important part of investing in the era of the digital economy to create new value for companies besides maintaining and enhance competitive advantage for companies and to maintain the governance system inside it through increase the investment in technology such as cybersecurity tools and other IT governance tools [21–24]. Public in 1998 developed a methodology to measure the company's efficiency related to the intellectual capital and financial capital, by using the concept of value-added [14]. This method has been known as the VAIC™, to create a measure of performance for a knowledge-based organization [21–24]. VAIC™ was widely used by researchers looking for a relation between intellectual capital and firm performance.

The VAIC™ model used by [25, 26] to measure the IC performance of 16 commercial banks in Malaysia through the period 2001 to 2003, and found IC is the key resource of value creation in the Malaysian banks, while the value creation capacity of the Malaysian banks is attributed to HCE. As for [25, 26] study, it was written in two parts to examine if VAIC™ has an impact on the market to book value (M/B), return on assets (ROA), assets turnover ratio (ATO), and return on equity (ROE) of organizational performance, the data from the companies of the Hong Kong Stock Exchange from 2001 to 2005. The results appear no evidence to support the association between IC, as measured by VAIC™, and financial performance. While [27] study, it was one of the earliest ones which examined the relation between VAIC™ and EPS. The study was applied to 150 publicly listed companies on the Singapore Exchange. The results of the study showed that there was a positive relation between VAIC™ and the EPS. In Jordan, [28] found a positive influence of VAIC™, HCE, and CEE on the firm performance, whereas SCE has no effect on the firm performance of Jordanian industrial companies. The VAIC™ model used by [29] to discover the effect of IC on firms' competitive advantage in Indian, the results showed that the firms' competitive advantage condition is relatively better explained by some of the individual IC components especially HC rather than by the VAIC™ model. In Turkey, [28] found VAIC™ shows a significant positive impact on firm performance before the crisis. SCE and HCE have a positive significant relationship with a firm performance before the crisis, but CEE show a negative significant impact on ATO after the crisis. The study of [29] found positive indications in the relation between VAIC™ and the firm performance of the Jordanian industrial companies.

In recent years, a few of studies have emerged aimed at developing a modified VAIC™ model. In Taiwan, [30] study a modified VAIC™ model by added innovation capital which measures by R&D expenses to the VAIC™ model and found that there was a positive relationship between IC and each of the firm performance. In Indonesia [19] suggested a modified VAIC™ model for measuring the value-based performance of the Indonesian banking sector by added RCE to the VAIC™, the results show that the M-VAIC™ can be used to measure the performance of all industries. In an extensive study, [31] propose a modified VAIC™ model by dividing the structural capital to customer capital, innovation capital, process capital. Whereas [32] modified the VAIC™ by added RC to the model and considered RC as an element of VAIC™ and measure it as all types of marketing and sales promotion expenses.

So, this study comes to examines the relationship between IC that measure by VAIC™ and EPS in the industrial sector in Jordan, it also examines whether RC moderates the relationship between VAIC™ and its components with the EPS.

3 Study Model and Hypothesis Development

Figure 1 displays the conceptual framework for the present study. It shows the relationship among the variables mentioned in the study to answer the questions of the present study. It shows the relationship between the VAIC™ as an independent

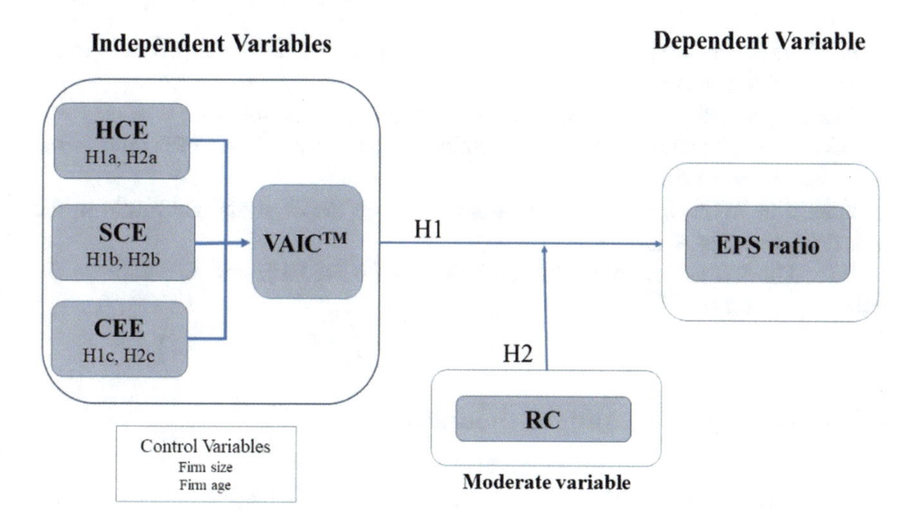

Fig. 1 Conceptual framework (Developed by chapter authors)

variable, firm performance as a dependent variable, and the RC as a moderating variable; to examine the relationship among the VAICTM and firm performance, and then to investigate the influence of RC on the relationship between VAICTM and firm performance.

The independent variable is VAICTM that consists of three components are: HCE, SCE, and CEE. These variables are expected to have an association with EPS for Jordanian industrial companies. Then, there is one hypothesis between VAICTM and EPS H1 that will be separated into three sub hypotheses in order to examine the relationship between HCE, SCE, and CEE and EPS. These hypotheses are illustrated in Fig. 1.

H1 developed to fill the gap in the previous studies that excluded examined the relationship between VAICTM and EPS ratio in the industrial companies in Jordan. Thus, it is expected that VAICTM should have an important relationship with the EPS.

H1: VAICTM has a significant relationship with EPS in the industrial sector in ASE.

These hypothesis separates into three sub hypotheses as follow:

H1a: HCE has a significant relationship with EPS in the industrial sector in ASE.

H1b: SCE has a significant relationship with EPS in the industrial sector in ASE.

H1c: CEE has a significant relationship with EPS in the industrial sector in ASE.

The second hypothesis developed to explained the indirect relationship of RC on the VAICTM and EPS. Besides, this study considered the RC is the market and advertising expenses [19, 31]. This hypothesis has assumed to explain the importance of VAICTM to be more efficient in explaining EPS and help the stakeholder to select the companies that have a sustainable VA.

The hypothesis that related the RC and its relationship with VAICTM and EPS:

H2: The RC as moderator has an influence on the VAIC™ and EPS ratio in the industrial sector in ASE.

These hypothesis separates into three sub hypotheses as follow:

H2a: The RC as moderator has an influence on the HCE and EPS ratio in the industrial sector in ASE.

H2b: The RC as moderator has an influence on the SCE and EPS ratio in the industrial sector in ASE.

H2c: The RC as moderator has an influence on the CEE and EPS ratio in the industrial sector in ASE.

4 Research Design and Methodology

4.1 Research Design

This research employs quantitative research approach. Panel data utilized to observe a number of variables in the companies, over a varied time [33]. An advantage of the panel data through cross-section analysis is that a model can be structured for estimating the impact that some time-varying variables (the values of which also vary across individuals) have on some dependent variable. Using secondary data is appropriate for this research as financial data are readily available from trusted authorities either in printed version or online database. For the purpose of this research, the unit of analysis is the companies in industrial sector listed on ASE.

4.2 Research Sample

The population comprises all companies in the industrial sector listed in ASE (77 companies). Sample data was collected from annual reports through the period 2008–2017 for companies in industrial sector listed on ASE. The selected companies are based on the following criteria:

1) The industrial company must be listed on the ASE within the period of this study.
2) The shares of the industrial companies must be actively published and traded in ASE during the period of study.
3) The financial year of the industrial companies in ASE must end in 31st of December every year.

Of the 77 industrial companies, 50 companies met the sampling criteria. Table 1 presents the sample for this study.

Table 1 The sample

Number	Sub-sector	N	percent
1	Pharmaceutical and medical industries	5	10%
2	Chemical industries	8	16%
3	Paper and cardboard industries	1	2%
4	Printing and packaging	1	2%
5	Food and beverages	8	16%
6	Tobacco and cigarettes	2	4%
7	Mining and extraction industries	11	22%
8	Engineering and construction	6	12%
9	Electrical industries	3	6%
10	Textiles, leathers and clothing	5	10%
	Total	50	100%

4.3 Measuring Variables and Study Models

The study has three main variables as discussed in the previous chapter. First, the independent variable is VAIC™ and it is component HCE, SCE, and CEE. Second, the dependent variable is EPS. Third, moderating variable proposed is RC. Moreover, proposed a control variable which includes company size and company age.

The VAIC™ model used to measure companies' IC, by uses companies' annual reports and estimates the total efficiency of the IC and asset value of the company [14]. There are six steps in arriving at the VAIC™ calculation, Table 2 summarizes the calculation steps of VAIC™ $_{it}$.

Where the VA according to [34, 35] is appeared the company capacity to create value from the resources investments including dividends, interests and salaries, taxes. While HCE estimates the VA created on an average profit per employee the amount of value in each monetary unit which is generated in HC that consider an indicator to create value added because it is estimated the average profit per employee or the contribution of the employee in the created value in the company [34, 36] SCE measures the amount of SC needed to generate a monetary unit of VA [14, 34, 35].

Table 2 Summary of calculation steps of VAIC™

Step	Variables	Equation
Step 1	VA	$VAit = OPit + ECit + Dit + Iit + Divit + Tit$
Step 2	HCE	$HCE_{it} = VA_{it}/HC_{it}$
Step 3	SC	$SC_{it} = VA_{it} - HC_{it}$
Step 4	SCE	$SCE_{it} = SC_{it}/VA_{it}$
Step 5	CEE	$CEE_{it} = VA_{it}/CE_{it}$
Step 6	VAIC™	$VAIC_{it}^{TM} = CEE_{it} + HCE_{it} + SCE_{it}$

However the CEE illustrates the amount of VA which created through one monetary unit of the physical and financial capital of the company [35], and the VAIC™ illustrates the amount of VA which is created per invested monetary unit in each resource in the company.

EPS ratio is one of the traditional firms' performance, usually used as a measure for the evaluation of companies by financial market analysts. It is a measure of profitability that includes the result of operating, investing, and financing decisions [25]. EPS ratio is defined as the amount of income earned on the common stock through an accounting period [37]. It is a condition for companies listed on the ASE to state EPS in companies' annual reports. It is a common indicator used in most analysts' reports in ASE. EPS ratio is calculated as follow:

$$EPS = \frac{Net\ Income - Preferred\ Dividends}{Weighted\ Average\ Numper\ of\ Shares} \quad [37]$$

Control variables are used to minimize external influences that may affect the relationship between intellectual capital and corporate performance [38]. This research uses company age and size as control variables, the company age is indicated by using the natural logarithm for the company age [39], while company size is the natural logarithm for book value of the total assets in the company [40, 41].

4.4 Study Models

The present study is using multiple regression analysis for testing the hypotheses. The specific test used is hierarchical regression to examine the interaction between independent and moderator variables on the dependent variable by determining the order of entry of the variables in the regression models. According to [42] the hierarchical model might include moderating variables and panel data.

On the other hand, the regression models for hypotheses and sub-hypotheses have been formulated through the models as below:

First model: To examine the effect of VAIC™ on EPS of the company i in year t.

Model 1: $EPS_{it} = \alpha 0 + \beta 1\ VAIC^{TM}_{it} + \beta 2\ logsize_{it} + \beta 3\ logage_{it} + (ui + \varepsilon it)$.

Second model: To examine the influence of VAIC™ and RCE on EPS of the company i in year t.

Model 2: $EPS_{it} = \alpha 0 + \beta 1\ VAIC^{TM}_{it} + \beta 2\ RCE_{it} + \beta 3\ logsize_{it} + \beta 4\ logage_{it} + (ui + \varepsilon it)$.

Third model: To examine the influence of VAIC™, RCE, and interaction of RCE on EPS of the company i in year t.

Model 3: $EPS_{it} = \alpha 0 + \beta 1\ VAIC^{TM}_{it} + \beta 2\ RCE_{it} + \beta 3\ RCE_{it} * VAIC^{TM}_{it} + \beta 4\ logsize_{it} + \beta 5\ logage_{it} + (ui + \varepsilon it)$.

Fourth model: To examine the influence of VAIC™ components on EPS of the company i in year t.

Model 4: $EPS_{it} = \alpha 0 + \beta 1\ HCE_{it} + \beta 2\ SCE_{it} + \beta 3\ CEE_{it} + \beta 4\ logsize_{it} + \beta 5\ logage_{it} + (ui + \varepsilon it)$.

Fifth model: To examine the influence of VAIC™ components and RCE on EPS of the company i in year t.

Model 5: $EPS_{it} = \alpha 0 + \beta 1\ HCE_{it} + \beta 2\ SCE_{it} + \beta 3\ CEE_{it} + \beta 4\ RCE_{it} + \beta 5\ logsize_{it} + \beta 6\ logage_{it} + (ui + \varepsilon it)$.

Sixth model: To examine the influence of VAIC™ components, RCE, and interaction of RCE on EPS of the company i in year t.

Model 6: $EPS_{it} = \alpha 0 + \beta 1\ HCE_{it} + \beta 2\ SCE_{it} + \beta 3\ CEE_{it} + \beta 4\ RCE_{it} + \beta 5\ RCE_{it} * HCE_{it} + \beta 6\ RCE_{it} * SCE_{it} + \beta 7\ RCE_{it} * CEE_{it} + \beta 8\ logsize_{it} + \beta 9\ logage_{it} + (ui + \varepsilon it)$.

Where; EPS it = Earning Pare Share of company i in year t. $VAIC^{TM}_{it}$ = Value-Added Intellectual Coefficient of company i in year t. HCE_{it} = Human Capital Efficiency of company i in year t. SCE_{it} = Structural Capital Efficiency of company i in year t. CEE_{it} = Capital Employed Efficiency of company i in year t. The moderating variable is RCE_{it} = Relational Capital Efficiency of company i in year t. Control Variables are Logsize = Natural logarithm for book value of the total assets in the company. Logage = Natural logarithm for company age Interaction Terms: RCE_{it}*VAIC™ = Interaction between Value-Added Intellectual Coefficient of company i in year t and Relational Capital Efficiency of company i in year t. $RCE_{it}*HCE_{it}$ = Interaction between Human Capital Efficiency and Relational Capital Efficiency of company i in year t. $RCE_{it}*SCE_{it}$ = Interaction between Structural Capital Efficiency and Relational Capital Efficiency of company i in year t. $RCE_{it}*CEE_{it}$ = Interaction between Capital Employed Efficiency and Relational Capital Efficiency of company i in year t. β0 = is the intercept of Regression Model. β1, β2, β3, β4, β5, β6, β7, β8, β9 = the Parameters of Regression Model (Regression Coefficients). (ui + εit) = the error items.

5 Results and Discussions

To test the effect of RCE as a moderator on the relationship between the VAIC™ and the EPS ratio, hierarchical regression is used for control variables according to models 1, 2, and 3.

Table 3 presents the results of hierarchical regression analysis in testing the effect of RCE on the relationship between VAIC™ and EPS ratio. The first step is to test the control variables in this study, which is the size and age of the company, and the results were as follows: Chi^2 value for control variables is significant at 5% with the value of 8.50, with a significant relationship between EPS and company size at Z-test 2.91 at level 1%, but insignificant relationship with company age.

Step two, show there is a positive and significant relationship between VAIC™ and EPS (coefficient = 0.1988612) at 1% level of significance (Z = 13.16). R^2 change (27.44%) is significant. This means that an additional 27.44% of the variation in

Table 3 Hierarchical regression results for indirect relationship of RCE on VAIC™ and EPS ratio

Variables	$EPS_{it} = \alpha_0 + \beta_1 VAIC^{TM}_{it} + \beta_2 RCE_{it} + \beta_3 RCE_{it} * VAIC_{it}^{TM} + \beta_4 logsize_{it} + \beta_5 logage_{it} + \varepsilon_{it}$							
	Step 1		Step 2		Step 3		Step 4	
	Control Variables		Model 1 Independent Variables		Model 2 Moderate Variable		Model 3 Interaction Variable	
	Coef	Z-test	Coef	Z-test	Coef	Z-test	Coef	Z-test
Constant	−0.4038895	−2.53**	−1.781613	−1.31	−0.1456804	−1.32	−0.1258299	−1.14
Control Effect								
Size	0.0274373	2.91***	−0.0044499	−0.53	−0.0052271	−0.77	−0.0059876	−0.88
Age	−0.0038272	−0.26	0.0237977	1.85*	0.0207389	1.91*	0.018139	1.65*
Main Effect								
VAIC™			0.1988612	13.16 ***	0.1900729	13.54 ***	0.1909562	13.61***
Moderate Effect								
RCE					−0.2826239	-8.33 ***	−0.1969227	−2.89***
Interaction Effect								
RCE*VAIC™							−0.0874954	−1.45
R^2	0.1090		0.3834		0.6258		0.6299	
R^2 change	–		0.2744		0.2424		0.0041	
Chi^2	8.50		184.95		286.09		289.11	
Significant Chi^2	0.0143		0.0000		0.0000		0.0000	

Notes: ***, **, and * significant at the levels 1, 5, and 10%
VAIC™: Value added intellectual coefficient. Size: company size. Age: company age. RCE: Relational Capital Efficient.
RCE*VAIC™: interaction between VAIC™ and RCE

EPS is due to the effectiveness of the VAIC™ and control variables. Therefore, H1 is supported.

In fact, this is an expected outcome because many previous studies support this hypothesis and found a significant relationship between VAIC™ and EPS such as [28, 43–47]. But the results of the present study are not consistent with some studies that found no relationship between VAIC™ and EPS [48–53].

Step three, when adding the moderating variable RCE, the R^2 increased to 0.6258 with R^2 change (24.24%) is significant. This means that an additional 24.24% of the variation in EPS by the effectiveness of the RCE. The VAIC™ has a positive and significant relationship with the EPS coefficient 0.1900729 at the 1% level of significance ($Z = 13.54$), and RCE is negative and significant with EPS coefficient −0.2826239 at the level 1% of significant ($Z = -8.33$). While company size not significant but company age has a positive relationship with EPS.

The fourth step, presents the impact of interaction terms RCE*VAIC™ on the EPS ratio. The Chi^2 value is 289.11 and significant. While, the R^2 of 0.6299 indicates the strong model explanatory power, this means 62.99% percent of the level of the EPS can be explained by model variables with R^2 change 0.41%. Regarding the interaction between RCE*VAIC™, Table 3 clarifies that beta coefficient for the interaction − 0.0874954 is negative and not significant at chi^2 −1.45. Thus, it can be decided that

RCE does not influence the relationship between VAIC™ and EPS. Therefore, H2 is not supported.

This means investors are not interested in investing in IC and they are not interested in RC, but rather that investors are more interested in tangible assets and finances. The concept of VAIC™ is a new conception, and not completely understood by industrial companies' managers in Jordan or the Arab world [50]. Figure 2 clarifies that there is no interaction effect of RCE on the VAIC™ and EPS nexus.

Table 4 presents the results of hierarchical regression analysis to test the sub-hypotheses for H1 which are H1a, H1b, and H1c, and H2 which are H2a, H2b, and H2c, to test the effect of components of VAIC™ on the EPS, and to examine the influence of RCE on the relationship between components of the VAIC™ and EPS ratio.

Step one comes to test the control variables and the results were as follows: the R^2 of 0.1090 indicates the strong model explanatory power, while, Chi^2 value for control variables is significant at 5% level with the value of 8.50. There is a significant positive relationship between company size and EPS ratio at 1% level (coefficient = 0.0274373, Z-test = 2.91). Meanwhile, there is no significant relationship between EPS ratio and company age (coefficient = −0.0038272, Z-test = −0.26).

Step two indicates that there is a significant relationship between HCE, SCE and EPS ratio, but insignificant relationship between CEE and EPS ratio. The R^2 change is 0.2719 and significant. This means that an additional 27.19% of the variation in the EPS is due to the effectiveness of the HCE, SCE, SCE, and control variables.

In model 4, two components of VAIC™, HCE (coefficient = 0.0912264, Z-statistic = 6.41) and SCE (coefficient = 0.3001, Z-statistic = 3.27) are found to be positively significant, therefore support H1a and H1b. On the other hand, there is no significant relationship between CEE and EPS, therefore H1c is rejected. Notably, the R^2 value of 38.09% shows that the variations in the EPS ratio of industrial companies in ASE are explained by the components of the VAIC™ (HCE, SCE, and CEE) and control variables. Furthermore, the probability F is 0.0000 which indicates the overall model is significant at a 1% level and considered fit.

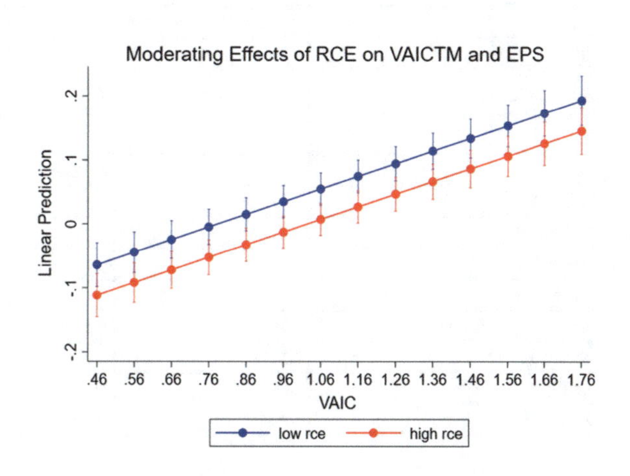

Fig. 2 Moderating Effect of RCE on VAIC™ and EPS (Developed by chapter authors)

Previous studies found a significant relationship between HCE, SCE, and CEE with EPS [45]. The results in present study vary with prior studies such as [43, 51] that found HCE has a significant relationship with EPS but the other components are not significant. In contrast [28] found SCE is the most effective factor than HCE and CEE before and after the subprime mortgage crisis, especially with EPS. The results of the present study also find SCE had a significant effect on the EPS, which is in line with [30].

In the third step, the Chi^2 value is 0.6412 and significant. The R^2 change is 0.2603. This means that an additional 26.03% of the variation in the EPS is due to effectiveness of the RCE. The company' size and age are insignificant. HCE, SCE, and CEE are positively significant with EPS ratio at Z-test 7.00, 2.48, and 3.18 at level 1%, 5%, and 1% respectively, with the coefficient 0.0934348, 0.2157243, and 0.1048973 respectively with EPS ratio. But the RCE is negatively significant with the EPS ratio at Z-test −8.00 at 1% level and coefficient −0.272002.

Step four tests the sub-hypothesis H2a, H2b, and H2c, The Chi^2 value for this step is 385.07 and significant at probability 0.0000. The R^2 is 0.6416 and R^2 change

Table 4 Hierarchical regression results for indirect relationship of RCE on VAIC™ components and EPS ratio

$EPS_{it} = \alpha_0 + \beta_1 HCE_{it} + \beta_2 SCE_{it} + \beta_3 CEE_{it} + \beta_4 RCE_{it} + \beta_5 RCE_{it} * HCE_{it} + \beta_6 RCE_{it} * SCE_{it} + \beta_7 RCE_{it} * CEE_{it} + \beta_8 logsize_{it} + \beta_9 logage_{it} + \varepsilon_{it}$

Variables	Step 1		Step 2		Step 3		Step 4	
	Control Variables		Model 4 Independent Variables		Model 5 Moderate Variable		Model 6 Interaction Variable	
	Coef	Z-test	Coef	Z-test	Coef	Z-test	Coef	Z-test
Constant	−0.4038895	−2.53	0.0022406	0.02	0.0021457	0.02	0.0466071	0.42
Control Effect								
Size	0.0274373	2.91***	−0.0083428	−1.03	−0.0078652	−1.15	−0.0093332	−1.40
Age	−0.0038272	−0.26	0.0152744	0.86	0.0124194	1.13	0.0053947	0.50
Main Effect								
HCE			0.0912264	4.21***	0.0934348	7.00***	0.0843902	6.64***
SCE			0.3001	2.57***	0.2157243	2.48**	0.2503722	3.03***
CEE			0.1071256	1.37	0.1048973	3.18***	0.166487	5.11***
Moderate Effect								
RCE					−0.272002	−8.00***	−0.0966541	−1.39
Interaction								
RCE*HCE							0.0760085	1.34
RCE*SCE							−0.0763446	−0.21
RCE*CEE							−0.877651	−7.25***
R^2	0.1090		0.3809		0.6412		0.6416	
R^2 change	–		0.2719		0.2603		0.0004	
Chi^2	8.50		88.32		294.37		385.07	
Prob Chi^2	0.0143		0.0000		0.0000		0.0000	

(continued)

Table 4 (continued)

$EPS_{it} = \alpha_0 + \beta_1 HCE_{it} + \beta_2 SCE_{it} + \beta_3 CEE_{it} + \beta_4 RCE_{it} + \beta_5 RCE_{it} * HCE_{it} + \beta_6 RCE_{it} * SCE_{it} + \beta_7 RCE_{it} * CEE_{it} + \beta_8 logsize_{it} + \beta_9 logage_{it} + \varepsilon_{it}$								
Variables	Step 1		Step 2		Step 3		Step 4	
	Control Variables		Model 4 Independent Variables		Model 5 Moderate Variable		Model 6 Interaction Variable	
	Coef	Z-test	Coef	Z-test	Coef	Z-test	Coef	Z-test

Notes: ***, **, and * significant at the levels 1, 5, and 10%. HCE: Human capital efficiency. SCE: Structural capital efficiency. CEE: Capital employed efficiency. RCE: Relational Capital Efficient. Size: company size. Age: company age

0.0004, this means that an additional 0.04% of the variation in the EPS is due to effectiveness of the interaction between RCE and HCE, SCE, and CEE.

According to Table 4 in step four, results for the companies' size and age are still insignificant as steps two and three, but the HCE, SCE, and CEE still positively significant with EPS ratio as the step three. But RCE becomes insignificant with the EPS ratio at Z-test −1.39 with coefficient −0.0966541.

The regression results in Table 4, show there is no significant effect of RCE on the relationship between HCE and the EPS ratio in the industrial sector in ASE (coefficient = 0.0760085, Z- statistic = 1.34). Therefore, H2a is not supported. Figure 3 explains that no effect of RCE on the HCE and EPS nexus.

The results in Table 4 do not support H2b because there is no significant effect of RCE on the relationship between SCE and the EPS ratio in the industrial sector in ASE (coefficient = −0.0763446, Z- statistic = −0.21). Figure 4 clarifies that there is no effect of RCE on the SCE and EPS nexus.

The results in Table 4 support H2c because there is a negative significant effect of RCE on the relationship between CEE and the EPS ratio in the industrial sector in ASE at 1% level (coefficient = −0.877651, Z- statistic = −7.25). Figure 5 explains that there is a negative interaction effect of RCE on CEE and EPS. However, if the company has low RCE even though it has a high CEE, it has a high EPS ratio. From

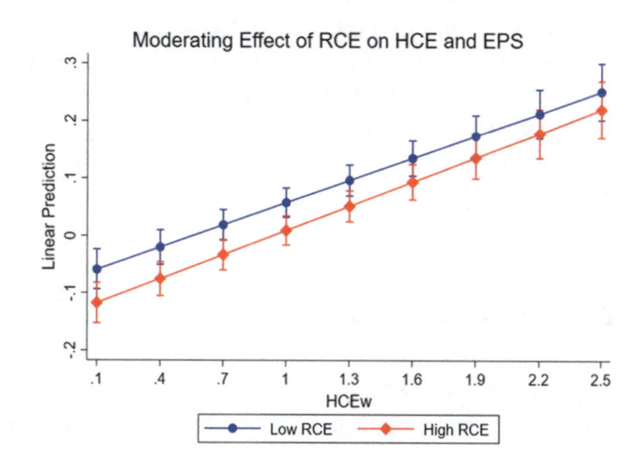

Fig. 3 Interaction effect of RCE on HCE and EPS (Developed by chapter authors)

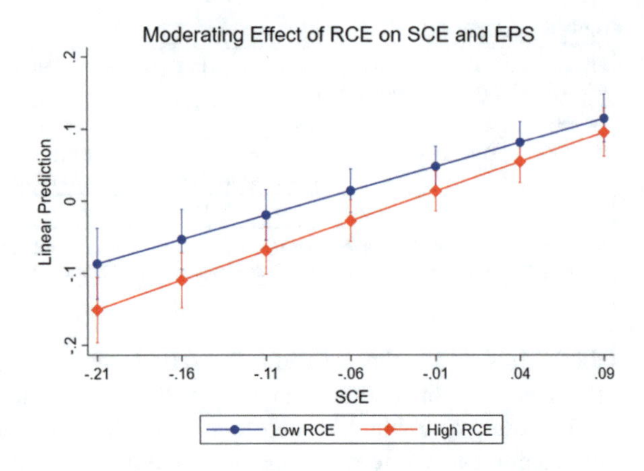

Fig. 4 Interaction effect of RCE on SCE and EPS (Developed by chapter authors)

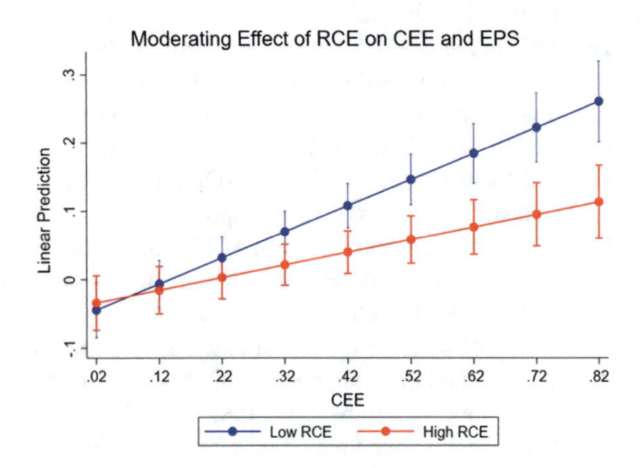

Fig. 5 Interaction effect of RCE on CEE and EPS (Developed by chapter authors)

the point of view of the EPS ratio, the RCE may negatively influence the relationship between CEE and EPS ratio. Therefore, the hypothesis H2c is supported.

These results may be due to several reasons, including the economic and political situation which the Middle East and Jordan region was exposed to the slowdown in Jordanian economic growth. Furthermore, the study was carried out during the subprime mortgage crisis 2008–2011, in addition to the Arab Spring crisis 2011 up now.

6 Conclusion

The main purpose of the present research is to determine the effect of RC on the relationship between IC and EPS in the industrial sector in ASE. Panel data for 50 industrial companies listed in ASE through the 10-year period (2008–2017) are used to test the hypotheses in this study.

This research presents two main contributions to research related to managerial accounting and the field of IC in the context of EPS. The first contribution has to do with the direct effects of IC on the EPS in the industrial companies in Jordan, which have been calculated by VAIC™. The second contribution, about the indirect effects of RC on the relationship between VAIC™ and EPS, through RCE as a moderator variable between VAIC™ and EPS in the industrial sector in Jordan during the period 2008–2017.

This research examines the RC measured by RCE in terms of its interaction with VAIC™ on the EPS in the industrial companies in ASE. Based on the arguments and findings of previous studies, the study proposes the first hypothesis H1 to test a direct relationship between VAIC™ and EPS and it is extended to three sub-hypotheses to examine the relationship between VAIC™ components (HCE, SCE, and CEE). Then, proposed the second hypothesis H2 to test the effect of RC on the relationship between VAIC™ and EPS, besides the sub-hypotheses (HCE, SCE, and CEE) to test the effect of RC on the relationship between VAIC™ components and EPS.

Generally, the results of this study indicate that the role of VAIC™ does matter in the EPS of the industrial companies in Jordan. Specifically, this research supports that the role of the VAIC™ increases the profit as a positive indicator that supports the EPS of the industrial companies listed in ASE. In addition, the role of HCE and SCE have a positive significant relationship on both EPS increase the profit as a positive indicator that supports the firm performance of the industrial companies through the period study, but CEE has a negative role with EPS.

Based on the hierarchical regression the results of this research indicate that RC does not find any significant influence of RC on the relationship between the VAIC™, HCE, SCE, and EPS. Additionally, the results reveal that RC plays a negative significant role as a moderating effect on the relationship between CEE and EPS.

The results suggest that industrial companies in Jordan must pay attention to the IC elements, especially the HC because it is the main item of VAIC™ and make new researches about the role of RC to enhance the effect of VAIC™ on the relationship with EPS. Moreover, the results suggest the policymakers need to find new export markets for products with a review of employment legislation and a minimum wage for employees. Furthermore, the present research shares new findings regarding the role of RC in VAIC™ and EPS. It is suggested that both factors of VAIC™ and RC are important elements for the success of industrial companies in Jordan.

Furthermore, future research should focus on the RC and the relationship between VAIC™ and firm performance especially EPS by paying attention to study all sectors of the ASE. In addition, studying a different period from the period of this study, as the current study period came within two financial and economic crises that affected Jordan.

This research has some limitations. First, the results of this study cannot be generalized to the Middle East, because the sample of the present study was restricted to Jordanian industrial companies listed in ASE. In addition, the findings are confined only to the industrial companies listed in ASE. Second, the time period for the study comes through two crises the Jordanian economy is still suffering from the effects of these two crises. Finally, the research utilized quantitative data and ignores the qualitative aspects of VAIC™ and RC.

References

1. Švarc J, Lažnjak J, Dabić M (2020) The role of national intellectual capital in the digital transformation of EU countries. Another digital divide? J Intellect Cap 823971. https://doi.org/10.1108/JIC-02-2020-0024
2. Chahal H, Bakshi P (2016) Measurement of intellectual capital in the indian banking sector. Vikalpa 41(1):61–73. https://doi.org/10.1177/0256090916629253
3. Momani KMK, Nour A-NI, Jamaludin N (2019) Sustainable universities and green campuses. In Global Approaches to Sustainability Through Learning and Education, no. September, IGI Global is prohibited, pp 17–27
4. Zhilenkova E, Budanova M, Bulkhov N, Rodionov D (2019) Reproduction of intellectual capital in innovative-digital economy environment. IOP Conf Ser Mater Sci Eng 497:012065. https://doi.org/10.1088/1757-899X/497/1/012065
5. Smithies J (2017) The challenge of the digital humanities
6. Gofman M, Jin Z (2019) Artificial intelligence, human capital, and innovation. Hum Capital Innov 20:1–55. https://doi.org/10.2139/ssrn.3449440
7. Momani KMK, Nour ANI (2019) The influence of intellectual capital on the return of equity among banks listed in Amman Stock Exchange. Int J Electron Bank 1(3):220. https://doi.org/10.1504/IJEBANK.2019.099613
8. Chen J, Zhu Z, Xie HY (2004) Measuring intellectual capital: a new model and empirical study. J Intell Capital 5(1):195–212. https://doi.org/10.1108/14691930410513003
9. Steward T (1997) Intellectual capital. The new wealth of organizations/tomas a. steward. newyork: doudleday¤cy, 278(2)
10. Bontis N (2003) Intellectual capital disclosure in canadian corporations. J Hum Resour Costing Account 7(1):9–20. https://doi.org/10.1108/eb029076
11. Al-Sartawi A (2020) Social media disclosure of intellectual capital and firm value. Int J Learn Intell Capital 17(4):312–323
12. Stewart TA (2007) The wealth of knowledge: intellectual capital and the twenty-first century organization. Crown Business
13. Schiemann F, Richter K, Günther T (2015) The relationship between recognised intangible assets and voluntary intellectual capital disclosure. J Appl Account Res 16(2):240–264. https://doi.org/10.1108/JAAR-11-2012-0076
14. Pulic A (1998) Measuring the performance of intellectual potential in the knowledge economy. In The 2nd World Congress on the Management of Intellectual Capital, pp 1–20
15. Fijałkowska J (2014) Value added intellectual coefficient (VAICTM) as a tool of performance measurement. Przedsiebiorczosc i Zarz 15(1):129–140. https://doi.org/10.2478/eam-2014-0010
16. Maditinos D, Chatzoudes D, Tsairidis C, Theriou G (2011) The impact of intellectual capital on firms' market value and financial performance. J Intellect Cap 12(1):132–151. https://doi.org/10.1108/14691931111097944
17. Cuganesan S (2005) Intellectual capital-in-action and value creation. J Intellect Cap 6(3):357–373. https://doi.org/10.1108/14691930510611102
18. Lenart R (2014) Relational capital as an instrument of increasing competitiveness. In 8th International Management Conference "Management Challenges Sustainable Development November 6th-7th, 2014, Bucharest, Rom., pp 14–26. https://conferinta.management.ase.ro/archives/2014/pdf/2.pdf

19. Ulum I, Ghozali I, Purwanto A (2014) Intellectual capital performance of indonesian banking sector: a modified VAIC (M-VAIC) perspective. Asian J Financ Account 6(2):103. https://doi.org/10.5296/ajfa.v6i2.5246
20. Sharabati A-AA, Nour A-NI, Shamari NS (2013) The effect of intellectual capital on jordanian tourism sector's business performance. Am J Bus Manag 2(3):210–221. https://doi.org/10.11634/216796061302370
21. Al-Sartawi A (2020) Information technology governance and cybersecurity at the board level. Int J Crit Infrastruct 16(2):150–161
22. Al-Sartawi A (2015) The effect of corporate governance on the performance of the listed companies in the gulf cooperation council countries. Jordan J Bus Adm 11(3):705–725
23. Al-Sartawi A (2018) Ownership structure and intellectual capital: evidence from the GCC countries. Int J Learn Intellect Capital 15(3):277–291
24. Iazzolino G, Laise D (2013) Value added intellectual coefficient (VAIC). J Intellect Cap 14(4):547–563. https://doi.org/10.1108/JIC-12-2012-0107
25. Gho PC (2005) Intellectual capital performance of commercial banks in Malaysia. J Intellect Cap 6(3):385–396. https://doi.org/10.1108/14691930510611120
26. Al-Sartawi A (2018) Corporate governance and intellectual capital: evidence from gulf cooperation council countries. Acad Account Financ Stud J 22(1):1–12
27. Chan KH (2009) Impact of intellectual capital on organisational performance: an empirical study of companies in the Hang Seng Index (Part 1). Learn Organ 16(1):4–21. https://doi.org/10.1108/09696470910927641
28. Chan KH (2009) Impact of intellectual capital on organisational performance an empirical study of companies in the Hang Seng Index (Part 2). Learn Organ 16(1):4–21. https://doi.org/10.1108/09696470910927650
29. Tan HP, Plowman D, Hancock P (2007) Intellectual capital and financial returns of companies. J Intellect Cap 8(1):76–95. https://doi.org/10.1108/14691930710715079
30. Haan AA-A, AL-Sakin SA-K, AL-sufy FJH (2016) The effect of value added intellectual coefficient on firms ' performance : evidence from jordanian industrial sector. Res J Financ Account 7(14):163–174
31. Tripathy T, Gil-Alana LA, Sahoo D (2015) The effect of intellectual capital on firms' financial performance: an empirical investigation in India. Int J Learn Intellect Cap 12(4):342–371. https://doi.org/10.1504/IJLIC.2015.072197
32. Nassar S (2018) The impact of intellectual capital on firm performance of the turkish real estate companies before and after the crisis. Eur Sci J 14(1):29–45. https://doi.org/10.19044/esj.2018.v14n1p29
33. Momani KMK, Jamaludin N, Zanani Wan Abdullah WZ@W, Nour A-NI (2020) The effects of intellectual capital on firm performance of industrial sector in Jordan. Humanit Soc Sci Rev 8(2):184–192. https://doi.org/10.18510/hssr.2020.8222
34. Chang WS, Hsieh J (2011) The dynamics of intellectual capital in organizational development. African J Bus Manag 5(6):2345–2355. https://doi.org/10.5897/AJBM10.1039
35. Nazari JA, Herremans IM (2007) Extended VAIC model: measuring intellectual capital components. J Intellect Cap 8(4):595–609. https://doi.org/10.1108/14691930710830774
36. Mondal A (2016) Application of modified VAICTM model for measuring intellectual capital performance. Int J Res Financ Mark 6(11):19–30
37. Baltagi BH (2005) Econometric analysis of panel data, Third edit. John Wiley & Sons Ltd
38. Pulic A (2004) Intellectual capital – does it create or destroy value? Meas Bus Excell 8(1):62–68. https://doi.org/10.1108/13683040410524757
39. Pulic A (2008) The principles of intellectual capital efficiency - a brief description. Croat Intellect Cap Center, Zagreb, no. February, pp 2–25. https://doi.org/10.1504/IJEIM.2004.005479
40. Abdulsalam F, Al-Qaheri H, Al-Khayyat R (2011) The intellectual capital performance of kuwaitibanks: an application of vaicTM1 model. iBusiness 03(01):88–96. https://doi.org/10.4236/ib.2011.31014

41. Gibson CH (2011) Financial reporting and analysis. 11th Editi. South-Western Cengage Learning
42. Hill R, Murphy G, Trailer J (1996) Measuring performance in entrepreneurship research. J Bus Res 36(1):15–23
43. Yasuda T (2005) Firm growth, size, age and behavior in Japanese manufacturing. Small Bus Econ 24(1):1–15. https://doi.org/10.1007/s11187-005-7568-y
44. Dang C, (Frank) Li Z, Yang C (2018) Measuring firm size in empirical corporate finance. J Bank Financ 86:159–176. https://doi.org/10.1016/j.jbankfin.2017.09.006
45. Harford J, Mansi SA, Maxwell WF (2008) Harford, mansi, maxwell - 2008 - corporate governance and firm cash holdings in the US☆.pdf. 87:535–555. https://doi.org/10.1016/j.jfineco.2007.04.002
46. Gelman A, Hill J (2006) Data Analysis Using Regression and Multilevel/Hierarchical Models. Cambridge University Press, Cambridge. https://doi.org/10.1017/CBO9780511790942
47. Ahmad M, Ahmed N (2016) Testing the relationship between intellectual capital and a firm's performance: an empirical investigation regarding financial industries of Pakistan. Int J Learn Intellect Cap 13(2/3):250. https://doi.org/10.1504/IJLIC.2016.075691
48. Amin S, Aslam S (2018) Relationship between intellectual capital and financial performance : the relationship between intellectual capital and financial performance : the moderating role of knowledge assets. Pakistan J Commer Soc Sci 12(August):521–547
49. Rehman W, Abdul Rehman C, Rehman H, Zahid A (2013) Intellectual capital performance and its impact on corporate performance: an empirical evidence from Modaraba sector of Pakistan. African J Bus Manag 5(20):8041–8049. https://doi.org/10.5897/AJBM10.1088
50. Suhendra ES (2016) The influence of intellectual capital on firm value towards manufacturing performance in Indonesia. Int Conf Eurasian Econ 10(4):438–445. https://doi.org/10.3923/ibm.2016.438.445
51. Ulum I, Kharismawati N, Syam D (2017) Modified value-added intellectual coefficient (MVAIC) and traditional financial performance of Indonesian biggest companies. Int J Learn Intellect Cap 14(3):207. https://doi.org/10.1504/IJLIC.2017.086390
52. Bakhsha A, Afrazeh A, Esfahanipour A (2017) A criticism on value added intellectual coefficient (VAIC) model a criticism on value added intellectual coefficient (VAIC) model . IJCSNS Int J Comput Sci Netw Secur 17(6):59–71
53. Sharabati A-AA, Jawad SN, Bontis N (2010) Intellectual capital and business performance in the pharmaceutical sector of Jordan. Manag Decis 48(1):105–131. https://doi.org/10.1108/00251741011014481

Machine Learning and Earnings Management Detection

Zakeya Sanad

Abstract Earnings management and the technological advancement of audit analytical procedures have become significant fundamental issues in the current digital economy. A rising number of studies criticized the traditional auditing procedure and discussed the new approaches in earnings manipulation detection. The aim of this paper is to discuss earnings management practices and then rereview the previous studies that discussed the role of machine learning in uncovering earnings management practices. The study also suggests future research directions.

Keywords Earnings management · Auditing · Machine learning · Data mining

1 Introduction

Earnings management (EM) is a serious concern for many countries. Many regulators around the world are continuously seeking to tighten their corporate governance regulations in order to limit earnings management practices and to attain investors' confidence. EM literature mainly classified EM methods into three categories: accruals-based management (ABM), real activities management (REM) and classification shifting (CS) [1, 3, 35, 50].

However, regardless of the method used to manage firms' earnings, all practices increase the information asymmetry between managers and interested parties and hide firm's actual performance, thereby, diminishing financial reporting reliability and credibility [4, 32].

Accordingly, in order to reduce the probability of manipulation and misleading the stakeholders, high attention was given to developing effective financial accounting fraud detection using technological approaches. The traditional audit has many limitations for instance, auditors have limited audit capabilities that could help them to reach the highest audit compliance at low costs [15, 40]. Hence, in todays' digital economy, auditors need new mechanisms for analyzing big data transactions.

Z. Sanad (✉)
Ahlia University, Manama, Kingdom of Bahrain
e-mail: Zsanad@ahlia.edu.bh

A. M. A. Musleh Al-Sartawi (ed.), *The Big Data-Driven Digital Economy: Artificial and Computational Intelligence*, Studies in Computational Intelligence 974,
https://doi.org/10.1007/978-3-030-73057-4_6

As a result, the application of technological techniques like machine learning are rising tremendously to expand the effectiveness and productivity of audit analytics. Unlike the widely used traditional audit where calculating statistics and ratios manually, the modern audit approach integrates sophisticated techniques such as data mining and machine learning technique [38]. Hence, this led to attracting the attention of scholars in testing how machine learning is used within accounting field [49].

2 Literature Review

The extant literature documented that managing opportunistically firms' earnings could lead to negative consequences in the long run. [29] defined ABM as when managers exercise their discretion and judgment in the financial reports to alter firms' earnings without cash flow consequences. The literature identified different ways which managers can engage in ABM, such as choosing different fixed asset depreciation methods, underestimating the expected bad debts amounts and loan loss provision, and delaying or accelerating asset write-offs [5, 12, 13, 43].

Although the majority of EM literature consisted of ABM studies, the last decades comprised a changing balance in favor of REM [47]. Real activities manipulation as an EM tool was not well understood due to the ambiguity associated with it [13, 17], and it remained a largely unexplored area in the literature until recent years. However, recent EM studies are increasingly paying attention to REM studies. Unlike ABM, REM is based on business decisions rather than accounting decisions, and it has a direct influence on cash flow generation [4, 16, 20].

As stated by [14], differentiating between REM and ordinary business decisions is considered as a challenge for investors. A rising number of EM studies tackled the issue of CS. Although CS might not sound as a significant EM concern since it does not deal with managing the actual bottom-line earnings, it is considered as an increasingly important EM issue.

Managers are motivated to manage income statement items because many interested parties such as investors, financial analysts, lenders, and other stakeholders are more likely to assess firms' earnings performance using core earnings metric rather than net income [10, 36]. Previous studies also agreed that unlike bottom line earnings, core earnings are conceived more value-relevant for market participants [6, 22]. Besides, [2] stated that the market participants perceive core earnings as a more reliable source for predicting future profitability than bottom line earnings.

Different researchers explained CS. [23, 45] defined CS as the manipulation of informational content, more precisely, the opportunistic misclassifying of income statement items. [35, 36] defined it as the misclassification of the income statements' items while the bottom-line remains the same. [42] explained CS as the intentional reporting of expenses or revenues on different lines on the income statement.

[39] provided more detailed definition by stating that CS refers to the deliberate misclassification of income statement items that result in increasing the gross profit

or core earnings while net earnings is unchanged. [2] claimed that CS is a subtle yet practical way to manage earnings, while [24, 35] described it as "recent form of earnings management".

In response to the previous accounting scandals that resulted mainly from opportunistic earnings management practices, the audit profession realized the weaknesses attributed with the traditional auditing process including manual verifications, inventory counts, and the application of simple ratios [22]. As stated by [15], the traditional audit methods are slow and could provide imprecise outdated audit reports.

Furthermore, prior researchers claimed that the traditional auditing method is associated with a number of drawbacks such as spending long time to analyze the huge volumes of data especially in the current digital economy [41, 46, 50]. However, the recent technological development in the business world raised the importance of adapting technology for fraud and opportunistic earnings management detection purposes.

A number of effective analytical techniques were introduced to integrate and improve the various current audit procedures and to make sure that the monitoring mechanisms are effective to avoid future accounting scandals. [33] mentioned that due to the advanced technology, auditors could not only enhance the reliability of the financial reports but also offer other beneficial services such as counseling managers.

In recent years, modern auditing techniques like continuous audit, real time audit and risk-based audit are extensively used to improve the traditional audit procedure and overcome the manual audit work issues [32]. Continuous audit has been discussed a lot in the literature. Continuous audit is beneficial because it is not only dealing with historical data, but predicting the future outcomes which help auditors to take decisions. However, sometimes auditors might take bad decisions because of data insufficiency [44], however, due to the advanced technology, the continuous auditing technique became more efficient.

Studies were motivated to test the impact of these techniques on earnings management practices to see if they can contribute in mitigating it. These studies mostly relied [9] model which is a statistical model that apply financial ratios obtained from firms' financial statements and then the M-Score is calculated to uncover the potential areas of opportunistic earnings management [26]. The Certified Public Auditors were required to use this model to make sure that financial reports are not opportunistically managed and free from errors [37].

Another more recent continuous audit technique that helps in uncovering earnings management practices is machine learning [38]. The machine learning method includes complicated techniques like data mining that goes in depth with regards to data analytics [26]. [16] highlighted that machine Learning help auditors to make reasonable assurance regarding potential earnings management practices within the financial statements. Studies also documented that machine learning is beneficial for justifying firms' accruals [19], selecting boards directors [18], and forecasting firms' stock returns [21].

Data mining has become a topic of interest in the audit field. [27] stressed that earnings management model cannot be applied as a simple linear model and suggested to apply the data mining method. Unlike statistical models, data mining has the ability of obtaining the related information from massive data sets and assist auditors in uncovering the hidden patterns and taking decisions.

[15] stated that data mining technique is a process of converting data into knowledge. Data mining made it possible to advance the Computer Assisted Audit Techniques [34] which includes various techniques such as the Support Vector Machines, Decisions Trees, Logic Base Algorithm, Bayesian Naïve Classifier, and Neural Network Based Machine Learning [8, 15].

Although there are various techniques available, researchers concurred that there is no specific technique that could accurately work with all kinds of data sets [8, 30]. Also, [40] suggested that usually the percentage of manipulators is low compared to the nanomanipulators which result in having imbalanced data, hence, leading to biased earnings prediction.

A number of studies responded to the importance of testing the effectiveness of the current models in detecting earnings manipulations. For instance, [25] tested the ability of Beneish M-Score models in uncovering fraud and found that the overall Beneish M-Score model could uncover financial fraud effectively. [15] used Bayesian Naïve Classifier to improve the decision-making procedure in the earnings manipulation detection. The findings revealed that mathematical models are better functioning than traditional audit.

Another study by [40] focused on the enhancement of earnings manipulation detection using again Bayesian Naïve Classifier as type of supervised machine learning. The study findings showed that Beneish model performs better than the manual auditors' method which indicates that manual auditors' method is less sufficient in uncovering earnings manipulation.

[28] recommended to use the non-parametric machine learning techniques because it is useful in different accounting predictions. A more recent study by [11, 31] conducted in Taiwan suggested an improved model for to uncover earnings management through applying hybrid machine learning methods. The study results showed that earnings management detection model using elastic net and C5.0 algorithm perform more accurately.

3 Conclusion and Future Studies

It is unsurprising that until today many auditors are still using the traditional manual auditing methods. Although earnings management literature includes a massive number of studies compared to the other accounting literature, most of the studies relied on traditional mathematical earnings management models methods and a limited but rising number of studies tackled the issue of the advanced technological audit methods such as machine learning role in monitoring EM practices.

Some researchers argued that the consequences of integrating technology with audit procedures still vague, hence, more studies are needed to explain how the machine learning and new data mining algorithms as an earnings manipulation detection tool could help forensic accountants, auditors and regulators in uncovering the hidden opportunistic practices.

Moreover, [48, 49] argued that few studies agreed on data patterns to uncover potential earnings manipulation and this might be due to the fact that studies do not rely on developing theory or expert information. Hence, future studies should take into consideration this issue. In addition, the attention on accrual manipulation in the literature is much more than the other manipulation methods including real earnings management and classification shifting.

Therefore, studies should develop models that could detect these kinds of earnings management especially that recent studies confirmed that it is essential to take into consideration more than one earnings management because managers use different EM practices to influence firms' earnings and studies confirmed that REM and CS are more likely to be used as substitutes when ABM is restricted [1]. Therefore, focusing on one type of EM would not provide a full picture of managers' opportunistic practices toward managing firms' earnings [7].

Furthermore, [40] suggested the use of supervised machine learning models by regulators because they are more likely to identify high risk firms at early stages, thus, qualitative studies are needed to know exactly the perception and awareness regarding the role of supervised machine learning in the audit field. Studies can also test advanced models for predicting manipulation.

References

1. Abernathy JL, Beyer B, Rapley ET (2014) Earnings management constraints and classification shifting. J. Bus Finan Account 41(5–6):600–626
2. Alfonso E, Cheng CSA, Pan S (2015) Income classification shifting and mispricing of core earnings. J Account Auditing Finan https://doi.org/10.1177/0148558X15571738
3. Al-Sartawi A (2015) The effect of corporate governance on the performance of the listed companies in the gulf cooperation council countries. Jordan J Bus Adm 11(3):705–725
4. Al-Sartawi A (2020) Does it pay to be socially responsible? Empirical evidence from the GCC countries. Int J Law Manag 62(5):381–394
5. Al-Sartawi A (2019) Assessing the relationship between information transparency through social media disclosure and firm value. Manag Account Rev 18(2):1–20
6. Al-Sartawi A, Sanad Z (2019) Institutional ownership and corporate governance: evidence from Bahrain. Afro-Asian J. Finan Account 9(1):101–115
7. Anagnostopoulou S, Tsekrekos A (2016) The effect of financial leverage on real and accrual-based earnings management. Account Bus Res 1–46
8. Anto S, Chandramathi S (2011) Supervised machine learning approaches for medical data set classification: a review. Int J Comput Sci Technol 2(4):234–240
9. Beneish M (1999) The detection of earnings manipulation. Financ Anal J 55:24–36
10. Black EL, Christensen TE, Taylor Joo T, Schmardebeck R (2017) The relation between earnings management and non-GAAP reporting. Contemp Acc Res 34(2):750–782
11. Chen S, Shen ZD (2020) An effective enterprise earnings management detection model for capital market development. J Econ Manag Trade 77–91
12. Chi W, Lisic LL, Pevzner M (2011) Is enhanced audit quality associated with greater real earnings management? Acc Horiz 25(2):315–335

13. Commerford B, Hermanson D, Houston R, Peters M (2016) Real earnings management: A threat to auditor comfort? Auditing: A J. Pract Theory 35(4):39–56
14. Commerford BP, Hermanson DR, Houston RW, Peters MF (2019) Auditor sensitivity to real earnings management: the importance of ambiguity and earnings context. Contemp Acc Res 36(2):1055–1076
15. Dbouk B, Zaarour I (2017) Financial statements earnings manipulation detection using a layer of machine learning. Int J Innov Manag Technol 8(3)
16. Dechow P, Schrand C (2004) Earnings quality. Charlottesville, VA: The Research Foundation of CFA Institute
17. DeFond ML, Park CW (2001) The reversal of abnormal accruals and the market valuation of earnings surprises. The Account Rev 76(3):375–404
18. Erel I, Stern LH, Tan C, Weisbach MS (2018) Selecting directors using machine learning (No. w24435). National Bureau of Economic Research
19. Frankel R, Jennings J, Lee J (2016) Using unstructured and qualitative disclosures to explain accruals. J. Acc Econ 62(2–3):209–227
20. Galdi FC, Johnson ES, Myers JN, Myers LA (2019) Accounting for inventory costs and real earnings management behavior. SSRN 3480492
21. Gu S, Kelly B, Xiu D (2018) Empirical asset pricing via machine learning (No. w25398). National Bureau of Economic Research
22. Gu Z, Chen T (2004) Analysts' treatment of nonrecurring items in street earnings. J Acc Econ 38:129–170
23. Haw IM, Ho SS, Li AY (2011) Corporate governance and earnings management by classification shifting. Contemp Account Res 28(2):517–553
24. Hematfar M, Hemmati M (2013) A comparison of risk-based and traditional auditing and their effect on the quality of audit reports. Int Res J Appl Basic Sci 4(8):2088–2091
25. Herawati N (2015) Application of beneish M-Score models and data mining to detect financial fraud. Procedia-Soc Behav Sci 211:924–930
26. Hogan CE, Rezaee Z, Riley R, Velury U (2008) Financial statement fraud: insights from the academic literature. Auditing: A J Pract Theory 27(2):231–252
27. Höglund H (2012) Detecting earnings management with neural networks. Expert Syst Appl 39(10):9564–9570
28. Hunt J, Myers J, Myers L (2019) Improving earnings predictions with machine learning. https://doi.org/10.1007/s10997-007-9015-8
29. Kothari SP, Lester R (2012) The role of accounting in the financial crisis: Lessons for the future. Acc Horiz 26(2):335–351
30. Kotsiantis SB (2007) Supervised machine learning: a review of classification techniques. Informatica 31(3):249–268
31. Krishnan GV, McDermott JB (2012) Earnings management and market liquidity. Rev Quant Financ Acc 38:257–274
32. Kuenkaikaew S, Vasarhelyi AM (2013) The predictive audit framework. The Int J Digital Acc Res 13:37−71
33. Lee TH, Ali AM (2008) The evolution of auditing: An analysis of the historical development. J Modern Acc Auditing 4(12):1–8
34. Malikov K, Manson S, Coakley J (2018) Earnings management using classification shifting of revenues. Br Acc Rev 50(3):291–305
35. McVay SE (2006) Earnings management using classification shifting: an examination of core earnings and special items. The Acc Rev 81(3):501–531
36. Nwoye UJ, Okoye EI, Oraka AO (2013) Beneish model as effective complement to the application of SAS No. 99 in the conduct of audit in Nigeria. Manag Adm Sci Rev 2(6):640–655
37. Omar N, Koya RK, Sanusi ZM, Shafie NA (2014) Financial statement fraud: a case examination using beneish model and ratio analysis. Int J Trade Econ Finance 5(2):184−186
38. Poonawala SH, Nagar N (2019) Gross profit manipulation through classification shifting. J. Bus Res 94:81–88

39. Rahul K, Seth N, Kumar UD (2018) Spotting earnings manipulation: using machine learning for financial fraud detection. In international conference on innovative techniques and applications of artificial intelligence, pp 343–356
40. Sanad Z, Al-Sartawi A (2016) Investigating the relationship between corporate governance and internet financial reporting (IFR): evidence from bahrain bourse. Jordan J Bus Adm 12(1):239–269
41. Sharma A, Panigrahi KP (2012) A review of financial accounting fraud detection based on data mining techniques. Int J Comput Appl 39(1):37–47
42. Skousen C, Sun L, Wu K (2019) The Role of Managerial Ability in Classification Shifting Using Discontinued Operations . Adv Manag Acc 31:113–131
43. Trejo-Pech CJ, Weldon RN, Gunderson MA (2016) Earnings management through specific accruals and discretionary expenses: evidence from US agribusiness firms. Can J Agr Econ/Revue Can d'agroeconomie 64(1):89–118
44. Utami I, Nahartyo E (2016) Audit decisions: the impact of interactive reviews with group support system on information ambiguity. AJBA 9(1):105–140
45. Al-Sartawi A (2018) Institutional ownership, social responsibility, corporate governance and online financial disclosure. Int J Crit Account 10(3/4):241–255
46. Verner J (2012) Why you need internal audit at the table. Published on Business Finance. https://businessfinancemag.com
47. Vladu AB, Cuzdriorean DD (2014) Detecting earnings management: insights from the last decade leading journals published research. Procedia Econ Financ 15:695–703
48. Whiting DG, Hansen JV, McDonald JB, Albrecht C, Albrecht WS (2012) Machine learning methods for detecting patterns of management fraud. Comput Intell 28(4):505–527
49. Gupta M, Sikarwar TS (2020) Modelling credit risk management and bank's profitability. Int J Electron Bank 2(2):170–183
50. Zalata A, Roberts C (2016) Internal corporate governance and classification shifting practices: an analysis of UK corporate behavior. J Acc Auditing Financ 31(1):51–78

Artificial Intelligence and IT Governance: A Literature Review

Anjum Razzaque

Abstract There is scant research which has assessed the role of Artificial Intelligence (AI) on IT governance, when reviewing literature independently within the field of AI and IT governance the scholar of this chapter was able to critique literature and blend two schools of thoughts: Artificial Intelligence and IT governance to comprehend the role of one over the other. Seldomly these two research domains have been addressed in literature: though an important area of discussion. Also, implications and future research agendas are also depicted in this paper. Such a research topic is unique and significant since this opens the doors to a new phenomenon in financial performance.

Keywords Artificial intelligence · IT governance · Internet of ideas

1 Introduction

The growth of the internet is at its fastest pace when it comes to comprehending its technical infrastructure. And this is an observation since the past two decades. The Internet launched as a revolutionary innovation and is now an essential creation utilizable by more than two billion individuals [57]. The business sector has reported numerous disruptive technologies such as cloud computing, social networks, and the next generation of m-commerce, which is constantly shifting the application of IT by companies for managing knowledge and information. Currently, roughly 70% of transactions are conducted online, thus demanding secured systems which assure transparent transactions [26]. As a result, the scope of Artificial Intelligence (AI) comes in a synchronous harmony with IT systems which are being utilized by organizations and their critical infrastructure. AI are autonomous systems, which operate without the interaction of human beings, and can learn and pinpoint decision-making patterns, to reach to different inferences based on by analyzing various circumstances [10].

A. Razzaque (✉)
Ahlia University, Manama, Bahrain

A. M. A. Musleh Al-Sartawi (ed.), *The Big Data-Driven Digital Economy: Artificial and Computational Intelligence*, Studies in Computational Intelligence 974,
https://doi.org/10.1007/978-3-030-73057-4_7

AI has started getting noticed in literature since the 1940s. This was when it was discussed whether machines could decide [8]. The 1970s shed literature which developed the idea of applying AI solutions pertaining to the management of knowledge since the past decade discussed AI for numerous public sector tasks [39]. In the healthcare sector [51], transport industry [30], etc. Just as Albert Einstein stated that problems cannot be solved at the similar consciousness level created by the creator [7], understanding the role of AI for IT governance, such a combined effort of IT can help support top management.

Smart phones and cloud computing mean that we are seeing a new set of different problems that are related to interconnection and it requires new regulation and thinking. Smart phones users are connecting mobile devices together many are in developing countries. The sheer number are likely to have social impact like flash mobs. A lot of politics is now migrating to the cyber space field. It has been 40 years with the outsourcing the filling of data. The new thing is the wide spread of the storage. Cloud computing is defined by [36, 37] as a rapid, on demand network to a shared pool of computing resources. Outsourcing means cost saving and it is used widely by companies for computation and data storage. The big names such as Amazon, eBay, Google, and Facebook are all outsourcing computation to cloud. Among other raised issues is the cost of process power and connectivity and net neutrality. The problem of security and jurisdiction has been raised with the new storage facilities [3].

The internet is highly secured currently through different private regulatory activities and many defensive products and strategies, international cooperation and regulation and many defensive strategies and products. Non-governmental entities play major roles in the cyber security arena. The technical standards for the internet (including current and next-generation versions of the Internet Protocol) are developed and proposed by the privately controlled Internet Engineering Task Force (IETF); the Web Consortium, housed at the Massachusetts Institute of Technology, defines technical standards for the Web. Other privately controlled entities that play significant operational roles on aspects of cyber security include the major telecommunications carriers, Internet Service Providers (ISPs), and many other organizations, including: The Forum of Incident Response and Security Teams (FIRST), which attempts to coordinate the activities of both government and private Computer Emergency Response Teams (CERTs) and is also working on cyber security standards; The Institute of Electrical and Electronics Engineers (IEEE), which develops technical standards through its Standards Association and in conjunction with the U.S. National Institute of Standards and Technology (NIST); The Internet Corporation for Assigned Names and Numbers (ICANN), which operates pursuant to a contract with the U.S. Department of Commerce (September 2009) transferring to ICAAN the technical management of the Domain Name System [2, 37].

2 IT Governance

Research has expressed the notion that organizations that follow an efficient IT governance plan end up producing returns which are about 40% greater than their relative investments pertaining to IT which have been made by their opponents [5, 45]. According to [45], IT governance stipulates the accountability of a framework which promotes the desirable results when applying ICT. IT governance and IT management differ; in the sense that organizations apply IT governance to distinguish who will make critical and held accountable for IT related decision-making. However, IT management pertains to the day-today decision-making and the activities that pertain to executing organization ICT. The phenomenon of governance recognizes individuals that will be the crucial and accountable decision-makers of a firm. What does it mean by good governance? The notion of a good governance means, the enablement and the reduction of bureaucratic layers as well as those nonsense political principles that formulate an organization and hence lead to critical and repeated mistakes [56]. When the same notion is applied using ICT, then IT governance then the same good governance is applied within the context of an organizational IT related investments.

To assure enterprise governance, and part of the organizational and leadership structure, the board of directors along with the executive body assume the responsibility of IT governance. The aim here is to sustains the strategies of an organization. The concept of IT governance is progressing swiftly since a decade. Practitioners are keen under researching this area [24]. The value of IT governance is most appreciated by the practitioners since IT governance facilitates organizational performance, i.e., by measuring profs. Enterprises also support IT governance for its tendency to control cost-effective business processes that lessen the risks: a governance approach, referred as COBIT. COBIT is an IT governance audit framework popular amongst Information Systems professionals [24]. COBIT is more than just a framework but a comprehensive Information Systems operation which offers prescriptive assistance for manages and IT auditors (Kerr and Murthy 2013). The education sector also applies the COBIT for teaching learners the IT governance principles [4, 44, 45].

The academic scholars researching within the field of IT have been observed to frequently linger after those IT experts when it comes to proposing and forecasting various IT phenomenon. This is since the academics and the practitioners have got to fulfill their own goals and follow their own set of working strategies. For academics to collaborate with IT when building and implementing Information Systems is not simple. ample disappointments get reported [14, 16]. However, Information Systems run organizations where the absence of information systems questions even the survival of an organization [31]. Even though the Information Systems professionals well regard the COBIT framework, such a framework is solely for the IT professionals [11, 12, 19, 47]. There is a dire need in academic research for a theoretical foundation to contribute to problem-solving contexts. And such a concern has been lifted by the COBIT IT professionals. The issue is the enormous set of intricate narrative standards, raise the need for a solid emphasis on the outlining of theoretical goals. Such a gap portrayed in academic and practitioner research, though caters to the "*what*," does not cater to the "*how*:" a need suggested by risk managers and IT auditors, but no concern for IT managers [43, 48].

When comprehending IT governance one meaning can shape out when thinking of the application of IT within the organizational operations. For instance, the application of IT governance within the government is referend as e-governance for increasing the citizens' reach to the operations/services of a government. According to Razzaque and Sohmen (2009) nations are applying e-government models for availing its platform as an efficient tool for tackling immense challenges which governments face on a long- and short-term bases. The whole idea is to enhance those communication infrastructures where the citizens (also referred as the end-users) are not entirely literate to using computers. Also, there are nations which still encounter limited Internet access. Following are the leading nations with the e-government initiative, as according to the United Nation Government Surveys (2020): Denmark, followed by Republic of Korea, then Estonia, Finland, Australia, Sweden, UK, New Zealand, USA, Netherlands, Singapore, Iceland, Norway, and lastly Japan [48, 52].

The concept of IT governance details an insight in the principles of corporate governance and constituents. Literature Epstein and Roy, which measures the performance of corporate boards, state that governance is practices of corporate boards and senior managers, where the concern is whether the made decisions and their processes are within the shareholders', employees', interests, or just executives' interests [53, 54]. The corporate governance framework encourages an efficient usage of resources, requiring accountability of such resources. Thus, it is important to align individual, corporate, and societal interests. IT governance is a taxonomized into three categories: decision-making structure (specifying the decision rights and responsibilities for using a firm's IT system), relational mechanisms (promoting a synergic communication between employees) and processes (guidelines for monitoring decision-making, and IT control) [41]. IT governance cultivates a favorable organizational atmosphere for improving its efficiency, a well-documented stance by numerous scholars [41, 45]. IT governance is a subset of corporate governance [32], and responsible for executives to allow IT governance to aid firms for managing technological risks [35]. According to various scholars, there is an increasing reliance on IT by firms, a global phenomenon, as IT is a major and a vital facilitator for the betterment of firm performance [55].

IT governance manages IT investments; therefore, firms are interested in IT governance: as a facilitator for monitoring and controlling IT effectiveness and investments [6]. The us why IT governance is vital where IT interconnects with various complex components, infrastructures, and architectures. Such interconnectedness is tremendous when considering the regulations resonating upon an organization, and when related to IT [37]. Recent literature is concerned over IT governance for IT controls, i.e., though IT controls are gaining significance when establishing IT controls: a challenge (Power, 2009). Such scholars tend to guesstimate the economic cost of ineffective IT security controls with mixed results.IT governance concerns pertain to IT practices within the senior managers, questioning whether IT structures, processes, and IT related decisions are within the interests of shareholders, or primarily in the executives' interests. IT governance relates to corporate governance, where the IT organization structure and its objectives are aligned with the business objectives. IT Governance is a process which controls organizational IT resources, plus ICT. According to the IT Governance Institute, IT governance is responsible for setting the direction of executives and the board of directors for optimizing organizational structure and processes to assure enterprise's IT is sustained and extends its organizational strategies and objectives.

3 Artificial Intelligence

AI, when viewed as a technology, has its roots for 60 years. It has gained its fame through its embeddedness within the various applications which significantly influenced many lives as AI replicated and enhanced the intelligence of human beings through artificially enabled technological machines 36. Various scholars emphasize on the notion that AI can act rationally as well as thinking like human beings. However, there are also scholars who believe otherwise [49]. AI is a well-established research area within the field of mathematics, philosophy, linguistics, computer science as well as various other sciences [15]. Research in AI has p[assed through three waves. The first wave observed AI research with computing capabilities that introduced the sense of AI and were very limited in their processing power as such computing power was scarcest few decades ago. In the second wave, AI development prevailed in the form of neural networks mimicking the human mind while showing off how the computing capabilities can enhance such a phenomenon of the human mind when implemented by a machine. The third wave, i.e., the current wave, sees AI implacable within the context of deep learning applicable within real-world problem-solving situations [15].

4 Artificial Intelligence Influences IT Governance

A scan of current literature led the scholar of this chapter to realize that there is hardly any literature which relates AI to IT governance. One possible reason is that the field of AI is new, though as stated earlier in this chapter that AI has been within scholarly discussions since the 1940s. AI's application is still considered new, and IT governance needs AI. Future research should also investigate this matter and take under consideration how AI can enhance IT governance. At this stage, the question is how AI relates with IT governance.

When taking IT governance under control, i.e., control IT governance, is a concept requiring a thorough agenda accounting for IT governance control. Considering that there are thousands of input connections which are composed of decision trees, there are numerous systemic calculations for comprehending best solution-paths. This is where AI, or in other words fuzzy logic based mathematical formulas, also referred as expert system, come forth as a viable option. However, management needs to be aware

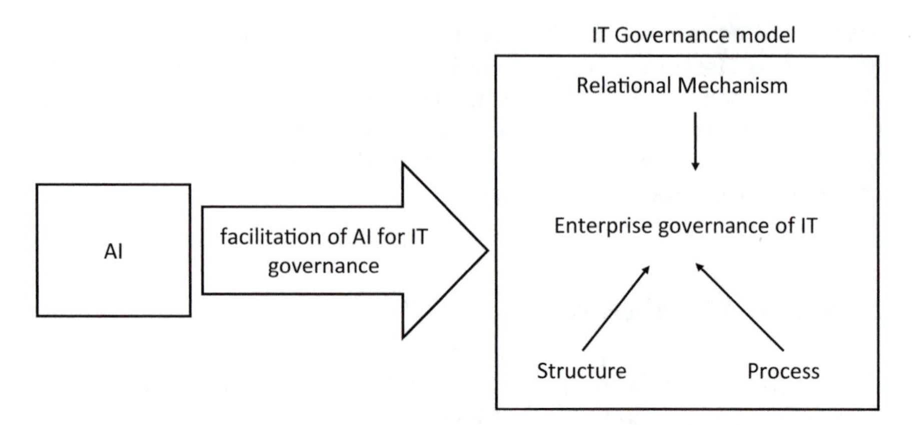

Fig. 1 AI for IT governance (Note: this model has been solely prepared by the author of this manuscript)

of operating such a system. IT governance is a model that offers decision-making mechanisms for supporting IT and thus, organizations. Consequently, devising a smart decision support system for IT governance is vital for capturing all opportunities for managerial decision-making. All the management needs to do is monitor the core embedded systemic criteria, to decide on the clearest way out [34].

For AI to support IT governance, first a model needs to be proposed. This chapter proposes IT governance model in three layers: (1) structure and processes, [49] enterprise governance of IT, and (3) relational mechanism. Structure refers to organizational units whose role is for making IT related decisions [41]. Process refers to the formulating of IT strategies-based decision-making or also when performing daily and routine monitoring IT procedures as per organizational policies [41, 42]. Enterprise governance of IT is reflective through Kaplan's and Norton's IT balance scorecard system [41, 42]. The relational mechanisms are active and collaborating relations between executives, IT stakeholders and mangers, reflecting announcements, advocacies, etc. [28, 41, 42]. Figure 1 depicts the role of AI for IT governance.

The model above depicted in Fig. 1, address to what [27] pointed out when it comes to the scant research which has yet to discuss how AI can be applied within various sectors. This is since there remain limited resources as well as the risk involved when applying AI, which scholars fear and hence to not embark in AI applications. The issue is that scholars need to notice the impact that AI can have on the governance of IT as in the same way as if AI is incorporated within wearable devices incorporate AI, and the same goes for objects like vehicles, smartphones, camera, or sensors, then AI catapults in a new dimension. There will be ample innovations that would better serve human beings and their enterprises [27]. But such innovations are possible when one recognizes the importance of AI for IT governance.

5 IT Governance Mechanisms

IT governance has primarily been driven by the need for the transparency of enterprise risks and the protection of shareholder value. The overall objective of IT governance is to understand the issues and the strategic importance of IT, so that the firm can maintain its operations and implement strategies to enable the company to better compete now and in the future. IT governance thus enables the enterprise to take full advantage of its information, thereby maximizing benefits, capitalizing on opportunities, and gaining competitive advantage [1]. The key IT governance mechanisms are: Business/IT strategic alignment, Value creation and delivery, Risk management (value preservation), Resource management and information systems auditing and performance measurement. Primarily of interest to business and technology management are the management guidelines tools and mechanisms to help assign responsibility, measure performance, and benchmark and address gaps between actual and desired capability. The guidelines help provide answers to typical management questions [50].

This study has previously expressed its interested in the COBIT 5 framework as a holistic mechanism for IT governance, COBIT has been publicized since 1996 and has been known for an IT auditing framework. The fifth version of COBIT was released in 2012 [24]. The COBIT is an IT governance and management framework has been developed by IT professionals who serve enterprises, such as the those from the insurance or banking or the consulting sector. Though there are academic scholars who have participated in the establishment of the COBIT framework, but the majority of the COBIT contribution comes from the IT practitioners [25, 29].

The COBIT framework supports organizations for accomplishing governance goals and managing IT part of an organization. The COBIT is a framework which is founded on five standards to represent the governance goals. Such a framework allows an organization to pursue efficient governance through an optimized investment scheme on IT [24]. The COBIT's five standards are: (1) meeting the needs of the stakeholders, (2) protecting an organization from one end to the other end, by the (3) utilization of one single, unified COBIT framework, though a (4) holistic methodology which (5) splits the governance from the management.

6 Financial Performance in Companies

Financial performance is a term that describes how well a company can use assets from its main mode of business and generate profits. It is also used as a wider measure of a company's overall financial strength within a specific period and can be used in comparing similar firms within the same industry or sectors, e.g., the education sector. The corporation itself as well as involved groups such as managers, shareholders, creditors, tax authorities and others search for answers to important questions. These questions can be answered using the financial analysis of a firm. It includes financial statements which is a structured collection of information according to rational and reliable accounting measures [21]. The purpose behind it is to spread a good understanding of some financial characteristics of a business firm. It may focus on Balance Sheet in some circumstances or may disclose a series of accomplishments over a specific period which refers to Income Statement. Therefore, the financial statements usually refer to two important statements which are: The Balance Sheet and the Income Statement [23].

- The Balance Sheet displays the financial position of the company at a certain point of time. It provides a snapshot and may be observed as a static image. BS is a summary of a company's financial condition on a specified date that shows total assets = total liabilities + owner's equity.
- The income statement indicates the performance of the corporation over a period. Income statement is a summary of two main parts which are revenues and expenses over a specific period, in conclusion it shows two results either net income or loss.
- Cash Flow Statement The cash flow statement includes both the IS income statement and BS balance sheet. For some forecasters, this statement is the most significant financial statement since it provides a settlement between net income and cash flow. This is where forecasters can see how much the company is paying on dividends, stock repurchases and capital expenditures. It also offers the source and usages of cash flow from processes, capitalizing and financing [23].

The main objective of recording, keeping, and analyzing financial statements is to make an improved business decision. Recognizing emerging issues and starting timely corrective act, as well as recognizing probable chances for better profit, are some of the clear benefits. Hopefully, continues analysis will assist the manager find previous errors and learn from them. The data taken from these three financial statements also can be used to make extra financial methods that disclose the strong points and weak points of the firm. These additional financial measures can be used to make some evaluations and comparisons [22].

The Firm Financial Values Assembly developed the (Financial Guidelines for Agricultural Producers), a set of suggested consistent financial aspects, measures and recording formats that can be used to improved understanding of a business. The suggested procedures for financial analysis are gathered into five broad classes: solvency, liquidity, financial efficiency, profitability, and repayment capability. These typical performance measures, occasionally mentioned to as the "*sweet 16*", historical and current financial information are not the only reasons affecting financial performance. Keep in mind that checking the "sweet 16" measures as a set are more significant than concentrating on only one or two measures [38].

7 Research Methodology

The aim of this study is to critique only specific literature which discusses the role of AI on IT governance. Such an aim was inspired through the need in such research pertaining to both disciplines of these two research areas, i.e., AI on IT governance. As a result, the author of this study predominantly depended on journal articles followed by conference papers from all accessible databases from his affiliating academic institution. This is a research in progress whose first phase is to conduct this literature review, which will follow into a deeper structured review of literature, followed by an identification of a more defined model, than the one depicted in Fig. 1 of this chapter.

8 Conclusion and Implications

At the beginning of our research, we listed three significant questions. The first questions were What is the level of applying Cyber security? To answer this question, we searched for more than 30 elements to determine whether the bank is applying cyber security system or not. Secondly, we collected 12 listed banks between Bahrain and Qatar to measure their appliance of cyber security. Based on us researches we found out that the most of these banks are applying a huge part of this system or even trying to develop this side in the bank itself. As the applying of cyber security increases the ROA will increase too resulting a noticeable improvement in the financial performance of the banks.

Government can free up its labor by implementing AI. This would aid in automating tasks and increasing the speed of real-time transactions, thus enhancing governmental services by improving citizens' satisfaction and enhancing the security and privacy of transactions, thus paving the way to improve policies that can drive AI for IT governance. [9, 40] AI and IT governance interrelate to facilitate various sector, such as education, finance, healthcare, just to name a few. It is important that experts implement AI for IT governance so to improves the quality of human life as well as the efficiency of any organization's governance [33].

9 Research Gap and Future Research

When it comes to the outline for future research, one must keep in mind the research gaps. Based on the literature review conducted through this study, literature seems scant on IT governance and ICT application within the healthcare sector as well as the education and the service sector that pertains to cultural or social affairs. However, research has highlighted various episodes where the pragmatic AI implementations were documented from various sectors, especially in the developed world. As far as other sectors go, scholars must explore the technological implementations of AI [27]. Ample research showcased technical facets when applying AI, with no focus on any model which could depict what the administrative or managerial implications AI has on IT governance perspective. There remains a necessity for scholars to pursue a deeper comprehension of the impact of AI on IT governance [46]. Scholars in the future could investigate the adoption of AI, and hence explore the related hazards when adopting AI in difference industrial segments. For instance, scholars could assess the adoption of AI within the various sectors of a government, with a comparative analysis on the results between the developing and the developed world [27].

10 Conclusive Recommendations

To guarantee that banks are working efficiently with improved financial performance level, they must apply cyber security system, since it provides a higher stage of security for both users and workers. Training employees is one of the most important aspects that increase their knowledge about day-to-day enhancement. On the other hand, IT governance is a must, because as we wish to get a higher security level, we would need members in the board of directors and workers with IT background. Neglecting these two independent variables could affect negatively in the performance of banks. For the future, researchers can research deeper into this this topic, to build on the knowledge that is accounting the IT background by adding more elements and variables such as: training, for more in-depth sectors this topic will be significant reference to banks and individuals who seek for innovation and creativity [9, 13, 17, 18].

References

1. Abdulrasool FE, Turnbull SI (2020) Exploring security, risk, and compliance driven IT governance model for universities: applied research based on the COBIT framework. Int J Electron Bank 2(3):237–265
2. Alhakimi W, Esmail J (2019) The factors influencing the adoption of Internet banking in Yemen. Int J Electron Bank 2(2):97–117
3. Al-Hashimi M, Hamdan A, Razzaque A, Al-Sartawi A, Reyad S (2020) Skill gaps in management information systems alumni. In: The international conference on artificial intelligence and computer vision (AICV'2020), pp 773–782, Cairo, Egypt. Springer, Switzerland
4. Alves V, Ribeiro J, Castro P (2012) Information technology governance—a case study of the applicability of ITIL and COBIT in a Portuguese private school. In: 7th Iberian conference on information systems and technologies. IEEE, Madrid

5. Blog.wallix.com (2019) Elements of an effective cybersecurity plan. https://blog.wallix.com/elements-of-an-effective-cybersecurity-plan. Accessed 26 July 2019
6. Bradley R, Byrd TA, Pridmore JL, Trasher E, Pratt RM, Mbarika VW (2012) An empirical examination of antecedents and consequences of IT governance in US hospitals. J Inf Technol 27:156–177
7. Brainly Quotes (2001) Albert Einstein Quotes. https://www.brainyquote.com/quotes/albert_einstein_130982. Accessed 26 July 2019
8. Buchanan A (2005) Handbook of emotional and behavioural difficulties. Child and Family Social Works. https://doi.org/10.1111/j.1365-2206.2005.00382_4.x
9. Cavoukian A (March 2009) Online privacy: make youth awareness and education a priority. Information and Privacy Commissioner of Ontario, Toronto. https://www.ipc.on.ca/wp-content/uploads/resources/youthonline.pdf. Accessed 26 July 2016
10. Čerka P, Grigienė J, Sirbikytė G (2017) Is it possible to grant legal personality to artificial intelligence software systems? Comput Law Secur Rev 33(5):685–699
11. Al-Sartawi A, Sanad Z (2019) Institutional ownership and corporate governance: evidence from Bahrain. Afro-Asian J Financ Account 9(1):101–115
12. Choi W, Yoo D (2009) Assessment of IT governance of COBIT framework. In: International conference on u-and e-service, science and technology, pp 82–89. Springer Link
13. Colorado Technical University (2019) The history of cybersecurity worms. Viruses. Trojan horses. Logic bombs. Spyware (26 July 2019)
14. Conboy K (2010) Project failure en masse: a study of loose budgetary control in ISD projects. Eur J Inf Syst 19(3):273–287
15. Darlington K (2017) The emergence of the age of AI. https://www.bbvaopenmind.com/en/technology/artificial-intelligence/the-emergence-of-the-age-of-ai/
16. Dwivedi YK, Henriksen HZ, Wastell D, De' R (2013) Grand successes and failures in IT. Public and private sectors. In: TDIT: international working conference on transfer and diffusion of IT. Springer
17. Federal Commnuications Commission (2013) Cyber security planning guide. https://transition.fcc.gov/cyber/cyberplanner.pdf. Accessed 26 July 2019
18. Gelnaw A (2018) Creating a cybersecurity awareness culture at financial institutions. https://www.bitsight.com/blog/creating-a-cybersecurity-awareness-culture-at-financial-institutions. Accessed 26 July 2019
19. Goldschmidt T, Dittrich A, Malek M (2009) Quantifying criticality of dependability-related IT organization processes in CobiT. In: IEEE 15th Pacific rim international symposium on dependable computing, pp 336–341. IEEE, Shanghai
20. Gunawan W, Kalensun EP, Fajar AN, Sfenrianto S (2018) Applying COBIT 5 in higher education. In: IOP conference series materials science and engineering, vol 420, no 1, p 012108
21. Gupta M, Sikarwar TS (2020) Modelling credit risk management and bank's profitability. Int J Electron Bank 2(2):170–183
22. Gupta N (2019) Influence of demographic variables on synchronisation between customer satisfaction and retail banking channels for customers' of public sector banks of India. Int J Electron Bank 1(3):206–219
23. Investopedia (2019) Financial performance. https://www.investopedia.com/terms/f/financial performance.asp. Accessed 26 July 2019
24. ISACA (2012a) COBIT 5 - a business framework for the governance and management of enterprise IT. Rolling Meadows, IL
25. ISACA (2012b) Cobit 5 - enabling processes. Rolling Meadows
26. ITU (2018) Critical Information Infrastructure Protection Role of CIRTs and Cooperation at National Level. Global Cybersecurity AgendaITU. https://www.energypact.org/wp-content/uploads/2018/03/Maloor_Day2_Critical-Information-Infrastructure-Protection.pdf. Accessed 26 July 2019
27. Kankanhalli A, Charalabidis Y, Mellouli S (2019) IoT and AI for smart government: a research agenda. Gov Inf Q 36(2):304–309

28. Kaplan RS, Norton DP (July–August 2005) The balanced scorecard-measure that drives performance. Harvard Business Review: https://hbr.org/2005/07/the-balanced-scorecard-measures-that-drive-performance. Accessed 1 Sep 2020
29. Kerr DS, Murthy US (2013) The importance of the CobiT framework IT processes for effective internal control over financial reporting in organizations: an international survey. Inf Manag 50(7):590–597
30. Kouzioka GN (2017) The application of artificial intelligence in public administration for forecasting high crime risk transportation areas in urban environment. Transp Res Procedia 24:467–473
31. Laudon JP, Laudon KC (2012) Management information systems: managing the digital firm, 16th edn. Pearson Education, Canada
32. Lunardi G, Becker J, Macada A, Dolci P (2014) The impact of adopting IT governance on financial performance: an empirical analysis among Brazilian companies. Int J Account Inf Syst 15(1):66–81
33. Marda V (2018) Artificial intelligence policy in India: a framework for engaging the limits of data-driven decision-making. Philos Trans R Soc A Math Phys Eng Sci 376(2133):1–19
34. Mårten S (2008) Predicting IT governance performance: a method for model-based decision making. KTH, Royal Institute of Technology, Stockholm
35. Mohamed N, Gian Singh JK (2012) A conceptual framework for information technology governance effectiveness in private organizations. Inf Manag Comput Secur 20(2):88–106
36. Mohasses M (2019) How AI-Chatbots can make Dubai smarter? In: 2019 Amity international conference on artificial intelligence (AICAI), pp 439–446. IEEE Xplore
37. National Institute of Standards and Technology (March 2006) Minimum security requirements for federal information and information systems. NIST Computer Security Division. https://csrc.nist.gov/csrc/media/publications/fips/200/final/documents/fips-200-final-march.pdf. Accessed 5 Jan 2019
38. nibusinessinfo.co.uk (2019) Measure your financial performance. https://www.nibusinessinfo.co.uk/content/measure-your-financial-performance. Accessed 26 July 2019
39. Pan J, Ding S, Wu D, Yang S, Yang J (2019) Exploring behavioural intentions toward smart healthcare services among medical practitioners: a technology transfer perspective. Int J Prod Res 57(18):5801–5820
40. PayTabs (3 July 2018) 7 tips for safe online transactions. Pay Tabs Blog. https://www.paytabs.com/en/7-tips-for-safe-online-transactions/. Accessed 26 July 2019
41. Peterson R (2004) Crafting information technology governance. Inf Syst Manag 21(4):7–23
42. Power M (2009) The risk management of nothing. Account Organ Soc 34(6/7):849–855
43. Razzaque A (2019) Multicultural teaching-learning model for western higher education institutions. Int J Electron Bank 2(1):358–373
44. Musleh Al-Sartawi AMA (2020) E-learning improves accounting education: case of the higher education sector of Bahrain. In: Themistocleous M, Papadaki M, Kamal MM (eds) Information systems. EMCIS 2020. Lecture Notes in Business Information Processing, vol 402. Springer, Cham
45. Razzaque A, Sohmen V (2009) A transcultural model of e-governance for the healthcare industry. In: Annual international conference on global business (ICGB 2009), Dubai, UAE
46. Reis J, Santo PE, Melão N (2019) Artificial intelligence in government services: a systematic literature review. In: Rocha Á, Adeli H, Reis LP, Costanzo S (eds) World conference on information systems and technologies, pp 241–252. Springer Link
47. Ridley G, Young J, Carroll P (2008) COBIT 3rd edition executive summary. Aust Account Rev 18(4):334–342
48. Ross JW, Weill P (2004) IT governance: how top performers manage IT decision rights for superior results. Harvard Business Press
49. Russell S, Norvig P (1995) Artificial intelligence: a modern approach. Prentice Hall, New Jersey. https://storage.googleapis.com/pub-tools-public-publication-data/pdf/27702.pdf. Accessed 2 Jan 2020

50. Al-Sartawi A (2020) Information technology governance and cybersecurity at the board level. Int J Crit Infrastruct 16(2):150–161
51. Sun Q, Medaglia R (2019) Mapping the challenges of artificial intelligence in the public sector: evidence from public healthcare. Gov Inf Q 36(2):368–383
52. United Nations Department of Economic and Social Affairs (2020) E-government survey 2020 digital government in the decade of action for sustainable development. New York, USA: United Nations. https://publicadministration.un.org/egovkb/Portals/egovkb/Documents/un/2020-Survey/2020%20UN%20E-Government%20Survey%20(Full%20Report).pdf. Accessed 1 Feb 2021
53. Van Grembergen W, De Haes S (2009) Enterprise governance of IT in practice. In: PauwelsBelgium S, Belgium M (eds) Enterprise Governance of Information Technology. Springer, New York, USA, pp 21–75
54. Weill P, Ross J (2004) IT governance: how top performer manage IT decision rights for superior results. Harvard Business School Press, Boston
55. Wilkin CL, Chenhall RH (2010) A review of IT governance: a taxonomy to inform accounting information systems. J Inf Syst 14(2):107–146
56. Al-Sartawi A (2018) Corporate governance and intellectual capital: evidence from gulf cooperation council countries. Acad Account Financ Stud J 22(1):1–12
57. World Economic Forum (2019) 1. Introduction: the digital infrastructure imperative. World Economic Forum. https://reports.weforum.org/delivering-digital-infrastructure/introduction-the-digital-infrastructure-imperative/. Accessed 26 July 2019

Blending of Physicians' Leadership and Decision-Making Style Within Virtual Platforms for Improving Service Quality

Anjum Razzaque and Magdalena Karolak

Abstract Literature demands for improving the quality of healthcare services, in response to high reported patient dissatisfaction from physicians' diagnostic errors. Social networks are a promising platform facilitating healthcare professionals for knowledge-shared-decision making (DM), in addition to leadership. Past research argues on the vitality of decisions based on appropriate leadership using knowledge management. Hence, this literature review paper proposes a conceptual framework making one proposition viable for future quantitative empirical evidencing, bearing theoretical and managerial implications.

Keywords Leadership style · Decision-making style · Physicians · Healthcare sector

1 Introduction

Literature widely reports a rising demand for improve the healthcare service quality, due to ample long-term/chronic diseases [6] and ample physicians' diagnostic errors causing poor medical decision-making [15, 18]. Hence, the healthcare sector undertook initiatives, e.g., Electronic Health Record, informed as expensive and a failure [12, 32]. On the other hand, the Health 2.0's social networks introduce knowledge management (KM) tools, like virtual communities (VCs), evoking better patient care through collective support between patient care stakeholders [8, 17, 22, 31, 36]

Knowledge sharing, which occurs within such virtual and social platforms facilitate an improvement in DM quality and patient care [4], when DM is achieved by leveraging knowledge assets [10]. Also, such a notion needs underpinning of effective leadership [16], such that leaders make regularly make decisions (Riaz and

A. Razzaque (✉)
College of Business and Finance, Ahlia University, Manama, Bahrain

M. Karolak
Zayed University, Dubai, United Arab Emirates

A. M. A. Musleh Al-Sartawi (ed.), *The Big Data-Driven Digital Economy: Artificial and Computational Intelligence*, Studies in Computational Intelligence 974,
https://doi.org/10.1007/978-3-030-73057-4_8

Khalili, 2014). Depending on a decision's contingency a leader's style could be decisive, consulting like with another individual or a group or delegate and facilitate a venture. A leadership style reflects a sense of a DM style being: (1) intuitive, (2) rationale, (3) dependency, (4) avoidance and (5) spontaneousness [34].

Medical leadership is gaining importance in the research arena. Within the clinical healthcare setting, a leader should be capable of motivating and managing teams to perform correct diagnostics. Hence, in a clinical setting, a leader will assume an un-given role is of a team player so to achieve a common goal efficaciously and cost-effectively [3, 24]. Based on this argument linking leadership with medical DM, which is aided by knowledge sharing using KM tools, can only succeed in improving healthcare service quality if the leadership, and their followers, jointly recognize and achieve with a common goal or vision, to ultimately improve healthcare service quality [16].

2 Problem Statement

While [9] reported that DM has been a neglected research area in leadership research they also critiqued literature to establish the link between DM and the content, context, and practice of leadership. [9] research was, extended by [35] pragmatic assessment assessing moderation of the KM processes (i.e., knowledge utilization, knowledge modification, knowledge creation, knowledge transference, knowledge storing, knowledge accessing, and knowledge disposing) to predict the role of transformational and transactional leadership on rational DM. Hence, [9]'s empirically conformed identifying that knowledge sharing enables DM in VCs, where a physician individually possesses his/her leadership style. [34] extended this study by comprehending the role of transformational leadership and the five DM styles, and empirically confirmed the importance of leadership-based DM. The issue is that though past literature empirically expressed the significance of how KM tools moderate for enhancing leadership-based DM; they neglect the holistic role of leadership style on decision making style for comprehending how these two variables interaction to bond a strong association within the Web 2,0 era [24, 29, 30].

3 Literature Review

3.1 Leadership

The research has shifted form human resource to human capital to achieve sustainable competitive advantage. Human capital is a combined knowledge base of skills, capabilities, and innovativeness. Leadership gained importance for the direct relation between leadership and organizational performance [1, 2]. Various theories tried the success of leader. These leadership theories are: Great man theory, Trait theory, "Behavioral theory", Contingency theory, Transactional leadership, and Transformational leadership.

According to the Great man theory, at leaders are born—not made thus, molded through personalities and attitudes. However, and unfortunately this theory does not pertain to the healthcare sector. As per the trait theory processes from great man theory, suggests that only those with are born with the skills are real leaders. Hence, even the trait theory is ill-representative in the healthcare sector. According to the behavioral theory, leadership is not based on trait of inbuilt skills but behavior. As per the contingency theory, a successful leader operates a style based on a given situation. An applicable theory in the healthcare sector. Also, the transactional leader is functional in the healthcare sector, where a leader either rewards or punishes depending on the outcomes of the follower. However, the role of transaction leadership is limited in the service section, such as the HC sector, though an asset to its administration side. A transformational leader is a change agent, someone who is inspires, leading to change. Transactional leaders are a preferred by healthcare professionalism of the UK's healthcare system [16].

Leaders set organizational vision, encouraging its adherence. Such a leader is a persuasive by interpersonally influencing followers in each situation by communicating goals, and thus accomplishing such goals Such a leader leaves follower alone until follower fails to perform, or a transformation leader (a role model figure who motivates employees through intellectual stimulation [10]. [3] argument is support by [4], according to whom leadership sets vision, motivates, organizational structure for sharing knowledge. The art of knowledge sharing, as per [4], is like [3], as per who leadership sets up and motivates followers for adhering to vision and fulfilling pre-set goals. Hence, an argument supported by [35] who assessed the moderating role of KM for predicting the effect of leadership style on DM style, though an assessment still needed within the healthcare sector and that too for physicians.

In conclusion, the leadership approaches help improve cost savings and reflect efficiency [9]. Managers differ from leaders. A distinction expressible through the invisible management approach portrayed in the social constructionist model, where leaders and follows act through their interpretation of how others would react [11]. Leadership is someone who motivates, is flexible, and inspired by encourage his/her subordinates [37]. A leader develops others, restructure to motivate, so to push a group of various personalities which are different from each other, thus dealing with those different personalities through the lens on their own leadership personalities, and cognitively diagnostic capabilities. However, it is the confidence, dominance, intelligence, and the assertiveness of a leader which plays a role to build his/her leadership qualities. Hence, a leader needs to be cognizant of his/her strengths, while also aware of his/her flaws. A leader must be culturally sensitive when creating a group identity, which is through the norms of a group for improving the productivity and performance of that group [39]. Besides, there are four types of leadership :

transformational, transactional, charismatic and team leadership [34] along with five leadership styles: (1) must be decisive, (2) one must be able to individually consult him or herself, (3) who is able to consult with the group, (4) facilitate others and (5) delegate tasks [34].

3.2 *Decision-Making*

The DM process occurs sequentially, where first a problem is identified followed by a classification of a solution/s, followed by reviewing of those solutions. A selection of a solution follows the formation of an action plan [23, 26, 27]. Within the healthcare sector, DM occurs during diagnoses. The quality of healthcare service suffers from uncertain diagnoses via poorly made decisions [22]. There are five types of DM styles, i.e. (1) avoidant DM, (2) dependent DM (3) intuitive DM, (4) spontaneous DM and (5) rationale DM.

Rationale DM is logical and based on reason/s. Intuitive DM is a gut feeling. Dependent DM is via peer support. Avoidant DM is postponing, consumes time and us spontaneous. In general, DM is a swiftly performing an impulsive choice [33, 34]. Out of the five DM styles, rationale DM, closely relates to the healthcare sector where clinicians sensibly perform, or process medical decisions supported through scientific evidence and clinical experimentation. Such decisions are under assumptions that a patient too, would make such a decision, given the circumstance [23].

3.3 *Leadership And Decision-Making*

The leader is known for his/her ability to perform daily (short-term) as well as the long-term DM roles. Leadership is a value each manager should possess. When a transformational leader uses his/her time to perform rationale DM, such a leader is comprehensively evaluating various alternatives, possible coordination, and the resource of knowledge at his/her disposal. Moreover, leadership style is based on governance system and the normality of information that must be considered for decisions and must be disclosed to internal and external users with voluntary or mandatory methods to communicate the past and the future events or circumstances (through the significance of his/her decision, leadership capability, commitment and commitment for group support, group expertise and group ability/efficiency) [32, 34, 38].

There seems to be a growing need for inspiring physicians for their DM capabilities over organizational policies and budgets [16, 33]. Within the healthcare sector, leadership is aligning vision with strategy where various leadership styles: coercive, collaborative, despotic and democratic leadership, help shift UK healthcare related policies. Medical leadership is now a buzz word, which migrated from the management towards a clinical leadership [3, 25]. [9] review the connection between leadership and medical DM through the Cynefin framework, in addition to past literature explaining this link using the Social Constructionist and the Social Constitutive approach. Also, leadership for DM is a budding research arena, gaining awareness in the medical world. DM was neglected in leadership research. That too in the healthcare sector. But leadership DM wise, clinicians from the NHS, show interest since 2008 [9].

The [9]'s Cynefin framework is a cognitive approach mirroring sense made by narrative and language, since medical DM is performed through models expressing story-told achievements. The Cynefin framework's four contexts relate cause and effect: simple, complicated, complex, chaos. The simple and complicated realms are ordered, (visible cause and effect), in contrast to complex and chaos (no relation to cause and effect: past actions are unable to predict present/future results). All the realms require leadership based medical decisional analysis. The middle of this model is the fifth domain, disorder, i.e., where DM situation is uncertain of what domain it is in. The best practice domains are simple and complex where order is needed in disturbances while the chaos domain is an opportunity for change through a problem–solution approach. Cynefin framework's cause—effect theory is like the experience-based informal problem-solving through trial and error within the healthcare sector. Here medical decisions are made through (1) patterns learnt in medical schools, (2) scientific research which pinpoints to a problem and solution, (3) analysis of collected data which expresses findings for suitable tests, (4) treatment proposition where a treatment is proposed by analyzing a diagnosis, (5) standard tests are made a prerequisite for DM and (6) treatment and diagnoses lead to DM that prescribe treatments. Further to [9]'s Cynefin framework, f leadership for DM is viewed from the social construction approach through a sense-making process, occurring for DM during and after the occurrence of a crisis [20].

4 Method

This research is in progress and it was conducted by reviewing literature as current as the last decade to assure that the review of the case of this study remains as concise as possible, and for an acquisition of a holistic knowledge of how medical leadership style facilities medical DM style for improving healthcare service quality. The preferred are journal articles and conference papers. But book and online reports were also considered to remain well-informed of current healthcare sector happenings. The critiqued literature furnished describe current scientific justification that led to a conversation demanding for comprehending the role between leadership and DM within the healthcare sector.

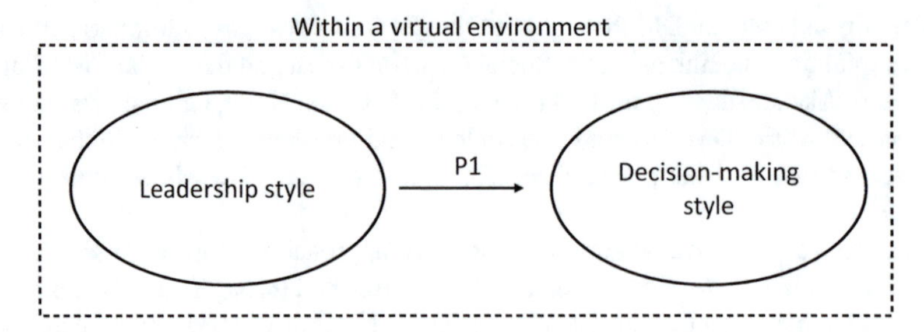

Fig 1 Conceptual framework (developed by the authors)

5 Proposed Solution

Figure 1 depicts one proposition. The question this study raises is what effect leadership style has on decision making style, particularly given the context of physicians, the healthcare sector, and for the mere improvement of healthcare service quality. Such a model is underpinned a critique of by current literature where such a model is viable for future quotative analytical assessment.

6 Conclusion

Considering that past research is scant on assessing what role leadership style plays on medical decision-making style, for improving medical DM quality for enhancing healthcare service quality, it is imperative that literature take under consideration the Fig. 1 model, not only for its literature-driven motive of existence, but also for future assessment of its true viable cause, bearing in mind implications. While there are theoretical implications which extend past research and pulls research to assessing this framework within the new medium of global choice, i.e., Web 2.0, this study also allows future research to take under consideration the managerial implications of empirically assessing the Fig. 2 framework. Managers within the healthcare sector, and those leaders who are involved in designing policies to evolve/improve the healthcare sector, would like to know how KM tools like virtual communities may prove cost effective, and efficient so that leadership can blend with the Industry 4.0 technology to, not only realize how leadership style can facilitate DM quality through what DM style, but also comprehend how policies can be re-engineered so to enhance the social fabric of online social platforms which physicians can adopt and utilize for the betterment of patient care and an improvement of healthcare service quality.

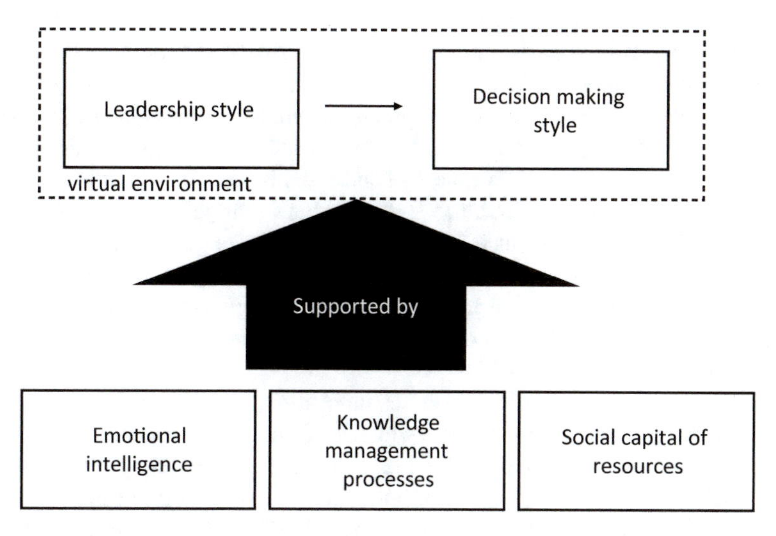

Fig 2 Holistic approach to improving service quality (developed by the authors)

7 Artificial Intelligence Led Innovations Important for Enhancing Leadership and Decision-Making in the Healthcare Sector

The world of the healthcare sector has seen ample innovations which have transformed medicine towards new heights. This has been evidenced in research since a century now [21]. The healthcare sector is a major concern point considering that it is a two and a half trillion-dollar service-oriented commerce [28]. The healthcare sector has advanced scientifically, such that it has reflected an achievement in such an extent that research has reported a longer rate of survival for patients who are suffering from illnesses. This is such, twenty years ago, patients would lose lives if they contracted such disease. Physicians now pioneered in thousands of new medicines and treatments, such that this has led the healthcare industry to also advance itself with its high-tech gadgets that have led the healthcare industry to improve its surgical tactics. This is how patients have also reported a rise in their satisfactions since there is an improved service of the healthcare delivery of services. This is particularly since the intervening procedures that enable the delivery of care, e.g., Artificial Intelligence enabled Internet of Things based medical systems. Still, ample work needs to be done, for instance—especially when incorporating Artificial Intelligence, Internet of Things, etc., since the healthcare sector continue to have its systems benchmarked at a level which practitioner report as remaining infantile. This is the reason why this study leaps into this review of literature to investigate how physicians' leadership affects their behavior to make valuable decisions in virtual environments. Virtual environments are where physicians (human beings) meet machines, to integrate and work together to solve healthcare service-related problems [5].

Though our world is at the advent of applying Artificial Intelligence, and state of the art technologies, the healthcare sector remains to prevail in the nineteenth century, and this is due to its backward practices which are forced to integrated with the twenty-first century healthcare devices. For instance, it is though unbelievable, but physicians continue taking their note by hand. This is just one example—observed and reported in literature. Literature further reports that patients get inspected in numerous CT scans simple since CT scan images seem to get lost during logistics. Such activities skyrocket healthcare costs, hence requiring innovative means for enhancing physicians' decision-making capabilities facilitated by their leadership qualities, while supported by innovative healthcare technologies like the Electronic Health Records (EHR) [13, 14].

Though the healthcare sector possesses innovation, its stakeholders do not seem to reflect sufficient leadership qualities which could reflect their sound abilities to be able to communicate with each another [13]. This is since even though the healthcare sector possesses innovations with an attitude that their role is optimistic; their adaption leads to a common consensus that such innovative technologies are complex, especially since the healthcare sector is shadowed by a web of ample governmental regulations that do not allow the physicians' leadership capabilities to decide properly within ample situations [5].

Physicians need Artificial Intelligence as a mediator to enhance the role of their leadership capabilities for making quality decisions since such a technology aids in the development and improvement of medical practices when handing patient records or when administering and managing chronic diseases. Such a just-described situation could further complicate the mutation of any healthcare innovation as complicated. However, those swiftly adapted inventions could reduce in value to others sluggishly adapting innovations. Rapidly adapted innovations, e.g., the role of Artificial Intelligent enabled innovations, can be applied by doctors, to appreciate them, without knowledge their consequences. Scholars may state that such innovations that get adopted rapidly are simply fads, due to well-marketed means, such innovations could also threaten the healthcare service quality [5, 7]. But then, this would not be really the case for Artificial Intelligence, which is a six-decade old discipline, a well-known general fact. But then, in the developed world, where the healthcare sector flourishes with ample hospital beds, then, physicians face an calmer time adapting innovative solutions [28].

Future research should assess this paper's model within the context of Bahrain's Ministry of Interior to empirical assess the research question of this study. There are also theoretical implications inviting future scholars to empirically tested model expressed in the Fig. 1 of this paper. If this model is empirical proven then there is an assurance for organizations to invest in virtual community environments where employees share knowledge through the resources of their social capital stored within their firms, and as a result achieve learner readiness, which is a remedy for competitive advantage for any firm. From a practical perspective, as per [20]'s Holistic model which proposes towards the improving of service quality, social capital, and knowledge management processes, bearing in mind the process of sharing knowledge,

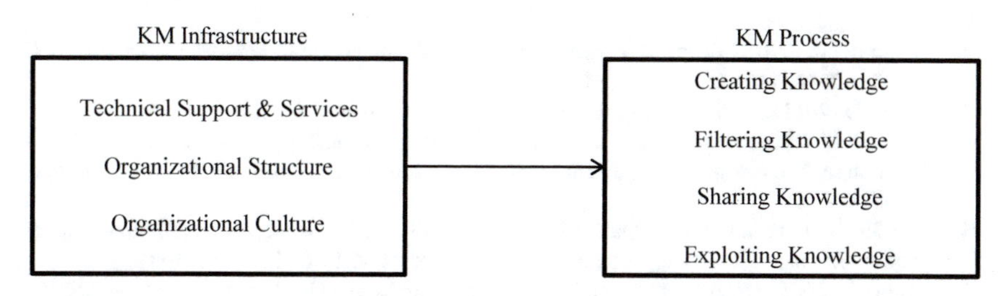

Fig 3 Knowledge management Infrastructure and processes interaction model (developed by the authors)

(Fig. 2) supports the leadership of any organization for supporting the achievement of quality decisions, in virtual environments.

For this study to value the functional ramifications of this present examination's experimentally confirmed model (Fig. 1), a sending of this model is the initial step to tackling its impact. Since the reliant variable of this model is information sharing quality, a knowledge management cycle, a benchmark of a knowledge management framework and design is vital. Supervisors, especially at Ministry of Interior, need to guarantee that their IT models have a knowledge management framework, in view of three measurements: technical sustenance and facilities, authoritative edifice, and hierarchical culture since knowledge management foundation encourages knowledge management measures [29, 30] as portrayed in Fig. 3. The board can execute the model (Fig. 1), if their association have a three-measurements based knowledge management framework. Knowledge management foundation likewise requires suitable knowledge management design, portrayed in Fig. 3.

Knowledge management architecture (Fig. 3) blends with any virtual environment, i.e., virtual community through a knowledge gateway endorsed by the service layer which is composed of the collaborating service, sharing service, and the personalizing services like tools employing an integrated network of knowledge bases [19]. Such bases come to be taken care of by between related data and information, spoken to as reported connections recorded in the information map layer. This data and information get put away in the substance the executive layer, which is thusly made from comprehend what, skill, know-why, and information data storehouses. Content administration layer underpins an information map layer. The foundation from where this knowledge management engineering achieves data and information is information systems like knowledge management devices like virtual networks [18, 26, 27].

References

1. Al-Sartawi A (2020) Social media disclosure of intellectual capital and firm value. Int J Learn Intellect Cap 17(4):312–323

2. Al-Sartawi A (2018) Ownership structure and intellectual capital: evidence from the GCC countries. Int J Learn Intellect Cap 15(3):277–291
3. Bhugra D (2011) Medical leadership in changing times. Asian J Psychiatry 4(3):162–164
4. Birasnav M, Rangnekar S, Dalpati A (2009) Enhancing employee human capital benefits through knowledge management: a conceptual model. Glob J e-Bus Knowl Manag 32(2):106–126
5. Birasnav M, Rangnekar S, Dalpati A (2011) Transformational leadership and human capital benefits: the role of knowledge management. Leadersh Organ Dev J 32(2):106–126
6. Bryant SE (2003) The role of transformational and transactional leadership in creating, sharing and exploiting organizational knowledge. J Leadersh Organ Stud 9(4):32–44
7. Chee MA, Choong P (2014) Social capital, emotional intelligence and happiness: an investigation of the asymmetric impact of emotional intelligence on happiness. Acad Educ Leadersh J 18(1):105
8. Coccia M (2012) Driving forces of technological change in medicine: radical innovations induced by side effects and their impact on society and healthcare. Technol Soc 34(4):271–283
9. Dixon-Woods M, Amalberti R, Goodman S, Bergman B, Glasziou P (2011) Problems and promises of innovation: why healthcare needs to rethink its love/hate relationship with the new. BMJ 20(Suppl_1): i47–i51
10. Dubé L, Bourhis A, Jacob R (2006) Towards a typology of virtual communities of practice. Interdisc J Inf Knowl Manag 1:69–93
11. Fulop L, Mark A (2013) Relational leadership, decision-making and the messiness of context in healthcare. Leadership 9(2):254–277
12. Hsia T-L, Lin L-M, Wu J-H, Tsai H-T (2006) A framework for designing nursing knowledge management systems. Interdisc J Inf Knowl Manag 1:95–108
13. Ismael K (2009) Mekerene university research repository item 123456789/628. Uganda Scholarly Digital Library. https://hdl.handle.net/123456789/628. Accessed 24 July 2011
14. Jalal-Karim A, Balachandran W (October 2008) Interoperability standards: the most requested element for the electronic Hea15lthcare records significance. In: 2nd international conference – e-medical systems, e-medisys 2008, pp. 29–31. IEEE, Tunisia
15. Jha A (5 February 2011) 21st Century Medicine, 19th Century Practices. Harvard Business Review. https://blogs.hbr.org/2011/02/21st-century-medicine-19th-cen/. Accessed 12 Nov 2013
16. Khurana R (2003) Invisible management: the social construction of leadership. Adm Sci Q 48(1):152–154
17. Kozer E, Scolnik D, Macpherson A, Keays T, Shi K, Luk T et al (2002) Variables associated with medication errors in pediatric emergency medicine. Pediatrics 110(4):737–743
18. Kumar RD (2013) Leadership in healthcare. Anaesth Intensiv. Care Med 14(1):39–41
19. Landro L (2006) Social networking comes to health care. https://www.post-gazette.com/pg/06363/749317-96.stm. Accessed 2 Dec 2010
20. Lin C, Chang S (2008) A relational model of medical knowledge sharing and medical decision-making quality. Int J Technol Manag 43(4):320–348
21. Lin C, Hung H-C, Wu J-Y, Lin B (2002) A knowledge management architecture in collaborative supply chain. J Comput Inform Syst 42(5):83–94
22. Maitlis S, Sonenshein S (2010) Sensemaking in crisis and change: inspiration and insights from Weick (1988). J Manag Stud 47(3):551–580
23. Omachonu VK, Einspruch NG (2010) Innovation in healthcare delivery systems: a conceptual framework. Innov J Publ Sect Innov J 15(1):1–20
24. Ohno-Machado L (2012) Informatics 2.0: implications of social media, mobile health, and patient-reported outcomes for healthcare and individual privacy. J Am Med Inform Assoc 19(5):683
25. Puschner B, Steffen S, Slade M, Kaliniecka H, Maj M, Fiorillo A et al (2010) Clinical decision making and outcome in routine care for people with severe mental illness (cedar): study protocol viewing options: abstract. BMC Psychiatry 10(1):90

26. Razzaque A (2015) Medical decision making aided by physicians' leadership: mediated by social capital in a social networking environment. In: KS global research international conference on emerging trends in multidisciplinary research (ETMR 2015). KS Global Research, Istambul
27. Razzaque A (2016) Social network based leadership decision making supported by social capital, knowledge management processes and emotional intelligence. In: 8th European conference on intellectual capital, ECIC 2016, 12–13 May 2016, p 224. ACPI, Venice
28. Razzaque A (2019) Multicultural teaching-learning model for western higher education institutions. Int J Electron Bank 1(4):358–373
29. Razzaque A (2019) Virtual learning enriched by social capital and shared knowledge, when moderated by positive emotions. Int J Electron Bank 2(1):77–95
30. Razzaque A, Eldabi T, Jalal-Karim A (2012) An integrated framework to classify healthcare virtual communities. In: European, mediterranean & Middle Eastern conference on information systems 2012, Munich, Germany
31. Razzaque A, Eldabi T, Jalal-Karim A (2013) Culture sustained knowledge management infrastructure and architecture facilitate medical decision making. In: 2013 IIAS-IASIA joint congress. IIAS-IASIA, Manama
32. Razzaque A, Eldabi T (2014) Link between social capital theory and innovation mediated by knowledge sharing and technology. In: European, mediterranean & Middle Eastern conference on information systems. Qatar University, Doha
33. Razzaque A, Karolak MM (2014) Significance of collaborative innovation for medical decision making in a virtual community. In: 5th health and environmental conference in the Middle East. HBMeU Congress 2014, Dubai, pp 2–10. https://congress.hbmeu.ac.ae/
34. Razzaque A, Al-Sartawi AM, Hamdan A, Al-Hashimi M (2019) Trusted physicians' virtual social capital facilitates decision-making during knowledge sharing. Int J Recent Technol Eng 8(4):2277–3878
35. Razzaque A, Eldabi T, Chen W (2020) Quality decisions from physicians' shared knowledge in virtual communities. Knowl Manag Res Pract 1–13. https://doi.org/10.1080/14778238.2020.1788428
36. Gupta N (2019) Influence of demographic variables on synchronisation between customer satisfaction and retail banking channels for customers' of public sector banks of India. Int J Electron Bank 1(3):206–219
37. Riaz MN, Khalili MT (2014) Transformational, transactional leadership and rational decision making in services providing organizations: moderating role of knowledge management processes. Pak J Commer Soc Sci 8(2):355–364
38. Sanad Z, Al-Sartawi A (2016) Investigating the relationship between corporate governance and Internet financial reporting (IFR): evidence from Bahrain Bourse. Jordan J Bus Adm 12(1):239–269
39. Wright A, Sittig DF (2008) A four-phase model of the evolution of clinical decision support architectures. Int J Med Inform 77(10):641–655

Social Capital Improves Virtually Shared Learner for Enhancing Their Readiness to Learn

Anjum Razzaque

Abstract In the previous literature there is ample proclamations of knowledge management frameworks but with seldom baring that are capable of applicable knowledge management strategies. In the Gulf Corporation Council region there are reports of hefty invests in knowledge management infrastructures, particularly when it comes to its sectors like sectors like banking, education, healthcare, etc. This study aims to express a review of current literature revealing a want to know the impact of participants' social capital for enhanced their virtually shared knowledge, in turn for improving their readiness to learn as they participate in an online (virtual) community. Furthermore, there is further elaboration on the theoretical and practical implications.

Keywords Knowledge sharing · Knowledge management · Social capital · Learner readiness · Identity · Trust · Shared vision · Shared language

1 Introduction

Historical studies on social capital and on learner readiness reveals the link between these two disciplines and the sharing of quality knowledge. But needs to address their all-encompassing integration, to get it how employees' SC of assets and their learner status impacts their sharing of quality information behavior whereas taking part in virtual communities. There appears to be a need for the organizational courses or procedures to bolster the sharing of quality information. The ostracized labor force that manages to create much of the knowhow in proficient positions within the Inlet locale may be difficult. As the danger of exile staff turnover can start a misfortune of information, numerous of them accept that their information is their control and incline toward not to share such information with others. Such conduct adversely influences sharing of quality information. This ponder points to survey whether employees' sharing of quality information conduct is esteemed all through

A. Razzaque (✉)
Ahlia University, Manama, Bahrain

A. M. A. Musleh Al-Sartawi (ed.), *The Big Data-Driven Digital Economy: Artificial and Computational Intelligence*, Studies in Computational Intelligence 974,
https://doi.org/10.1007/978-3-030-73057-4_9

organizations, and in case employees' sharing of quality information conduct drives choice making, efficiency, and advancement. This finding of this investigate may well be utilized to enhance the existing neighborhood inquiries about and cause an activity for encourage and more thorough inquire about within the future. The presence of social capital in virtual communities is to some degree constrained although a past inquire about theme, but not within the conte. For instance: [7, 10]. Section 2 and 3 critiques literature. Section 4 talk about the research methodology which led this study to propose its Fig. 1 model. Section 5 concludes upon the study's summer, highlighting its limitations and implications which are theoretic and are workable, along with future research recommendations.

2 The Quality of Virtually Shared Knowledge Affects the Readiness to Learn

To adjust e-learning (i.e., learning in virtual communities: as the considered context of this research) with learner interface, learners should be prepared to memorize through one's own directed learning as week as the one's own motivation towards discovering. The learner availability is propelled when some instruction method becomes supplemented with electronic or e-learning frameworks. A learner's availability, i.e., availability, could be a learner's intrigued, pivotal for the victory of e-learning when motivated by obtaining and sharing information. Intrigued is a person (individual intrigued through intelligent with peers) and situational (self-inspired transitory intrigued). Intrigued makes a difference accomplish wanted learning results. Since the sharing behavior of knowledge is based within a social environment, for learners to actively learn; e-learning setting shapes a reasonable stage to propel learning for committed cooperation, to memorize inside e-learning virtual communities. The need for the sharing of knowledge comes about when research has reported a decline in the interest for online participation, subsequently a drop in the learning culture, consequently the disappointment for any e-learning framework [6, 17, 25].

Learner readiness is a general notion very much applicable within flipped learning. This is a very efficient teaching–learning method by millennial learners, who oppose to conventional lectures; but instead choosing by participating in social media platforms which are mixed with conventional lectures. Their intolerant performance has indorsed researchers to re-imagine conventional lecture-based approaches by proposing effective teaching/learning policies [15]. [15] has evaluated the concept of flipped education by means of a questionnaire which was distributed to eighty-four undergrad-students, thus indicating that flipped learning seems central for teaching–learning, when thinking of a student cantered approach for achieving learning outcomes. Here, during flipped learning, learning occurs through the attainment of information before classes, therefore preparing at home before arriving to the class [26]. And, during a class session the instructor customized his/her directions to benefit learner through case-studies that lead to students doing problem solving sessions. Various academics have also reported that students better perform during flipped learning sessions. Although, other scholars remain concerned about the lack of readiness of students to flip learn, which reflects that e-learning tools are not yet efficient enough, as there remains scant empirical evidence in this matter. Empirical inquire ought to be conducted to affirm upon the case of flipped learning and its readiness of learners, as this can aid make strides intelligent for made strides e-learning when mixed with conventional course lessons [15]. This study aims to inquire upon this research gap, to investigate its role within the higher education sector of the Gulf Corporation Council (GCC) [3, 4, 12].

Studies have appeared that numerous work execution lacks are due to a need for spoken experience [1, 23]. Consequently, the sharing of knowledge behavior leads to the readiness and the availability to memorize or discover after other peers is acknowledged as having an impressive effect on work execution [9]. The sharing of knowledge is a knowledge management process which occurs when participants express their eagerness and their readiness to pursue learning through their peers who they are participating with. Or else it could also be since they are aiding others to pass through a learn process to aid them in contributing to the comprehension of their new understandings (Wei, Choy, Chew & Yen, 2012; 34]. Though, for the most part of the information that is required to support commercial competing is implied in nature meaning it is settled in individuals and is not self-evident to others. Tacit knowledge is implicit information; it cannot be easily codified and is not promptly transferable from one individual to another. Be that as it may, considers have demonstrated that the sharing of implied information is an imperative quality for team-oriented learning organizations. Furthermore, companies that lock-in in persistent learning are further aped to attain better work-based performance. Although it is accepted to be one figure that recognizes effective directors from others: there's a need of information from the perspective of the of the affect that tacit knowledge influences on a procedure [24].

It is vital for the Gulf region, meaning the GCC countries, to recognize and therefore evaluate the position of the sharing behavior of quality knowledge for inspiring learner-readiness. This is since the GCC seems to be blending face-to-face learning with e-learning, and this is a current trend in this region [8]. Examining the participation of online students through their readiness to learn helps design creative programs organized through e-learning portals. And this is fruitful only, when the scholars can realize how e-learning events can link the learners outside the classrooms. [20] interviews on undergrad students, to examine the scope for acquiring and applying knowledge in e-learning; revealed that students feel the challenges of attaining and utilizing knowledge as they interact socially. Nevertheless, [20] assessed the mediation technologies for cultivating friendly networks; however not to acknowledge the quality of their shared knowledge, which would have been necessary for learning why such an examination needs empirical verification, i.e., empirically assessing the Fig. 1 model of this study. Based on the argument this study proposes its first proposition:

P1: Employees' virtual sharing of quality knowledge positively and significantly affects the learner readiness.

3 Relation Between Learners' Social Capital and Their Virtually Shared Quality Knowledge

To comprehend the notion of e-learning one needs to appreciate the philosophy of a virtual situation, where knowledge is communicated. The term virtual network (virtual community) originated from the term network of training: depicted utilizing social capital theory and the social cognitive theory [7, 10], etc. social capital theory means participation within networks through a network of relations which hence exploit the social capital of resources. The resources from the social capital, hence the social capital, fits within the group whose social structure is being utilized and within which an environment of a learning cantered virtual community is generated within the education context. Previously scholars have reported that the social capital theory reflects bridging and bonding through this theory's three dimensions: structural, relational, and cognitive, e.g. [7, 10, 34]. The structural element is members' association is through the know-who which shapes the communication links.

The relational aspect indicates the ties of the participants which are laid out through the bonds developed by their trust for their peers within the group and for the social protocol, i.e., the norms of reciprocity, once they have identified themselves with each other. The trust supports the sharing of knowledge and the norms of the participants collectively reflect such a knowledge sharing behavior. Meaning that the sharing of knowledge occurs during the participations which must occur amidst the pre-set norms. The cognitive dimension reveals the members' universal portrayal of their mutual principles and traditions. It is the stories and the myths which actually help the participants to indulge in the sharing of their knowledge behaviors, which gets further expedited when their communicated through a common language, i.e., shared language and also when they communicate while they hold a common interest i.e., shared vision, [10, 16]. By using the social capital theory, scholars can portray social networks, such as WhatApp, Facebook, etc., to analyze those interactions which occur virtually. Even Though the participation within virtual communities is currently a trendy topic [11]; few scholars have come about to describe virtual communities through the lens of its members' participation, and that too through the lens of the social capital theory. And that too, by analytically examining how the social capital of a virtual community facilitates the sharing of quality knowledge [7, 10, 33].

By definition, a virtual community us a casual stage esteeming both sorts of knowledge (implied and unequivocal), composed through encounter, making it underpinned by an education structures [5]. The auxiliary measurement of the social capital hypothesis communicates collective intuitive relations are the quality of community relationships making social capital of assets through enthusiastic back and concepts distributing via members organize ties. Such assets made from the social capital of members' resources, is made during team partnerships, and usually conceivable through the sharing of knowledge [7, 13, 18, 22]. The communications encourage knowledge exchange. So, such a sharing behavior happens in virtual areas,

Fig. 1 Model

thus extending associates' knowledge and, as a result, their skill [2, 14]. From the viewpoint of the social aspect strengthen members' knowledge [2, 14].

On Or After the viewpoint of the social measurement of the social capital theory; the sharing behavior of knowledge involves communal ties to empower the judgment for charitable sharing of interacted knowledge [7, 10, 11] where the distribution of knowledge conduct is by replying to inquiries, by offering or proposing opinions. Computer-generated (virtual) communities are considered as weakly linked up systems where individuals with confrontational (face to face) meeting. And these are considered as strong-ties systems. And this also is the case within the conventional classrooms, i.e., the actual higher instruction classes happen face-to-face, and e-learning is just supplementary to facilitate the instructs during teaching–learning. Most of the past inquire about cantered on strong-network-ties; whereas there seems scant investigation on the weak-tie communities, consequently there is a must be wonder why members still volunteer to take part in weak-tie virtual communities [19, 21, 27, 31, 32]. A few researchers question on the off chance that there is any shared quality information inside frail ties [28–30]. Based on this argument this chapter recommends its second proposition:

P2: Employees' social capital of reserves (social capital theory) positively and significantly affects their self-directed e-learning.

4 Research Method

This study is a analysis of the existing literature. It reviewed past works to detect a gap/s in studies. The research gap/s led to proposing three propositions. As a result, this is a deductive approach, answering its research question: what is the impact of members' social capital on their learning readiness, in the absence and presence of their knowledge sharing behaviors in online environments? This is when this study was able to propose its Fig. 1 research model.

5 Discussion and Conclusion

This chapter is meant to evaluate the position of learners' social capital when virtually sharing quality knowledge, and hence the effect of such a virtual behavior on their readiness to learn. The social capital is assessed through the social capital theory, which is comprised of identification, social interaction led ties, norms of reciprocity, trust, mutual vision, and mutual language. According to by [10, 25], this study merely conveys trust, identification, mutual language, and mutual vision; as variables of the social capital theory (Fig. 1 model). Future research should assess this paper's model within the context of Bahrain's Ministry of Interior to empirical assess the research question of this study. There are also theoretical implications inviting future scholars to empirically tested model expressed in the Fig. 1 of this paper. If this model is empirical proven then there is an assurance for organizations to invest in virtual community environments where employees share knowledge through the resources of their social capital stored within their firms, and as a result achieve learner readiness, which is a remedy for competitive advantage for any firm. From a practical perspective, as per [25].

References

1. Chang HH, Chuang S-S (2011) Social capital and individual motivations on knowledge sharing: participant involvement as a moderator. Inf Manage 48(1):9–18
2. Alwis RS-D, Hartmann E (2008) The use of tacit knowledge within innovative companies: knowledge management in innovative enterprises. J Knowl Manage 12(1):133–147
3. Aubert J-E, Reiffers J-L (2003) Knowledge Economies in the Middle East and North Africa Toward New Development Strategies. World Bank Institute, Eashington DC
4. Baehr C (2012) Incorporating user appropriation, media richness, and collaborative knowledge sharing into blended E-learning training. IEEE Trans Prof Commnu 55(2):175–184
5. Bentley C, Browman G, Poole B (2010) Conceptual and practical challenges for implementing the communities of practice model on a national scale - a Canadian cancer control initiative. BMC Health Serv. Res 10(3):1–8
6. Blayone TJ, Mykhailenko O, vanOostveen R, Barber W (2018) Ready for digital learning? A mixed-methods exploration of surveyed technology competencies and authentic performance activity. Educ Inf Technol 23:1377–1402
7. Akdere M, Schmidt SW (2007) Measuring the effects of employee orientation training on employee perceptions of organizational learning: Implications for training & development. Bus Rev 8(1):172–177
8. Musleh Al-Sartawi AMA (2020) E-learning improves accounting education: case of the higher education sector of Bahrain. In: Themistocleous M, Papadaki M, Kamal MM (eds) Information Systems. EMCIS 2020. Lecture Notes in Business Information Processing, vol 402. Springer, Cham
9. Chow IH-S (2012) The role of social network and collaborative culture in knowledge sharing and performance relations. SAM Adv Manage J 77(2):24–37
10. Chiu C-M, Hsu M-H, Wang ET (2006) Understanding knowledge sharing in virtual communities: an integration of social capital and social cognitive theories. Decis Support Syst 42(3):1872–1888
11. Edelenbos J, Klij E-H (2007) Trust in complex decision-making network. Admin Soc 39(1):25–50
12. El-Khoury G (2015) Knowledge in Arab countries: selected indicators. Contemp Arab Affairs 8(3):456–468
13. Ellison NB, Steinfield C (2010) Connection strategies: social capital implications of Facebook-enabled communication practices. New Media Soc 13(6):873–892
14. Alhakimi W, Esmail J (2019) The factors influencing the adoption of internet banking in Yemen. Int J Electron Banking 2(2):97–117

15. Hao Y (2016) Exploring undergraduates' perspectives and flipped learning readiness in their flipped classrooms. Comput Hum Behav 59:82e92
16. Al-Sartawi A (2018) Corporate governance and intellectual capital: evidence from gulf cooperation council countries. Acad Account Fin Stud J 22(1):1–12
17. Hwang Y (2016) Understanding social influence theory and personal goals in e-learning. Inf Dev 32(3):466–477
18. Jansen RJ, Curseu PL, Vermeulen PA, Geurts JL, Gibcus P (2011) Social capital as a decision aid in strategic decision-making in service organizations. Manag Decis 49(5):734–747
19. Khorsheed MS (2015) Saudi Arabia: from oil Kingdom to Knowledge-based economy. Middle East Policy 22(3):147–157
20. Al-Sartawi A (2020) Social media disclosure of intellectual capital and firm value. Int J Learn Intellectual Capital 17(4):312–323
21. Lin C, Hung H-C, Wu J-Y, Lin B (2002) A knowledge management architecture in collaborative supply chain. J Comput Inf Syst 42(5):83–94
22. Gupta M, Sikarwar TS (2020) Modelling credit risk management and bank's profitability. Int J Electron Banking 2(2):170–183
23. Peroun DL (2007) Tacit knowledge in the workplace: the facilitating role of peer relationships. J Eur Ind Train 31(4):244–258
24. Al-Sartawi A (2020) Information technology governance and cybersecurity at the board level. Int J Critical Infrast 16(2): 150–161
25. Yang C, Chen L-C (2007) Can organizational knowledge capabilities affect knowledge sharing behavior? J Inf Sci 33(1):95–109
26. Razzaque A, AlAlawi M (2016) Role of positive emotions for knowledge sharing in virtual communities like knowledge management tools. In: 2nd international conference on emerging trends in multidisciplinary research (ETMR 2016), Bangkok, Thailand. KS Global Research, Bangkok
27. Razzaque A (2019a) Knowledge management infrastructure for the success of electronic health records. In: Al-Sartawi A, Hussani K, Hannon A, Hamdan A (eds) Global approaches to sustainability through learning and education. IGI Global Publications
28. Razzaque A (2019) Multicultural teaching-learning model for Western Higher Education Institutions. Int J Electron Banking 1(4):358–373
29. Razzaque A (2019) Virtual learning enriched by social capital and shared knowledge, when moderated by positive emotions. Int J Electron Banking 2(1):77–95
30. Razzaque A (2020) M-learning improves knowledge sharing over e-learning platforms to build higher education students' social capital. SAGE Open 10(2)
31. Razzaque A, Eldabi T, Jalal-Karim A (2013) Culture sustained knowledge management infrastructure and architecture facilitate medical decision making. In: 2013 IIAS-IASIA joint congress. IIAS-IASIA, Manama, Bahrain
32. Razzaque A, Eldabi T, Jalal-Karim A (2013) Physician virtual community and medical decision-making: mediating role of knowledge sharing. J Enterp Inf Manage 26(1)
33. Xiao ZX, Tsui AS (2007) When brokers may not work: the cultural contingency of social capital in Chinese high-tech firms. Adm Sci Q 52(1):1–31
34. Razzaque A (2016) Social network based leadership decision making supported by social capital, knowledge management processes and emotional intelligence. In: 8th European Conference on Intellectual Capital, ECIC 2016, 12–13 May 2016, p 224. ACPI, Venice, Italy

Game-Based Learning: Recommendations Driven from Literature

Azam Abdelhakeem Khalid, Adel M. Sarea, Azzam Hannoon, and Abdalmuttaleb M. A. Musleh Al-Sartawi

Abstract The aim of current study is to critically review some of the recently published articles tackling the concept and the importance of Game-based learning to come up with some practical recommendation that can be used in developing the mechanism of applying the concept and how we can use it during the COVID-19. The study recommends that to develop teaching methods and students' skills as well, by investing more in AI techniques which will create more usable games in teaching and learnings during and after COVID-19. Furthermore, educational institutions need to adopt and foster digital literacy to enhance the level of using the technology inside and outside the classroom which will enhance the learners' skills and decrease the risk of spreading the virus.

Keywords Game-based learning · AI · COVID-19 · Education

1 Introduction

During the 1950s, after World War 2, the research on Artificial Intelligence (AI) started emerging when scholars tested the extent to which machines could compete with the processes of human beings [14]. The 1960s brought about a decade of intensified research on AI, reflected in projects pertaining to chess games and robotics. In the current era, AI systems have rapidly spread in many of our life aspects. Research studies have introduced the Expert Systems and Neural Networks which imitate human behavior, such as learning, cognitive logic rationalizing, problem-solving, and computational intelligence, and using mathematical tools mimicking the natural surroundings. Despite AI systems are not being able to replicate human intelligence,

A. A. Khalid (✉)
Universiti Pendidikan Sultan Idris, Perak, Malaysia

A. M. Sarea · A. M. A. M. Al-Sartawi
Ahlia University, Manama, Kingdom of Bahrain

A. Hannoon
American University in the Emirates, Dubai, UAE

A. M. A. Musleh Al-Sartawi (ed.), *The Big Data-Driven Digital Economy: Artificial and Computational Intelligence*, Studies in Computational Intelligence 974,
https://doi.org/10.1007/978-3-030-73057-4_10

it can provide accurate outputs that can far replace human efforts [17, 20]. In education, we need to analyze the strengths and the limits of AI systems to make it useful in solving problems and to determine the training and skills needed to allow students to easily control intelligent systems. When students start learning about the AI, they will be capable to enhance their competences and will effectively participate in improving the business which will be part of its future intellectual capital [19, 21].

The technology is playing a vital role in today economy. Many countries around the world have developed its own plans to increase the investment as they considered it as a clean resource which can develop the economy and have high contributions on the sustainability goals [16]. In the age of Artificial Intelligence (AI), societies are leaning on social media, knowledge management, and data science to survive and achieve the sustainability goals. Accordingly, the rapid expansion of intelligent systems will increase the quantity of education, the demand for new teaching methods, and will increase demand on well-educated and skilled students who can operate the based artificial intelligence systems and enhance the future technological governance [18, 22].

Since the outbreak of Coronavirus (COVID-19), never has technology been more important than it is today. However, the effectiveness of social distancing as a mitigator depends on the extent to which it is maintained by individuals and organizations using technology. Consequently, COVID-19 has been a wake-up call to the education regulators, forcing scholars and professional to quickly adapt and deal with emerging changes and to find practical solutions to carry on normally during extraordinary circumstances while maintaining the confidence of learners and the future employee. In the age of Artificial Intelligence (AI), societies are leaning on social media, knowledge management, and data science to survive and fight the pandemic. Accordingly, the rapid expansion of intelligent systems will increase the quantity of education by creating new educational systems and tools and the large firms in any country will be interested in supporting such changes as part of its social responsibilities [15, 23].

Electronic learning can be considered as a kind of developing programs by specialist to support students in understanding their courses and connecting with the real outside market's needs [1, 20]. E-learning could be a form of educational technology to facilitate learning using a set of prepared or live online courses and content [2, 14, 24]. There are variety of applications or software that can be used to achieve like learn-smart, Kahoot! QuickBooks for Accounting and a lot of many examples can be used as training portals for the students depending on the remote education or in class bases. Consequently, [4] noted that computer-based techniques can be used both as an adjunct to existing training methods and as a replacement for conventional approaches to instruction that do not result in learning. Despite the shortcomings, the use of computers often opens new worlds of learning which go beyond the traditional classroom activity and stimulate the intellectual climate and social interaction among pupils [7, 24]. They help teachers to do better what they already know and act as amplifiers of existing practice [3]. Applying the games- based E-learning plays a vital role in developing the students' ability in understanding the subjects in very smooth ways and to develop their expectations about the external world beyond the classroom [5, 24]. Accordingly, the current study is to try to review the previous studies to develop some practical recommendations that can be used by the regulator of education sectors to develop the education system by formally adopting the gaming based education. This paper is arranged as follows. Section 2 is about published literature review. Section 3 provides the conclusion and recommendations.

2 Literature Review

Game-based learning is a new approach to training students by imitating real-world actions to provide a better understanding of the content. Accordingly, gamified content is used in training and there are so many games that have been developed to reach different audiences and industries. [6] explains that game learning is founded on different principles that include constructivist learning, which underlines that there is a need to offer students the required tools to enable them to build their individual procedures for them to solve a problem. [6] noted that this requires students to participate in the learning, as they interact with their environment to solve challenges that they are given. Another principle of game-based learning is experience, practice, and interaction. Through game learning, learners can safely practice various concepts and experiment on these concepts in non-threatening situations, and gain knowledge and skills through social interaction and practice.

[11] conclude that game-based learning can draw the attention of the learners easily to maximize the effectiveness of the education process by emphasizing creativity and motivation. Having such an approach in teaching will enhance the abilities of the countries to invest more in the digital economy which will enhance the countries' competitive level. Many countries in the world adopted the concept of game learning by easing the accessibility of electronic and web-enabling technologies, which have tremendously contributed to the successful implementation of game learning. [8] add that the internet and networks provide a strong basis for game learning allows for traditional learning approaches to be integrated into electronic technologies to create a new dynamic learning approach. The significance of advanced electronic technologies like the internet in education continues to increase as more people seek to advance their learning. However, game-based learning must be designed and developed with a lot of care, implementing a scientific approach that applies well-designed techniques and procedures. [10] asserts that in the current information era, learning opportunities last a lifetime, from childhood to even old age. People need to continuously refresh and improve their skills and knowledge to keep up with new market trends and technologies. Game-based learning empowers individuals to know, to learn more, learn faster, and utilize the power and resources of information and knowledge. [10] adds that game-based learning results in higher productivity and improved skills and knowledge.

Technology has changed the way of teaching and learning, enhanced the students' skills, fostering innovation among the learners, and make them self-learners which reflected positively on their performance [9]. Nowadays, especially with the arrival of COVID-19, the students and the parents preferred to use iPads and tablets, mobile phones, and other mobile devices instead of using traditional notebooks and textbooks [24].

[13] contend that students are using digital devices and technologies to connect with their colleges and teachers even outside their countries. At the same time, students engage themselves in self-directed learning in various areas such as personal and professional. [12] argue that this kind of learning has helped these students to improve their learning skills leading to significant improvement of their grades. [13] suggested that game-based learning improved the students' grades by 80% and saving the time that they spend in preparation for the exams and classes by 81% which can be a clear indicator of the importance of game-based learning in transferring the education process to be more efficient which will increase the students' productivity. Nonetheless, when learning from home, a lot of students are more motivated to use their mobile devices, which is considered as very important since it is easy and feasible. Beside the mobile phones, students as well chose to study using their devices like tablets and laptops. Many students are dazed with new technologies and wish to integrate them into their learning. Acknowledging the use of digital learning technologies underlines the fact that students have embraced digital learning to improve their grades. In a study carried out by [12] they reached to a conclusion that game-based learning had a positive effect on academic achievements of students. Similar findings were reported by [1, 12] who also concluded that using game-based learning significantly improved the attitudes and motivation of students and their academic achievement. Game-based learning has also been found to improve the learning process and creativity of students. Additionally, [13] mentioned that game-based learning improved the recalling and learning process for the students. Moreover, game-based learning in teaching and learning effectively enhanced the education quality by transforming education to be student-oriented rather than teacher-oriented. Integrating e-learning in traditional learning approaches enhances learners' performance and understanding. However, this must be done in focusing on the needs of the students; other approaches can be used to improve game-based learning applications by integrating internship with the course, to increase the knowledge and competencies of the students [13]. Students at college and university should as well be given a chance to have an internship to get a better idea about what the industry expects from them upon entering the employment world [12], which can be achieved by applying the simulation and conducting remote workshops by senior managers. Students and educators need to have more training about game-based education to allow them to learn more about the actual market's needs and to obtain a good understanding of where and how to apply and integrate game-based learning applications. Universities and the regulators must develop a system that allows the students to learn depending on real-life based problems.

3 Conclusion and Recommendations

Game-based learning is not a very new term for many students. Game-based learning has many advantages over traditional methods of learning such as it allows students to learn from mistakes and learn through doing. Games provide students with instant feedback as well as any negative outcomes because of any mistakes made, which cannot be possible with the traditional methods of learning. Therefore, game-based learning enables students to recognize mistakes almost instantly through the provisioning of feedback, hence it becomes easier to quickly correct them. After analyzing the previous literature, the researcher finds that game-based learning has a significant effect on performance through enhancing the students' ability to develop their own model of learning, manage their time, connecting them with real-life problems, motivate them, enhance creativity, creating innovative thinking, and shape their employability skills.

Games offer students with experimental learning by placing them in the position of the decision makers which allow them to reflect on what they learned to solve complicated problems extracted from real life.

Accordingly, it is highly recommended to develop the teaching methods to develop students' skills by investing more on the AI techniques. As it is shown that educational games provide an opportunity for students to develop and practice physical skills like coordination of hand and eye. Through game-based learning, students can develop, social, cognitive as well as physical skills at the same time something that improves basic life skills such as teamwork and cooperation.

Furthermore, educational institutions need to adopt and foster digital literacy to enhance the level of using the technology inside and outside the classroom which will enhance the learners' skills and decrease the risk of spreading the COVID- 19 virus. Through digital literacy, students will develop key skills such as accountability, strong communication in addition to ethical skills. One more recommendation can be taken into consideration that educational centers such as universities should motivate and encourage students to be a part of game-based learning, because it included not only learning but also entertaining. Therefore, since games consist of measurable objectives, rules, and goals this will provide students with an interactive experience that enhances the sense of achievement among learners.

References

1. Carenys J, Moya S (2016) Digital game-based learning in accounting and business education. Acc Educ 25(6):598–651
2. Hamari J, Shernoff DJ, Rowe E, Coller B, Asbell-Clarke J, Edwards T (2016) Challenging games help students learn: An empirical study on engagement, flow and immersion in game-based learning. Comput Hum Behav 54:170–179
3. Hermanto SB (2018) The role of sharing of accounting learning materials in the use of e-learning in higher education. Am Sci Res J Eng Technol Sci (ASRJETS) 40(1):252–272

4. Hussin AH, Tahir WMMW, Noor IHM, Ismail N, Daud D (2018). Learning accounting via game based approach. In: 2018 ICBMATH, p 26
5. Hwang GJ, Chiu LY, Chen CH (2015) A contextual game-based learning approach to improving students' inquiry-based learning performance in social studies courses. Comput Educ 81:13–25
6. Qian M, Clark KR (2016) Game-based Learning and 21st century skills: a review of recent research. Comput Hum Behav 63:50–58
7. Shah KA (2017) Game-based accounting learning: the impact of games in learning introductory accounting. Int J Inf Syst Serv Sector (IJISSS) 9(4):21–29
8. Shukla AK, Sharif MI (2017) Technology vs. traditional teaching in accounting education: a case study from Fiji National University. Pac J Educ 1(2):41–50
9. Silva R, Rodrigues R, Leal C (2019) Play it again: how game-based learning improves flow in Accounting and Marketing education. Acc Educ 28(5):484–507
10. Watty K, McKay J, Ngo L (2016) Innovators or inhibitors? Accounting faculty resistance to new educational technologies in higher education. J Account Educ 36:1–15
11. Canela AM, Alegre I, Ibarra A (2019) Quantitative methods for management: a practical approach. Springer, Cham
12. Morgan LD (2013) Integrating qualitative and quantitative methods: a pragmatic approach. SAGE Publications, Thousand Oaks
13. AlJeraisy MN, Mohammad H, Fayyoumi A, Alrashideh W (2015) Web 2.0 in education: the impact of discussion board on student performance and satisfaction. Turkish Online J Educ Technol TOJET 14(2):247–258
14. Al-Sartawi A (2020) Social media disclosure of intellectual capital and firm value. Int J Learn Intell Capital 17(3). Accepted article
15. Al-Sartawi A (2020) Does it pay to be socially responsible? Empirical evidence from the GCC countries. Int J Law Manage 62(5):381–394
16. Al-Sartawi A (2020) Information technology governance and cybersecurity at the board level. Int J Crit Infrastruct 16(2):150–161
17. Al-Sartawi A (2019) Assessing the relationship between information transparency through social media disclosure and firm value. Manage Account Rev 18(2):1–20
18. Al-Sartawi A, Sanad Z (2019) Institutional ownership and corporate governance: evidence from Bahrain. Afro-Asian J Finance Account 9(1):101–115
19. Al-Sartawi A (2018) Ownership structure and intellectual capital: evidence from the GCC countries. Int J Learn Intell Capital 15(3):277–291
20. Al-Sartawi A (2018) Institutional ownership, social responsibility, corporate governance and online financial disclosure. Int J Crit Account 10(3/4):241–255
21. Al-Sartawi A (2018) Corporate governance and intellectual capital: evidence from Gulf Cooperation Council Countries. Acad Account Financial Stud J 22(1):1–12
22. Sanad Z, Al-Sartawi A (2016) Investigating the relationship between corporate governance and internet financial reporting (IFR): evidence from Bahrain Bourse. Jordan J Bus Admin 12(1):239–269
23. Al-Sartawi A (2015) The Effect of Corporate Governance on the Performance of the listed companies in the Gulf Cooperation Council Countries. Jordan J Bus Admin 11(3):705–725
24. Al-Sartawi A (2020) E-learning improves accounting education: case of the higher education sector of Bahrain. In: European, mediterranean, and middle eastern conference on information systems, pp 301–315. Springer, Cham

Utilizing Blockchain Technology to Manage Functional Areas in Healthcare Systems

Esha Saha, Pradeep Rathore, and Ashna Gigi

Abstract A blockchain is a public ledger that is secured and trusted which automatically records and verifies huge amount of digital transactions. This study highlights the importance of blockchain technology, and provides an overview on how blockchain can be implemented in the healthcare sector. It mainly focuses on applying blockchain technology in pharmacy, blood banks, laboratory and radiology department. The paper also mentions the various challenges and limitations of implementing blockchain technology in healthcare systems. In addition, it investigates how blockchain can be used as an important tool in fighting against pandemic situations. The present study will help and support decision makers for effective implementation of blockchain technology in the healthcare sector.

Keywords Healthcare systems · Blockchain technology · Applications

1 Introduction

Healthcare is of national importance worldwide. According to the Global spending on health report in 2019, global health spending increased by 3.9% between 2000 and 2017, and is expected to rise at a rate of 5% by 2023 according to 2020 Global health care outlook. It has been reported that considerable sources of these spending are due to the high administrative costs, over-priced medical tests, unwanted treatments, protection of patient data, increasing medical frauds, inefficient operational activities. Also, there is a huge boom in the amount of money spent on duplicating services due to miscommunication between the hospital management and doctors. All these

E. Saha (✉)
Institute of Management Technology, Hyderabad 501218, Telangana, India

P. Rathore
Symbiosis Centre for Management Studies, Nagpur, Constituent of Symbiosis International University (Deemed University) Pune, Nagpur 440008, Maharashtra, India

A. Gigi
Rajagiri College of Social Sciences, Kochi 682039, Kerala, India

A. M. A. Musleh Al-Sartawi (ed.), *The Big Data-Driven Digital Economy: Artificial and Computational Intelligence*, Studies in Computational Intelligence 974,
https://doi.org/10.1007/978-3-030-73057-4_11

factors have ultimately led to poor cost savings. With the emergence of blockchain technology, the ability to organize, synthesize and share complicated health information led to enhanced coordination of care, resulting in proper cost savings and better health outcomes [1, 2].

According to a report from International Data Corporation, 2019, 1 in each 5 therapeutic and healthcare organizations will begin utilizing blockchain for data administration and quiet character purposes by 2020. By 2025, 55 percent of all healthcare applications will have conveyed blockchain for commercial purposes. By the same year, the valuation of blockchain in healthcare will bounce to $5.61 billion from the present $170 million.

Thinking about blockchain's practicality, administration related regions, health care division are starting to change and adjust to coordinate their present status with the blockchain innovation [3]. The business esteem add is required to increment by more than $176 billion by 2025, and surpass $3.1 trillion by 2030. As indicated by the World Economic Forum report, 10% of worldwide total national output will be put away utilizing the blockchain innovation. The worldwide blockchain advertise compound yearly development rate is foreseen to rise 71.46% somewhere in the range of 2017 and 2022 to arrive at an aggregate advertise size of US$4.401 billion by 2022, expanding from US$0.297 billion of every 2017. These evaluations do not astound speculators who put resources into blockchain new businesses; however, the capability of blockchain is as yet obscure all through society. This innovation is as of now being embraced by businesses, for example, banking, minerals and mining, and energy sector [4, 5]. It is currently on the process to be used in the medical services space to alter wellbeing data innovation and its installment models [6–9].

The healthcare segment is enormous and spread over the globe. It includes patients, specialists, clinics, laboratories, pharmacies, and supply chain. Until presently, there is no well-structured method to put through different components of the healthcare industry. To give a compelling framework, blockchain has been put into utilize. Blockchain is capable of changing healthcare offices with patients as the central line. Binding together blockchain innovation in healthcare optimizes restorative data, in this way expanding the security and interoperability [10]. It more over makes a difference to convey an inventive show for trading wellbeing information by bringing together electronic therapeutic records in a proficient and secure way. This dispersed record innovation tool plans to revolutionize the healthcare industry.

Blockchain in healthcare moves forward by and large security of patients' electronic therapeutic records, settle the issues of drugs genuineness and drugs supply chain traceability, and empowers secure interoperability between healthcare organizations [11, 12]. Therapeutic information should be controlled, worked out and allowed to be used by subjects other than medical clinics. It is a core concept of patient-focused interoperability that contrasts historically structured interoperability. Blockchain innovation allows patients to dole out access rules for their restorative information. For instance, enabling specific experts to access parts of their information within a specified timeline. With the invention of blockchain, patients may interact with different clinics and ultimately gather their medicinal knowledge. Given the common size and affectability of medicinal information, it is widely accepted that the label data of therapeutic information will be placed in information squares, not the restorative information itself. In any case, some critical data can be distributed in an open blockchain, for example, on medicate hypersensitivity data. Blockchain innovation is constantly improving as opposed to finishing, and it has some potential difficulties that need to be targeted for applications in biomedical and medical services.

With outlines for the healthcare information administration and therapeutic investigate, and sedate falsifying in pharmaceuticals, this study points to exhibit conceivable impacts and possibilities associated to application of the groundbreaking blockchain innovation within the healthcare segment. Whereas the selection of blockchain in healthcare has been comparatively moderate all these a long time, a major worldview move is in course.

The research question to be addressed in this study is how the blockchain technology can be utilized in different functional areas of a healthcare system? Along with that we will also highlight the challenges and limitations of implementing blockchain technology in healthcare.

The organization of the paper is as follows: Sect. 1 includes the Introduction with an overview of blockchain technology, the research issues along with research question and objective of the study. Section 2 reviews the literature. The blockchain technology in healthcare system is described in Sect. 3. Section 4 includes discussions. Finally, Sect. 5 concludes the paper.

2 Review of Literature

The blockchain platform is a shared database that can execute a transaction through a peer-to-peer network without the central authority's permission. In the healthcare sector, everybody wants to trust in their business, from patient to doctor, data analyst to clinical data providers. Blockchain technology will carry the digital trust among big data dealers in healthcare.

Blockchain has various potentials for healthcare sector. The three factors needed for digital trust that can be achieved by blockchain technology in hospitals are security, identifiability and traceability [13]. The healthcare industry handles many confidential records and documents and are stored in a centralized database [14]. Using blockchain in healthcare can, first, ensure the confidentiality of the data stored. Reliability is calculated by verifying every record from multiple sources. Second, over time, the data will remain unchanged. Nobody may make adjustments to the record without consenting to other outlets, let alone removing the data. Lastly, there will be sufficient data protection because no one can access it without permission [15] Blockchain technology are decentralized and compatible with digital payment systems and hash chain event structure [16].

Blockchain technology creates a platform for the secure exchange of data by addressing various challenges due to patient driven interoperability. In patient-driven interoperability, patient's electronic health data are available through standard mechanisms [17]. There are many applications in the field of healthcare sector. Patient Master Identifier is a type of blockchain in healthcare where distinct identifier is used for all healthcare providers seamlessly. Another feature of the blockchain is fraud detection with smart contracts without the help of third parties. Supply chain management in healthcare can also utilize blockchain with the use of smart contracts. Another sensitive and most important issue is drug counterfeiting that can be minimized if verification is done through the blockchain network [18].

It has been identified that the various applications of blockchain in radiology are authentication and verification, applications in administrative and governance like billing, supply chain tracking, etc. and in the field of research and machine learning [19]. This method performs a distributed verification of the information transiting over the hospital network, i.e. when a new set of MRIs are generated, a check is performed with the aim to guarantee the MRIs integrity by verifying the similarity [20]. It is found that the inclusion of blockchain technology in healthcare happens usually in four stages: all clinical details in current health IT networks are monitored and processed. Various patient data is transmitted to the blockchain network through API utilizing Patient IDs. A smart contract is then used to perform inbound transactions. If the individual decides to disclose his / her name with the health care professional, ultimately, they can share their private key and then the provider can access patient information and provide solutions or care for any symptoms identified [21].

The next breakthrough in the field of blockchain in the medical field is the exchange of data related health. Electronic Health Records (EHR) help attain a secure and encrypted search scheme without any framework for authentication. Taking advantage of blockchain, this scheme achieves feasibility and efficient justice in the sense that honest users are compensated [22]. For a cloud-based EHR distribution network, the search scheme allows the rapid retrieval of EHRs by a medical professional whose symptoms include the keywords queried. The ABE algorithm also helps users to authenticate based on their attributes. In this way, a blockchain based solution can benefit both patients and healthcare institutes to manage health records securely and quickly [23]. The blockchain is used as a peer-to-peer network in a blockchain-based SSE scheme to store user data in a pay-per-use manner. Every user in the decentralized framework has equal status and demands that other users store the data by making a transaction. The person who attaches these data to the blockchain as a block will earn bonuses, from which these data are stored in a public database. This scheme however only supports one keyword quest [24].

Because of its largely open nature, securing patient data for patient access storage and interoperability in the health care system is a challenge for blockchain implementation. As mentioned earlier, one solution to this is the use of a hybrid approach, but the interoperability issue remains with these models. Omni PHR is a model geared towards the distribution and interoperability of personal health records (PHR). Omni PHR model holds PHR in encrypted blocks of data that are spread over its network nodes. Each block is signed by the entity that inserts the information into the data block, which may be a health care professional, a patient, a caregiver or a medical device [25].

It has been noted that records can be preserved and modified in a permanent, verifiable manner, like employment, licenses, and other certificates. The exchange and authentication of licenses would become more effective by utilizing a blockchain-based database to store medical certificates and licenses. The ledger can be used as the only source of truth for current credentials.

MedRec offers a record of their medical history to patients, which is not only detailed but also open and reliable. This restores the agency for patients, as participants are now more fully informed about their medical history and any changes. We allow for patient- initiated data exchange between medical jurisdictions through permission management on the blockchain. MedRec provides for different authorizations to support the need for confidentiality on a granular scale. Specific metadata fields may be exchanged separately within a single record, which may contain additional constraints such as an expiry date for viewership rights. The blockchain ledger keeps an auditable history of medical interactions for patients, providers and regulators. MedRec enjoys a powerful model of failover, depending on the system's multiple participating organizations to prevent one single point of failure. Medical records are stored locally in separate provider and patient databases; on each node in the network, copies of authorization data are stored. In addition, since the medical data remains distributed, the system does not create a new, central target for content attack [26].

Challenges in data processing, privacy and security are rising rapidly day by day. In addition, safety is about protecting delicate data from intruders [27]. By providing the required tool to create consensus among spread entities without relying on a single reliable group, the blockchain technique can ensure data protection, control of sensitive data and encourage the monitoring of health care data for the patient and various actors in the medical field.

A record reminder for screening electronically based medical records was improved in [28] and this planned recall concept recognizes double incremental screening rates, in particular, for older patients. A privacy preservation platform through the use of third-party mobile services was put forward in [29] to modify the authorisations only if they follow the access control policies kept on the blockchain. The management of access controls within the healthcare data sharing systems was identified in [30]. The authors proposed a one-purpose, centric access control model. In this, two data operator classes were recognised: r-users and p-users. It is suggested that r-users read raw data while it is suggested that p-users acquire data about the results obtained. For each data request, a requester must make a transaction to gain access to data for a particular category within a limited period of time, depending on their need.

Blockchain also helps to verify financial transactions in the health care sector and validate records of patients that include test results, medications and medical notes. The Health-Coin is provided for health-related financial transactions, and its contribution to a type 2 diabetes cure proposed a new storage and security strategy based on the blockchain decentralization network, proposing a modern, decentralized EHR program to address third-party dependence. The latest program aims at providing patients a way to warn healthcare professionals to the smallest adjustments in the database. The program reduces the amount of medical errors and helps patients obtain straightforward consultations.

A blockchain system is proposed in [31] which provides patients, health care professionals and third parties with easy as well as secure access to medical information in EHRs, while maintaining private patient data. The research seeks to discuss third party, vendor and patient protection and safety issues in EHRs.

A decentralized attribute-based signatures scheme is used to keep patient data confidential and to retain the anonymity of signer identity within the EHR framework [32, 33]. An innovative on- and off-chain information partnership management can provide reliable information and monitoring of data sharing across multiple health care providers. The open blockchain-based database system ensures unmanipulated, critical and verifiable medical information stored or exchanged in EHR. A blockchain-based framework can be for processing and maintaining EHR in a cloud setting, containing three core elements: doctors, patients, and health insurance providers [34].

A new blockchain-focused program is introduced in [35] that allows patients, healthcare professionals and other stakeholders to access medical information safely and efficiently while maintaining sensitive data confidentiality.

A secure cryptosystem and blockchain-based EHR system is proposed in [36] to support fine-grained access control and ensure systematic, reliable, and confidential authentication of medical data in the EHR cloud. Developments in the program use the simple blockchain decentralization model to ensure the transparency and traceability of EHR medical data records.

A framework based on data sharing to secure EHR data sharing privacy is introduced in [37] that focused on a blockchain network that manages cloud medical details and database databases are inserted into (tamper-proof) blockchain ledgers that resolve possible protection threats from centralized data storage.

MeDShare system is developed for data authenticity, auditing, and protection in the exchange of medical data among multiple organizations, such as research and medical institutions, in an untrustworthy environment. MeDShare was developed using smart contracts on blockchain technology to effectively determine data behavior and detect cyberattacks of offending behavior of the entities [38]. A framework to exchange and maintain medical data and render the platform safe, confidential and auditable is built by [39]. This program is improved by asking data agencies with the aid of a blockchain network to verify the validity of the medical records. A new data preservation system to use blockchain as a reliable storage solution to ensure that the data stored is primitive and verifiable, while maintaining user privacy is discussed in [40].

Apart from that, an important of application of blockchain technology is waste management. There are various problems in waste management as discussed in [41], which also includes hospital wastes, and [42] applied a delegated issue proof consensus algorithm (DPoS)-based blockchain to manage, coordinate, and monitor the disposal of these wastes. Inventory management of pharmaceuticals is also an important issue [43, 44] and blockchain technology can provide complete transparency of the pharmaceutical items from the source to the shelf.

Thus, from the critical appraisal of the existing literature we highlight the importance of blockchain technology in healthcare systems along with various tools and techniques applied in healthcare to utilize blockchain technology effectively.

3 Blockchain Technology in Healthcare Systems

3.1 Utilization in Different Functional Areas of a Healthcare System

Blockchain Technology in Pharmaceutical Supply Chain. There are various challenges and issues where blockchain can progress the capability of passing on a secured and fruitful medicine to the patients. Medication track and trace, product verification, notification about illegitimate drugs, counterfeit drug prevention, manage drug development cycle are the major activities possible with the blockchain technology. Blockchain addresses the challenges of the entire pharmaceutical supply chains from sourcing, production to distribution to pharmacies and hospitals.

Sourcing. Blockchain technology in procurement and sourcing can be applied in smart contracts that can be used for secured multiparty agreements. Also, the overall procure-to-pay process can be done through the application of blockchain technology. The supplier credentials, evaluation criteria, selection process, and performance measures can be shared among the supply chain parties but at the same time will be secured by the use of the blockchain technology.

Production. Blockchain technology in the production can include securing the documentation regarding the entire manufacturing process starting from the sourcing, producing, quality check, package and delivery. The documentation generally includes regulatory requirements, detailed product specifications, packaging components, certifications, and collaborations. Besides, with transparency achieved from blockchain technology, manufacturers are better able to know the demand in real time, and plan the manufacturing of pharmaceutical items.

Distribution. With blockchain, drug shipment can be tracked. Also, with the data, the shortest routes can be obtained to minimize the delivery time, remove unnecessary steps, and enhance the delivery process. Also theft, missing cargo, spoilage can be prevented, and secure payment can be ensured. Ledger-based smart contracts can also increase transparency and profits while decreasing delivery time and costly errors.

Blockchain Technology in Blood Banks. Blockchain could be a secure, self-updating record, giving all clients in its arrange real-time perceivability into each exchange. For blood administrations, this would be required that each sack of blood is checked and labeled, checked all through transportation, and followed as it is handled into extra items counting plasma, cells and platelets. Blockchain makes it conceivable for it to reach possible beneficiary, driving potential advancements in effectiveness and fetched reserve funds for the blood benefit organization, and way better results for patients.

Blood Banks will be associated to distinctive blood banks using an arrangement of distributed-decentralized ledgers. Information are put away in squares. Blood banks can efficiently direct the blood. Excess blood can be quickly transferred to the nearest blood bank requiring blood. Blood bank will be in a position to supervise blood donors and receivers.

- Blood bank can analyze real-time blood status report by querying the blockchain.
- Donors and patients can inquiry the blockchain, after login to the framework utilizing one of a kind recognizable proof number (UID—Aadhar) and the patients will receive message.

Each blood donation is required, and in the long run each sack of blood might be matched with somebody who needs it. As each pack moves through the supply chain, the information gathered will be able to nourish into manufactured insights and machine learning, making modern bits of knowledge that can be put to great utilize—to driving superior outcomes. For blood administrators universally, it is required to see how developing innovation can change the way blood comes to those who require it. For donors, it gives a clear see into how they are making a difference in people's lives. For patients, it guarantees way better and conceivably life-changing results.

Blockchain Technology in Laboratories. Therapeutic research facilities and anatomic pathology bunches are distinctly mindful that associate, secure, interoperable wellbeing records are basic to smooth, productive work. Be that as it may, the current regularly broken state of wellbeing data innovation in health-care of framework regularly disturbs the security and usefulness of data trade between clinic and subordinate hone persistent record systems. One arrangement to this may be blockchain innovation. With its enormous information and inexhaustible touch points (regularly: safety net providers, research facility, physician, hospital, and domestic care), the healthcare industry may well be ready for blockchain data trades. Blockchain might empower secure and trusted linkage of payer, supplier, and patient information.

The laboratory can be associated to the blockchain technology using conveyed ledgers. Therapeutic lab records will be stored in individual's blockchain, thus lab record will be maintained in dispersed records within the frame of pieces. The user can see their reports utilizing their interesting identification number (UID – Aadhar).

Blockchain Technology in Radiology. Over the years blockchain has observed applications in the authentication, verification and administration in the field of radiology.

Authentication and Verification. Blockchain software can be utilized to validate details across a broad variety of uses, from patient-centered exposure to monitoring of health records to visual exchange and cyber protection upgrades. Checked information may be exchanged safely at the patient's alert, counting a list of critical health details, such as differentiating information about allergies and closeness or the need for metal inside the body. If this information was secured off chance on a patient's blockchain health record, the exchange of information would be more consistent across the institutions. A crucial technical enhancement is that current plans and use cases for blockchain applications include holding all touchy wellness details off-chain and off the ledger, whereas, the special hashes ensuring the validity and sharing of information between users are registered. Blockchain will complement existing patient data management strategies within the Electronic Health Care Record (EHR).

This could lead to more notable efficiency and allow for improved auditing of these systems. One of the possible uses cases is within the picture sharing and tele-radiology field. Applications for picture sharing will give patients the right to have full control and get information about their images, as allowed by law. This will encourage patients to require consent to access their information about healthcare, leading to more patient- driven healthcare. Because of its disseminated existence, these use cases can also be extended to include tele-radiology. Within the blockchain, no safe health data is shared or stored. Through using the blockchain to store a specific examination ID and understanding linked to the test, patients can get to whoever they want to share their images, thereby taking advantage of their test's protection and portability. Blockchain can also be used in image sharing and tele-radiology to enhance information security, ensuring the images and commitment distributed have not been modified. While most radiologists and alluding suppliers may not see the advantages of unchanging knowledge in the midst of trade instantly, the software we use today does not defend against these cyber assaults.

Administration. In radiology, settling and administration charges can be applied in regions to mitigate medical billing-related extortion. Shrewd contracts help speed up approval of claims with the potential of streamlining ways of pre-authorisation. Therefore, the blockchain can be used for supply chain follow-up and restorative drug administration. It can also be used to preserve a record of all image equipment assessments and maintenance.

Blockchain During Pandemics. Blockchain is a very helpful tool that could be applied in healthcare industries for tracking drug supply chains, protecting medical data and identifying symptoms of infection. Implementing an interoperable blockchain solution will dramatically boost day-to-day management as well as addressing crisis situations, like COVID-19. The world is fighting hard to get rid of this virus. Imagine a situation when a state provides funds and medical material to all its districts except one which is considered as the hotspot because of the lack of enough medical materials and sufficient amount of fund. Blockchain technology solve this problem by having a record of each district and examine which district did not receive enough funds and medical materials. After examining, it helps donors by providing them with sufficient fund and medical materials. This technology thus allows users to track demand and supply chain for medical supplies. Hashlog is a tool used in blockchain technology which visualize data that helps us to know total number of medical cases worldwide, mortality and recovery rates from the disease. The technology also makes sure that the data that being shared is not modified or manipulated. Blockchain also makes the health insurance claim process much easier. For an individual to claim healthcare insurance, the person need to approach the company to receive the paper work for the insurance but with the help of blockchain technology, documents are provided by the clinic. This makes it possible to process the transactions without human intervention thus reducing the chances of infection from direct contact.

With the support of blockchain, organizations, without any third party, can easily handle centralized capital across a variety of organizations. Using this, businesses will have the potential to build a centralized archive for all work surrounding Covid-19 and could share the chance to cooperate and discover the solution with the entire planet. In addition, the World Health Organization (WHO) has joined hands with major block- chain firms, government entities and global health organizations to create a blockchain- driven monitoring and communications network to track Covid-19 carriers and infection hotspots rapidly and more reliably. The aim is to solve one of the major problems clinicians, scientists and researchers face, i.e., lack of integration of verified data sources that can be used with confidence. Hence, blockchain is an emerging platform that is able to help government, healthcare industries and businesses around the world.

3.2 Challenges and Limitations

The challenges and limitations of implementing blockchain technology in health care, the following observations are made.

Legislation. There are complexities in coordinating appropriate avenues for ownership administration rights over restorative transactions around the world with the planned and blockchain-based healthcare system. Because of the presence of numerous partners, information ownership and existing therapeutic law of the conventional healthcare framework are critical issues that need to be adequately tendered to. It would be difficult to adjust the existing regulatory structure according to new organizational strategy priorities that manage the carefully defined, decentralized and widespread existence of blockchain. Proprietorship of documents, issued get to rights, and blockchain's distributed capability structure should be carefully explained.

Sustainability. Encryption key plays a major part in blockchain. There is no way to recoup the corrupted private key. Because of its long-lasting existence this adds uncertainty to healthcare knowledge. Missing pieces of a patient's health record diminish dramatically its unwavering consistency and importance. In addition, hacking or removing the private encryption key from a user allows access to all data that is placed away in connection with the system.

Transparency and Privacy. Blockchain innovation demonstrates convenience that does not attract wellness space in some situations. Given the fact that it offers protection through encryption, the availability of a database is seen as an imperative problem for healthcare partners, even in scrambled picture. In this way, it would be tendentious to get to power inside the blockchain environment properly.

Governance. For the proposed blockchain-based healthcare system interoperability be- tween open and private blockchain is needed. This raises the need for all-inclusively coordinated interventions negotiations across uniform boundaries and jurisdictions.

Scalability. When customers incorporate information, blockchain grows—in this case, by putting away all the hashes associated with the information being added. It raises the demands of resources and processing power, which means that the organize would have fewer hubs with enough computing power to plan and accept data on the blockchain. In the event that wellness experts struggle to meet demands for resources and technical power, the potential for increased centralization and slower acceptance and confirmation of knowledge is increasing.

Operational Cost. However, while the cost of setting up and working such a framework as well as moving from traditional wellness data frameworks is not known, it can be reduced by open-source innovations and the conveyed nature of blockchain. The un- ceasing operation and maintenance of the proposed blockchain-based co-ordinates health system needs continuous accessibility of assets for purposes of inquiry, updating, improving, and reporting.

4 Discussion

Blockchain technology is not widespread around the world because only few use this technology. The reason behind this is that many people are not aware of the technology and at the same time there are various challenges associated with implementing blockchain technology in healthcare. Cross-border exchange of health data where there are multiple and sometimes overlapping jurisdictions may impede the advantage of data exchange through blockchain. Moreover, individual privacy standards vary from country to country based on government regulations. Another possible challenge under-researched is the blockchain's ability to manage and handle large transfers with access to data in a timely manner. Medical records of a patient consist of images, documents and laboratory reports which need a significant amount of storage space but the blockchain technology does not have that much storage space.

As transaction volumes rise, there will be delay in the mining blocks which further increase exponentially. However, it also creates a problem if the number of blocks in a blockchain network is too less, because this makes the mechanism more centralized and thus less stable. Hence, blockchain is not an easily scalable technique. There are also complexities associated with the rules and regulations for the development of blockchain in healthcare. Even though there are many applications for blockchain technology in healthcare sector, it has not been fully utilized in any of the functional areas of a hospital. So, the health care industry is now under extreme pressure to regulate cost and provide high quality services to patients. The main problem that blockchain technology is facing is lack of experts and specialists to operate it. This problem can be solved by making more people aware about the opportunities of blockchain technology. Many industries have started adopting blockchain technology, so there will be a lot of career opportunities for people who are experts in this technology. We also discussed about the delay due to increase in the volume of transactions. So, there is a need for innovative mechanisms or algorithms to reduce this delay which would increase the performance of the technology. We have also addressed the topic of medication tracking; it is evident that blockchain technologies would be an important resource for pharmacists and healthcare professionals to efficiently and quickly authenticate the distribution of legal medications and their transfer to the patients. Additional studies into reliable monitoring systems that control the Product Registration is required. Current tracking systems that rely on RFIDs and barcodes are not immune from tampering as these codes are sent in the supply chain process as fixed values that can be modified/copied by counterfeiters. Blockchain can truly shape the future of industrial sectors such as healthcare, financial institutions, banking, agriculture, supply chains, education, the Internet of Things, etc. But there is a lot of work to do in minimizing the power consumption while making a network safer.

5 Concluding Remarks

Blockchain provides various preferences that can be used within the healthcare application to illuminate specific record sharing, protection and privacy issues. Blockchain could not be the best scheme to use in any circumstances. Instead, a thorough review of common blockchain problems and how they impact the healthcare application should be evaluated. Mining rewards which are the central tool of blockchain and specific blockchain attacks which can halt the entire system are not completely considered in healthcare applications. Blockchain gives various points of interest, such as decentralized innovation in combination with transparent and unchanging structure, which can prove to be useful for the health sector. Although it offers crucial opportunities, it brings with it a few challenges in the form of administration problems, the need for implementation, belief and stability, scalability, support acceptance, and taking an operational toll. At the same way, this study explores the opportunities and complexities of implementing blockchain engineering at healthcare space from an all-encompassing perspective. The healthcare sector is distinguished by the world's highest estimation of projects and investments, inefficient activities and unpredictable legislation. Blockchain helps us to tackle some of these issues by directing the contact between stakeholders and pushing unused advanced business models and wellness activities. To this end, blockchain has the ability to produce exciting arrangements in the immediate future and such arrangements will lead to disruptive changes within the healthcare industry.

References

1. Angraal S, Krumholz HM, Schulz WL (2017) Blockchain technology: applications in health care. Circ Cardiovasc Qual Outcomes 10(9):e003800
2. Daniel J, Sargolzaei A, Abdelghani M, Sargolzaei S, Amaba B (2017) Blockchain technology, cognitive computing, and healthcare innovations. J Adv Inf Technol 8(3)
3. Ahram T, Sargolzaei A, Sargolzaei S, Daniels J, Amaba B (2017) Blockchain technology innovations. In: 2017 IEEE technology & engineering management conference (TEMSCON), pp 137–141
4. Andoni M, Robu V, Flynn D, Abram S, Geach D, Jenkins D, McCallum P, Peacock A (2019) Blockchain technology in the energy sector: a systematic review of challenges and opportunities. Renew Sustain Energy Rev 100:143–174
5. Chitchyan R, Murkin J (2018) Review of blockchain technology and its expectations: case of the energy sector arXiv preprint arXiv:1803.03567
6. Linn LA, Koo MB (2016) Blockchain for health data and its potential use in health it and health care related research. In: ONC/NIST use of blockchain for healthcare and research workshop, Gaithersburg, Maryland, United States, pp 1–10
7. Engelhardt MA (2017) Hitching healthcare to the chain: an introduction to blockchain technology in the healthcare sector. Technol Innov Manage Rev 7(10):22–34
8. Hölbl M, Kompara M, Kamišalić A, Nemec Zlatolas L (2018) A systematic review of the use of blockchain in healthcare. Symmetry 10(10):470
9. Khezr S, Moniruzzaman M, Yassine A, Benlamri R (2019) Blockchain technology in healthcare: a comprehensive review and directions for future research. Appl Sci 9(9):1736
10. Brodersen C, Kalis B, Leong C, Mitchell E, Pupo E, Truscott A, Accenture L (2016) Blockchain: Securing a new health interoperability experience. Accenture LLP 1–11
11. Rifi N, Rachkidi E, Agoulmine N, Taher NC (2017) Towards using blockchain technology for eHealth data access management. In: International conference on advances in biomedical engineering, pp 1–4. IEEE
12. Pandey P, Litoriya R (2020) Securing and authenticating healthcare records through blockchain technology. Cryptologia 1–16
13. Onik MdMH, Kim C-S, Lee N-Y, Yang J (2019) Privacy-aware blockchain for personal data sharing and tracking. Open Comput Sci 9:80–91
14. McGhin T, Choo KR, Liu C, He D (2019) Blockchain in healthcare applications: research challenges and opportunities. J Netw Comput Appl 135:62–75
15. Koshechkin KA, Klimenko GS, Ryabkov IV, Kozhin PB (2018) Scope for the application of blockchain in the public healthcare of the Russian Federation. Proc Comput Sci 126:1323–1328
16. Roman-Belmonte JM, De la Corte-Rodriguez H, Rodriguez-Merchan EC (2018) How blockchain technology can change medicine. Postgrad Med 130(4):420–427
17. Gordon WJ, Catalini C (2018) Blockchain technology for healthcare: facilitating the transition to patient-driven interoperability. Comput Struc Biotechnol J 16:224–230
18. Ahmad S, Kamal M (2019) What is blockchain technology and its significance in the current healthcare system? A brief insight. Curr Pharm Des 25(12):1402–1408

19. Abdullah S, Rothenberg S, Siegel E, Kim W (2020) School of block-review of blockchain for the radiologists. Acad Radiol 27:47–57
20. Brunese L, Mercaldo F, Reginelli A, Santone A (2019) A blockchain based proposal for protecting healthcare systems through formal methods. Procedia Comput Sci 159:1787–1794
21. Tanwar S, Parekh K, Evans R (2020) Blockchain-based electronic healthcare record system for healthcare 4.0 applications. J Inf Secur Appl 50:102407
22. Chen Y, Bellavitis C (2020) Blockchain disruption and decentralized finance: the rise of decentralized business models. J Bus Ventur Insights 13:e00151
23. Liu Z, Weng J, Li J, Yang J, Fu C, Jia C (2016) Cloud-based electronic health record system supporting fuzzy keyword search. Soft Comput 20(8):3243–3255
24. Li M, Weng J, Yang A, Lu W, Zhang Y, Hou L, Liu J (2017) CrowdBC: a blockchain-based decentralized framework for crowdsourcing, Report, IACR Cryptology ePrint Archive, 444
25. Roehrs A, da Costa CA, da Rosa RR (2017) OmniPHR: a distributed architecture model to integrate personal health records. J Biomed Inform 71:70–81
26. Azaria AEA, Vieira T, Lippman A (2016) MedRec: using blockchain for medical data access and permission management. In: International conference on open and big data, pp 25–30. IEEE
27. Omar A, Rahman S, Basu A, Kiyomoto S (2017) MediBchain: a blockchain based privacy preserving platform for healthcare data. In: Wang G, Atiquzzaman M, Yan Z, Choo KK (eds.) Security, privacy, and anonymity in computation, communication, and storage, SpaCCS 2017. Lecture Notes in Computer Science, vol. 10658, pp 534–543. Springer, Cham
28. Kershaw C, Taylor JL, Horowitz G, Brockmeyer D, Libman H, Kriegel G (2018) Use of an electronic medical record reminder improves HIV screening. BMC Health Serv Res18:14
29. Zyskind G, Zekrifa DMS, Alex P, Nathan O (2015) Decentralizing privacy: using blockchain to protect personal data, pp 180–184. IEEE
30. Yue X, Wang H, Jin D, Li M, Jiang W Healthcare data gateways: found healthcare intelligence on blockchain with novel privacy risk control. J Med Syst 40:218
31. Vora J, Nayya A, Tanwar S, Tyagi S Kumar N, Obaidat MS, Rodrigues J (2019) BHEEM: a blockchain-based framework for securing electronic health records. In: IEEE Globecom workshops
32. Guo R, Shi H, Zhao Q, Zheng D (2018) Secure attribute-based signature scheme with multiple authorities for blockchain in electronic health records systems. IEEE Access 6:11676–11686
33. Ramani V, Kumar T, Bracken A, Liyanage M, Ylianttila M (2018) Secure and efficient data accessibility in blockchain based healthcare systems. In: IEEE global communications conference, pp 206–212
34. Kaur H, Alam MA, Jameel R, Mourya AK, Chang V (2018) A proposed solution and future direction for blockchain-based heterogeneous medicare data in cloud environment. J Med Syst 42(8):156
35. Dagher GG, Mohler J, Milojkovic M, Marella PB, Ancile, (2018) Privacy-preserving framework for access control and interoperability of electronic health records using blockchain technology. Sustain Cities Society 39:283–297
36. Wang H, Song Y (2018) Secure cloud-based EHR system using attribute-based cryptosystem and blockchain. J Med Syst 42:152
37. Liu B, Yu XL, Chen S, Xu X, Zhu L (2017) Blockchain based data integrity service framework for IoT data. In: International conference on web services, pp 468–475
38. Xia Q, Sifah EB, Asamoah KO, Gao J, Du X, Guizani M (2017) MeDShare: trust-less medical data sharing among cloud service providers via blockchain. IEEE 5:14757–14767
39. Theodouli A, Arakliotis S, Moschou K, Votis K, Tzovaras D (2018) On the design of a blockchain-based system to facilitate healthcare data sharing. In: 17th IEEE international conference on trust, security and privacy in computing and communications/12th IEEE international conference on big data science and engineering, pp 1374–1379
40. Li H, Zhu L, Shen M, Gao F, Tao X, Liu S (2018) Blockchain-based data preservation system for medical data. J Med Syst 42(8):141

41. Rathore P, Sarmah SP (2018) Allocation of bins in urban solid waste logistics system. In: Harmony search and nature inspired optimization algorithms, pp 485–495
42. Kassou M, Bourekkadi S, Slimani K, Chikri H, Kerkeb M (2021) Blockchain-based medical and water waste management conception. In: E3S web of conferences, p 234
43. Saha E, Ray PK (2018) Inventory management and analysis of pharmaceuticals in a healthcare system. Healthc Syst Manage Methodol Appl 71–95
44. Saha E, Ray PK (2019) Modelling and analysis of healthcare inventory management systems. OPSEARCH 56(4):1179–1198

Fintech and Entrepreneurship Boosting in Developing Countries: A Comparative Study of India and Egypt

Hebatallah Adam

Abstract Financial Technology or what is so-called "Fintech" is one of the prominent sectors that have been introduced by the Fourth Industrial Revolution at the beginning of the second millennium. The Fintech sector is providing a variety of innovative digitized financial activities in the aims of improving and facilitating traditional financial services as they are often faster, cheaper, easier to access. The rise and development of Fintech sector in today's world is playing a crucial role in increasing the financial inclusion levels among individuals as well as among businesses, especially the Micro, Small and Medium Enterprises (MSMEs). Higher financial exclusion has a major negative impact on poverty and income distribution in societies. About 2.5 billion people and over 200 million businesses are still excluded from the formal financial system which is restraining economic growth, job creation and employment opportunities [13]. Similarly, in India, the figures of the unbanked population counted for 190 million adults without a bank account [13]. For a population size of more than 98 million people, 77% of adults in Egypt are not having a formal bank account [30]. The main objective of this study is to understand how the digitization of financial services can play a major role in increasing inclusive growth and entrepreneurship with a special emphasis on the Indian and Egyptian cases. This study is based on a theoretical and conceptual framework. The study examines various archival, reports of national and international organizations. The findings of the study are showing that the Fintech sector is having immense opportunities for enhancing entrepreneurship and economic growth in developing countries. The Fintech sector in Egypt is still new but rapidly expanding. In comparison with the successful Indian Experience, the digital financial ecosystem in Egypt is facing a major shortfall.

Keywords Fintech · Digital financial services · Entrepreneurship · MSMEs · Financial inclusion · Ecosystem · India · Egypt

H. Adam (✉)
Jindal School of International Affairs, O.P. Jindal Global University,
Sonipat 131001, Haryana, India
e-mail: dhadam@jgu.edu.in

A. M. A. Musleh Al-Sartawi (ed.), *The Big Data-Driven Digital Economy: Artificial and Computational Intelligence*, Studies in Computational Intelligence 974,
https://doi.org/10.1007/978-3-030-73057-4_12

1 Introduction

Bringing the Bottom of the Pyramids population into the country's formal financial system remains the main challenge for most developing countries. Entrepreneurship, investment, production and economic growth can be increased by the introduction and development of innovative digital financial infrastructure and services. Digital financial services play an important role in speeding up the financial inclusion goals in developing countries by narrowing the gaps between genders, as well as between income levels, education and degrees of urbanization. [2] defines Fintech as a "*broad range of financial services accessed and delivered through digital channels, including payments, credit, savings, remittances and insurance*". [25] underlines the importance of the Fintech sector explaining that it represents a relatively new, low-cost means of digital access to traditional financial services.

Emerging financial technology and innovations in traditional financial services play an important role in entrepreneurship growth in developing countries by expanding the access of MSMEs to financial services (i.e. credit, insurance, payments and savings). In a study of Boston Consulting Group (BCG) examining the impact of mobile financial services across multiple countries, it has concluded that mobile financial services alone have the power to increase gross domestic product by up to 5% in the countries examined. The same above-mentioned study has estimated that India's GDP could rise 5% by 2020 through the addition of mobile financial services in society. Increased access to credit, greater investment opportunities, and the creation of new businesses could result in an additional four million jobs for India's total workforce [26].

The main objective of this study is to identify the role played by the Fintech sector in increasing the financial inclusion levels and boosting inclusive growth and entrepreneurship in developing countries. We start at first place by introducing the concept of Fintech and the key market players and the ecosystem. Then in second place, we explain the importance of the Fintech sector for the growth of entrepreneurship in developing countries. And lastly, we explore the Digital Financial Services Ecosystem in India and Egypt to end up at last by the conclusion and recommendations.

2 Literature Review

2.1 Fintech Conceptual Framework

Since the beginning of the Second Millennium, the world is witnessing major developments and disruptions in the ICTs sector that has been accompanied by a tremendous growth of global networks connections. The digital revolution has introduced to the world a globally digitalized economy where new technologies are widely used by governments, businesses and markets system. The intensified interconnectivity of networks and interoperability of digital platforms across all sectors of the economy has created new sectors where technology is the driven factor. Financial Technology (Finance + Technology) or what is also called "Digital Financial Services" is one example of these new services that been brought out by the digital revolution.

Fintech is defined as "the broad range of financial services accessed and delivered through digital channels, including payments, credit, savings, remittances and insurance" [2]. The Fintech sector is introducing a relatively new, low-cost means of digital access to traditional financial services [25].

The use of digital technologies to access financial services is covering a wide range of financial service providers, financial services types and financial services users. [20] illustrates that Fintech services are covering:

1. All types of digital financial services, such as payments, savings accounts, credit, insurance, and other financial digital solutions.
2. All types of users, including individuals at all income levels, businesses of all sizes, and government entities at all levels.
3. All types of providers of financial services, including banks, payment providers, other financial institutions, telecoms companies, financial technology start-ups, retailers, and other businesses.

Furthermore, the Fintech services are incorporating all types of business models: G2C, B2C, B2B, C2B, and C2C models [27]. The Digital Financial Services ecosystem involves four main players (Fig. 1) that are all working together to achieve the financial inclusion goals and the stability and integrity of the national financial systems. [24] explained that the four main players of the digital financial services ecosystem are market users, market services providers, ITCs infrastructures, and government:

1. Market users: Include all kind of users like consumers, businesses, government agencies and non-profit organizations who have needs for digital and interoperable financial products and services.
2. Market services providers: Include all kind of digital financial services providers like banks, other licensed financial institutions, and non-banks (i.e., Fintech companies) who supply those products and services through digital means.

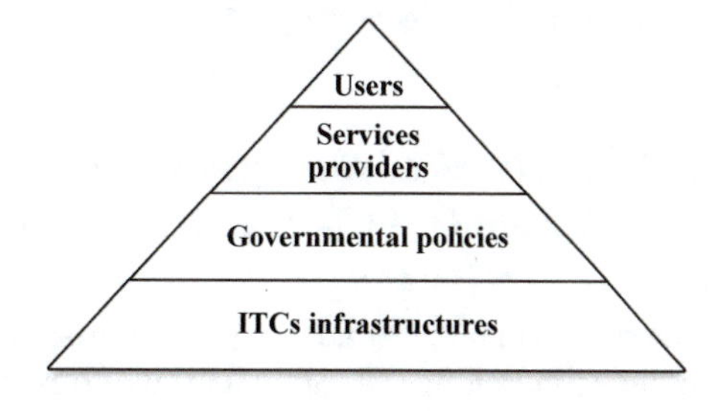

Fig. 1 Digital financial services ecosystem main players. **Source** Author's construction

3. ITCs infrastructures: All the needed ITCs infrastructures (financial, technical, and other infrastructures) that make the provision of digital financial services faster, easier, cheaper, and secure of any fraud or cybercrimes.
4. Government: Includes governmental policies, laws and regulations which set up the rules of the play in the market to ensure that digital financial services are provided in an inclusive, accessible, affordable, and safe manner and that they are serving the national sustainable development strategy.

The effectiveness of the Digital Financial Services ecosystem depends on two fundamental factors: Enabling environment and infrastructure readiness [24]. The enabling environment covers all the regulatory laws and organizing institutions that are organizing the market and set codes of conducts, like:

- Sector organizing laws and regulations that define the authority of financial regulators, permissions are given to non-bank financial services providers and ICTs providers including the telecommunication specific regulations.
- Consumer protection law and regulation.
- The standard-setting enabling environment on different levels: national, regional and international.
- Industry unions and interest groups, NGO's and development organizations that are working on implement digital financial services ecosystems (e.g. World Bank, and CGAP).

The efficiency of the Digital Financial Services ecosystem is also determined by the country's efforts in building a solid level of infrastructure readiness that incorporates the technical systems needed to enable Digital Financial Services such as:

- Secured payments systems available for the transaction between and among sector players (individuals, businesses, Fintech companies, governments).
- Identity systems capable of identifying end-users and their providers, and authentication systems capable of recognizing and validating these identities. Identity systems may be national ID's, sectorial ID's (e.g. financial industry identifiers, bank account numbers, mobile phone numbers) or private sector ID's (e.g. PayPal identifiers).
- Voice and Data Communication Networks. The quality and security of these networks are a necessary component of infrastructure readiness (Table 1).

2.2 Theoretical Background: Fintech Importance for the Growth of Entrepreneurship

Micro, Small and medium-sized enterprises (MSMEs) are a playing a key enhancer of the economic activity around the world as they are considered as a central source of job creation, productivity growth, and innovation. Despite their crucial role, MSMEs

Table 1 Factors affecting the effectiveness of a digital financial services ecosystem

Factors affecting the effectiveness of DFS Ecosystem	Enabling environment	Infrastructure readiness
Sector organizing laws and regulations	X	
Consumer protection law and regulation	X	
Sector national and international standards	X	
Industry unions and interest groups, international organizations	X	
Secured payments system		X
Authentication systems		X
Voice and data communication networks		X

Source Author's construction based on [24]

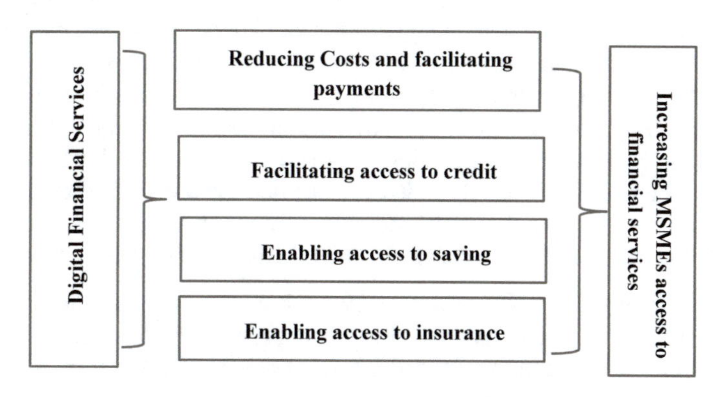

Fig. 2 Digital financial services channels to boost MSMEs. **Source** Author's construction

are still getting a small share of credit from the traditional financial system in developing countries. Emerging financial technology and innovations in traditional financial services play an important role in entrepreneurship growth in developing countries by expanding the access of MSMEs to financial services (i.e. credit, insurance, payments and savings). Financial digital innovation is paving the way for MSMEs to have access to better financial services through several channels (Fig. 2):

1. Reducing Costs and facilitating payments: Digitizing payments have significantly expanded access and reduced costs for all users and especially for the financially excluded ones. Digital finance has reduced the costs of financial services by making it self-service or automated.
2. Facilitating access to credit: Digital Financial Services has made it much easier to offer loans especially to the Bottom of the Pyramid group. Time and costs related to the loan applications have been reduced. The use of digital data

analysis and machine learning for better assessment of existing structured and unstructured data sources (such as financial history, mobile usage data and e-commerce transaction data) have made it possible to expand credit access to the poor and financially excluded small businesses.

3. Enabling access to savings: Digital financial solutions has made it also easier to get access to deposit and investment products by eliminating the high monetary (i.e. minimum-balance requirements and related fees to open an account) and non-monetary (i.e. geographic distance from a bank branch) costs that prevent individuals and businesses from opening deposit accounts.
4. Enabling access to insurance: Digital Financial Services enables the financially excluded segment by enabling them to get access to tailored insurance products that match their specific conditions and needs (i.e. insurance services via mobile apps, ensuring automobiles, electronics, crops, micro and small businesses).

In the Global Fintech Adoption Index report [16], a study made on senior decision-makers at 1,000 SMEs in five countries (the UK and the US, China, Mexico and South Africa), has shown that across the five countries, SMEs choose Fintech solutions because they provide a good range of functionality and features, have services available around the clock, and are easy to set up, configure and use.

Fintech services play an important role in enhancing the financial inclusion of MSMEs in developing countries by providing innovative solutions to traditional MSMEs financing constraints including lack of credit information, little competition and, more generally, the relatively high cost of servicing MSMEs' financing needs. In this perspective, Blancher, [7] underlined different innovative solutions Fintech firms have implemented to overcome conventional MSMEs financing problems, including:

1. Credit Information: New technologies such as big data analytics and cloud computation facilitate the gathering and processing of large amounts of consumer credit performance and behavioural data (for example, from social media, psychometric information, and retail receipts).
2. Real-time Credit Scores: For SMEs, credit analysis can rely on artificial intelligence and machine learning that combine SME registration and accounting information with geographic and socioeconomic information which is helping in generating real-time credit scores and profiles and a stronger credit risk management. Blockchain technology can improve also the information management related to collateral registry and ownership by providing security, privacy, and transparency in decentralized open-source platforms.
3. Enhancing Competition: By using digital finance platform, borrowers can compare between credit cards, insurance, leasing, and other SME banking products.

The Fintech sector has created new channels to finance MSMEs, in both developed and developing markets. New electronic platforms (Table 2) have emerged to offer different financing solutions to MSMEs including crowdfunding, peer-to-peer lending, and other channels.

Table 2 Fintech solutions for financing MSMEs

Types of fintech solutions to finance MSMEs	Examples
P2P/marketplace business lending: online platform collects contributions from investors towards a loan to the business	Lending Club (USA), Beehive (USA)
Equity-based crowdfunding: online platform allows individuals or institutional investors to purchase equity issued by a business	Crowdfinance (UK), Eureeca (UAE)
Reward/donation-based crowdfunding: online platform allows individuals or institutional investors to provide funds in exchange for non-monetary rewards/products/philanthropic motives	Kickstarter (USA), Zoomaal (Lebanon)
Balance sheet business lending: online platform lends directly to the business from its balance sheet	OnDeck (USA), CAN Capital (USA)
Invoice lending: online platform provides liquidity to businesses in form of (discounted) payments for outstanding customer invoices	BlueVine (USA), MarketInvoice (UK)
Merchant and e-commerce finance: online platform that does not have lending as its core business but has rich information about its customer base that it could potentially use to provide credit products	Amazon (USA), Alipay Financial (China)

Source Author's construction based on [7]

3 Research Methodology

This research focuses on India and Egypt. Secondary data in the form of published reports have been commentated to highlight the Digital Financial Services Ecosystem in both countries and their challenges with the different possible opportunities.

This study analyzes archives of national and international organizations like Alliance for Financial Inclusion, Alexandria Bank, the Central Bank of Egypt, the Financial Regulatory Authority in Egypt, the Government of India (GOI), McKinsey Global Institute, UNCDF, World Bank and Bank Negara Malaysia, and World Bank [1–4, 8, 9, 14, 17, 18, 20, 28–31].

4 Discussion

4.1 Digital Financial Services Ecosystem in India

The government of India via India Stack and the Jan Dhan-Aadhaar-Mobile Trinity is supporting the digitization of payments, amending KYC requirements, and customers digital onboarding, and enabling automated access to data from various digitized government systems in the country.

India has started since 2009 to create multiple innovative digital platforms as public goods. Each platform, designed within the regulatory system, solves a single need such as identity, payments or data sharing. Together, these platforms, create a powerful "stack" of applications that is reinforcing the Fintech sector by supporting open, free and contestable markets in digital finance [12].

India Stack, the largest open API in the world, is "*a set of (Application Programming Interface) APIs that allows governments, businesses, startups and developers to utilize a unique digital Infrastructure to solve India's hard problems towards presence-less, paperless, and cashless service delivery*" [17]. It has been implemented in stages, starting in the year 2009 by the introduction of Unique Identification numbers (UID) called the "Aadhaar" which was the first digital public good in India that has been designed for the specific purpose of authenticating individual identity. Starting from 2011, the Aadhaar number is a central key for electronically channelizing the government benefits and subsidies. The next important stage was in 2012 where the eKYC (electronic Know Your Customer) has been launched to allow businesses to perform Know Your Customer verification process digitally using Biometric or Mobile OTP. The introduction of eKYC has reduced significantly transaction costs for businesses and enabled greater access to bank accounts. The government could now also directly transfer subsidies to bank accounts rather than conduct these transactions in cash, reducing fraud and leakages.

Other stages have been implemented to India Stack, in 2015 digital signature (eSign) and digital repository (DigiLocker). In 2016, the Unified Payments Interface, has been launched as the most advanced public payments system in the world to revolutionize digital payments in India. In three years since 2016, it has handled a total of more than 12 billion transactions compared with 5 billion credit card transactions over the same period [12].

The GOI established a regulatory environment in the country and encouraged new businesses to take the lead and make a mark in the finance industry. The GOI has launched initiatives such as the National Payments Council of India (NPCI), Digital India Programme and Jan Dan Yojana.

Another initiative to digitizing financial services and speeding the movement to a cashless economy is the JAM (Jan Dhan-Aadhaar-Mobile) trinity. It is a financial inclusion initiative introduced by the Government of India in 2014 to link Jan Dhan accounts, Mobile numbers and Aadhar cards of Indians to directly transfer subsidies to intended beneficiaries and eliminate intermediaries and leakages.

India is amongst the fastest growing Fintech markets in the world. India ranked the highest globally in the Fintech adoption rate with China. The overall transaction value in the Indian Fintech market is estimated to jump from approximately $65 billion in 2019 to $140 billion in 2023. India has overtaken China as Asia's top Fintech funding target market with investments of around $286 million across 29 deals, as compared to China's $192.1 million across 29 deals in Q1 2019 [18]. The launch of India stack (Unique Identification numbers—UID, Aadhar), the eKYC, eSign, DigiLocker, and UPI) along with the high level of banking penetration through the Jan Dhan Yojana (more than one billion bank accounts) have played a significant role in the rise of the

Fintech sector in India. Besides, there were key factors that have fasten the speed of the growth in the Fintech sector in India during the last 10 years:

- High smartphone penetration (1.2 billion mobile subscribers)
- The growing disposable income
- The expansion in the wide middle-class (By 2030, India will add 140 million middle-income and 21 million high-income households which will drive the demand and growth on the Indian Fintech sector).

The National Association of Software and Services Companies (NASSCOM) revealed that 400 Fintech firms are currently operating in India, and the number is expanding every quarter. The key segments within the Fintech sector in India are:

1. Remittance services: both outbound and inbound remittance transaction are being taken up by Fintech start-ups like FX, Instarem, Remitly and others, posing challenges to giants such as MoneyGram and Western Union.
2. Personal finance and loans: Like Loanbaba website, that is helping people access quick loans within 24 to 72 h.
3. Payment services: web and mobile apps for accepting and transferring payments from businesses and individuals saw a rise after the demonetization drive in 2016. Examples of leading Fintech payment services firms are Paytm, Mobikwik and Oxigen Wallet.
4. Peer-to-peer (P2P) lending: a P2P lending platform allows borrowers and lenders to communicate with each other for lending and borrowing cash, regulated by the Reserve Bank of India (RBI) norms (example: Lendbox).
5. Equity funding: crowdfunding platforms like Start51 and Wishberry.

4.2 Digital Financial Services Ecosystem in Egypt

Banks are the leading services providers in the Egyptian financial market. They hold the biggest share of the financial assets and flows in the market. The Central Bank of Egypt annual report of the FY (2017–2018) [8] shows that the number of working banks has reached 39 banks with more than 2,800 branches across Egypt. Sixteen of these banks offer full-service e-banking and mobile financial services with 133,651 mobile payment agents [4]. Thirty-two of Egypt's 39 banks are providing internet banking services which represent 1.4 million registered accounts and EGP 128 million EGP (approximately USD 7 million) in transaction volumes. Moreover, 4.5 million cards were issued by 2,800 governmental institutions for payroll and seven million cards for pensions [4]. The other non-banking institutions that are providing financial services in Egypt are (statistics available end of 2017):

- The Egyptian National Post Organization (ENPO) with 3,900 branches located all over Egypt.
- Microfinance institutions and non-governmental organizations, 873 licensed with 848 branches.

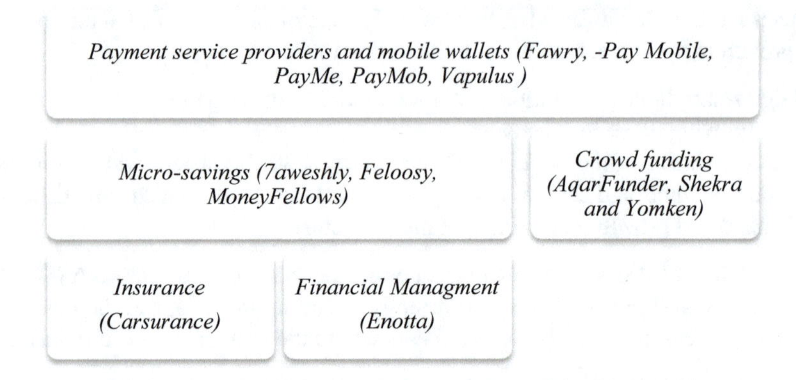

Fig. 3 Egyptian Fintech companies by services type. **Source** Author's construction

- Brokerage companies (140), insurance companies (37), leasing companies (226) and mortgage companies (13).

There is a fast-growing number of Fintech startups in Egypt, driven by the Egyptian Government and the Central Bank of Egypt's initiatives to digitize the payment systems and to attain financial inclusion goals. The most mature sector in the Egyptian Fintech is the provision of payment services, mobile cash and smart wallets [10].

The Egyptian Fintech startups are growing in number (16 startups), value, and specialization [15]. The most mature sector is the provision of payment services, mobile cash and smart wallets [11]. Other sectors are also covered by the Egyptian Fintech including Savings and investments, insurance, financial management, crowdfunding, and blockchain (Fig. 3).

1. Mobile wallets: Following the new regulations for cashless payments using smartphones issued in 2016 by the Central Bank of Egypt, only licensed banks can apply to provide mobile wallets and to act as an issuing bank to take cash deposits in exchange for issuing electronic money. Mobile wallets services, according to the new regulations, include cash-in/cash-out, person-to-person (P2P), person-to-merchant (P2M), merchant-to-merchant (M2M), ATM cash-in/cash-out, international money transfers (IMT), virtual card number (VCN) and account value load (AVL) from bank to wallet accounts.
2. Payment service providers: With collaborate in partnership with banks, Fintech startups provide payments services to banked and unbanked customers to transfer money, pay telephone and other utility bills and different payments use. Fawry is the most prominent Egyptian Fintech companies providing payments services. Other digital payments Fintech companies in Egypt like T-Pay Mobile, PayMe, PayMob, and Vapulus.
3. Micro-savings: Egyptian Fintech startups specialized in micro-savings are: 7aweshly which offers to unbanked customers a tool to save small amounts of money. Feloosy assists people in saving money for a particular investment.

MoneyFellows is allowing users to organize money circles securely and it classifies users according to their income and other factors.

4. Crowdfunding: Some Fintech startups in Egypt have been specialized in crowdfunding, like AqarFunder, Shekra and Yomken. They are offering different ways to support local innovation with their crowdfunding platforms where they collect funds via the internet for a specific project or venture from a large group of supporters.
5. Micro-Insurance: Other Fintech companies aim to increase access to insurance to the Bottom of the Pyramid of the Egyptian population through e-platforms and mobile apps. Carsurance is an example of an Egyptian Fintech company providing insurance quotes.
6. Financial management: Enotta is an example of an Egyptian Fintech startup offering cash management tools for companies.

The Fintech Ecosystem in Egypt involves in addition to banking, non-banking institutions and Fintech companies, local Fintech accelerators that give technical support to existing and new companies in the market. Examples of the Fintech accelerators, incubators and hubs in Egypt: 1864 and Flat6Labs as well as AUC's venture lab Fintech accelerator. Nile University (NU)—in collaboration with the Academy of Scientific Research and Technology (ASRT)—has launched Egypt's first blockchain-focused incubator at NU Tech Space [22].

Banking institutions in Egypt are created under Law No. 88 of 2003 which is also known as Central Bank law. Article 31 of that law requires that any banking activity must have the prior authorization of the Central Bank of Egypt [9]. Banking activity is described by the law as "any activity which essentially and habitually entails the acceptance of deposits, obtaining financing, and investing these monies in the provision of financing and credit facilities, participating in capitals of companies, and all that which the banking custom considers a banking activity" [6].

Egyptian Financial Regulatory Authority (FRA) was established in accordance with the law No. 10 of the year 2009. It is the responsible regulator of the non-banking financial institutions in Egypt. Non-banking financial institutions working in Egypt in sectors: financial leasing, real estate finance, insurance, and microfinance are having each specific regulatory law [14].

Many important steps have been taken by the Egyptian government and the Central Bank of Egypt to accelerate the transformation towards a cashless economy and to increase the financial inclusion main indicators. Alexandria Bank, 2017, denoted four important steps. First, the establishment of the National Council for Payments in 2017, to reduce the use of banknotes outside the banking system, stimulating the use of electronic payments, and developing the national payments systems. Second, the issuing of new version regulations for the Mobile Payment Services by the Central Bank of Egypt in November 2016. The new regulations permit banks' customers to transfer or receive funds and remittances through their mobile accounts. Third, the signature of a Memorandum of Understanding (MoU) between the Egyptian government and Visa, to enable digital payment of government subsidies to 22 million Egyptian families. Earlier, Visa signed an MoU with the Federation of Chambers of

Table 3 Challenges facing the digital financial transformation in Egypt

Challenge type	Problems persisting	Comment/Explanation
Regulatory infrastructure	• Full KYC/CDD[a] requirements are not yet applied in Egypt	• Opening an account is much complex because of the required documents to verify a customer's identity and agents which interrupt the KYC process
	• Lack of consumer protection and cybersecurity	• Expected to be solved under the newly issued regulation by the Central Bank of Egypt under which banks must establish consumer protection measures and systems that provide customers with confidence and protect them against risks, fraud, loss of privacy, threats and criminal activities
	• Small daily and monthly transfer limits	• The transfer limits were fixed at EGP 3,000 daily and EGP 25,000 monthly
Under-development of the current digital infrastructure	• Lack of stable mobile and internet connectivity	• Although mobile penetration in Egypt stands at 109.45%, only 29.42% of these subscriptions have internet access • Overall, internet penetration is relatively low at 37.8% which is considered as a potential barrier to the adoption and usage of mobile financial services
	• Interoperability problem	• Interoperability across different networks in Egypt (e.g., telecommunication, banking, educational, health networks and other sectors/networks)
Formal financial transactions	• High costs and related risks to formal financial transactions	• Despite that 44 million Egyptians are qualified to get access to the formal banking and financial services according to the World Bank, problems like the high cost of formal financial services have weakened consumer accountability on the formal financial institutions

(continued)

Table 3 (continued)

Challenge type	Problems persisting	Comment/Explanation
Technical and financial assistance	• Lack of funds and technical assistance	• Insufficiency in funds available to start new digital infrastructure projects and to implement new market reforms and regulation efficiently • Service providers need additional assistance to improve product design, increase financial literacy and awareness

Source Author's construction
[a]KYC or "Know Your Customer" regime is the due diligence and bank regulation that financial institutions and other regulated companies must conduct to recognize their clients and ascertain relevant information relevant to doing financial business with them. KYC regulating rules aim to prevent identity theft fraud, money laundering and terrorist financing. Similarly, Customer Due Diligence (CDD) is a requirement for the bank to get enough information to verify the customer's identity and assess the related risk.

Commerce in Egypt to encourage the acceptance of debit cards and other electronic payments among merchants and SMEs. Similarly, the Ministry of Communications and Information Technology and MasterCard signed an MoU aiming to extend financial services to 54 million Egyptians. Finally, the creation of the Financial Inclusion Unit by the Central Bank of Egypt to support and enhance financial inclusion. Additionally, the CBE announced a plan to establish an independent central administration unit to protect consumers rights of financial market risks.

Although many market reforms and reorganization have been taken in the last two years in Egypt, still Digital Financial Services meeting numerous regulatory and market infrastructure challenges [4] as it is shown in Table 3.

5 Conclusion and Recommendation

The implementation of an efficient Digital Financial Services ecosystem is crucial to boost the economic welfare of the Bottom of the Pyramid Population in developing countries. An inclusive Digital Financial Services ecosystem includes a range of digitized financial services that provide opportunities for accessing and moving funds, growth capital, and reducing costs and risks of financial transactions.

The analysis of India's experience in the digital finance transformation is showing that the Government of India has implemented an inclusive digital financial ecosystem that is having key features including the widespread coverage of mobile services, quickly authenticate identity, and real-time payment services for low-value transactions for hundreds of millions of customers at the same time. Four main pillars are making from the digitization of financial services in India a unique experience globally: (a) providing digital financial infrastructure as a public good; (b) encouraging private innovation by providing open access to this infrastructure; (c) creating a level playing field through the regulatory framework; and (d) empowering individuals through a data-sharing framework that requires their consent [12].

During the last two years, the Fintech market in Egypt is witnessing many developments like the creation of new innovative Fintech startups, expansion of services coverage of already existing companies. However, the digitization of financial services in Egypt still significantly facing many challenges that are related mostly to the regulatory framework and the currently available market infrastructures. Besides other challenges like strong cash culture, expansive geography and limited accessibility and consumers acceptance of cashless transactions. In respect of the Indian experience, we underline the following recommendation to increase the effectiveness of the digital financial ecosystem in Egypt:

1. Apply full KYC/CDD market requirements and to start implementing the eKYC to allow businesses to perform Know Your Customer verification process digitally.
2. Enforce by market regulation the consumer protection and cybersecurity.
3. Increase the daily and monthly accounts transfer limits to meet individuals and business's needs.
4. Increase investment in mobile and internet infrastructure.
5. Implement competent interoperability across different networks in Egypt.
6. Decrease costs and related risks to formal financial transactions.
7. Provide financial and technical assistance (national/international) to market providers.

References

1. Alexandria Bank (2017) Financial Inclusion in Egypt. https://www.alexbank.com/document/documents/ALEX/Retail/Research/Flash-Note/AR/Financial-Inclusion_Dec17.pdf
2. Alliance for Financial Inclusion (2016) Digital financial services basic terminology, August. https://www.afi-global.org/sites/default/files/publications/2016-08/GuidelineNote-19 DFS-Terminology.pdf. Accessed 20 Oct 2019
3. Alliance for Financial Inclusion (2018a) Digital transformation of microfinance and digitization of microfinance services to deepen financial inclusion in Africa. https://www.afi-global.org/sites/default/files/publications/2018-08/AFI_AfPI_SpecialReport_AW_digital.pdf
4. Alliance for Financial Inclusion (2018b) Financial inclusion through digital financial services and Fintech: the case of Egypt, August 2018. https://www.afi-global.org/sites/default/files/publications/2018-08/AFI_Egypt_Report_AW_digital.pdf
5. Arner DW, Buckley RP, Zetzsche DA (2018) Fintech for financial inclusion: a framework for digital financial transformation. UNSW Law Research Paper No. 18-87; University of Hong Kong Faculty of Law Research Paper No. 2019/001; University of Luxembourg Law Working Paper No. 004-2019, 4 September. https://doi.org/10.2139/ssrn.3245287
6. Bälz K, Rizk L (2018) Client alert: banking without banks? The regulation of Fintechs in Egypt. https://amereller.com/wp-content/uploads/2018/04/180410-Egypt-Fintech-Regulation.pdf

7. Blancher, Mr Nicolas R, et al (2019) Financial inclusion of small and medium-sized enterprises in the Middle East and Central Asia. International Monetary Fund
8. Central Bank of Egypt (2018) Annual report 2017/2018. https://www.cbe.org.eg/en/EconomicResearch/Publications/AnnualReportDL/AnnualReport2017-2018.pdf
9. Central Bank of Egypt (n.d.) Digital financial services regulations. https://www.cbe.org.eg/en/PaymentSystems/Pages/Regulations.aspx. Accessed Nov 2019
10. Chance C (2017) Fintech in the Middle East – an overview, October 2017. https://www.cliffordchance.com/content/dam/cliffordchance/briefings/2017/11/Fintech-in-the-middle-east.pdf
11. Chance C (2019) Fintech in the Middle East - developments across MENA, January 2019. https://www.cliffordchance.com/briefings/2018/12/Fintech_in_the_middleeast-developmentsacros.html
12. D'Silva D, Filková Z, Packer F, Tiwari S (2019) The design of digital financial infrastructure: lessons from India. BIS Paper (106)
13. Demirguc-Kunt A, Klapper L, Singer D, Ansar S, Hess J (2018) The global Findex database 2017: measuring financial inclusion and the Fintech revolution. The World Bank
14. Financial Regulatory Authority (FRA) (n.d.) About FRA. https://www.fra.gov.eg/content/efsa_en/efsa_pages_en/main_efsa_page_en.htm. Accessed Nov 2019
15. Fitch Solutions (2019) Start-ups to drive fast-developing Egyptian Fintech market, 20 November 2019. https://www.fitchsolutions.com/corporates/telecoms-media-technology/start-ups-drive-fast-developing-egyptian-Fintech-market-20-11-2019.
16. GFAI (2019) https://www.ey.com/en_gl/ey-global-fintech-adoption-index. Accessed 05 June 2020
17. India Stack (2020) Indiastack.org. Accessed 05 Feb 2020
18. Invest in India (2020) https://www.investindia.gov.in/sector/bfsi-Fintech-financial-services. Accessed 05 Feb 2020
19. Klapper L, El-Zoghbi M, Hess J (2016) Achieving the sustainable development goals. The role of financial inclusion. https://www.ccgap.org. Accessed 23 May 2016
20. McKinsey Global Institute (2016) Digital finance for all: powering inclusive growth in emerging economies. www.mckinsey.com/mgi
21. McKinsey Global Institute (2016) Digital finance for all: powering inclusive growth in emerging economies. https://www.mckinsey.com/~/media/McKinsey/FeaturedInsights/EmploymentandGrowth/Howdigitalfinancecouldboostgrowthinemergingeconomies/MG-Digital-Finance-For-All-Full-report-September-2016.ashx
22. Odhiambo R (2018) Egypt's first blockchain-focused incubator to launch in april, 29 March. https://bitcoinafrica.io/2018/03/29/egypts-first-blockchain-focused-incubator-to-launch-in-april/. Accessed Nov 2019
23. Ozili PK (2018) Impact of digital finance on financial inclusion and stability. Borsa Istanbul Rev 18(4):329–340. https://doi.org/10.1016/j.bir.2017.12.003
24. Perlman L (2017) Competition aspects of digital financial services. ITU-T Focus Group Tech Rep (03):17–50
25. Perlman L (2018) An introduction to digital financial services (DFS). SSRN Electron J. https://doi.org/10.2139/ssrn.3370667
26. Prasad MVNK (2019) Financial inclusion: emerging role of Fintech. Fintechs Evolv Ecosyst 85
27. Sadłakowski D, Sobieraj A (2017) The development of the Fintech industry in the Visegrad group countries. World Sci News 85:20–28
28. UNCDF (2016) How to succeed in your digital journey: a series of toolkits for financial service providers, DFS Toolkit #1: use mobile as a too. MicroLead Programme, 11 November 2016. https://www.uncdf.org/article/2046/digital-financial-services-toolkit-1-use-mobile-as-a-tool. Accessed Nov 2019
29. World Bank and Bank Negara Malaysia (2017) Global symposium on microfinance: revolutionizing microfinance: insight, innovation, inclusion (2017, May 22–23, Kuala Lumpur, Malaysia). https://pubdocs.worldbank.org/en/332301505318076916/GSM2017-Synthesis-report-draft-August-9th-2017-Final.pdf. Accessed 25 Oct 2019

30. World Bank (2018) Global financial inclusion (Global Findex) database, 19 April 2018. https://datacatalog.worldbank.org/dataset/global-financial-inclusion-global-findex-database
31. World Bank (2019) Financial inclusion global initiative (FIGI), 18 July 2019. https://www.worldbank.org/en/topic/financialinclusion/brief/figi. Accessed 29 Oct 2019

Role of Artificial Intelligence During the Covid-19 Era

Husain Alansari, Oksana Gerwe, and Anjum Razzaque

Abstract Artificial' ability the term artificial intelligence (AI) was first coined by John McCarthy in 1956 when he had the initial theoretical meeting on this topic. But this journey to see if machines will really believe started much before this. Five years later Alan Turing published the article on this idea of machines being able to imitate human race and the ability to do intelligent things, e.g., play chess. Yet, most people be uncertain about the purpose of AI. What if he produces the AI that has the only aim of estimating the amount of Pi, but has decided that by getting earthly death and giving human lives, it could gain another digit to its collection of digits efficiently? Therefore, the entire idea of knowing in machines is unlikely. The only plausible explanation of the power to philosophize, emotionally interact with humans and reason motivation is artificial awareness.

Keywords Artificial Intelligence · Covid-19 · Social Capital · Trust

1 Introduction

1.1 Artificial Intelligence

Financial sectors use artificial intelligence (AI) nowadays to manage transactions, invest in capitals, to manage housing to achieve wellbeing. AI may respond to changes overnight or when job is not taking home [51]. In August 2001, robots can win versus human at the simulated business market contest. AI has eases crime and business offenses by watching behavioral patterns of users for any anomalous modifications

H. Alansari · O. Gerwe
Brunel University, London, UK
e-mail: husain.alansari@brnel.ac.uk

O. Gerwe
e-mail: oksana.grew@brunel.ac.uk

A. Razzaque (✉)
Ahlia University, Manama, Bahrain

A. M. A. Musleh Al-Sartawi (ed.), *The Big Data-Driven Digital Economy: Artificial and Computational Intelligence*, Studies in Computational Intelligence 974,
https://doi.org/10.1007/978-3-030-73057-4_13

or anomalies. (financial assistance conference. 2 April 2015). The ideological point that John Serene has named AI as powerful has posted that the programmed machine with the right inputs and outputs could attain the knowledge and make similar sense as human beings would cognitively' [48]. According to [48] the statement with his Chinese room argument, which takes us to see inside the machine and try to see where this "*thought*" might be. Are there limitations to how smart machines—or human–machine hybrids: will take? The super intelligence, hyper intelligence, or superhuman ability is a hypothetical agent that could have power far exceeding that of the brightest and most talented human brain. Super intelligence may also relate to the shape or level of information owned by such an agent. [52], in fact some people exaggerations about the truth about AI, as well as the media distort and mislead the concept of AI and represent it as a power that will fight humanity to add some interesting action, that drives the people into believing and considering AI as something we should be cautious form, which leads to an idea that we cannot trust AI. Despite the development and growth of the AI sectors, it is not really satisfied with the real concept that satisfied with AI is [36], follow by emphasizing that what he called real-world clinical is not fully AI operated [68] stated that AI are spread out vastly, but the society reaction and prospect should be taking in consideration, to avoid social resistance due to Trust reasons, which will slow down the development of the desired sections.

The term AI has been used in many ways, but one of the most common uses is for computers to perform tasks that humans are incapable of doing, such as job replacement with AI that will resulted increase the unemployment to a high numbers as some scholars, [31] predict that that AI will create an unemployment and economic crises, what he called technological unemployment, this and many scholars and scientist are creating panic wave against AI, in media many science movies talks about scenario that someday AI will decide that people are useless and some predicted that as a result one day AI will decide that humans are not important, then it will start to eliminate humans and will begin the era of human extinction, moreover in his book Do Androids Dream of Electric Sheep?, Philip K. Dick, says The sci-fi tale about Androids and humans collision in The futuristic world. Components of AI allow this sympathy box, mood organ, and the androids themselves. Dick depicts the idea that human judgement is changed by technology made with AI [31, 32]. All the above were used to media consuming or scholars without trying to work parallelly with AI and the employment that can use AI toward developing humanity prosperity and protection.

The concept of AI has become increasingly involved the area of computer science that highlights the creation of intelligent machines, which function and react like humans. Some of the activities and functions that computers with AI carry out include the following: knowledge, learning, planning, speech recognition, problem solving, deep learning, machine learning, ability to manipulate and move objects, computer programming, reasoning, medical field, and knowledge engineering. In other words, AI is a branch of computer science. It seeks to create intelligent machines to do many tasks and activities. Therefore, it has become an indispensable part of the technology industry (Habeeb, 2017). In fact, machines can act and react in the same way humans do in case they have plentiful information associated with the world around them. Thus, AI needs to have access to objects, properties, categories, and the relations existing amongst all of them to apply knowledge engineering. Starting reasoning, common sense and problem-solving ability in machines is rather a difficult approach. Learning without any type of supervision necessitates a capability of identifying patterns in streams of inputs, while learning with suitable supervision includes classification and numerical regressions.

Side by side with the increasing interest in AI, knowledge management is gaining increasing interest by both industry and governments. Nowadays, there is a move towards building knowledge organizations. In such knowledge organizations, knowledge management is planned to play an essential role in the success of transforming individual knowledge into organizational knowledge. One of the main building blocks for developing and enhancing this knowledge management is AI. Many knowledge management specialists and theorists are overseeing this AI to utilize in in knowledge management and knowledge sharing within organizations [1]. Both AI and knowledge management rotate around the same concept of "knowledge", which represents a fundamental component in both. In addition, AI provides all mechanisms to machines enabling them to accumulate knowledge and to learn. Moreover, AI helps machines to obtain knowledge from various sources, process information by means of systematic rules, and then apply this attained knowledge in the best places, where there is a need to share it. To make direct live decisions, machines need to unlock the knowledge stored in their system. In this respect, decision-making represents a key challenge in both fields of knowledge management and AI. Thus, experts decisively believe that AI and knowledge management represent two essential sides of the same coin. They also believe that unless there is a dependable online knowledge base, machines will not be capable of enlarging, creating, or even using knowledge in ways that man has never imagined [61].

The unique association between AI and knowledge management has led to the presence of cognitive computing. This cognitive computing utilizes several computerized models, which stimulate the same way the human brain processes things. It encompasses the utilization of two main components. These are deep learning and self-learning neural networks. The software programs used to carry out cognitive computing make use of pattern recognizers, natural language processors, and data/text mining methods to imitate the activities carried out by human brain [16]. In the field of social media and its association with both AI and knowledge sharing, there the Chatbot, which software that conducts a conversation via auditory or textual methods. This kind of software program is designed to persuasively imitate how a human would behave as a conversational partner, however far the distance is. These Chatbots are typically used in dialog systems for many practical purposes such as information acquisition or customer service [9, 10]. Today, most chatbots are accessed by means of virtual assistants including Google Assistant and Amazon Alexa, via messaging apps like Facebook Messenger or WeChat, or via individual organizations' apps and websites [53]. Chatbots can be classified into usage categories that include conversational commerce (e-commerce via chat), entertainment, finance, education, health, news, and productivity. This chapter tries to investigate the literature available in relation to social capital. Therefore, it casts the light on the definitions given to the concept of social capital, dimensions of social capital, of social capital, as well as investigating social capital in social media, factors affecting social capital, and the significance of social capital. Moreover, this chapter overviews the concept of civic engagement with its definitions, theory of civic engagement, civic engagement and social media, factors affecting civic engagement, and the significance of civic engagement. Additionally, it discusses knowledge sharing, significance of knowledge sharing, knowledge sharing and social media, knowledge sharing in the context of AI, knowledge sharing and social capital, knowledge sharing and civic engagement. Finally, the chapter ends with presenting the research gap and rounds of with a conclusion of the points discussed in this chapter.

1.2 Civic Engagement

The concept of "civic engagement" means "working to make a difference in the civic life of our communities and developing the combination of knowledge, skills, values and motivation to make that difference". In other words, it refers to promoting the quality of life in a certain community or society by means of both political and non-political procedures and processes [26, 27]. In addition, civic engagement includes activities wherein citizens take part in these activities of personal and public interest. Such activities are individually life enriching and socially useful to the society, as well. Another definition states that civic Engagement "is a process in which people take collective action to address issues of public concern" and is totally instrumental to democracy [17].

A report prepared for the Carnegie Corporation confirms that there is the scant agreement of civic engagement [58]. While there is not a singular agreed upon universal definition given to the term civic engagement, many scholars associate the term to the clue of public work or activities that influence public matters [43, 59]. For example, [60] claims that political participation mostly refers to a collection of activities citizens utilize to affect the government structure, the choice of government officials, or the governmental policies. Likewise, civic engagement includes voting in political elections as a kind of social activity [3]. Recently, some studies have started to associate civic engagement with social justice and social work. For example, [55] investigated several civic behaviors of the university students ranging from voting and political engagement to participating in protests and joining social action groups. Many studies sought to provide definitions of civic engagement that limit the concept to a certain kind of activity. These definitions are broader and more inclusive.

Civic engagement as community service. Some definitions, given to the term civic engagement, focus on the participation in voluntary service in the local community. This participation could be by a single individual working independently or working as a member in a group. In this sequence, "Civic engagement is an individual's duty to embrace the responsibilities of citizenship with the obligation to actively participate, alone or in concert with others, in volunteer service activities that strengthen the local community" [27]. Civic engagement as collective action. While the previous definition mentions that activities could be individually or in groups, other definitions restrict the concept of civic engagement to include only activities taken place collectively or in groups to improve community. In this respect, "Civic engagement is any activity where people come together in their role as citizens" [27]. Moreover, it is defined as "the means by which an individual, through collective action, influences the larger civil society" [23].

Civic engagement as political involvement. Other definitions bound the meaning of the concept to actions and activities, which are not only cooperative but specifically political, as well. Therefore, "Civic engagement differs from an individual ethic of service in that it directs individual efforts toward collective action in solving problems through our political process" [26, 27]. Civic engagement as social change. [26], in his definition of the term, concentrates on the component of social change existing in civic engagement. Here, civic engagement is the participation of an active citizen within a community which hence shapes the future. Ultimately, civic engagement involves social change [21]. The emergence of social media has made a breakthrough in almost every aspect of our modern life. Noticeably, social media provide users with high potential for facilitating and enabling civic engagement. For example, when there was a sharp decline in citizens' participation in politics, representing one of the most difficult problems facing democracy in the Western hemisphere, this potential of social media provided new hopes [23]. It has proved that it can help reinvigorate such political raising the levels of voting in the parliamentary elections, and therefore strengthening democratic responsibility at national and international levels [16]. The numerous accounts of users on social media have emphasized the presence of such new possibilities for self-organizing participation like direct democracy and for avoiding mass media gatekeepers and taking action to address the different issues of concern directly. At the same time, others believe that these accounts have highlighted the supremacy of individualization, commercial interests, non-committal participation and safety and censorship [47, 65–67].

For most academics and researchers, social media represent a new and promising arena for civic engagement, for all the social media users especially youth who come at the forefront with their extensive usage in large numbers as the highest segment using the different social media websites [13, 14]. Recent studies and reports show the so-called "Twitter revolutions" or "Facebook revolutions" in the Middle East area, known as "The Arabian Spring," as well as, the Occupy Wall Street movement, and many other similar movements, have arisen because of the role played by these social websites. They played a significant role in mobilizing people to take part in such civic actions. Additionally, social media provide users with new opportunities for having social interaction and active participation via the online meetups, chat rooms, blogs, video sharing sites like YouTube, and social networking sites such as Facebook, Instagram, Whatsapp, Twitter, Tumblr, Linked In, and Google+ [13, 62–64]. In this way, social media provides users with the ability to become collaborative and active participants instead of being passive viewers [57]. Therefore, social media can provide a suitable arena for enhancing participatory democracy in relation to enhanced civic engagement. Consequently, government entities, non-profit institutions such as humanitarian organizations and political parties, as well as the industrial enterprises with a social mission are nowadays investigating such recent social movements with greater concern to utilize them.

Nowadays, there is significant intensive effort made by research community to explore the impact of social media sites like Facebook, Twitter and MySpace on users' social capital and civic engagement [11, 12].

In terms of the impact of social media upon social capital, Jenkins, and co-authors (2006) describe the participatory culture existing on such websites as having comparatively low barriers facing artistic expression and civic engagement. They provide strong support and assistance to single individuals to create and share their own creations with others. Moreover, with some sort of informal mentorship, what most experienced people in various fields of specialization is passed along to others to benefit from in some sort of collective collaboration enriching social capital. On such online sites, members feel that their contributions matter and have some sort of significance. Additionally, these websites of social media provide members with opportunities to feel some degree of social connection with others who share them the same interests and concerns (Haller 2018). Further academic studies on online discussion have indicated that online conversations moderated by government officials lead to obtaining more respectful behavior from participants than that behavior coming out of the spontaneous, unofficial online deliberations. This strengthens the assumptions that the upcoming design of civic engagement in social media ought to enhance and support an informal method, over which users can join intensive deliberations in a flexible but decent fashion [24].

Many factors can affect the presence of effective civic engagement. These include the following:

Trust: Local democracy of the community is not a fundamentally legislative process. Yet, it considers the different cultural and systemic elements existing in the community. Consequently, it is important to think of the existence of mutual relation between associations and citizens. The legislation cannot exist without paying attention to a continuous process of information, discussion, and exchanges with individuals. This will create trust from both sides, allowing laws and legislations to be implemented. In this sequence, building trust is crucial and the absence of such trust between authorities and citizens hinders the effective civic engagement [66]. Awareness: The processes of decentralization within the society address the organization of competences, powers, and responsibilities. Two main parties participate in this process: institutions and citizens. Both parties need to pass through their own process of capacity building, development, and training. The improvement should include these two integrated parties whether the governance, institutions, or civil society. The training of both parties increases the levels of civic engagement in the community. The lack of awareness and understanding in this process could hinder effective civic engagement, represents another challenge for concerned parties.

Joint Approach: Clearly, these two constituents of the governance (citizens and public institutions) could never realize development and improvement in parallel paths without meeting and crossing to each other. There should be collaborative joint processes, where the two develop and cooperate side by side. Such developed abilities are liable to continuous and discussion between them to achieve highest engagement in the civic actions, which benefits the whole society. Mainly, having a perfect legislative system should be built side by side while working on governance and citizenship at the same time. Such joint work guarantees better civic participation from the side of citizens. Attitude to participation: Participation in social activities is not a process that goes gradually and rule by rule when required. Instead, it is an inherent "attitude" towards participative actions for the sake of the whole community. Those who are concerned in schools, universities and organization should work on develop such an attitude to be deep rooted within citizens. When there is a little recognition and knowledge of the required skills of cooperation, participation, negotiation, conflict resolution and team building, this affects engagement. Normally, the attitude to engagement as a regular process is associated with the cultural elements related to dealing with public policies. Legislative and institutional limitations: Sometimes, a community suffers the existence of legislative and institutional limitations that impose restrictions of the activities of civic engagement. These restrictions should be seriously revised and modified, in a way that paves the way to full civic participation in different social activities including politics, education, charity works and the like. The presence of such limitations creates a challenge for developing civic engagement.

Financial and Structural Limitations. In most countries concerned with the improvement of civic engagement, municipalities, and local bodies responsible for this are financially week and this affects their competences. They only get funding from the central government. Lack of capital, necessary for initiating activities related to community such as a real center of decision-making, represents a challenge. In such a case, existence of fruitful dialogue with citizens can be useful, where civil entities can participate in carrying out different projects for the benefit of the community, relieving the pressures and burdens from the shoulders of the central government and the local authorities. Lack of transparency and corruption. One of the major factors that affect civic engagement is lack of transparency and presence of corruption. Public institutions are heavily affected by corruption that hinders a virtuous growth in democratic, social, and economic terms. In turn, the presence of such corruption refrains citizens from participation in civic activities, where they do not know whether their efforts go to those deserving it or not. Additionally, transparency encourages citizens to further share in the social activities where they get acknowledged with everything openly. Lack of transparency and existence of corruption, undoubtedly, refrain citizens from participating effectively in civic duties and actions. Lack of Role Models: The feeling of citizenship in made and not born with. Deep rooting citizenship in the souls of citizens needs the collaboration of both the governmental entities and the social ones including schools, and home. Orders, teachers, and parents should provide good examples of citizenship to their young. Such models teach citizens to follow the news or talk about public affairs raising their devotion to public service. This increases levels of civic participation in the community.

Nowadays, we live in the age of globalization and the revolution in information and technology, where people tend to be more interconnected. However, people are often away from taking part in the policy decisions that affect their own lives. Most governments do not possess the sufficient capability or resources necessary to address all the political, economic, and social gaps existing in the society, whatever advanced it is. Therefore, it is vital to engage individual citizens in the improvement of the societies. In this respect, civic engagement is one assured way where citizens can help effectively and successfully in developing better future for their communities collectively. Civic engagement covers a widespread collection of activities where individuals share in the formal and informal political practices, which deal with the needs of the community. Thus, civic engagement seeks to improve and enhance the quality of life for all of individuals, groups, and whole communities.

Civic engagement provides both individuals and governments with chances to change communities to better. By means of active civic engagement, citizen can recognize and addressing the different social challenges facing the community. In this respect, Civic minded people see themselves as parts or members of a larger entity and their collective responsibility and sense of belonging drive them to participate in shaping the society for the better. With or without the interference of the government, they approach problems, work collectively, and find solutions to them. Additionally, civic engagement helps in preparing for and responding to emergencies and disasters that take place in the community and the whole country. Civic engagement provides the community with its precious social capital, which underlines all positive activities and effects of the created interactions amongst individuals in the community. This social capital benefits many fields such as education provided for the community citizens, increased levels of safety within the society, employment offered to graduates, and decreased crime, illiteracy, and health and socioeconomic differences [34]. The level of this civic engagement in the different civic activities determines whether the community social capital can either beneficial or not. Generally, Civic engagement works as a strategy that targets encountering the challenges found in the society and the whole nation at large. For instance, it strengthens the performance of the schools found in the community and tackles the dropout crises that might occur in the educational community. Undoubtedly, this improves the levels of education in the community enjoying high levels of effective civic engagement, which, in turn, benefits education in the whole country, through preparing students who are ready to participate in serving their country willingly. Moreover, civic engagement assists in improving energy competence, preserving the available energy sources in the community. This facilitates the process of safeguarding the environment as well as all is natural its resources. In turn, this helps in refining health care for all the community members. In addition, it enlarges the economic opportunities in the areas with low income, where new chances of getting job opportunities are offered for the unemployed. Thus, the social capital of the nation can guarantee its progress and advance.

Academics and practitioners have shown greater interest in researching the field of knowledge management since it plays a very important role in enabling the success of institutions and organizations [33, 38]. The shorter product life cycles, rapid transformation of customer requirements and the rising costs of technologies have encouraged institutions and organizations to look at knowledge management willingly as promising sphere worth investment [46, 47, 49]. In literature, there are different definitions of knowledge management. Most of them accounts on the researcher and the academic discipline. One of these definitions given to Knowledge Management (KM) identifies it as processes at maximize outcomes of business units' produce knowledge [30]. As an intangible asset of the organizations, management of knowledge is more multifaceted than any other physical asset. The main aim of Knowledge management in organizations is to make knowledge obtainable to the right person at the right time and in the correct form [39]. In this respect, Knowledge sharing plays an essential role in stating the success of knowledge management initiatives within institutions and organizations [34, 48, 56].

Knowledge sharing represents one of the most significant elements of knowledge management [19]. Knowledge sharing refers to "the exchange of different types of knowledge between individuals, groups, units, and organizations" [40, 41]. It is seen as the process of exchanging knowledge inside a group or an organization [45]. Additionally, Knowledge sharing is "the process of capturing knowledge or moving knowledge from a source unit to a recipient unit" [8, 9]. Previous literature has studied the antecedents and investigated consequences of the knowledge sharing within organizations. Wang and Noe (2010) have recognized several antecedents. These include organizational context such as management support, rewards, organizational structure, organizational culture and climate, and incentives. They also include interpersonal and team features such as team features, diversity, and processes. There is also the cultural characteristics, individual characteristics, and motivational factors such as interpersonal trust and justice, as well as the individual attitudes. All these affect the process of knowledge sharing [39, 70]. For example, [18] examined the individual, organizational, and technological factors influencing the readiness to share knowledge. They came out with the results that personal expectations and incentive systems represent two foremost factors in enticing academics to involve in knowledge sharing activity. They reached the result that "Forced" participation does not represent an effective way in relation to knowledge sharing amid academics.

In a second study, [38] indicated that the applied systems of rewards as well as performance appraisal proved to positive impact upon knowledge sharing between the individual members of the academic staff. Concerning the consequences of knowledge sharing, many of the theoretical and empirical studies carried out recently indicated that knowledge sharing has numerous implications for single individuals, [25], teams [68, 69], teams [35, 46] and organizations [28]. Concerning the way of sharing knowledge in organizations, [48] explained that there is not one single correct way to make people share knowledge among them, but there are many ways. These ways rely on the style and values of this organization. Additionally, [67] made a differentiation between two main different forms of knowledge sharing. These two forms are knowledge donating, which refers to communicating to others what one's personal intellectual capital is, and knowledge collecting, which refers to consulting colleagues to urge them to share their intellectual capital. Then, they concluded that the knowledge required can be obtained through the usage of these two different forms [38, 40].

2 Problem Statement

It is necessary that the winners of future want to use AI to better sales and marketing campaigns. But, at the end of this day, they would even want to learn how to relate to and interact with different humans. The most powerful teams can learn how to provide the best of both AI and human ability to develop a revenue strategy that can improve customer participation and increase exchange of resource [28–32]. Many scientists warned that there is a need for humans to be worried that AI will be able to replace human, Scientist Stephen Hawking, Microsoft founder Bill Gates, and SpaceX founder Elon Musk have revealed his fear about this chance that AI would grow to this fact that humanity would not keep it, With Hawking hypothesized that they would "mean the ending of the human beings" (Rawlinson, Kevin), which lead to human disengagement. as a result, will play an effect on AI development because the lack of civic engagement and participation are not enough to make people aware of the need for AI. Nowadays, AI, which was when believed to go purely in the domain of the human imagination, is a very serious and looming potential. In the fact of time imitating creation, we are confronted with this question of whether AI is harmful and if its benefits far outweigh its potential for severe effects to all humanity. It is no longer the topic of it [2], but when, moreover, scholars emphasize that the medical sector needs serious demand for real time AI technology that are able take decision that will help us handle the pandemic from spread around the world [55, 56]. A question of trust is not condoned discussion about how some confidence humans take in machines, and what functions people are willing to outsource [36].

Furthermore, civic engagement in the modern era is facing challenges. Especially today's youth who are growing in an era of social media network do not seem to have interest in the civic engagement activities, which is posing problems to many organizations involved in civic engagement. This is one side of the problem. On the other side are participants engaged in civic activities but are challenged with advancing technologies like social media network characterized by AI. This breed of participants faces the daunting task of learning to use modern techniques in civic engagement failing which there is a danger of these participants get disinterested in civic engagement leading to their disengagement. The overall result either way shows that civic engagement could be threatened with lack of participants who could volunteer and engage in civic activities. This in turn leaves a gap in the society due to the absence of civic engagement in the long run affecting those people who are benefited by civic activities.

3 Trust

A Question is if how AI is trustworthy, was raised as an issue that could potentially affect the future of AI. Here is the great news: Some tourniquet we must learn to believe humans we will have to learn to believe AI [36, 37], this meaningful sentence is important to shed the light into Trust in AI will be certainly the future of AI. The prospect of AI should be more developed that will even change humanity relationship with machines ever, over time the confidence in the person should change. You will get the sense of what the person is better in, and where he or she might require a moment of supervision. Perhaps, but perhaps, you will see that because the person is now part of the team and has gained the confidence, you will go on to focus on more important and interesting things [37].

This emerging period of partnerships between humans, machines, and AI brings with it. The recent rise of business opportunities. A recent study from Forrester Research argues that automation will not ruin American businesses, but instead change the workforce. This study findings suggest that automation and AI can produce approximately 15 million new jobs in the USA over the next ten years. These current businesses are required to be technologically oriented and involve new skillets. And the force change is not specific to the United States. But a necessary question needs to be asked, since the trust in AI is low, what about AI is within the mobile also! Siri, Google today, Cortana are all smart assistants using AI technologies designed to help you make answers to your questions or do activities with sound power. For instance, you may tell: what is the weather today? And the worker can search for this data and refer it to you.

Since trust is identified as a component of civic engagement [23], and there is a gap in the literature that concentrated on trust as a sub-variable of civic engagement, the researcher decided on investigating its relationship with the independents in this research and the mediator which is knowledge sharing. Assessment such a relationship can guide researchers to conclude that knowledge sharing. And trust. If such a relationship is significant then it would compliant the revelations of [57] who indicated that 'higher the trust in or perception of value of social media culture that is characterized by AI, lower will be the disengagement of participants from civic engagement". Additionally, the same research indicated that knowledge sharing is an important variable that enables transfer or sharing of knowledge about social capital mediates between social capital and organizational performance. [2] assured that there is a direct linkage between knowledge as a construct to civic engagement. Nevertheless, the researcher did not assure that knowledge sharing is correlated to components of civic engagement.

3.1 Trust vs Covid-19

Despite the low indication of Trust, scholars mentioned that since the beginning of COVID-19 pandemic accrued the Trust has indicators has changed to positive in many sectors, for instance, if we look closer how AI is used in healthcare, we will be able to determine if the indications are justified from their point of view: Scientists succeeded in using AI algorithms to differentiate between chest disease and Covid-19 between 2060 patients [71], the benefit is not only on the human and healthcare sectors, but it extends to reach urban monitoring and management [6, 7].The diagnostic of Covid-19 can be quicker with AI, using mobile devices to cite an infected town or area will help the insolation process [Rao.. 2020], but more effort needs to be done in data classification and that come up with Covid-19 data set discovered also more applications and models needed [18–21]. Due to diagnostic reasons experiments was conducted to measure the public opinion on social media discussion that analyze sentiments as well as graphic analysis on the most devastated on that period, the result was three key words 'Trump, mask, hospital' that gives a hint about the negative in public concerns (Hung. 2020). According to [43] there is high impact of AI toward improving development, of medicine, vaccine, monitoring daily cases, tracking the cases and monitor their behavior change as well as case recovery as well as cases follow up, as well as decrease human interference in medical cases, nevertheless, most of these AI models are not shown to be the public communities the real participation of the AI, it is important to know that physicians and health care workers will not be able to treat all cases. Thus, computer scientists need to participate by start working to solve COVID-19 by presenting more intelligent possible solutions to get rapid control of severe acute respiratory syndrome, the virus that causes the disease [4]. Practical AI, a.k.a. Sophisticated data processing, aims to develop commercially viable "*intelligent*" *systems—for information, "expert*" medical identification systems and stock-trading systems. Practical AI has enjoyed significant success, as reported in the section Expert organizations [4]. While AI can help the healthcare sector, where the professionals decide the best diagnosis or care for the patient, AI can also make a change in conventional drug discovery methods, resulting in intelligent drug design. Using AI and other instruments, scientists and researchers would be able to more efficiently determine compounds that are possible to succeed in processing the given consideration. Benevolent AI is a good example of the company spearheading campaigns at applying AI in drug discovery. Elders amongst others are engaging with other Covid-19 patients without knowing, are assumed to have Covid-19 therefore they need to be examined, people with considered genetics that are more effected by Covid-19 needs to be on the top list, spical enzime in the lung was discovered related to Covid-19 symptoms indicates its more sensitive and more need medical attention such as diabetes and hypertension [43].

During the pandemic of Covid-19, many researchers got working shoulder to shoulder with computer scientist to collect and analyze specific data that could detect Covid-19. Therefore, an early accurate detection, that is so hard to screen, was a problem that got solved by AI. As a result, AI differentiated between who is already living with a lung disease versus who has contracted the Covid-19 [44], AI plays an essential role in disseminating robust tool that aid in powerful impacting the reshaping of the workflow as well as clinical analysis and decisions [16,64,65]. While it is such the case, there is a lot of work to be done to improve producing medical services and patient help, the effect of AI is already evident there, and physicians believe how cooperating with AI is advantageous to a point where it is considered crucial. Elsie Ross, a vascular surgeon at Stanford University in Palo Alto, California, prefers if doctors adopt AI to help in aiding patients. This is since AI simulates human intellect that can recognize proposed a proper vaccine development for covid-19, appropriate monitoring, scan, predict and follow the cases and determine possible patients [56].

Machine learning, high system virus detection with high tense and speed, respiratory pattern, as well as automated to acquire more capacity diagnostic, that indeed will protect the healthcare worker [43, 66]. Moreover [49] mentioned that AI has enabled rapid diagnostic of Covid-19 patients, AI has reuse drugs that enable to treat Covid-19 [5, 72] stated that AI can treat structured and unstructured neural networks diagnostic. AI has successfully approved in scanning chest detection using multinational datasets, after conducting AI based on deep learning experiment on 1280 patient the result was 90% accuracy [36], moreover the Brazilian and US AI models that are based on temperature and perception variable has 3,08% error the writer concluded this model is the best to forecast and predict 6 days ahead [22].as well as protecting the workers and stopping the spread (The Lancet Digital Health). Some issues occurred such as the phone call models that keep the line busy all the times, [18, 46] chatBot is easy way to diagnose the cases without waiting guaranteed the diagnostic and report 100% and distinguished 96.32% of the cases of 20,000 clinical cases. [46, 57] natural language processing and AI have significant clinical influence when used to data stream, that will be more vital in response. Mapping landscape tackle many aspects and the main reasons that crisis stopped, and it could be worked more effectively if there is more data provided that could be implemented to the project which will reflect on the AI performance and know in which way to direct the community [15, 16], however there is need to balance between data privacy and public health the more data the more lives will be saved but all has to be under supervision that will guarantee to manage the data security or not being lost [48].

4 Conclusion

Since there is still no cure o the COVID-19 pandemic, the best way to avoid the healthcare workers contracting this virus and the same goes for the patience, is through social distancing. And if this notion can be first facilitated by AI and machine learning, this would be a greater good [43]. Clinical trials that are properly planned are important before these emerging methods are applied in a real clinical environment [57]. From the healthcare perspective it is obvious that AI is vital in ample critical cases that needs proper consideration, it becomes clear that there are important benefits but not announced to the public in a way that they would embrace it. From the above we can notice the discussion indicating the trust issue with AI, hence with application of AI in dealing with covid-19 has raised these indicators, however according to the above scientist, the discipline needs to be developed but before that jobs have to be upgraded to fit and worked to the purpose of AI so that the public and humanity will make AI serve and comply by working to the interest of humanity prosperity, therefore Trust will that will reflect and raised the performance of AI.

References

1. Abbas M, Sajid S, Mumtaz S (2018) Personal and contextual antecedents of knowledge sharing and innovative performance among engineers. Eng Manage J 30(3):154–164
2. Adams M (2016) Is Artificial Intelligence Dangerous? Forbes. https://www.forbes.com/sites/robertadams/2016/03/25/is-artificial-intelligence-dangerous/
3. Adler RP, Goggin J (2005) What do we mean by civic engagement? J Transf Educ 3(3):236–253
4. Alafif A, Adly A, Adly M (2020) Faculty of Computers and Artificial Intelligence. In: Adly C (n.d.) Approaches Based on Artificial Intelligence and the Internet of Intelligent Things to Prevent the Spread of COVID-19: Scoping Review. Retrieved November 29, 2020. https://www.jmir.org/2020/8/e19104/
5. Alafif T (2020) Machine and Deep Learning Towards COVID-19 Diagnosis and Treatment: Survey, Challenges, and Future Directions. https://doi.org/10.31224/osf.io/w3zxy
6. Alguezaui S, Filieri R (2010) Investigating the role of social capital in innovation: sparse versus dense network. J Knowl Manage 14(6):891–909
7. Allam Z, Jones D (2020) On the coronavirus (COVID-19) outbreak and the smart city network: universal data sharing standards coupled with artificial intelligence (AI) to benefit urban health monitoring and management. Accessed 29 Nov 2020. https://www.mdpi.com/2227-9032/8/1/46/htm
8. Bircham-Connolly H, Corner J, Bowden S (2005) An empirical study of the impact of question structure on recipient attitude during knowledge sharing. Electron J Knowl Manage 32(1):1–10
9. Bradeško L, Mladenić D (2017) A survey of chabot systems through a loebner prize competition. Artificial Intelligence laboratory, Jozef Stefan Institute, Ljubljana Slovenia
10. Britannica (2018). Artificial intelligence. www.britannica.com/technology/artificial-intelligence
11. Brandtzæg PB, Følstad A, Mainsah H (2012). Designing for youth civic engagement in social media. In: Proceedings of the IADIS International Conference of Web Based Communities and Social Media, pp. 65–73, July 2012
12. Brandtzaeg PB, Følstad A (2017) Why people use chatbots. In: International Conference on Internet Science, pp. 377–392. Springer, Cham, November 2017
13. Brooks R (2014) Artificial intelligence is a tool, not a threat, 10 November 2014. Archived from the original on 12 November 2014
14. Boulos MNK, Resch B, Crowley DN, Breslin JG, Sohn G, Burtner R, Chuang KYS (2011) Crowdsourcing, citizen sensing and sensor web technologies for public and environmental health surveillance and crisis management: trends, OGC standards and application examples. Intl J Health Geograph 10(1):1–29
15. Bullock J, Luccioni A, Pham K, Lam C, Luengo-Oroz M (2020) Mapping the landscape of artificial intelligence applications against COVID-19. Accessed 29 Nov 2020. https://arxiv.org/abs/2003.11336
16. Castrounis A (2017) Artificial intelligence, deep learning, and neural networks, explained. Kdnuggets.com. https://www.kdnuggets.com/2016/10/artificial-intelligence-deep-learning-neural-networks-explained.html

17. Checkoway B, Aldana A (2013) Four forms of youth civic engagement for diverse democracy. Child Youth Serv Rev 35(11):1894–1899
18. Chen NDJ, Nokes RME, Kruijshaar JK-H, Yu A, Diekmann OH, Pastor-Satorras A, McCall B (1970) A review of mathematical modeling, artificial intelligence and datasets used in the study, prediction and management of COVID-19. https://doi.org/10.1007/s10489-020-017 70-9. Accessed 29 Nov 2020
19. Chong AYL, Ooi KB, Bao H, Lin B (2014) Can e-business adoption be influenced by knowledge management? An empirical analysis of Malaysian SMEs. J Knowl Manage
20. Chong CW, Yuen YY, Gan GC (2014) Knowledge sharing of academic staff. Libr. Rev. 63:203–223
21. Correction to Lancet Digital Health 2020 published online May 6. https://doi.org/10.1016/S2589-7500(20)30104-7. (2020). The Lancet Digital Health **2**(6). https://doi.org/10.1016/s2589-7500(20)30110-2
22. Da Silva R, Ribeiro M, Mariani V, Coelho L (2020). Forecasting Brazilian and American COVID-19 cases based on artificial intelligence coupled with climatic exogenous variables, October 2020. https://www.ncbi.nlm.nih.gov/pmc/articles/PMC7324930/. Accessed 29 Nov 2020
23. Dahlgren P (2011) Young citizens and political participation: online media and civic cultures. Taiwan J Democracy 7(2):11–25
24. Davies T, Chandler R (2012) Online deliberation design. Democracy in motion: Evaluation the practice and impact of deliberative civic engagement 103–131
25. Dee J, Leisyte L (2017) Knowledge sharing and organizational change in higher education. Learn Organ 24:355–365
26. de Paz JP, Caujapé-Castells J (2013) A review of the allozyme data set for the Canarian endemic flora: causes of the high genetic diversity levels and implications for conservation. Ann Bot 111(6):1059–1073
27. Diller EC (2001) Citizens in service: The challenge of delivering civic engagement training to national service programs. Corporation for National and Community Service, Washington, DC
28. Ehrlich T (Ed.). (2000). *Civic responsibility and higher education.* Greenwood Publishing Group.
29. Ekbia HR, Nardi BA (2017) Heteromation, and other stories of computing and capitalism. MIT Press, Cambridge
30. Fatemi F (2017) Artificial Intelligence and Human Intelligence: The Essential Codependency. Forbes, October 2017. https://www.forbes.com/sites/falonfatemi/2017/10/24/artificial-intelligence-and-human-intelligence-the-essential-codependency/
31. Forsythe DP (2017) Human rights in international relations. Cambridge University Press, Cambridge
32. Galvan J (1997) Entering the posthuman collective in Philip K. Dick's, "Do androids dream of electric sheep?" Sci Fiction Stud 24(3):413–429
33. Gerwe O, Al-Ansari H (2020) Impact of artificial intelligence based social capital on civic engagement in an environment of changing technology: development of a theoretical framework. SSRN 3659186
34. Habeeb A (2017) Introduction to artificial intelligence. University of Mansoura, Egypt
35. Harmon S, Sanford T, Xu S, Turkbey E, Roth H, Xu Z, Yang D, Myronenko A, Anderson V, Amalou A, Blain M, Kassin M, Long D, Varble N, Walker S, Bagci U, Ierardi AM, Stellato E, Plensich GG, Franceschelli G, Girlando C, Irmici G, Labella D, Hammoud D, Malayeri A, Jones E, Summers R, Choyke P, Xu D, Flores M, Tamura K, Obinata H, Mori H, Patella F, Cariati M, Carrafiello G, An P, Wood B, Turkbey B (2020). Artificial intelligence for the detection of COVID-19 pneumonia on chest CT using multinational datasets, 14 August 2020. https://www.nature.com/articles/s41467-020-17971-2. Accessed 29 Nov 2020
36. He J (2019) The practical implementation of artificial intelligence technologies in medicine. Nat Med 25:30–36

37. Hollis C (2018) Let Artificial Intelligence Earn Your Trust. Forbes. https://www.forbes.com/sites/oracle/2018/03/26/let-artificial-intelligence-earn-your-trust/
38. Inkinen H (2016) Review of empirical research on knowledge management practices and firm performance. J Knowl Manage. 20:230–257
39. Jennex M (2017) Re-examining the Jennex Olfman knowledge management success model. In: Proceedings of the 50th Hawaii international conference on system sciences., January 2017
40. Kang M, Lee MJ (2017) Absorptive capacity, knowledge sharing, and innovative behaviour of R&D employees. Technol Anal Strat Manage 29(2):219–232
41. Khalil OE, Shea T (2012) Knowledge sharing barriers and effectiveness at a higher education institution. Int J Knowl Manage (IJKM) 8(2):43–64
42. Kiesa A, Orlowski AP, Levine P, Both D, Kirby EH, Lopez MH, Marcelo KB (2007) Millennials talk politics: a study of college student political engagement. In: Center for information and research on civic learning and engagement (CIRCLE)
43. Lalmuanawma S, Hussain J, Chhakchhuak L (2020) Applications of machine learning and artificial intelligence for Covid-19 (SARS-CoV-2) pandemic: a review. https://www.ncbi.nlm.nih.gov/pmc/articles/PMC7315944/. Accessed 28 Nov 2020
44. Li L, Qin L, Xu Z, Yin Y, Wang X, Kong B, Bai J, Lu Y, Fang Z, Song Q, Cao K, Liu D, Wang G, Xu Q, Fang X, Zhang S, Xia J, Xia J (2020). Using Artificial Intelligence to Detect COVID-19 and Community-acquired Pneumonia Based on Pulmonary CT: Evaluation of the Diagnostic Accuracy, August 2020. https://www.ncbi.nlm.nih.gov/pmc/articles/PMC7233473/. Accessed 28 Nov 28, 2020
45. Lilleoere AM, Hansen EH (2011) Knowledge-sharing enablers and barriers in pharmaceutical research and development. J Knowl Manage 15:53–70
46. Martin A, Nateqi J, Gruarin S, Munsch N, Abdarahmane I, Zobel M, Knapp B (2020, November 04). An artificial intelligence-based first-line defence against COVID-19: Obeid, J, Davis, M, Turner, M, Meystre, S, Heider, P, O'Bryan, E, Lenert, L. (2020, July 04). Artificial intelligence approach to COVID-19 infection risk assessment in virtual visits: A case report, 04 November 2020. https://academic.oup.com/jamia/article/27/8/1321/5843795. Accessed 29 Nov 2020
47. McCurdy P, Uldam J (2014) Connecting participant observation positions: toward a reflexive framework for studying social movements. Field Methods 26(1):40–55
48. McDermott R, O'dell C (2001) Overcoming cultural barriers to sharing knowledge. J Knowl Manage 5(1):76–85
49. Mei X, Lee H, Diao K, Huang M, Lin B, Liu C, Xie Z, Ma Y, Robson P, Chung M, Bernheim A, Mani V, Calcagno C, Li K, Li S, Shan H, Lv J, Zhao T, Xia J, Long Q, Steinberger S, Jacobi A, Deyer T, Luksza M, Liu F, Little B, Fayad A, Yang Y (2020). Artificial intelligence–enabled rapid diagnosis of patients with COVID-19, 19 May 2020. https://www.nature.com/articles/s41591-020-0931-3. Accessed 29 Nov 2020
50. Naudé W (2020). Artificial Intelligence Against Covid-19: an early review, 06 April 2020. https://papers.ssrn.com/sol3/papers.cfm?abstract_id=3568314. Accessed 29 Nov 2020
51. O'Neill E (2016). Accounting, automation and AI. www.icas.com. Archived from the original on 18 November 2016. Accessed 18 Nov 2016
52. O'Keefe B (2015) Can artificial intelligence stop hackers? Fortune. https://fortune.com/2015/07/14/artificial-intelligence-hackers/
53. Orf D (2016) The 8 best chatbots of 2016. https://venturebeat.com/2016/12/21/8-top-chatbots-of-2016/
54. Patel SJ, Sanjana NE, Kishton RJ, Eidizadeh A, Vodnala SK, Cam M, Gartner JJ, Jia L, Steinberg SM, Yamamoto TN, Merchant AS, Mehta GU, Chichura A, Shalem O, Tran E, Eil R, Sukumar M, Guijarro EP, Day CP, Robbins P, Feldman S, Merlino G, Zhang F, Restifo NP (2017). Identification of essential genes for cancer immunotherapy. Nature, 548(7669):537–542
55. Pritzker S, Springer M, McBride AM (2015) Learning to vote: informing political participation among college students. J Commun Engage Schol 8(1):8
56. Roberts J (2016) Thinking machines: the search for artificial intelligence. Distillations. 2(2):14–23
57. Rameesh A (2004) Artificail intelligence in medicine

58. Rao A, Vazquez J (2020) Identification of COVID-19 can be quicker through artificial intelligence framework using a mobile phone–based survey when cities and towns are under quarantine: Infection Control Hospital Epidemiology. https://www.cambridge.org/core/journals/infection-control-and-hospital-epidemiology/article/identification-of-covid19-can-be-quicker-through-artificial-intelligence-framework-using-a-mobile-phonebased-survey-when-cities-and-towns-are-under-quarantine/7151059680918EF9B8CDBCC4EF19C292. Accessed 29 Nov 2020
59. Salvatore S, Fini V, Mannarini T, Veltri GA, Avdi E, Battaglia F, Castro-Tejerina J, Ciavolino E, Cremaschi M, Kadianaki I, Kharlamov N, Krasteva A, Kullasepp K (2018). Symbolic universes between present and future of Europe. First results of the map of European societies' cultural milieu. PloS one 13(1):e0189885
60. Sanchez GR (2006) The role of group consciousness in political participation among Latinos in the United States. Am Pol Res 34(4):427–450
61. Singh N, Koiri P (2016) Understanding social capital. Soc Sci Spectr 2(4):275–280
62. Solaimalai A, Ramesh RT, Baskar M (2004) Pesticides and environment. Environ Contaminat Bioreclamat 345–382
63. Shklovski I, Valtysson B (2012) Secretly political: civic engagement in online publics in Kazakhstan. J Broadcast Electron Media 56(3):417–433
64. Sweeney L (2002) Achieving k-anonymity privacy protection using generalization and suppression. Int J Uncert Fuzz Knowl-Based Syst 10(05):571–588
65. Vaishya R, et al Artificial Intelligence (AI) applications for COVID-19 pandemic. In: IEEE reviews in biomedical engineering. https://doi.org/10.1016/j.dsx.2020.04.012
66. Van Benshoten E (2001) Civic engagement for people of all ages through national service. Unpublished manuscript
67. Van den Hooff B, de Leeuw van Weenen F (2004) Committed to share: commitment and CMC use as antecedents of knowledge sharing. Knowl Process Manage 11(1):13–24
68. Vellido A (2018) Societal issues concerning the application of artificial intelligence in medicine. Kidney Dis 5:11–17
69. Weber K, Dacin MT (2011) The cultural construction of organizational life: Introduction to the special issue. Organ Sci 22(2):287–298
70. Yeşil S, Hırlak B (2013) An empirical investigation into the influence of knowledge sharing barriers on knowledge sharing and individual innovation behaviour. Int J Knowl Manage 9(2):38–61
71. Zhang R, Tie X, Qi Z, Bevins N, Zhang C, Griner D, Song T, Nadig J, Schiebler M, Garrett J, Li K, Reeder S, Chen GH (2020). Diagnosis of COVID-19 Pneumonia Using Chest Radiography: Value of Artificial Intelligence, 24 September 2020. https://pubs.rsna.org/doi/full/10.1148/radiol.2020202944. Accessed 29 Nov 2020
72. Zhou Y, Wang F, Tang J, Nussinov R, Cheng F (2020) Artificial intelligence in COVID-19 drug repurposing. https://www.sciencedirect.com/science/article/pii/S2589750020301928. Accessed 29 Nov 2020

Improved Safety On-board Using Augmented Reality Technology as a Training Tool

Dimitrios Frossinis, Nikos Anaxagora, and Elena Chatzopoulou

Abstract The current study explores the advanced technology of Augmented Reality (AR) as a digital tool of Industry 4.0. As such, we investigate the potential use of AR for the on-board training of seafarers, as a new technological tool aiming to increase their safety and operational effectiveness. Based on existing research of training through technology, we have performed a gap analysis between current training methods whereas, we are discussing the adoption maturity level and the technological obstacles. Our research is qualitative in nature, by means of in-depth interviews. The study consisted of interviews with twenty-four people from the shipping industry aiming to provide fresh insights about the augmented reality in the industry. Augmented Reality in shipping is currently at a very early stage; apparently, around half of the interviewees are not welcoming the technology in general and especially the use of AR in on-board training. According to our findings, the market seeks for the adoption of AR in on-board operations rather than training. Practical implications are also outlined.

Keywords Augmented reality technology · On-board training · Safety onboard · Shipboard operations · Industry 4.0 · Maritime · Shipping

1 Introduction

In shipboard operations, there are three major factors which need to be considered concerning the on-board personnel: safety of the crew, safety of the ship and cargo and environmental protection. However, under all circumstances, safety comes first.

D. Frossinis (✉)
Hotel Feedback, Cyprus

N. Anaxagora
University of Frederick, Limassol, Cyprus

E. Chatzopoulou
Kent Business School, University of Kent, Canterbury, Kent, UK

A. M. A. Musleh Al-Sartawi (ed.), *The Big Data-Driven Digital Economy: Artificial and Computational Intelligence*, Studies in Computational Intelligence 974,
https://doi.org/10.1007/978-3-030-73057-4_14

Shipping companies support the human safety at sea, and thus, rank it as their very top priority.

It is a known fact that to achieve a high level of safety, on-board personnel need to undertake extensive training, possess the right safety equipment and operate in a safe working environment. However, working on-board is considered a dangerous workplace. The nature of the seafaring job, the characteristics of vessels and cargoes, the psychology of the seafarers while being at sea for months, is what causes lack of attention on safety procedures. Human failures are also common as they work under pressure and it is unlikely to underestimate potential risks [14].

Another existing gap has been the lack of communication among crew members. A multicultural crew means that people from several countries and/or cultures interact regularly [13]. An extensive literature review of the impact on crews on maritime safety, concluded that intercultural cooperation, communication, fatigue and language skills of a seafarer are the most important issues that contribute to maritime safety on an individual level [13].

The continuous crew changes and the fast operation cycle do not allow time effective safety training prior to boarding [25]. Judging by the current efficiency of training on-board, development of new technologies will result in a more effective, rather than a costlier method of training [23].

The International Transport Workers Federation [16] supports the idea that the key to maintain a safe shipping environment and keeping our oceans clean, lies on all seafarers across the world observing the highest standards of competence and professionalism in the duties they perform onboard. The International Convention on Standards of Training Certification and Watch-keeping for Seafarers [35] (established in 1978, amended in 2010), sets these standards, governs the award of certificates and controls watch keeping arrangements.

A seafarer needs to learn the working environment thoroughly, and competently analyse any risks. This knowledge and these skills need to be gained through proper training and appropriate practice in the use of equipment, tools, systems, safety procedures and emergency plans.

The high level of collaboration among individuals across departments and the trust between crew and on shore managers is a matter of safety. Lack of collaboration may lead to dire consequences. The uncertainty between crew and on shore managers will impede knowledge sharing at all levels. Moreover, horizontal communication is essential for individuals to build trust and understanding, therefore a requisite for collaboration and knowledge sharing.

This research investigates a new approach in on-board training to explore the potential of advanced technology usage and to improve safety. A theoretical hypothesis is formulated in a way that a rapid and substantial technological progress of industry 4.0 occurs today. The main objective of this research is to identify the challenges of implementing AR technology as an educational tool for on-board training.

"AR, can be applied in the training of qualified seafarers by combining real world data (technical indicators of different aggregates from the ship's power system, location and development fire, flood water, etc.) with computer generated data from the simulation complexes" [3]. AR systems can communicate with various sensors in real-time, which can offer a broad range of training affordances. This technology allows learners to examine all information in a stimulating and exciting setting that combines a traditional learning content with the world's most innovative virtual objects [17].

It is intriguing to investigate whether there is a possibility of improving safety procedures in shipboard training by using AR. Today, AR is used in the navigation of ships, mainly by companies from Scandinavia, and as a pilot or prototype model in marine engineering. No one has ever used it for training seafarers on-board ship.

2 Literature Review

Safe Training On-board

The science behind AR shows that it can make training interactive, more productive and gratifying. Several studies have shown that AR systems can provide motivating, entertaining, and engaging environments conducive for learning [20]. Thus, AR is capable to enhance currently used media in training such as booklets, posters and videos. By virtue of such a rich training content, assisted with AR tools, the owners of shipping companies can boost the effectiveness of on-board training.

Seafarers hold a high standard of expertise in their profession, where specific knowledge and skills are required to operate this sea bound mode of transportation [1]. A good quality training is required to ensure that a vessel maintains a high standard of operation [8]. Similarly, having well-trained seafarers as being essential to any maritime company that wishes to demonstrate their responsibility, while at the same time to be perceived by the industry as of having a high quality and competitive operation [8].

There have been numerous changes in methods of maritime training over the last two decades. With the assistance of modern technology, several companies provide innovative software solutions for on-board training. Maritime Education and Training (MET) has used Computer-Based Training (CBT) as an asset to achieve its training objectives [39].

Augmented Reality as a Training Tool

Advanced technologies always play a significant role in the education system, since they offer better opportunities for creating interactive, personalised learning materials and activities that are compliant with the specific needs and characteristics of learners [19].

AR for training and education can be highly effective, time-efficient and guarantees that trainees acquire the necessary skills [11]. Furthermore, it welcomes trainee's and students' feedback, alerts the trainee, provides instructions and different levels of guidance which are key for the learning process. AR training can combine real experiences with virtual instruction and guidance. Existing studies which applied AR in Chemistry classes by using AR applications managed to gain higher levels of student engagement in the classroom [6]. Moreover, in an experimental study of physics teaching to students, the findings suggest that the learning experience of the students is better when it is enhanced with AR technology [38].

AR has dramatically shifted the location and timing of education and training [20]. In AR, the environment is real, but extended with information and imagery from the system. In other words, AR bridges the gap between reality and the virtual environment in a seamless way [20]. Moreover, by using AR in higher education even students' attitude is improved as student seem to have a stronger motivation for learning while their academic performance is much improved [26]. There is also an evident potential for AR to provide helpful contextual, on-site learning experiences and serendipitous exploration and discovery of the connected nature of information in the real world [20].

Therefore, AR not only has the power to engage a learner in a variety of interactive ways, that have never been possible before, but can also provide to each individual their own unique discovery path, containing rich content from computer-generated three-dimensional environments and models.

Augmented Reality in the Shipping Industry

Digitalisation and new technologies are transforming the shipping industry, because they provide efficiency and flexibility. One of these technologies is the AR which can be used for many different tasks such as navigation, engineering and simplifying inspections.

In navigation, AR can provide to the navigator a direct view of the ship's movement and a path with the current rudder position and speed orders. Digital information projected to head-glasses is appealing, as most information is available and collected into one system, thus allowing the bridge team to spend more time looking out of the window than down into the bridge's scattered resources. The implementation also encompasses areas of conservation, like marine protected and environmentally sensitive areas [29].

In the engine room of a ship, the engineers can wear the head-glasses with which the virtual information about an equipment combined with the real environment, may aid engineers to view both the virtual information and the real equipment operational efficiency while controlling the virtual data through their sight, gestures and voice. They can receive additional three-dimensional information when it comes to repairing equipment or even to diagnose and solve a mechanical issue. Apart from the real environment, engineers will be capable to inspect animated components regarding the part which needs to be replaced and the tools to be used, while an audio instructor may guide them through each of the steps via the integrated speakers with head-glasses. The virtual process of disassembling is available through the head-glasses at anyplace, anytime [21].

3 Research Methodology

The current study explores the development of AR technology on-board ship operations and on maritime personnel training. In this qualitative research, the use of interviews was implemented to gather fresh insights about the implementation of AR as the topic is new and on a very early stage of development, interviews were required for in-depth research aiming better explanation of the phenomenon and higher validity of the findings.

The 'interpretivism' approach as a school of thought concentrates on the meaning of social interactions. Interpretivism guided our qualitative approach through which researchers and interviewees interpret and discuss in-depth their experiences and the environment in which they live [37]. Qualitative interviewing, using semi-structured questions, makes use of open-ended questions to encourage meaningful responses [28]. Moreover, by conducting semi-structured interviews there is a clear focus on the research topic and the aspects to be discussed without compromising the freedom of the interviewee to elaborate on issues and aspects relevant to him/her.

In the semi-structured interviews, there was a list of themes and questions to be covered, although these may vary from interview to interview [40]. This means that some interviews had to be adapted aiming to give a specific organisational context that relates to the research topic. This depends on the knowledge about AR and the background of the interviewee i.e. if the interviewee is a shipowner, captain, engineer, investor or director of a related shipping company (see Table 1 and Fig. 1). The order of questions also varied depending on the flow of the conversation. On top, confirmatory questions were sometimes required to guarantee the understanding of the meanings and the variation of events within particular organisations. Data were audio recorded during the interviews as long as participants agreed to be recorded.The twenty interview questions were divided in two parts. The first part concerned on-board training and the second part addressed the use of digital technologies and AR implementation. In the second part, interviewees listened carefully to the presentation

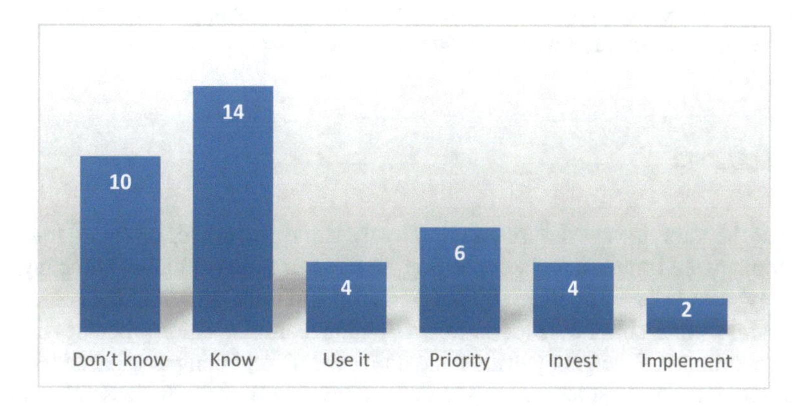

Fig. 1 Participants' prior knowledge about AR (Developed by the chapter authors)

of AR, they asked questions, made suggestions and finally expressed their opinions in the form of feedback.

The authors have significant experience in the development of new technologies and products. The current study sets the ground for a long-term plan to break down the complexity of this innovative and complex idea (the use of AR for on-board training) into easy to comprehend ideas and thereby increasing its efficiency and functionality. Qualitative data analysis approach relies on the processes and procedures that are used to analyse the data and provide some level of explanation, understanding, or interpretation [4].

The purpose of our analysis was to become familiar with the data, to identify the meaning and determine which pieces of data should be kept. Then, we grouped into themes relevant data and created a framework. As such, we identified patterns, we made connections, we identified relationships between themes or data sets, we interpreted data and explained in-depth the findings [14]. Finally, the authors developed a mapping and interpretation, involving the analysis of the key characteristics as these are depicted in the Figures.

Analysis began early and simultaneously with the data collection in order to improve the interview guide [5]. Audio records from interviews were transcribed and then edited with the use of NVIVO 12. Thematic analysis enabled the reduction of the amount of collected data by considering only data from specific sub-themes rather than all possible cases or elements [33]. The interviewee sample was taken from a range of viewpoints and job roles from the shipping industry (10 from shipping companies' offices, 6 active seafarers, 6 provider companies and 2 training companies). The participants' profile is outlined in Table 1 (Appendix).

Reliability and validity are two factors which we, as qualitative researchers, considered while we were designing the study, analysing results and evaluating the quality of the research [34]. To improve the internal validity of the research, only the most appropriate interviewees from different companies and positions of the shipping industry were taken into account. Finally, we finalised the data collection process when data saturation was reached meaning, data saturation determined how many interviews were enough to acquire rich and thick data [7]. This was reached when no new information was emerging from the data [15].

4 Findings

Most of the interviewees described virtually the same process for on-board training. Variations were found though, depending on the size of the shipping company, the vessel type and class. The main similarity in on-board training, across the board, was that seafarers don't have time for on-board training. Familiarisation processes take up to 3 h maximum. Emergency drills done on weekends and computer-based training were noted as the most tedious aspects of the training process. Successful training that has a major impact on the safety on-board ship relates to the qualifications and responsibility of the Second Officer that carries out the training on-board.

It has been clear that in shipping there is no consensus on the use of technology in on-board training. Half of the respondents believe, trust and use technology while the other half are sceptical of its benefits, especially when related to the expense that such technology entails.

AR technology is very new, not only for the shipping industry but other industries and also for the public. The attractiveness of AR lead to technological and academic proponents to continuously raise expectations, to come up with new ideas, which sometimes sound like science fiction. However, nowadays, the technology required for AR applications has matured and, as a consequence, the number of commercial AR applications and especially consumer AR applications is growing exponentially. When considering AR and on-board training today, research findings deliver the following pros & cons:

Advantages of AR

The interviewees argue that implementing AR in on-board training would bring only slight benefits. Only three out of twenty-four said that AR would be helpful in the improvement of training. Two interviewees have used AR and know what it is. This shows that the majority of the industry do not yet see any actual benefit from the use of AR in on-board training. As such, AR technology is not a priority for the time being (Fig. 2).

Disadvantages of AR

The biggest concern about AR usage is the price. The respondents argue that AR would not be of any use for on-board training. They feel that it is a very complex technology and distracting to use. They view it as inefficient, along with the concern

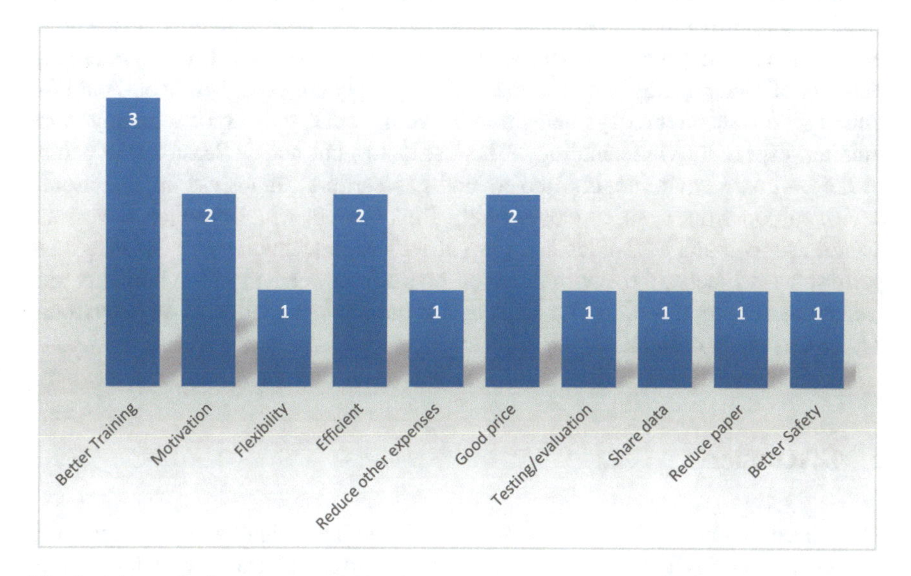

Fig. 2 Advantages from the potential implementation of AR (Developed by the chapter authors)

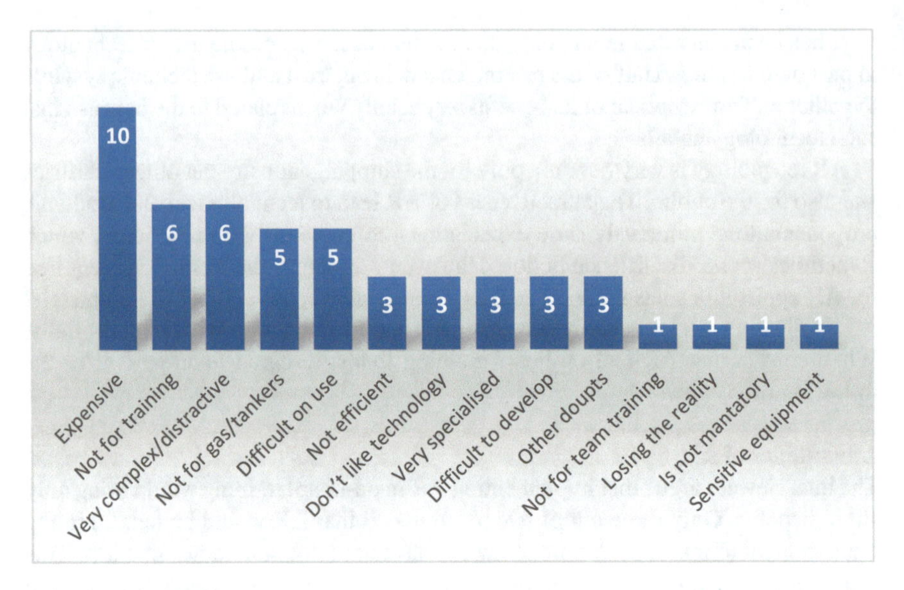

Fig. 3 Disadvantages from the potential implementation of AR (Developed by the chapter authors)

that it would be both difficult and expensive to develop an AR application that could be universally used for any vessel type (Fig. 3).

Major Concerns About Digitalisation and New Technologies
For shipping companies, due to the narrow profit margins, the major concern is cost. Business developers believe that an investment in new technologies must have a significant return of their investment in both money and time. Seafarers, on the other hand, argue that they are already using effective training methods. The overwhelming majority of seafarers hold the view that there is simply not enough time on-board for training, with seafarer's days being dominated by work, they don't want any more training, especially via technology. The last factor, but not the least, is the nature of the on-board environment. Tankers and gas carriers cannot use any electronic device outdoors, due to strict rules of safety. Bulk carriers, which are cheaper vessels, do not use as much as tankers and gas carrier's high technology to enhance their performance. Finally, the environment on-board (bad weather, saline humidity and dangerous structural setting of a ship) makes the implementation of any electronic device extremely difficult.

5 Discussion

The current study explores the application of learning technologies by expanding further the existing literature which supports the usage of learning technology in contrast to didactic styles [32]. Our findings suggest that there is a need for AR technology on shipboard operations but, not for additional on-board training. This is due to the fact that the shipping companies are physically training seafarers before taking them on-board. The manning standards of ships are leaning toward employing fewer and fewer crew on-board, therefore there is now less time for the crew to train. For this reason, advanced training technologies, implemented with simulators and VR technology, stands to give the seafarers a decent and proficient training experience.

Likewise, the Maritime Labour Convention (MLC) utilises the time from resting hours of seafarers for drills. Under these circumstances, the MLC shall take training more seriously by considering training time the same as working time, so seafarers can give the correct attention to training. Interviewees that were close to technology, and understood AR, argue that an AR solution is more appropriate for shipboard operations and not for training. AR has a great potential to help in maintenance, remote assistance, machinery diagnostics, work tasks, work checklists and finally exporting reports in PDF form or video.

Interviewees highlighted that, one of the biggest challenges which the industry is facing is the familiarisation of staff with new technologically advanced equipment previously unknown to operators and to seafarers. As such, there is a need for training in the use of this new equipment and machinery and this is the challenge which AR may address. We propose an integration of sensational pedagogy with training [12] aiming to aid the on-board crew to visualize processes and concepts resulting to learn in depth [30]. Sensational pedagogies enforce higher learners' engagement with the learning material in comparison with the one-way communication teaching [12] which will lead to deep learning. While, marine equipment is traditionally large, cumbersome and extremely technologically complex [27] also differs from vessel to vessel, while engineers have to move from vessel to vessel whereas, they need to familiarise with the equipment. As such, our study proposes that AR is not only a matter of efficiency but also a matter of safety via horizontal communication as for example trust and understanding is built between seafarers and on shore managers.

Contrary to the existing literature review which describes the perspective of AR in training, our findings suggest that there is much more potential for the AR tool and this format of AR is welcomed by the market. For instance, a 3D model of the vessel can be uploaded to the AR device, and by using the glasses the seafarer can 'see' behind bulkheads and/or underfoot. Every pipeline will be displayed to the seafarer for what it is and also request guides for specific processes or controls. In the engine room the engineer can see 'inside' the machinery and make a diagnosis of a potential problem by following steps and filling in the relevant checklist. A video recording can be made of the procedure and can be sent to the manufacturer. Similarly, a faulty part on machinery can be changed on-site with remote assistance from the manufacturer in real-time, thus eliminating the costly presence of a service engineer.

Everything can be controlled and checked easier and faster, reducing costs, paper usage and time. Checklists can be completed via AR smart glasses for every task and can instantly upload the information to a cloud. As such, all parties concerned, on-board and onshore will be informed about any work that will be done on-board. Companies' training costs are high, especially if manufacturers visit them on-board.

However, with AR smart glasses, training can be done remotely in real-time with a connection to the manufacturing company. This solution, enables the development of new technologies over time to produce an increased integration, interoperability and intermobility [23].

AR smart glasses are the extension of the land offices' eyes. Staff at the land offices (i.e. IT, trainers and managers) can connect in real-time with the vessel anywhere in the world. Nowadays, offices support the vessel not only by phone, but with real-time eye contact. Using the AR remote assistant smart glasses in this way saves both time and expense.

AR smart glasses can be an innovative communication channel through which the vessel will be connected with the land (land office, manufacturers, providers, etc.). Moreover, AR can bring vessels one step closer to autonomy. In the near future, machines and equipment will communicate and exchange data via internet networks. AR smart glasses will be the primary tool which will receive and send informative data for the operator. Other software which can connect with the glasses are digital maps (i.e. ECDIS). Integrated displays on the bridge have relative advantages over scattered displays [31]. By taking these into consideration we can appreciate the benefits of having all the information in front of the eyes of the navigational officer with the AR smart glasses, enhancing situation awareness (SA).

By using Artificial Intelligence (AI), the AR smart glasses can be utilised as the receiver of data, recorder of data, and as an exporter of newly analysed information based on AI to assist in better decision making. Even in cases where unmanned vessels will dominate the seas, as it is expected to happen in the following 15 to 20 years, sailors will be less but more vital [18]. Therefore, any additional help provided by technology to the crew will be helpful and valuable.

6 Conclusion

The digital revolution is shaping not only shipping but our everyday life. AR is capable to transform the shipping industry, and there are already a few obvious applications for it in training with the creation of real-world cases and skills development. Applications, too, can be envisioned in maintenance by providing engine diagnostic information and remote support to engineers at sea. AR technology can also provide improved situational awareness and decision-making support for operations.

Moreover, this research argues that AR is not appropriate for training on-board, but can be a powerful tool for the engine room: offering guided maintenance and/or assistance for machinery with steps and tasks, check-list control, inspections and training in advanced technology equipment. Engineers need to have a tool which enables training on new equipment, manages maintenance and has remote assistance with the provider on land. This tool can be the AR headset. AR technology can guide the engineers and train them: while engineers will be looking at the equipment AR will be adding the crucial information in digital format and it will also complete the checklist and any inspection.

The industry is based 100% on the human factor along with low profit margins and constant efforts to increase these by reducing costs. It is clear that ship-owners are financially orientated and cost-driven making every new investment of theirs to be affected by profitability and the requirements of current legislations. It is also widely recognised that the future is going to be even more digital. However, in shipping, new technologies have always been regarded with some scepticism. The shipping industry has a traditional, more conservative approach to the operation of ships.

7 Future Research and Limitations

Further research is strongly recommended on the necessity of integrating AR technology with appropriate bridge equipment and machinery. AR technology, can enhance the interaction of the crew with operations and maintenance. Radars, AIS, ECDIS, main and auxiliary machinery, can all be integrated with AR technology. What remains to be discovered is how AR affects shipboard operations and the needs of shipping companies.

Most of the interviewees were unaware of AR technology and its usability (see Fig. 1). Half of them were not supporters of technology and any of the new tools that it brings. It is recommended that an experiment should be carried out which will test a prototype AR application. An AR headset would add value to the experiment, giving a more realistic approach with participants. In order to test the validity of the research, it is recommended that an additional survey should be conducted.

Furthermore, the research was based mainly on 2 countries: Cyprus and Greece. Countries from Northern Europe and other continents may well produce different results. Future research may also need to consider how AR applications could be used for onboard operations and in industry in general. The impact of sales and revenue for providers, the added value and customer satisfaction for all the users should also be thoroughly researched. How all necessary information will appear in an AR headset, how the headset will be designed and how easily it can be adopted by seafarers, shows the way for future research.

Appendix

Company	Job Title	Years of experience
Shipping Management Company 600 vessels, 20,000 employees	IT Manager	5
VR training provider Running in 1000 vessels	CEO, Ex. Captain	12

(continued)

(continued)

Company	Job Title	Years of experience
LNG Shipping Management Company 31 LNG vessels	3rd officer	7
Shipping Management Company 600 vessels, 20,000 employees	1st Engineer	15
Family Shipping Management Company 10 vessels	IT Manager	27
Shipping Management Company Tankers & Bulk Carriers 21 vessels	Health Safety Quality Manager	25
Shipping Management Company 600 vessels, 20,000 employees	Fleet Manager	25
Shipping Management Company 70 owned vessels, 100 on a crew management	Captain / Head of Training Department	23
The biggest industry e-learning provider More than 10,000 vessels	Business Development Manager	10
Leading provider of global marine support services. Network of over 44,000 seafarers	HSEQ & Marine Superintendent	20
LNG Shipping Management Company 16 LNG vessels	Chief Officer, Onboard Trainer	10
International provider in maritime electronics equipment	Chief Engineer - Area Sales Manager	6
Training of seagoing personnel, Safety & Quality assurance, Crewing and incident investigation	HSEQ & Marine Trainer, consultant	32
Shipping Management Company 400 vessels, 15,000 employees	Master Mariner (Oil, Chem and Gas) – Pilot – QSHE – Marine – Crewing – Training	26
Training solutions for maritime security services for oil & gas industry	Training Development Manager	13
Crew Management company with worldwide network of offices and training centres. 13,000 crew onboard	Crew Training Manager	30
LNG Shipping Management Company 31 LNG vessels	3rd Officer	5
Crew Management company with worldwide network of offices and training centres. 11,000 crew onboard	Global HSEQ Manager	22

(continued)

(continued)

Company	Job Title	Years of experience
Shipping Management Company 100 vessels, 4,000 employees	Group Fleet Personnel Manager	30
Shipping Management Company 50 shore-based & 700 seafarers	HSE Manager	15
Shipping Management Company 100 vessels	Seafarer	4
Shipping Management Company 14 vessels	2nd Officer	10
Mechanical & Industrial marine engineering 19,000 employees, 200 locations, 80 countries	Technical Expert	12
Crew Competence Digital Management Solutions Training, E-Learning, ECDIS, Maritime Training, Human-Resource, E-Assessment, E-Certification, ETC training Centre, Shipping Industry, Crewing	Product Manager	12

References

1. Albayrak T, Ziarati R (2010) training: onboard and simulation-based familiarisation and skill enhancement to improve the performance of seagoing crew. TUDEV (Turkish Maritime Education Foundation - Institute of Maritime Studies)
2. Ari A (2018) Augmented reality in marine engine field service. Aalto University School of Business Information and Service Management. https://aaltodoc.aalto.fi/bitstream/handle/123456789/30486/master_Ahonen_Ari_2018.pdf?sequence=1&isAllowed=y. Accessed 4 Sept 2020
3. Bakalov I, Lutzkanova S, Kalinov K (2018) Applying the augmented reality concept in maritime engineering personnel training. Scientific Bulletin of Naval Academy, Constanta, Rumania
4. Berg B, Lune H (2017) Qualitative Research methods for the social sciences, 9th ed. Pearson
5. Charmaz K (2011) The SAGE handbook of qualitative research. SAGE
6. Chen SY, Liu SY (2020) Using augmented reality to experiment with elements in a chemistry course. Computers in Human Behavior, 106418
7. Dibley L (2011) Analyzing narrative data using McCormack's lenses. Nurse Res 18(3):13–19
8. Dragomir C, Utureanu S (2016) Drills and training on board ship in maritime transport. Emergency 3:1–4
9. Easy Marine website (2020). https://www.easymarine.it/augmented-reality/. Accessed 10 Aug 2020
10. Fei J (2011) An empirical study of the role of information technology in effective knowledge transfer in the shipping industry. Marit Pol Manage 38(4):347–367
11. Ferrati F, Erkoyuncu JA, Court S (2019) Developing an Augmented Reality based training demonstrator of manufacturing cherry pickers. In: 52nd CIRP conference on manufacturing systems

12. Gallagher M et al (2017) Listening differently: a pedagogy for expanded listening. Br Edu Res J 43(6):1246–1265
13. Gausdal AH, Makarova J (2017) Trust and safety onboard. WMU J Marit Aff 16(2):197–217
14. Genc YU (2019) Improvement of onboard crew safety by applying personal tracking and monitoring technologies (Master's thesis, University of South-Eastern Norway)
15. Guest G, Bunce A, Johnson L (2006) How many interviews are enough? An experiment with data saturation and variability. Field Methods 18(1):59–82
16. International Transport Workers' STCW A Guide for Seafarers (2020) Taking into account the 2010 Manila amendments. ITF Revised 2017. Accessed 5 Aug 2020
17. Ismaeel DA, Al Mulhim EN (2019) Influence of augmented reality on the achievement and attitudes of ambiguity tolerant/intolerant students. Int Educ Stud 12(3):59–70
18. Jo S, D'agostini E (2020) Disrupting technologies in the shipping industry: how will MASS development affect the maritime workforce in Korea. Mar Pol 120:104139
19. Kiryakova G, Angelova N, Yordanova L (2018) The potential of augmented reality to transform education into smart education. TEM J 7(3)
20. Lee K (2012) Augmented reality in education and training. TechTrends 56(2):13–21
21. Limbu BH, Jarodzka H, Klemke R, Specht M (2018) Using sensors and augmented reality to train apprentices using recorded expert performance: a systematic literature review. Educ Res Rev 25:1–22
22. Liarokapis F, Mourkoussis N, White M, Darcy J, Sifniotis M, Petridis P, Basu A, Lister PF (2004) Web3D and augmented reality to support engineering education. World transactions on engineering and technology education 3(1):11–14
23. Mazzarino M, Maggi E (2000) The impact of the new onboard technologies on maritime education and training schemes in Europe: some findings from the 'METHAR' project. Marit Pol Manage 27(4):391–400
24. Maritime Labour Convention 2006 – ILO, The general conference of the international labour organization
25. Markopoulos E, Luimula M (2020) Immersive safe oceans technology: developing virtual onboard training episodes for maritime safety. Future Internet. MDPI
26. Martín-Gutiérrez J, Fabiani P, Benesova W, Meneses MD, Mora CE (2015) Augmented reality to promote collaborative and autonomous learning in higher education. Comput Hum Behav 51:752–761
27. Millar IC (1980) The need for a structured policy towards reducing human-factor errors in marine accidents. Marit Pol Manage 7(1):9–15
28. Patton MQ (1990) Qualitative evaluation and research methods. SAGE Publications, Inc.
29. Procee S, Borst C, van Paassen MM, Mulder M, Bertram V (2017) Toward functional augmented reality in marine navigation: a cognitive work analysis. In: Proceedings of COMPIT, pp. 298–312
30. Ruiz-Primo MA, Briggs D, Iverson H, Talbot R, Shepard LA (2011) Impact of undergraduate science course innovations on learning. Science 331(6022):1269–1270
31. Sauer J, Wastell DG, Hockey GRJ, Crawshaw CM, Downingd J (2003) Designing micro-worlds of transportation systems: the computer-aided bridge operation task. Comput Hum Behav 19(2):169–183
32. Saunders FC, Gale AW (2012) Digital or didactic: using learning technology to confront the challenge of large cohort teaching. Br J Edu Technol 43(6):847–858
33. Saunders M, Lewis P, Thornhill A (2009) Research methods for business students, 5th edn. Prentice Hall
34. Suri H (2011) Purposeful sampling in qualitative research synthesis. Qualit Res J 11(2):63
35. STCW (2020) International convention on standards of training, certification and Watch-keeping for seafarers (1978, updated in 2010). https://www.imo.org/en/OurWork/HumanElement/Pages/STCW-Convention.aspx. Accessed 4 Sept 2020
36. Tamburini D (2018) Augmented Reality becomes mainstream in Manufacturing, changes the face of the industry. https://azure.microsoft.com/en-us/blog/augmented-reality-becomes-mainstream-in-manufacturing-changes-the-face-of-the-industry/. Accessed 5 Sept 2020

37. Thanh NC, Thanh TT (2015) The interconnection between interpretivist paradigm and qualitative methods in education. Am J Educ Sci 1(2):24–27
38. Thees M, Kapp S, Strzys MP, Beil F, Lukowicz P, Kuhn J (2020) Effects of augmented reality on learning and cognitive load in university physics laboratory courses. Comput Hum Behav 106316
39. Triand M (2018) Benefits of computer based training in the Maritime Industry. https://triandmaritime.com/benefits-computer-based-training-maritime-industry/. Accessed 10 Aug 2020
40. William A (2010) Conducting semi-structured interviews. handbook of practical program evaluation. Wiley, San Francisco
41. Wright I (2017) What Can Augmented Reality Do for Manufacturing? https://www.engineering.com/AdvancedManufacturing/ArticleID/14904/What-Can-Augmented-Reality-Do-for-Manufacturing.aspx

Online Paying Happily to Buy Pain in Electronic Games: The Marketing Mystery

Hanin Al Balushi, Vinay Gupta, Araby Madbouly, and Sameh Reyad

Abstract It is very ironical that thousands of people in our world spend millions of dollars to get rid of pain and there are those who pay hundreds for experiences that cause them pain. These buyers take the decision to buy goods or services based on their emotions only. Simply, these buyers buy their pain. This study is an attempt to understand this phenomenon. The main objective of this chapter is to identify the extent of emotions impact on the consumer of electronic games procurement decision.

Qualitative research method has been followed in this research via case study strategy. Interviews were conducted with a group of respondents who are used to play electronic games in Omani malls. This study has been conducted for participants from different nationalities. The main finding of this research is surprisingly 82% respondents were buying these painful experiences for fun and enjoyment and 89% of the respondents were satisfied after toiling trying to win when playing the electronic games. These results bring useful learning for experiential marketing professionals and academicians and pave way for further researches to develop this domain of knowledge. The main limitation faced researchers is the difficulty to get approval from participants who are interested to be interviewed. However, researchers successfully conducted the interviews with a diversified sample with varying age, gender, nationality and experience.

H. Al Balushi
Alumni of Muscat College, Muscat, Sultanate of Oman

V. Gupta
Infobeans Ltd, Pune, India

A. Madbouly (✉)
Department of Business and Accounting, Muscat College, Muscat, Sultanate of Oman
e-mail: araby@muscatcollege.edu.om

S. Reyad
College of Business and Finance, Ahlia University, Manama, Bahrain
e-mail: sreyad@ahlia.edu.bh

A. M. A. Musleh Al-Sartawi (ed.), *The Big Data-Driven Digital Economy: Artificial and Computational Intelligence*, Studies in Computational Intelligence 974,
https://doi.org/10.1007/978-3-030-73057-4_15

Keywords Online payment · Electronic games · Emotions in purchasing decisions · Oman

1 Introduction

With the increased knowledge and advanced researches our environment has improved to increase our safety against wars, illness, terrorism, starvation etc. Though there are still new risks which we are not being aware of and experience involuntarily, such as pleasurably experience playing electronic games. Many reasons can be behind taking such type of risk. Therefore, these electronic games are considered a sever example of buying the pain as they are considered painful experiences while they are marketed as challenging courses for participants. This cahapter analyses how emotions can play a major role in consumer decision making. The research questions are

1. What are the reasons of participation in the challenge electronic games?
2. What are the feelings of participants in electronic games?
3. How the participants deal with the pain during the challenge?
4. How participants feel after the completion of the challenge?

The main aim of this chapter is to bring a clear image of the painful experiences and why people like to pay to participate in these events. The research insights are drawn by doing the study on a group of respondents and asking them what makes them seek this kind of experiences as well as observing them and using netnography. The research is going to improve ethnography that attends to the psychological, emotional dimensions and physical material of pain. The results of the research will help in designing marketing strategies for products that cause pain to buyers, especially the electronic games.

This introduction is followed by the literature review in section two while the third section has the research methodology. Section four shows the data analysis and findings, then the discussion, conclusion and future scope & marketing implication presented in section five.

2 Literature Review

2.1 Experience Marketing

"Experiential marketing" has introduced first time by Pine and Gilmore [23] as a part of their job in the experience economy. They explained the experiential marketing as "when an individual buys a service it means he purchases a collection of intangible activities that are carried out on individual's behalf". The experience is described as consumer's consumption of and interaction with services or products that involve important affection [15]. Experiential marketing consequently is about taking the core of a product and expanding it into a set of physical, tangible and interactive experience that supports the offer [30]. Holbrook and Hirschman addressed the experiential aspects of consuming: feelings, fantasies, and fun. Experience marketing describes marketing that provides customers with depth, tangible experience with enough information to make the purchasing decision [28]. There is a growth of using experience marketing and this growth is expected to be more and more with time. We can say that Experience is the key factor in marketing in the future [18]. Individuals are looking for new forms of core values, fulfillment, happiness, and sensations in market offerings [10]. The varied personal cases may lead an individual into an extraordinary experience which can bring an increased level of emotional intensity and is made by an unusual event. According to Same and Larimo [25], an individual can go for an extraordinary experience when the event offers joy, value, absorption, and newness of process and perception.

Until 2004 there was no formal attempt to acquaint what frames an experience in marketing terms. The lack of clarity lies in various ways in which the concept "experience can be understood". It used in different ways to convey the process, the impact in which thought, emotion or object is felt through the mind or the sense, even the result of experience by way of learning or skill for example. Thus, it is unclear whether to experience is passive or active for the participants or the outcome must be a skill development or learning or whether it needs interaction or not [24]. So, consumer experience is a complicated whole that cannot be rotten into separated parts. It can be expanded to other consuming contexts, for example, art, entertainment, leisure activities, concerts, and museum. Emotions play a big role in business as overall, and it is an important aspect for the marketers and the next part of literature review will show how emotions are important in experience marketing.

2.2 Experience Marketing and Emotions

According to Ahuvia [1], customers relationships to "loved products" may mark stories that involve a social group, family members, and traditions. Cultures can be added to this as it plays a big role in individual's emotions [19]. Experience is supposed to have a new sense of the process and perception, uncommon events should trigger it and the interaction between people is an important trigger [7]. Within the context, moving to anthropology, it shows how emotional, cultural scenarios contribute in the conversion of consumers, that means they could return to the routine world as renewed people. Also, people can seek extraordinary experiences when they feel the need of new and different relationships or to accomplish new goals.

According to Tumbat and Belk (2011), since modernity excludes passion, mystery, soul and magic; people participate in extraordinary experiences. However, they argue that consumers may join different relationships out of need and may want to exceed everyday structural standard and participants can be more curious to expanding their accomplishments. Consumers should be expected to deliberate between different duplicity of the extraordinary enchanting and everyday mundane. In the experiential perspective, consuming experience in no longer related to be certain before purchasing activities (search for information, stimulation of need, assessments, etc.) or to some after purchasing activities. For example, the estimation of satisfaction involves some other activities that can affect consumers purchasing decisions and their future actions [5].

Consuming experience can be divided into four stages. Pre purchasing experience, purchasing experience, the core consuming experience and the remembered and nostalgia consuming experience activate photographs to re-live a memory which based on past augments and stories with friends and which moves to the tabulation of memories [3]. The consumer, therefore, is seen as a person who emotionally participates in purchasing process in which the imaginary, emotive and multi-sensory in particular are appreciated. Which means that the enjoyment of shopping does not come from the wanting, buying or desiring products, but from shopping that is considered to be a socio-economic process in which the consumer enjoys himself being with people while purchase [6]. People do follow their emotions to take some decisions. The next part will discuss how these emotions are used by the marketers and how emotions can change consumers purchasing decision making and will clarify the reason behind people buying the painful experience products.

2.3 Emotions and Purchasing Decision Making

For marketing, a perfect experience is memorable which allows the individual to exploit all his senses over the staging of the activity support/physical/social interaction. This kind of experience brings emotions in marketing, it turns on engaging consumers memorably and it offers them an experience or guiding them through the experiences [4]. Experts see that economic value gets high as offering shifts from merchandise to emotional transformations. In the offering, marketers concentrate on experience and making it memorable. Offering experience is an effective solution to avert the commodity trap, for any business. The goal is to build strategies those offer customers emotional and physical sense during their shopping experience. Getz and Fairley [12] have shown in their studies that around 70% of the target market for a special event in the United States has relied on their past experiences for their decision—making process regarding attending the special events instead of the traditional marketing communication processes that were used by the event organisers (quoted in [13]. The above research has clarified that emotions can play an important role on people's decision making. The results from this research indicate that emotions can be one of the reason people participate in obstacle events. Next part of this reports focuses specifically on obstacle courses, and it clarifies why people participate in such type of events.

3 Research Methodology

The qualitative method has been followed in this research, as qualitative data has been collected using more than one technique. Structured interviews and observation and netnography techniques. **Structured interview:** this type of interview allows the researcher to get all the needed data along with details that can value the research [21]. The interview schedule had 12 questions related to the different aspects of the research questions. **Netnography:** To triangulate the data collection in order to increase reliability, netnography was used to get more information about the event and participants feelings. The Netnography procedure included analysis of documents, photography, and other online data. Gathering additional information from the participant's social media accounts and web pages where they write about their experience, event, team, training and their stories increased credibility of data collected through interview and observation.

The population of this research is the people who play electronic games in Omani malls. Collecting data from these participants is the best choice for this study since they are the ones who are participating on their own free will, and leading themselves to painful experiences. The study covered 5 Malls (Three in Muscat, one in Barka and one in Suhar). 32 participants were interviewed who participated in different painful electronic games. 12 participants were in their twenties, 17 participants were in thirties and three participants in forties. Gender wise, there were 20 males and 12 females. The number of females interviewed were lesser since they were a bit conservative about being interviewed for this research. Before the interview, full explanation has been given to each respondent about the research and its' objectives. The language used throughout the interview was English. The questions were asked during the session and these questions and the answers were written by researchers. Letter of consent has been signed from each interviewed respondent. Further, names of respondents kept anonymous to maintain confidentiality.

Qualitative research heavily relies on analysis of opinions and observations. Content analysis was used in this study which helped in the analyses of social phenomena like reason behind people participating in painful electronic games [14]. The analysis of all the qualitative started by develop transcript for each interview. To make the transcription process easier, researchers referred to the question number in the transcription which made the later search faster and easier. Collected data has been categorized into groups. This process followed by give a code for each answer to be easier for categorization. The process of data categorization went into many levels started identify similarities and differences among the responses of respondents and categorised them into number of categories given labels or phrasal descriptors considered as an array. This categorization followed by 2nd-order theoretical level of themes, dimensions, and the larger narrative, which is called 2nd-order theme.

The first limitation faced researchers is that although medical instruments to assess emotional changes in mind one technique usually used to collect the data related to the buying pain giving electronic games. Using this method was difficult to be used due to financial constraints and reluctance of participants. However, the selected instruments were considered sufficient to achieve the research objectives. The second limitation was that the inability to record the interviews with respondents. This has been solved by taking notes for each answer during the interview.

4 Findings

The findings have levels of analysis. First, the researchers investigated personal pain the participants felt while playing electronic games. Second, the research focuses on how the participants see the pain and how they describe it. It shows whether this pain is meaningful to them or it is just a feeling they do not give attention to. Lastly, the feeling participants get after spend some time play electronic games and are these different than what they felt during playing the electronic games, it is difficult for them as event gets tough and sometimes the pain could suspend the participants from feeling what they are seeking.

Following are the results analysed for various interview questions.

What Did Participants Felt While Playing Electronic Games in Malls?
Researchers describe these events as a painful event, but the question was asked to the participants to see their perspective on electronic games. People have a different way of seeing things. Many of them said, electronic games are there to give them the chance to feel competition, train them and help them accomplish their goals and what they want. When they see people achieve their goals it brings joy and the feeling of satisfaction. Electronic games can bring so many great opportunities to an individual as a chance to know them self-better by going through tough experiences and the way they can deal with different situations in the electronic games. Others these games are so fun, and stress relief as some of the participants explained where it could bring different passions. It is an overall painful experience some of them said but yet a very satisfying challenge.

Netnography Analysis: Participants Comments
"I always encourage people to follow something beyond their comfort zone, be it learning a new language, painting, mastering a new skillset, or any other endeavour that's not within their day-to-day work basis". This was one of the comments from participant who already echoed same thing during interview. The responses were validated with the comments posted by participants on their social media accounts. Almost all the comments through netnography were in sync with interview and observation data.

Staying in the comfort zone will create a life full of boredom, laziness, and having an unsuccessful life. However, when trying new things in life like participating in new challenges or something different than what a person can do in their daily bases will improve and develop an individual personality as well as creating a happier successful life.

"This electronic games like a life track you have to imagine it like this, and you have to overcome your life obstacles and beat them to succeed" (Respondent no. 5).

"It feels really good to see myself when playing electronic games and win" (participant 12).

These responses are in line with results of a research which concluded that "Providing people with the help they need, being social, helping people and seeing them happy will reflect that happiness to the individual himself" [26].

Writing About Their Experiences in Social Media Platforms
As is in Fig. 1, more people are now documenting their experiences in playing electronic games on social media platforms. Some of the participants said that they are planning to open their own YouTube channel to share their tricks and lessons with others. Moreover, some other said they used to have a blog writing their experiences, but they shopped that since it is time consuming so now, they post directly on social media platforms. This reflects the social motive of participation.

The majority of the participants have said that there is pain while playing electronic games. Dealing with the pain differs from one to another. Some of them said that "being strong from inside" helps them to overcome the pain that they feel during the game. They see the challenge as it is between themselves before it is with anyone else.

Some respondents "found the joy" when play electronic games. Feeling uncomfortable is the most rewarding feeling to them. Some of them said that they do not exhaust themselves all at once. However, they divide their energy and power during play electronic games for long time. Others said that they "ignore the pain" and doesn't give it their attention until they complete the game or at least reach advance level. Pain feel comes after the game. During the game they mentally block it (Fig. 2).

Out of 32 participants, two of them were not that sure about their feeling after the game. However, with an overall rate of satisfaction, all of the participants were happy and satisfied. People who were explaining these games in a bet different way claiming that it is always pain being in these challenges, putting themselves in pain of full concentration during playing which is very exhausting and requires high level of concentration.

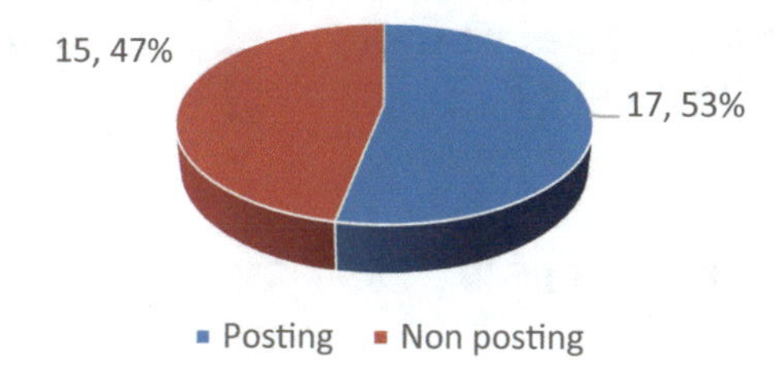

The way respondents dealt with the pain while playing electronic games (Developed by authors)

Fig. 1. Documenting their journeys

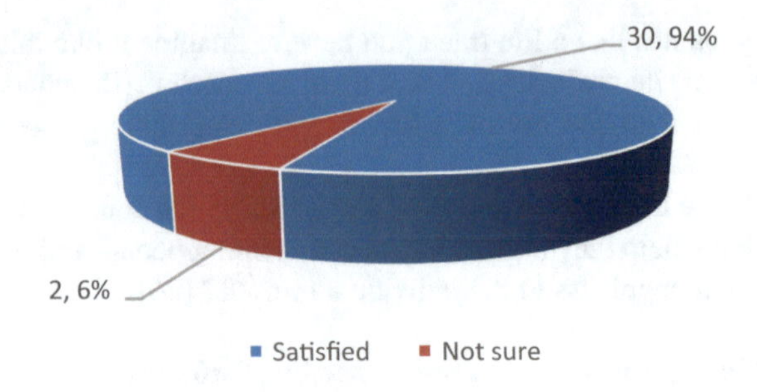

The feeling after the electronic game: Developed by authors

Fig. 2. Paticipants' satisfaction level

Participant's way of explaining their feelings showed how thrilled they were about this participation. It is like owning the world as they feel they are the best in this whole world the feeling of helping other seeing them achieve their goal and helping them improve their accomplishing new goals makes them feel unique. Some of them say as they get rid of all the negative energy after the playing they become fresh.

Moreover, the conviction that every person within him has a great power that he has not yet recognised. Once they get to the finish the electronic game, they feel happiness and proud comes to them, they forget the tough, hard work that they have been through. The more the participant finish this kind of games the more they get flexible and more energetic.

These games give people the chance to keep their stress and thinking of their problem and work away for a good period. Researchers found these games as a source of joy every time they play. The feeling after the game is an amazing feeling of accomplishment. Even though they participate so many times, the pride and joy of the finish line never go away.

"Before the electronic game, I am nervous, and I have anxiety. After the game, I feel relieved and happy, especially if I win any prize or get any more points" (respondent 3). The feeling of excitement and nervousness are common for players since they know what they will face and the goal that they want to achieve what makes them get worried. However, the feeling of being the best at the end will shut down all the negative feelings.

5 Discussion and Conclusion

5.1 Discussion

The aim of this study is to bring a clear image of the painful experiences and why people like to pay to participate in electronic games. As discussed in the literature review, painful experiences are not only a way to make people realise their physical power. It is a way that provides people time to "self-shattering" that is when an individual gets the opportunity to unravel the reflective self (respondent 12). The result of this study is in congruence with other studies that painful experiences are considered to be as special or unusual experiences hence, it is well protected in an individual's mind [11]. People seek for such experiences to know themselves better, particularly among young people who are trying to gain the clarity of their self-concept.

Once people go through several experiences in their life, they get to know themselves better, and their self-definition gets more clarified to them [17]. Researchers found that participants found electronic games as unique challenges and something funny to do, doing something new in life will change individuals' way of thinking and makes them think more positive. One of the well-known aspects of painful electronic game are seeing the participants shouting and screaming during playing; this pain makes them forget their entire world and thinking about their everyday routine [9]. In the recent years, the marketplace has been offering uncountable services that can ease role performance. By using scripts, props, stages, and services, it results in experiences that occupy customers in a personal way. Researchers found that when customers use individual competitive mindset, negotiate structural and antistructure characteristics carry out important theoretical aspects in understanding the interaction between the producer and consumer or between the marketer and consumer in different forms of market extraordinary (painful) services consumption [29].

Researchers covered this ethnography as the solving of the puzzling dilemma. In a time where people spend amount of money on pain relievers, Extreme tough events like Spartan race and other events are getting more popular in the recent years. Past studies have examined in extraordinary experiences and the way they are related to researcher's findings.

Electronic games can bring to an individual very special feelings; painful experiences can bring people to try new things. "I live by the mantra that "you only live once, but if you work it right, once is enough". I do play these games because I hate to stay within what the society would call "Normality". To me, Normality is what average people call living. I call it Death" (Participant 18).

From the most beneficial and emotional experiences are those that mentioned in the literature review as "extraordinary". Experiences linked with strong positive emotions like high excitement and pleasure, which are rare or temporary (Allen and Johnson, 2008). There are three factors which have been identified by Arnould and Price [3] for extraordinary experiences, which had a strong impact on the overall satisfaction. First of all, people experienced involvement with nature that included both escape from their duties and an emotional feeling brought on by diving in uncommon places like natural beauty. Second, the extraordinary experiences included connection feeling with family, strangers, and friends. Third individuals have a renewal of self as well as personal growth. In psychology, there is something that could flow experience when self- awareness is lost, and it involves a balance of challenge, enjoyment, a balance of skill and full concentration on an activity [8].

According to Caprariello and reis (2010), there are two approaches when purchasing experiential services. First, experiences of life can satisfy the psychosocial desires, which can raise the feeling of liveness which in turn can lead to happiness. Individuals are more into experiences that can share with other people instead of purchasing material items. Researchers found that athletes love being in groups and sharing their experiences together since that what makes this kind of experiences more meaningful to them.

Studies have mentioned that consumers are more likely to rely on the "word of mouth", or others opinions when making a purchasing decision. Consumers influence each other while making a decision [22]. Which this can be applied even in events participation, the researcher has found that most of the participants had at least one of their friends or family members with them in the same event.

The motive behind people participating in painful experience differs. However, there is one common answer which is to enjoy and have the greatest time of their lives. Painful experiences defined as rare and unusual, and it goes beyond what people experience in their daily life bases. People do not rely on others when it comes to valuing their performance; individuals judge their performance when they reach the finishing positions. They determine their achievement from their performance and comparing it to others [20].

5.2 Conclusion

The results of the research are largely centered around emotions of doing something extraordinary as main reason behind people buying such pain, happily. Extraordinary experiences allow the individual to free themselves temporally from the everyday life thinking and their usual obligations by trying a new experience which engages them for a good period. People seek for unusual experiences that can lead them to pain since it allows them to escape from their real live duties [27]. While imagination and escaping to the fantasy world can be powerful forms of escaping reality. Painful experiences challenges that are being conducted can provide the individuals with a different form of newness in their lives [16]. According to Csikszentmihalyi (2014) extraordinary experiences as "flow" it is when people lose their self-consciousness and they give their full attention to the activity, as the flow concept is needed and used especially in exorcise fraternity and sports.

5.3 Future Scope and Marketing Implication

For marketers, this chapter helps to determine consumer's thinking and what attracts their attention. According to the findings of this study, people are more into experiences that lead them to happiness, being with the social group, enjoying their moments, staying away from their daily concerns. Being in painful experiences can make participants know how they can face new challenges and deal with such kind of situations, tough and hard experiences help them to know more about themselves as well as being in social groups and helping others with what they do can bring joyful feeling to them as also suggested by Argyle [2]. Thus, marketers can take advantage of these aspects to market such kind of events. This kind of painful events helps the business in creating a unique brand experience as well as helping the business to discover the bodies by the preparation of the social material setups.

References

1. Ahuvia AC (2005) Beyond the extended self: Loved objects and consumers' identity narratives. J. Consum Res 32(1):171–184
2. Argyle M (2013) The psychology of happiness. Routledge, London
3. Arnould EJ, Price LL (1993) River magic: extraordinary experience and the extended service encounter. J Cons Res 20(1):24–45
4. Britton T, LaSalle DD (2003) Priceless: turning ordinary products into extraordinary experiences
5. Caru A, Cova B (2003) A critical approach to experiential consumption: fighting against the disappearance of the contemplative time. Crit Mark 23:1–16
6. Carù A, Cova B (2003) Revisiting consumption experience: a more humble but complete view of the concept. Mark Theory 3(2):267–286
7. Chronis A, Arnould EJ, Hampton RD (2012) Gettysburg re-imagined: the role of narrative imagination in consumption experience. Consumpt Mark Cult 15(3):261–286
8. Decloe MD, Kaczynski AT, Havitz ME (2009) Social participation, flow and situational involvement in recreational physical activity. J Leisure Res 41(1):73–91
9. Dietrich A, Audiffren M (2011) The reticular-activating hypofrontality (RAH) model of acute exercise. Neurosci Biobehav Rev 35(6):1305–1325
10. Fortezza F, Pencarelli T (2011) Experience marketing: specific features and trends. The wish days case study. J Mark Trends 1(6):57–69
11. Gershoff AD, Kivetz R, Keinan A (2011) Consumer response to versioning: how brands' production methods affect perceptions of unfairness. J Cons Res 39(2):382–398
12. Getz D, Fairley S (2004) Media management at sport events for destination promotion: Case studies and concepts. Event Manage 8(3):127–139
13. Gitelson R, Kerstetter D (2000) Anew perspective on the decision-making process of arts festival visitors. Events Beyond 2000: setting the agenda, 179–
14. Hsieh HF, Shannon SE (2005) Three approaches to qualitative content analysis. Qual Health Res 15(9):1277–1288
15. Ismail AR (2011) Experience marketing: an empirical investigation. J Rel Mark 49) 10(3):167–201
16. Jenkins R, Nixon E, Molesworth M (2011) 'Just normal and homely': the presence, absence and othering of consumer culture in everyday imagining. J Cons Cult 11(2):261–281
17. Lambert-Pandraud R, Laurent G (2010) Impact of age on brand choice. The Aging Consumer: perspectives from psychology and Economics 191–208
18. LaSalle D, Britton TA (2002) Priceless: turning ordinary products into extraordinary experiences
19. McCracken LM, Gutiérrez-Martínez O (2011) Processes of change in psychological flexibility in an interdisciplinary group-based treatment for chronic pain based on Acceptance and Commitment Therapy. Behav Res Ther 49(4):267–274
20. Mullins-Sweatt SN, Glover NG, Derefinko KJ, Miller JD, Widiger TA (2010) The search for the successful psychopath. J Res Personal 44(4):554–558

21. Neuman WL (2013) Social research methods: qualitative and quantitative approaches. Pearson education
22. Phelps JE, Lewis R, Mobilio L, Perry D, Raman N (2004) Viral marketing or electronic word-of-mouth advertising: Examining consumer responses and motivations to pass along email. J Advert Res 44(4):333–348
23. Pine BJ, Gilmore JH (1998) Welcome to the experience economy. Harvard Bus Rev 76(4):97–105
24. Poulsson SH, Kale SH (2004) The experience economy and commercial experiences. Mark Rev 4(3):267–277
25. Same S, Larimo J (2012). Marketing theory: experience marketing and experiential marketing. In: 7th international scientific conference business and management, pp 10–11, May 2012
26. Shapiro SL, Carlson LE (2017) The art and science of mindfulness: integrating mindfulness into psychology and the helping professions. American Psychological Association
27. Sherry Jr, JF, Kozinets RV, Borghini S (2013) Agents in paradise: experiential co-creation through emplacement, ritualization, and community. In: Consuming experience. Routledge, pp 31–47
28. Skavronskaya L, Moyle B, Scott N (2017) Experiential decision choice. Visitor Exp Des 68
29. Tumbat G, Belk RW (2010) Marketplace tensions in extraordinary experiences. J Cons Res 38(1):42–61
30. Tynan C, McKechnie S (2009) Experience marketing: a review and reassessment. J Mark Manage 25(5–6):501–517

The Role of Higher Education Institutions in Building the E-innovation System in the Sultanate of Oman

Araby Madbouly and Sameh Reyad

Abstract The purpose of this study is to analyze the Role of Higher Education Institutions on the E-Innovation of Oman. It was found that there are several conditions for the expansion of the contribution of universities to regional development of E-Innovation system. These conditions are predominantly related to a broad set of factors that relate to characteristics of HEIs, characteristics of the regional firms, aspects of the collaborative relationship, and characteristics of environmental context in which HEIs and firms are embedded.

Keywords E-innovation system · Higher Education Institutions · Oman

1 Introduction

Innovation, simply, is a "wellspring of growth" or "continuous process for producing new ideas". The terminology of "E-innovation" has been used by industry for last three decades [8]. E-innovation is related to "planning, scenario building, technology forecasting, market intelligence, new product development and creative business development is converging into a new, proactive response to change" [1]. Innovation and E-innovation are considered key drivers for the sustainable growth in countries. They depend, mainly, on the surrounding environment which includes a "system". Therefore, forming an effective innovation and E-innovation systems play a significant role in achieve long run economic development objectives for countries.

Higher Education Institutions (HEIs) are key players of innovation and E-innovation in the era of Knowledge economy as they create, share and transfer the knowledge to the industry to create new products, enhance the current products and uses better technologies [9, 10, 11].

A. Madbouly (✉)
Department of Business and Accounting, Muscat College, Muscat, Oman
e-mail: araby@muscatcollege.edu.om

S. Reyad
College of Business and Finance, Ahlia University, Manama, Bahrain
e-mail: sreyad@ahlia.edu.bh

A. M. A. Musleh Al-Sartawi (ed.), *The Big Data-Driven Digital Economy: Artificial and Computational Intelligence*, Studies in Computational Intelligence 974,
https://doi.org/10.1007/978-3-030-73057-4_16

Sultanate of Oman and other GCC countries have taken good steps towards increase the capacity for innovation [9]. The Omani national strategy focus on building a strong educational and economic infrastructure to be able to participate in technological developments. This will foster the will foster the innovation specially via build strong connection between HEIs and the industry. Although the taken steps towards building National Innovation System (NIS) and having innovation economy, the is ambiguity about role of the HEIs in building this system. Furthermore, there is no clear model of reflects the role of HEIs in having an E-Innovation system. Therefore, there is a need to have a model includes the mechanism of the contribution of Omani HEIs in Building Omani E-National Innovation System. Therefore, this chapters aims to elaborate a model shows how the Omani HEIs can contribute in building Omani E-innovation system.

2 E-innovation

The terminology "E-innovation" begun to be used by late 1970s and 1980s accompanied with the start of design software and having the open source software which created the potential for distributed innovation on a global basis. Therefore, E-innovation can be considered the raises significant implications for expansion of creativity in a wide range of domains. Sawhney and Prandelli [13] suggest a third way: "community of creation", which manages distributed innovation in turbulent markets.

Rayport and Jaworski [5] Examine innovation frameworks and processes in the Networked Economy. They identify five variables: investment required, time-to-market, flexibility, decision-making mechanism and innovation constraints. E-innovation is a new idea- a new label for a group of activities whose aim is to create new tools for innovation, product development and business development in the present climate of continuous change. These activities include competitive intelligence, strategic corporate planning, new product development.

E-innovation is also the computerization of the innovation processes and capabilities are integrated into the electronic work. In such cases, computers carry out most of the innovative activities, working products are digital and their diffusion costs are minimal. And they put forward a technology for realizing E-innovation: workflow technology and develop a NPDWF system which facilitate the standard process of NPD with information technologies in order to increase shared knowledge and information, and decrease the cycle of NPD. Lan ping summarize that E-innovation can be interpreted as using the internet to plan, initiate, conduct, run, facilitate, and/or promote innovation. It is defined as innovations closely tied to a digital platform, no matter whether the platform is used for delivery or to commercialize an innovation. He also established three dimensions to measure the E-innovation with the aspects of software-based outlets, distributed delivery, integrated toolkits (Lan ping 2004).

Finally, we conclude that E-innovation is a new way of working. It is the integrated usage of new and emerging technologies, which can greatly improve existing working methods to better deliver services and better sharing of knowledge. It comes from the Internet Era and success use of open source software development.

In recent years, E-innovation became more popular. Searching popular Internet search engines and directories such as Goolge for the keyword ‘E-innovation’, we find 18,700 results. In order to gather deeper understanding of E-innovation, Lan ping and Du examined 1060 samples of searching findings from Yahoo and draw a Found that about half the industrial E-innovation has been used by industry since the E-commerce stage, and the term is quietly incorporated into some enterprises’ operations. They also found that half the industrial usages of the term e-innovation are related to labeling new products or services and most usages emphasis on the new features of product or services such as ICL’s online smart card, Bournemouth Council in the UK providing ‘ask job’ service and Cell Consulting E-spin service which consists of E-vision, E-portfolio, E-conception and E-implementation.

Some companies apply E-innovation to a special field such as Arthur D. Little, one of the leading consulting companies, which develops an E-innovation hierarchy in the management of supply chain based on the combination of market innovation, channel innovation, product/service innovation, and business model innovation. Some companies apply E-innovation to indicate an area to the Internet such as e-government, e-service and so on. Others use E-innovation to show a capability. For example, GE use E-innovation not only refer to new service or product, but also to capabilities, progress and results based on the section “GE’s E-innovation in 2000’. Trifinit Networks, an American ISP, present E-innovation as the core of its three key E-businesses: Trifinit Service, Trifinit Venture and Trifinit Online. E-innovation is also used as a new company’s name or reflected in creating intellectual property.

Some researches discuss E-innovation on the relationship between innovation and other business activities such as manufacture, biotechnology, telecommunication, semiconductor and retail industries. Philip Rosson examine electronic marketplaces as one “digital economy” innovation based on the Canadian experience; Shigemi Yoneyama, Ingyu OH show how knowledge-integration capabilities underpin the success of the process of technology commercialization in Japanese companies; Jennifer Frahm, Prakash Singh indicate that telecommunication is the basic electronic infrastructure for E-innovation after investigating six biotechnology area in Australia; Hyuk-rae Kim, Shigemi Yoneyama trace the formation and usage of flexible innovation networking used by Taiwanese semiconductor manufacturing company; Rudy L. Ruggles focus on one dynamic driving force of E-innovation: connectivity; Bob cotton mention the definition, evolution, E-dimension, the global dimension of E-innovation and ten steps to making the E-innovation work.

2.1 The Framework of E-innovation

Examining the efforts of industry and academic in promoting E-innovation, it is apparent that many enterprises are positioning themselves in the emerging field of

E-innovation, which could bring new functionalities and new channels for them to compete in a new business environment.

Therefore, a theoretical working framework has to be developed for E-innovation. It has developed into a hot point about how to construct the E-innovation platform in recent years.

In order to seek such a framework, a retrospect of academician's study on innovation was conducted. Schumpter's [14] initial study builds the first interface between innovation and business systems. He explains how innovations cause economic evolution by linking two phases and two types of innovations with a wavelike business circle. He reveals how key components of an economic system such as entrepreneurs, money and banking affect the carrying out of innovation. Kogut, B. and Turcanu, A. [7] develop two models to illustrate the process of innovation. One is waterfall model which minimizes the need for coordination and communication because it sees the process of software development as a cascade of phases, the output of one being the input to the next. The other is iterative model, which recognizes the substantial overlap between various phases of software development. It is an integrated, parallel system that designs new products and related processes. Bob Cotton (2002) presents a bottom-up approach to E-innovation, which attempts to marry strategic planning, product portfolio management, and a responsive product innovation process with tactical, entrepreneurial opportunism. Lan Ping (2004) suggested a three-dimensions framework to accommodate current efforts for understanding and promoting new innovation based on Schumpter's framework. The three dimensions are digital innovation outlet, distributed innovation delivery and simplified innovation management. Address the changes of innovation following the thread of new product development and put forward a new model of New Product Development Workflow, which realizes E-innovation.

Therefore, a five–dimensions framework is suggested here to accommodate current efforts for understanding and promoting E-innovation based on the above framework. The E-innovation platform consists of five parts: distributed nature digital form decoupling activities dedicated to channel change deployment of new rules. Digital form means E-innovation focuses on the information and its controls over these physical works. Distributed nature means that E-innovation is open and decentralized to a much higher degree. Decoupling activities means the basic fundamental construction unit. E-innovation focuses on the change from traditional channel to electronic channel. Deployment of new rules emphasis on integration of rules and principles used in the innovation management to make great changes to the organization structure, incentive mechanism and intellectual.

3 Omani Effort Towards Having Innovation and E-innovation Economy

Oman started to pay more attention to innovation many years back which has been translated into establishing two entities focus mainly on promote innovation, they are: The Research Council (TRC) and Innovation Park Muscat. TRC aims at develop most R&D programs within the scope of the National Research Strategy. Joint efforts by Sultan Qaboos University and state-owned Shell-led Petroleum Development Oman are an example of such a fruitful partnership between academia and industry. TRC has many initiatives in the way of stimulating R&D and innovation in the sultanate. Innovation Park Muscat is Oman's latest and most ambitious science and technology creation. It is one of the main initiatives of the Research Council to foster scientific research, innovation and cooperation in local and international communities in the academic, private and diverse sectors of industry.

One of the initiatives to promote innovation and E-innovation is EJAAD which has been launched in January 2018 aims to collaborate and support research and innovation, and establish efficient R&D partnerships towards Oman's knowledge-based economy. EJAAD focuses on applied research and commercialization opportunities and continues to evolve with the support of technical focal points within institutions, steering committee and leadership directions. EJAAD aspires to considerably impact Oman's ranking in Global Innovation Index, Global Competitiveness Index (EJAAD 2015).

The move towards having innovation economy is not limited on the given initiatives only, it's a long term direction in the sultanate and has been highlighted in Oman vision 2040. Having "A diversified, integrated and competitive economy that is future-oriented and is driven by innovation and entrepreneurship" is one of the goals of "Economic Diversification and Fiscal Sustainability" national priority (Oman 2040 2017). To move the Omani economy towards a robust economic diversification base that relies on technology, information and innovation. The goal is to build a diversified, dynamic, globally interactive and competitive economy. Innovation will be the new engine of growth but will be closely dependent on appropriate infrastructure and educational system that encourages entrepreneurship. This pillar focuses on the achievement of Economic diversification in a way that ensures sustained economic growth in the next century, with a reduction in oil demand. It also aims to improve economic sectors' upstream and downstream integration. (Oman 2040 2017).

4 Omani HEIs and Their Role in Promote E-innovation

A dramatic change has been shown in the education in Sultanate Oman since the accession of His Majesty Sultan Qaboos bin Said in 1970 where the education played a vital role in the country's socioeconomic development alongside with many changes that have reshaped the Sultanate of Oman today. Since that time, general education (pre-tertiary) has been available to all Omani nationals which yielded a large number of secondary school graduates. The change, consequently, include the higher education where the Sultante developed a diversified model between public and private HEIs [3].

Actually, few literatures investigated the role of Omani HEIs in foster the national innovation. Abdul Madhar [1] Investigated the Knowledge Management in Higher Educational Institutions with Special Reference to College of Applied Sciences (CAS), he found that many Supporting factors for successful knowledge management strategies are available in CAS, they are: "A knowledge-based and knowledge-oriented culture, the cooperation of the top-management, an adequate technical and organizational infrastructure, a clear vision, motivating elements, a certain amount of knowledge structure and multiple channels for knowledge transfer". Therefore, the effective implementation of KM would improve CAS from the perspectives of teaching, learning and general working conditions as well.

Other study conducted by [6] to develop a conceptual framework of KM especially to Higher College of Technology of Ministry of Menpower (MOMP). The study suggested a KM center for the MOMP to be able to continuously store, maintain and redesign the knowledge base with a continuous contact with the labour market and to have better control over the academic and administrative knowledge. The stored data can be effectively used by all colleges equally and to develop a standardized system which supports in strategic and operational decision making.

Al-Hemyari [2] developed a KM model for HEIs is and proposed as an appraising mechanism that can be used to identify and monitor the quality and performance areas that are important to Omani HEIs. The conceptualisation and operationalisation of the KM model is also articulated extensively in this paper and tested rigorously using a sample of HEIs in the form of College of Applied Sciences (CASs) in Oman. The study concluded that there is a need to support the KM model as an appraising mechanism for identifying and monitoring the quality and performance of Omani HEIs.

Majority of Omani private HEIs are involved in partnership with international HEIs. Elezi and Bamber [4] found argued that "formulation of HE partnerships is strongly relied on aspects of trust, honesty and communication and the ability of educational institutions to share, transfer and absorb tacit and explicit knowledge effectively".

5 Actions for Omani HEIs Can Undertake to Stimulate E-innovation

From the above empirical studies, we can distil several deliberate actions that can be undertaken by HEIs to stimulate regional development. Essentially these actions pertain to the different roles of the HEIs in the regional innovation system. Thus, each of the traditional roles of research and education will have to be reinterpreted and expanded, in order for the HEIs to become system builders.

The HEIs in the region create social capital, attract students who spend money in the regional economy, contribute to social, political and cultural life, community development, architecture and integration in international society.

The first domain in which HEIs can actively contribute to regional system building is can be regionally-focused. Furthermore, the form of research projects can be such that it involves regional actors. We can distinguish between different forms of research agreements that vary in terms of the party that is undertaking the research and whether the project generates new insights from an academic point of view, or not.

On one side of the spectrum are contract research agreements that are commissioned by industry or policy makers and are undertaken by university researchers. Projects in this category usually pertain to original academic research that creates new knowledge. Also the collaborative development of innovative instruments and engineering design tools falls into this category of research agreements. A middle group consists of joint research agreements that involve research directed to generating academically new insights and that is undertaken collaboratively by several parties. Collaborative centres where university staff and personnel from companies do joint research in so-called third spaces or academic workplaces are a good example of projects in this category (Harloe and Perry 2004).

At the other side of the spectrum are consultancy projects that are commissioned by industry and do not involve original research. In the latter category HEIs are essentially exploiting and capitalizing on existing knowledge (Etzkowitz et al. 2000). Communication channels between HEIs and firms can be kept open by creating regional learning centres or Lernladen (Gnahs et al. 2008), which are specialized knowledge institutes that are easy accessible for firms.

The traditional role of education brings forward the second domain of activities that can be adapted in the light of the new role of regional development. Training relationships between HEIs and industry can be built (Musleh, 2020). Professors may identify opportunities for students to serve as interns in firms (Etzkowitz et al. 2000). 'Placements' within the region can be created for alumni in order to prevent bright graduates from leaving. In this way HEIs act as attractor, educator and retainer of students, shaping them into knowledge-based graduates for firms in the region (Boucher et al. 2003), and providing the research. Research can be employed in several ways. The content of the research can be directed to areas that underpin the region's economic base, that is, part of the research undertaken by HEI can be regionally-focused. Furthermore, the form of research projects can be such that it involves regional actors. We can distinguish between different forms of research agreements that vary in terms of the party that is undertaking the research and whether the project generates new insights from an academic point of view, or not.

On one side of the spectrum are contract research agreements that are commissioned by industry or policy-makers and are undertaken by university researchers. Projects in this category usually pertain to original academic research that creates new knowledge. Also the collaborative development of innovative instruments and engineering design tools falls into this category of research agreements. A middle group consists of joint research agreements that involve research directed to generating academically new insights and that is undertaken collaboratively by several parties. Collaborative centres where university staff and personnel from companies do joint research in so-called third spaces or academic workplaces are a good example of projects in this category (Harloe and Perry 2004; Wetenschappelijke Raad voor het Regeringsbeleid 2008). At the other side of the spectrum are consultancy projects that are commissioned by industry and do not involve original research. In the latter category HEIs are essentially exploiting and capitalizing on existing knowledge (Etzkowitz et al. 2000). Communication channels between HEIs and firms can be kept open by creating regional learning centres or Lernladen (Gnahs et al. 2008), which are specialized knowledge institutes that are easy accessible for firms.

The traditional role of education brings forward the second domain of activities that can be adapted in the light of the new role of regional development. Training relationships between HEIs and industry can be built. Professors may identify opportunities for students to serve as interns in firms (Abdulrasool et al. 2020 and Etzkowitz et al. 2000). 'Placements' within the region can be created for alumni in order to prevent bright graduates from leaving. In this way HEIs act as attractor, educator and retainer of students, shaping them into knowledge-based graduates for firms in the region (Boucher et al. 2003), and providing the region with a pool of skilled labour (Power and Malmberg 2008). Firm employees can enrol in university courses, or specific in-house training programmes targeted to firm employees can be set up by university personnel. Another type of interaction that is associated with a very intensive flow of knowledge is the temporary appointment of university members to the business sector (Schartinger et al. 2001). Existing education programs can be adapted to meet regional skills needs. Furthermore, student recruitment could adopt a strong regional focus.

Finally, several activities can be distinguished that enhance HEI-industry relationships and that cannot be categorized under research or education and that mainly originate from formal and informal participation as an institutional actor with other regional actors in linkages networks of learning, innovation and governance (Boucher et al. 2003). We label this type of activity as active collaboration with (regional) public and private actors. Attendance of university researchers at industry sponsored meetings and conferences is an example, as well as making the university premises available for local activities. Staff participation on external bodies can induce regional networking between HEIs, industry representatives and policymakers. It also supports decision-making and brokering networking between national and international contacts and key regional actors (Benneworth and Hospers 2007). Other collaborations between HEIs and industry can be in the form of academic entrepreneurship or spin-off firms, that generate new economic activity, revenues and jobs in the region (Etzkowitz et al. 2000; Lockett et al. 2003; Power and Malmberg 2008; Erdös and Varga 2009). Yet another type of collaboration that is associated with knowledge flows is creating or renting university facilities or equipment to industry (Ramos-Vielba et al. 2010).

Whether a certain mechanism for deliberate action is suitable to adopt by a certain HEI depends on a number of factors. These factors influence the ability of HEIs to engage in regional system building in which regional stakeholders work together to develop the overall capacity of human resources in the region (Boucher et al. 2003, p. 895). Studies on the contribution of HEIs to the regional innovation system have revealed several conditions for the expansion of the contribution of universities to regional development. These conditions are predominantly related to a broad set of factors that relate to characteristics of HEIs, characteristics of the regional firms, aspects of the collaborative relationship, and characteristics of environmental context in which.

HEIs and firms are embedded.

References

1. Abdul Madhar M (2010) Knowledge Management in Higher Educational Institutions with Special Reference to College of Applied Sciences (CAS). Ministry of Higher Education, Sultanate of Oman, August https://doi.org/10.2139/ssrn.1663543
2. Al-Hemyari ZA (2019) A knowledge management model for enhancing quality and performance of higher education institutions: insights from Oman. Int J Qual Innov (IJQI) 4(½). https://doi.org/10.1504/IJQI.2019.101435
3. Al-Lamki SM (2002) Higher Education in the Sultanate of Oman: the challenge of access, equity and privatization. J High Educ Policy Manag 24(1):75–86
4. Elezi E, Bamber C (2018) Understanding the role of knowledge management in higher education partnerships through experts. In 19th European Conference on Knowledge Management, 06–07 September, Padua, Italy
5. Jeffrey FR Bernard JJ (2001). Introduction to e-commerce. Boston: McGraw-Hill/Irwin MarketspaceU
6. Kumar DM, Kumar P (2010) Knowledge management and technology knowledge management model for the higher technology college of ministry of manpower, Oman. J Bus Admin 1(1):24–40
7. Kogut B Turcanu A (1999) The emergence of E-innovation: insights from open source software development, Wharton School paper, University of Pennsylvania (November 15). Accessed 25 February 2001. Online Available HTTP: http://jonescenter.wharton.upenn.edu/events/software.pdf.
8. Lan P, Du HH (2002) Challenges ahead E-innovation. Technovation 22:761–767
9. Madbouly A, et al (2020) The Economic Determinants for the Competitiveness of the Sultanate of Oman. NOVYI MIR Res J 5(10):135–151. https://novyimir.net/gallery/nmrj2627%20f.pdf
10. Madbouly A, et al (2020) The impact of knowledge management on the HEIs' innovation performance: an analytical study of Omani HEIs. NOVYI MIR Res J 5(11), 124–143. https://novyimir.net/gallery/nmrj%202649%20f.pdf
11. Madbouly A, Reyad S, Chaomancy G (2020) Analysis for the Knowledge Economy in GCC Countries. Kacprzyk, J. Stud Comput Intell. https://www.springer.com/series/7092
12. Musleh Al-Sartawi AMA (2020) E-learning improves accounting education: case of the higher education sector of Bahrain. In: Themistocleous M, Papadaki M, Kamal MM (eds.) Information Systems. EMCIS 2020. Lecture Notes in Business Information Processing, vol 402. Springer, Cham
13. Mohanbir S Emanuela P (2001) Communities of creation: managing distributed innovation in turbulent markets. California Management Review. 42. https://doi.org/10.2307/41166052
14. Schumpeter JA (1939) Business cycles: a theoretical, historical, and statistical analysis of the capitalist process. McGraw-Hill, New York

Foreign Currency Exchange Rate Prediction Using Bidirectional Long Short Term Memory

Rony Kumar Datta, Sad Wadi Sajid, Mahmudul Hasan Moon, and Mohammad Zoynul Abedin

Abstract This chapter aims to predict the foreign currency exchange rate over twenty-two different currencies based on the US dollar. This chapter proposes three machine learning algorithms, such as ridge regression, lasso regression, decision tree, and a deep learning algorithm named Bi-directional Long Short-Term Memory (Bi-LSTM) to predict the foreign currency exchange rate. Technical analysis of foreign currency exchange is also discussed in this chapter. The authors use mean absolute error (MAE), mean square error (MSE), root mean squared error (RMSE), and mean absolute percentage error (MAPE) to measure the performance of the algorithms. Empirical findings indicate that overall performance of all the algorithms is satisfactory, but Bi-LSTM performs better than others. This study is beneficial for the stakeholders in setting a range of strategies for the foreign exchange market.

Keywords Currency exchange rate · Machine learning · Deep learning · Time series analysis

R. K. Datta · M. Z. Abedin (✉)
Department of Finance and Banking, Hajee Mohammad Danesh Science and Technology University, Dinajpur 5200, Bangladesh
e-mail: abedinmz@hstu.ac.bd

R. K. Datta
e-mail: rony.datta@hstu.ac.bd

S. W. Sajid
Department of Electronics and Communication Engineering, Hajee Mohammad Danesh Science and Technology University, Dinajpur 5200, Bangladesh

M. H. Moon
Department of Computer Science and Engineering, Hajee Mohammad Danesh Science and Technology University, Dinajpur 5200, Bangladesh

A. M. A. Musleh Al-Sartawi (ed.), *The Big Data-Driven Digital Economy: Artificial and Computational Intelligence*, Studies in Computational Intelligence 974,
https://doi.org/10.1007/978-3-030-73057-4_17

1 Introduction

The Foreign Currency Exchange (FOREX) means trading of one currency for another currency [1]. It is a global market for currency trading. FOREX determines foreign exchange rates for each currency. The foreign exchange rate means in which rate one currency is exchanged for another. The FOREX market is open 24 h a day and this market is really complex, volatile, and almost impossible to predict. FOREX market trading happens following four major time zones: Asian Zone, Australian Zone, European Zone, and the North American Zone [2]. FOREX market is the most liquid, and the largest market in the currency world, with billions of dollars changing hands daily. As FOREX is complex to predict and this market is so volatile, it has been an interesting field for researchers over the last few decades. Researchers are repeatedly trying to crack the FOREX exchange rate prediction. They are using fundamental analysis and technical analysis to predict the FOREX market [3–5]. Those analyses include many different factors such as the industrial and economic conditions of a country and the technical analysis tries to forecast the market using previous time-series data. To predict the foreign currency, researchers used different types of technical approaches previously. But among all these methods, neural network-based analysis has proven to provide the best result. Many researchers have used Long Short-term Memory (LSTM), Gated Recurrent Unit (GRU), Ridge Regression, or Lasso Regression to predict the FOREX market with the help of Machine Learning (ML) and Deep Learning [6]. Most of them have used one or two algorithms and predicted only a few currencies. But we intend to do something better. In this research, we predict twenty-two different currency exchange rates against the USD (United States Dollar). As well as we utilize four different Machine Learning approaches: Bi-directional Long Short-term Memory (Bi-LSTM), Lasso Regression, Decision Tree Regression, Ridge Regression. The traditional RNN cannot handle the data when the input and output have the different sizes and Bi-LSTM overcome these problems. Our result is promising and it is supposed to provide an idea of upcoming currency rate changes to all money related organizations. The later part of this chapter has the application of machine learning, related works review, methodology, comparison, conclusion, and discussion for future research direction.

2 Literature Review

Many techniques have been applied to predict the Foreign exchange market in last few years. Islam and Hossain propose to predict the FOREX rate using GRU-LSTM Hybrid Network. They develop a hybrid model combining two promising neural networks, LSTM and GRU. Furthermore, they predict the FOREX price 10 min and 30 min before the actual time [7]. To recognize programmed change, devices have been applied to two unique capacities chains, time-stretch dependent on the phantom record picture contrast strategy, it was created. These devices are conceivable, have little effect on the local and mainland scale, and do multi-scale investigation [8]. Now-a-days many security patches and management systems are installed on gadgets and smart devices. These have made them really difficult to predict and related organizations are implanting complex algorithm. It is continuously a booking calculation of the general assignment [9]. From different input/output process with machine learning financial changes can be analyzed through a software interface. By using advanced inserted and Artificial Intelligence (AI), the expectation conversion standard can be examined. AI and Machine Learning work on rapid changes of financial transactions and predict the next change [1]. To forecast the gross domestic product (GDP) growth rate, artificial neural network and extreme machine were introduced. Root means square, coefficient of determination and Pearson coefficient indicators compare the predictive performances of the proposed ANN models. Extreme Machine Learning and Back Propagation learning algorithm develop GDP growth rate prediction accuracy [10]. Machine Learning (ML) approaches construct the training, test and off-sample data sets for the financial predictions. Low complexity binary machine learning can give good predictions on financial trading over an extensive period of time. Financial forecasting includes expert or rule based systems, decision trees, neural networks and genetic computing [11]. In a wide range of computational intelligence, economical time series prediction is a popular choice for researchers which affects the academia's implementation and regional financial industry significantly. Machine Learning researchers have created various models as well as published them accordingly. Consequently, many surveys are located in the financial time series prediction to cover the Machine Learning research [12]. Yuanyuan Pu proposes machine learning methods for rockbrust prediction which is related to sudden changes in currency rate. This work includes a complex and nonlinear procedure that is influenced by uncertainty parameters [13]. Air quality degradation for traffic has been a major issue; these pollution rates can be predicted using machine learning. Wang proposes regression models, Mean square error analysis, and root mean square analysis error for the neural network engine. The authors develop Land use regression model, Artificial neural network, XGboost models for different data segment machine learning [14]. It has embraced the mix's parity hypothesis to predict the conversion scale in this examination. The Global Financial Crisis is needed to consider for this investigation. The goal of the examination is to gauge the model and acquire explicit stun. The authors utilize the outcomes to the principal stage's swapping scale execution [15].

3 Research Methodology

3.1 Dataset

The dataset that we use for the analysis in this chapter is open-source data from Kaggle [www.kaggle.com]. This dataset comes from the Federal Reserve system. Brunotly make few changes. We have a total of 22 currency exchange rates against the US dollar in this dataset. We apply machine learning approaches on this dataset. Thus, we can predict the Foreign Currency Exchange rates against US dollar for 22 different countries.

3.2 Performance Measures

The authors measure and compare the performance of the different regression algorithms used in this study by computing the Mean Absolute Error (MAE), Mean Square Error (MSE), Root Mean Squared Error (RMSE) and the Mean Absolute Percentage Error (MAPE).

MAE: Mean Absolute Error is simply the arithmetic average of the absolute Errors. It is a common measure of forecast error in time series analysis and calculated as:

$$\mathrm{MAE} = \frac{\sum_{i=1}^{n} |y_i - x_i|}{n}$$

Where y_i is the predicted exchange rate on day i, x_i is the actual exchange rate on the same day, and n is the total number of days.

MSE: Mean Squared Error measures the average of the squares of the errors, i.e. average squared difference between the estimated values and actual values. It is calculated as:

$$\mathrm{MSE} = \frac{\sum_{i=1}^{n} (y_i - x_i)^2}{n}$$

RMSE: Root Mean Squared Error is a standard metric to compute the errors of a model for numerical prediction. It is the square root of the mean of the square of all the error and calculated as:

$$\mathrm{RMSE} = \sqrt{\frac{\sum_{i=1}^{n} (y_i - x_i)^2}{n}}$$

MAPE: Mean Absolute Percentage Error is usually used as a loss function for regression problems. It is a measure of prediction accuracy and found by calculating as:

$$\text{MAPE} = \frac{\sum_{i=1}^{n} \frac{|y_i - x_i|}{y_i}}{n} \times 100\%$$

3.3 Methodology

To calculate the prediction error of exchange rate movements of the selected countries, different algorithms have been used to select the best method. Here we utilize four different methods namely Ridge, Lasso and Decision tree as OLS regression algorithms and Bi-LSTM as neural network algorithm to estimate the prediction accuracy.

Since over-fitting and under-fitting may arise in simple linear regression, Ridge and Lasso regressions are used to reduce model complexity and prevent over-fitting. As we know that linear regression aims to optimize slope and intercept in such a way that it minimizes the cost function. The general cost function can be written as follows:

$$\sum_{i=1}^{M}(y_i - \hat{y}_i)^2 = \sum_{i=1}^{M}\left(y_i - \sum_{j=0}^{p} w_j \times x_{ij}\right)^2$$

Ridge Regression: In Ridge Regression the general cost function is transformed by adding a penalty term (lambda) equivalent to square of the magnitude of the coefficients.

$$\sum_{i=1}^{M}(y_i - \hat{y}_i)^2 = \sum_{i=1}^{M}\left(y_i - \sum_{j=0}^{p} w_j \times x_{ij}\right)^2 + \lambda \sum_{j=0}^{p} w_j^2$$

This is equivalent to minimizing the general cost function under the following condition:

$$\text{For some} \quad C > 0, \ \sum_{j=0}^{p} w_j^2 < C$$

The penalty term λ (Lambda) which is also known as regularization parameter, regularizes the coefficients in such a way that the optimization functions penalized when coefficients take large values. In this way ridge regression can shrink the coefficients and help to prevent the model from multi-collinearity. Here $\lambda \geq 0$ is used as a complexity parameter that controls the amount of shrinkage and the larger the value of λ, the greater the amount of shrinkage. The coefficients are shrunk toward zero. So when $\lambda \rightarrow 0$, the cost function of ridge regression becomes similar to the general linear regression cost function. The lower the value of λ on the features, the model will resemble the linear regression model.

Lasso Regression: The lasso is a shrinkage method like ridge with subtle but important differences. The cost function for the lasso regression can be defined as follows:

$$\sum_{i=1}^{M}(y_i - \hat{y_i})^2 = \sum_{i=1}^{M}\left(y_i - \sum_{j=0}^{p} w_j \times x_{ij}\right)^2 + \lambda\sum_{j=0}^{p}|w_j|$$

$$\text{For some } t > 0, \quad \sum_{j=0}^{p}|w_j| < t$$

The main difference in the cost function equation between ridge and lasso regression is instead of taking the square of the coefficients, magnitudes are taken into account in the lasso regression. This type of regularization (L1) can lead to a zero coefficient that is some of the features are completely neglected for the output evaluation. So lasso regression not only helps in reducing over-fitting but also helps in selecting features which makes the model easier to interpret.

Decision Tree Regression: Decision tree is a common practical approach for supervised learning. It is used for both classification and regression estimation. Decision tree is a tree-structured classifier consists of three types of nodes namely root node, interior node and leaf node. The root node is the initial node which denotes the whole sample and may be split into further nodes. The interior nodes represent the features of a data set and the branches denote the decision rules. Lastly, the root nodes signify the outcome. For a particular data point, decision tree is run by answering True/False questions until it reaches the leaf node. The final prediction is calculated by the average of the value of the dependent variable in that particular leaf node. In this way the tree is able to predict a proper value for the data point through multiple iterations. Decision tree is advantageous as it is simple to understand and requires lesser data cleaning. The model's performance is not affected by the non-linearity and the number of hyper-parameters to be tuned is almost null. Likewise ridge and lasso regression, the decision tree regression may have over-fitting problem which can be solved by the Random Forest algorithm.

BiLSTM Method: Deep learning based on recurrent neural network such as LSTM, BiLSTM works efficiently in analyzing any time series data better than any traditional model such as ARIMA, SARIMA and ARIMAX due to its bidirectional nature input pattern. When LSTM works with only preceded data pattern, BiLSTM model considers both previous and future data during training time that makes the BiLSTM more effective than LSTM. This behavior of BiLSTM helps to learn present status of data both from past data and future data through its forward layer and backward layer. It can capture not only local feature but also extract global feature in the time series data. Again in BiLSTM layer, there is no hidden-to-hidden connections between forward and backward layers. This helps to understand information both from backward layer and forward layer in each BiLSTM unit. That is the reason; Bi-LSTM works more accurately than LSTM on time series data analysis.

We compute and merge the forward and backward hidden states using the following equations: Where h is the hidden layer state, W is the weight, b is the bias term and σ is the activation function.

$$h_f = \sigma(W_f * X + h_f + b_f) \tag{1}$$

$$h_b = \sigma(W_b * X + h_b + b_b) \tag{2}$$

$$y = (h_f W_f + h_b W_b + b) \tag{3}$$

We employ LSTM units to construct the Bi-LSTM unit for interpreting the features across time steps in both directions in the proposed network.

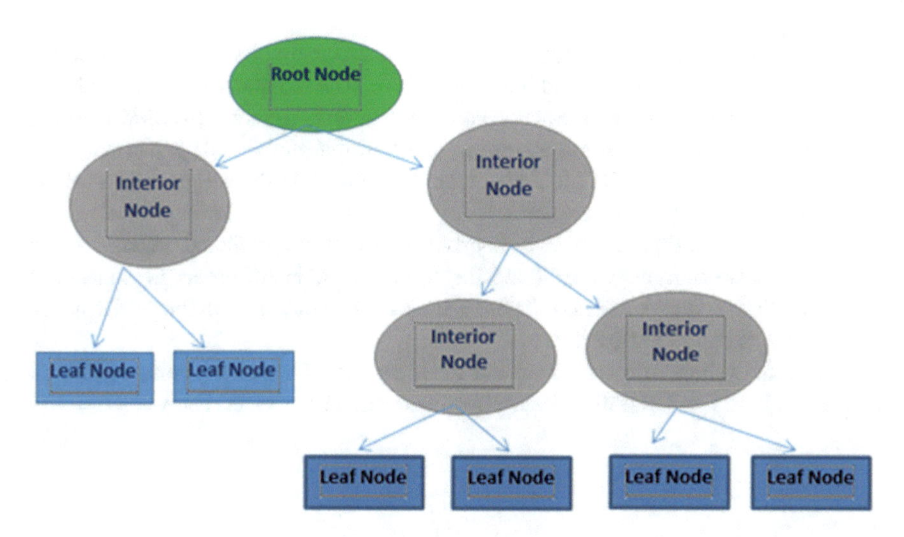

Fig. 1 Decision tree Nodes flowchart (Developed by authors)

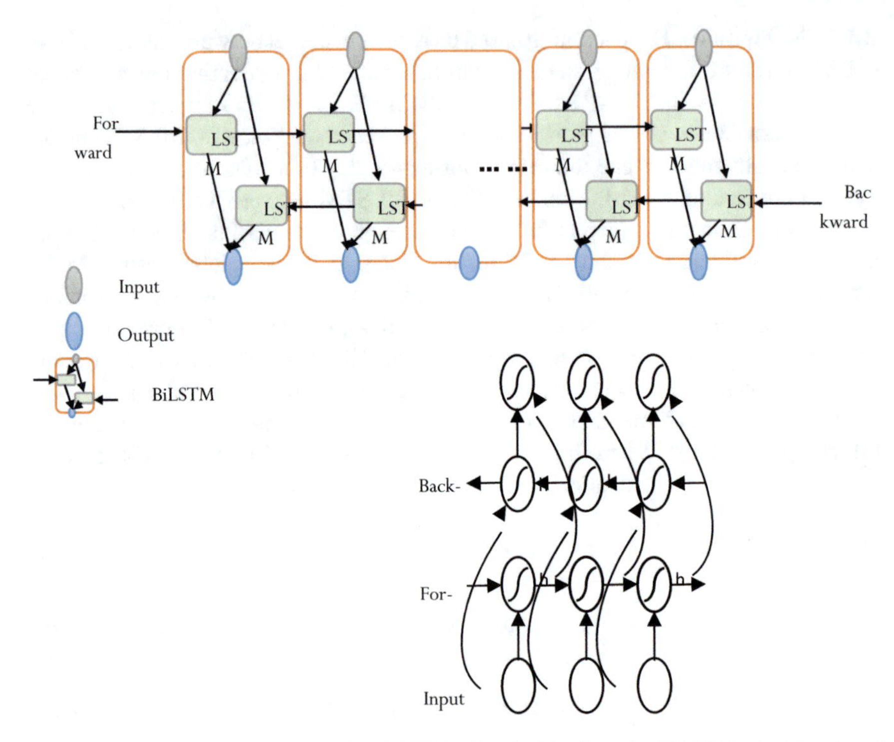

Fig. 2 shows the pictorial diagram of Bi-LSTM and the inside view of 1-BiLSTM unit. (Developed by authors)

4 Result and Discussion

The authors train the models using the dataset and create graphs that show the actual and predicted value using red and blue color curve. The red one shows the predicted value and the blue one shows the actual value. The authors use three machine learning algorithms and a deep learning algorithm and find that the overall performance of the deep learning method Bi-LSTM is better than others. In the below, authors show the graphs for twenty-two currencies.

From the above Figs. 1 and 2, it clearly shows that the predicted values and the actual values are nearer to each other. Most of the cases, the difference is too low but in case of New Zealand, Mexico, Japan, Canada, UK, South Africa the performance is not same as others but good.

The authors show the performance of the algorithms for the twenty-two different currencies based on US dollar. The performance Table 1 is given below:

Table 1 Different types of error for Currency exchange rate prediction

Variable	Algorithms	MAE	MSE	RMSE	MAPE
AUSTRALIAN DOLLAR/US$	Ridge	0.0039	0.00024	0.0049	0.85
	Lasso	0.0185	0.00038	0.0194	3.95
	D.Tree	0.0126	0.00035	0.0187	2.97
	BiLSTM	0.0047	0.00033	0.0057	0.32
EURO/US$	Ridge	0.0033	0.00019	0.0043	0.73
	Lasso	0.0192	0.00039	0.0198	4.12
	D.Tree	0.0065	0.00071	0.0084	1.40
	BiLSTM	0.0020	0.00068	0.0026	0.22
NEW ZELAND DOLLAR/US$	Ridge	0.0037	0.00023	0.0047	1.35
	Lasso	0.0054	0.00045	0.0067	2.03
	D.Tree	0.0071	0.00081	0.0090	2.57
	BiLSTM	0.0144	0.00025	0.0158	0.94
UNITED KINGDOM POUND/US$	Ridge	0.0085	0.00012	0.0112	0.98
	Lasso	0.0882	0.00805	0.0897	9.99
	D.Tree	0.0124	0.00025	0.0158	1.41
	BiLSTM	0.0042	0.00029	0.0055	0.54
BRAZIL - REAL/US$	Ridge	0.0083	0.00012	0.0108	0.92
	Lasso	0.0908	0.00842	0.0917	10.10
	D.Tree	0.0156	0.00042	0.0203	1.74
	BiLSTM	0.0023	0.00085	0.0292	0.57
CANADIAN DOLLAR/US$	Ridge	0.0042	0.00030	0.0054	0.71
	Lasso	0.0206	0.00046	0.0214	3.49
	D.Tree	0.0076	0.00095	0.0097	1.29
	BiLSTM	0.0098	0.00011	0.0105	0.74
CHINA - YUAN/US$	Ridge	0.0048	0.00054	0.0074	1.22
	Lasso	0.0099	0.00014	0.0117	2.72
	D.Tree	0.0068	0.00091	0.0095	1.77
	BiLSTM	0.0111	0.00028	0.0166	0.16
HONG KONG DOLLAR/US$	Ridge	0.0178	0.00072	0.0269	2.19
	Lasso	0.1161	0.01515	0.1231	12.69
	D.Tree	0.0249	0.00229	0.0478	2.96
	BiLSTM	0.0034	0.00018	0.0043	0.04

(continued)

Table 1 (continued)

INDIAN RUPEE/US$	BiLSTM	0.0118	0.00010	0.0130	0.81
	Ridge	0.0054	0.00051	0.0071	0.61
	Lasso	0.0789	0.00620	0.0793	8.88
	D.Tree	0.0096	0.00012	0.0121	1.08
KOREA - WON/US$	Ridge	0.0056	0.00056	0.0075	1.38
	Lasso	0.0305	0.00118	0.0344	7.26
	D.Tree	0.0100	0.00018	0.0135	2.47
	BiLSTM	0.0037	0.00025	0.0050	0.32
MEXICO - MEXICAN PESO/US$	Ridge	0.0060	7.04520	0.0083	0.76
	Lasso	0.0838	0.00710	0.0843	10.52
	D.Tree	0.0112	0.00021	0.0154	1.39
	BiLSTM	0.1014	0.01670	0.1295	0.53
SOUTH AFRICA - RAND/US$	Ridge	0.0079	0.00010	0.0103	0.99
	Lasso	0.0850	0.00730	0.0857	10.69
	D.Tree	0.0148	0.00030	0.0191	1.87
	BiLSTM	0.1336	0.027500	0.1658	0.91
SINGAPORE - SINGAPORE DOLLAR/US$	Ridge	0.0029	1.46330	0.0038	1.17
	Lasso	0.0213	0.00040	0.0217	8.55
	D.Tree	0.0048	4.12900	0.0064	1.91
	BiLSTM	0.0020	7.07730	0.0026	0.15
DENMARK - DANISH KRONE/US$	Ridge	0.0033	0.00019	0.0044	0.72
	Lasso	0.0200	0.00044	0.0206	4.26
	D.Tree	0.0090	0.00014	0.0111	1.94
	BiLSTM	0.0146	0.00031	0.0191	0.22
JAPAN - YEN/US$	Ridge	0.0046	0.00039	0.0063	0.84
	Lasso	0.0093	0.00011	0.0110	1.64
	D.Tree	0.0084	0.00011	0.0111	1.53
	BiLSTM	0.3499	0.20021	0.4474	0.32

(continued)

Table 1 (continued)

MALAYSIA - RINGGIT/US$	*Ridge*	*0.0036*	*0.00025*	*0.0050*	*0.46*
	Lasso	*0.0589*	*0.00350*	*0.0592*	*7.57*
	D.Tree	*0.0070*	*0.00012*	*0.0107*	*0.91*
	BiLSTM	*0.0055*	*6.22160*	*0.0078*	*0.1344*
NORWAY - NORWEGIAN KRONE/US$	*Ridge*	*0.0069*	*0.00071*	*0.0088*	*0.83*
	Lasso	*0.0605*	*0.00371*	*0.0615*	*7.19*
	D.Tree	*0.0100*	*0.00011*	*0.0130*	*1.21*
	BiLSTM	*0.0341*	*0.00180*	*0.0430*	*0.3869*
SWEDEN - KRONA/US$	*Ridge*	*0.0060*	*0.00064*	*0.0080*	*0.86*
	Lasso	*0.0638*	*0.00410*	*0.0645*	*8.96*
	D.Tree	*0.0119*	*0.00020*	*0.0146*	*1.68*
	BiLSTM	*0.0373*	*0.00230*	*0.0480*	*0.39*
SRI LANKA - SRI LANKAN RUPEE/US$	*Ridge*	*0.0022*	*0.00011*	*0.0033*	*0.23*
	Lasso	*0.1346*	*0.01810*	*0.1347*	*14.03*
	D.Tree	*0.0050*	*0.00044*	*0.0066*	*0.53*
	BiLSTM	*0.3812*	*0.24050*	*0.4904*	*0.21*
SWITZERLAND - FRANC/US$	*Ridge*	*0.0024*	*0.00010*	*0.0031*	*1.01*
	Lasso	*0.0283*	*0.00080*	*0.0285*	*11.81*
	D.Tree	*0.0037*	*0.00024*	*0.0049*	*1.54*
	BiLSTM	*0.0035*	*0.00018*	*0.0043*	*0.36*
TAIWAN - NEW TAIWAN DOLLAR/US$	*Ridge*	*0.0065*	*0.00012*	*0.0106*	*1.79*
	Lasso	*0.0220*	*0.00061*	*0.0249*	*6.67*
	D.Tree	*0.0086*	*0.00015*	*0.0125*	*2.39*
	BiLSTM	*0.0569*	*0.00661*	*0.0818*	*0.1841*
THAILAND - BAHT/US$	*Ridge*	*0.0034*	*0.00023*	*0.0048*	*2.58*
	Lasso	*0.0398*	*0.00160*	*0.0404*	*32.31*
	D.Tree	*0.0043*	*0.00032*	*0.0056*	*3.27*
	BiLSTM	*0.0786*	*0.00970*	*0.0986*	*0.25*

The performance of Bi-LSTM is better than other algorithms. Sometimes the Ridge regression shows the less error in terms of MAE, but most of the time MAPE is less than Bi-LSTM and overall performance of Bi-LSTM is shown in the Fig. 3.

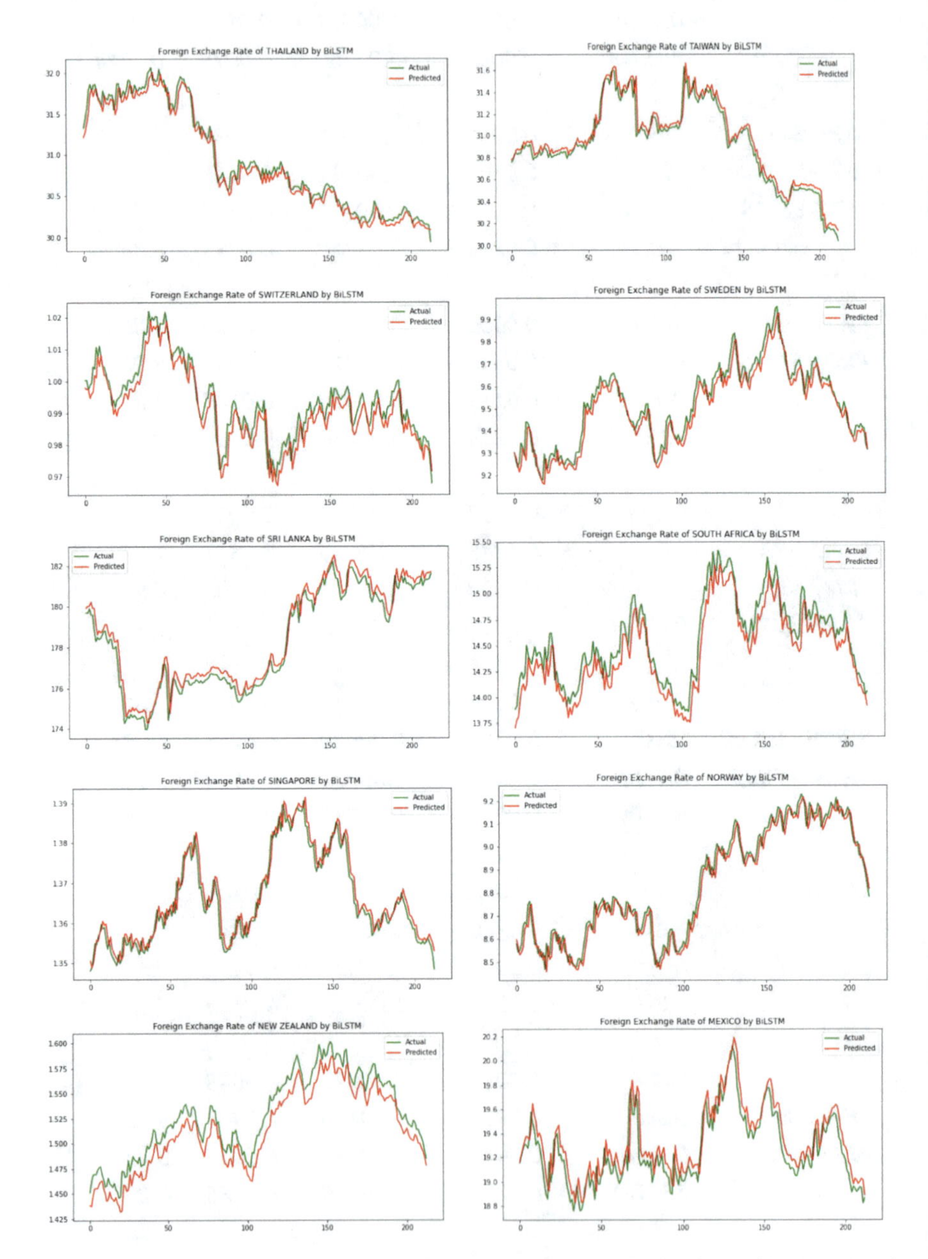

Fig. 3 The relation between actual and predicted values are shown in the above table. (Developed by authors)

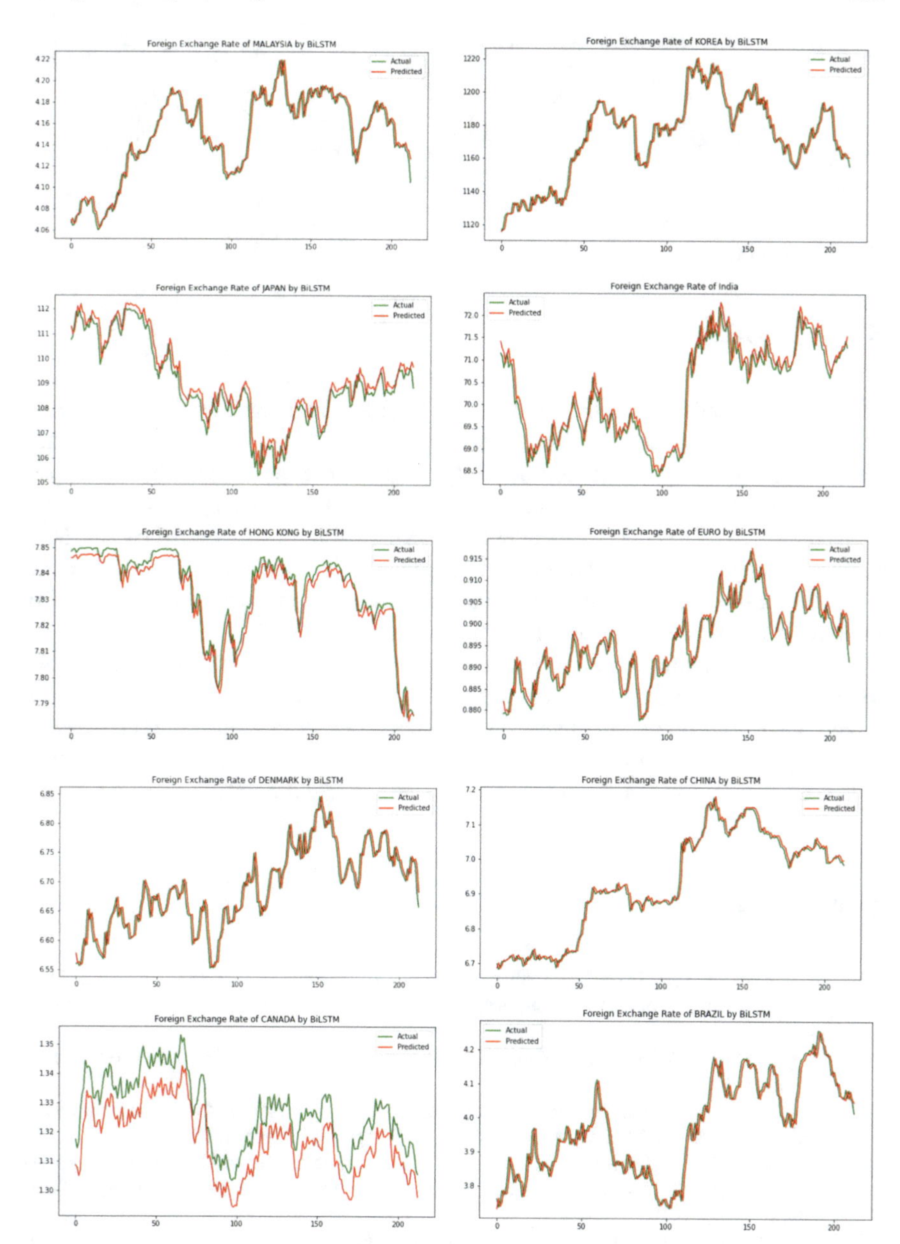

Fig. 3 (continued)

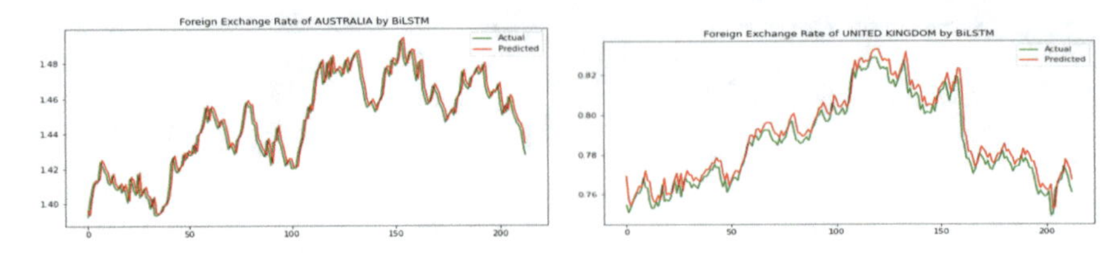

Fig. 3 (continued)

5 Conclusion

This chapter clearly shows that Bi-LSTM a deep learning algorithm shows the less errors to predict the foreign currency exchange rate. The performances of all other algorithms are also good. Sometimes Ridge regression shows a better performance than others, but to analyze a large amount of data, deep learning performs better than others. Bi-LSTM performs better because the working principle of Bi-LSTM is really amazing than others. The errors to predict the foreign currency exchange rate prediction are less than others in this chapter. It indicates that our proposed model is good for predicting foreign currency exchange rate.

In future, the authors want to add a converter that can predict the exchange rate of any currency based on any currencies. For more accurate prediction, the authors would like to add more parameters on the input and also minimize the time, space, and complexity for the models. The authors try to make an application software that can be used by any kinds of people and they get the real time exchange prediction.

References

1. Tao X, Haoxian Yang H (2010) Analysis of the real-time changes in financial exchange rates based on machine learning and complex embedded systems. Microprocess Microsys ISSN 0141-9331
2. Masry S, Dupuis A, Olsen R, Tsang E (2013) Time zone normalization of FX seasonality. 13(7):1115–1123
3. Sezer OB, Gudelek MU, Ozbayoglu AM (2020) Financial time series forecasting with deep learning: a systematic literature review: 2005–2019. Appl Soft Comput 90(106181)
4. Gonce A (2019) Prediction of exchange rates with machine learning. In: Proceedings of the International Conference on Artificial Intelligence, Information Processing, and Cloud Computing – AIIPCC
5. Tsaih RH, Kuo BS, Lin TH, Hsu CC (2018) The use of big data analytics to predict the foreign exchange rate based on public media: a machine-learning experiment. 20(2):34–41
6. Hailee M, Nasser AOM (2016) Financial depth and exchange rate volatility. Am Econ 62(1):19
7. Islam MS, Hossain E Foreign exchange currency rate prediction using a GRU-LSTM Hybrid Network. Soft Comput Lett SOCL 100009
8. Wada T, Shikishima A (2020) Real-time Detection system for smartphone Zombie based on machine learning. IEICE Commun Express 9(7):1–6
9. Zhou J Real-time task scheduling and network device security for complex embedded systems based on deep learning networks. Microprocess Microsyst MICPRO 103282

10. Milaci L, dan Jovi S, Vujovi T, Miljkovi J Application of artificial neural network with extreme learning machine for economic growth estimation. Physica A, PHYSA 17457
11. Abedin MZ, Chi G, Uddin MM, Satu MS, Khan MI, Hajek P (2021) Tax default prediction using feature transformation-based machine learning. IEEE Access 9:19864–19881
12. Huang K, Kelly PJ, Zhang J, Yang Y, Liu W, Kalalah A, Wang, C (2019) Molecular detection of bartonella spp. in China and St. Kitts. Canad J Infect Diseas Med Microbiol 2019 Article ID 3209013, 9 p. https://doi.org/10.1155/2019/3209013
13. Pu Y, Apel DB, Liu V, Mitri H (2019) Machine learning methods for rockburst prediction-state-of-the-art review. Int J Mining Sci Technol 23:565–570
14. Wang A, Xu J, Tu R, Saleh M, Hatzopoulou M (2020) Potential of machine learning for prediction of traffic related air pollution. Transp Res Part D 88:102599
15. Abedin MZ, Chi G, Moula FE, Azad S, Khan MSU (2019) Topological applications of multi-layer perceptrons and support vector machines in financial decision support systems. Int J Finan Econo 24(1):474–507

Board of Directors Composition and Social Media Financial Disclosure: The Case of the United Arab Emirates

Azzam Hannoon, Yousif Abdelbagi Abdalla, Abdalmuttaleb M. A. Musleh Al-Sartawi, and Azam Abdelhakeem Khalid

Abstract Among the most significant governance issues currently faced by the modern corporation are those relating to diversity, such as gender, and independence of directors. Another modern issue facing managers is social media disclosure. The paper aimed to address several research questions, mainly the association between boards of directors' composition: board diversity, and board independence, and the level of financial disclosure on social media by UAE firms. The research data is collected from 103 firms listed in the UAE financial markets for the period 2018–2019. The research result indicates that the relationship between board composition and the social media financial disclosure is significant and positive. This study provides contributions to UAE's government, policymakers, and regulators with regards to two important issues, particularly gender diversity on the board as well as the modern disclosure tools that could be used by firms to reduce the agency problem.

Keywords Social media · Gender · Corporate governance · UAE

1 Introduction

Corporate governance (CG) can be viewed as a system of law, contracts, and social norms that govern the structure by which firms make decisions. This system is important, as it reduces the information gap among the managers and the potential firm investors who have less access to firm information than the firm managers [8, 10].

A. Hannoon (✉)
American University in the Emirates, Dubai, UAE

Y. A. Abdalla
University of Sharjah, Sharjah, UAE

University of Khartoum, Khartoum, Sudan

A. M. A. Musleh Al-Sartawi
Ahlia University, Manama, Kingdom of Bahrain

A. A. Khalid
Universiti Pendidikan Sultan Idris, Perak, Malaysia

A. M. A. Musleh Al-Sartawi (ed.), *The Big Data-Driven Digital Economy: Artificial and Computational Intelligence*, Studies in Computational Intelligence 974,
https://doi.org/10.1007/978-3-030-73057-4_18

Therefore, to decrease the asymmetry of information and the agency cost associated, firms should disclose transparent, timely, and accurate information to the public.

Moreover, CG facilitates access to external finance, especially when the corporate governance structure is transparent through disclosure, and thereby insists on investor rights [14, 20].

Disclosure is a fluctuating term that frequently changes among researchers. [6, 7] define that disclosure is the revelation of the previously unknown so that it becomes shared information. Based on [53], numerous empirical studies have recognized that corporate governance is a key determinant of a firm's voluntary disclosure strategy and. It influences several disclosure practices of the firm, such as disclosing the management forecasts of earnings, adopting earnings press release, and disclosing corporate restructuring plans [45], the composition of the board of directors is one of a number of internal governance mechanisms that are intended to ensure that there is no conflict between the interests of shareholders and managers, and to discipline or remove ineffective management teams.

Nevertheless, among the most significant governance issues currently faced by the modern corporation are those relating to diversity, such as gender, and independence of directors. [12, 16, 25] argue that one of the major governance issues challenging managers and stakeholders, including shareholders and investors, of a modern firm, is the gender composition of the board of directors. As early as [50] issued a policy statement on CG that states the board should be composed of competent individuals who reflect diversity of experience, gender, race, and age. Diversity is a crucial criterion for investment because it is believed that a diverse board will be less obliged to follow management, thus reducing the agency problem [28]. These policies are usually supported on the ground that the participation of women has a positive effect on the operations of boards and managing different types of risks by adopting new technologies which will turn on the performance of firms [2, 19, 22].

Another modern issue facing managers and stakeholders of firms is the advancement of technology, and in turn the firms' readiness to keep up with such advancements. Recently, social media channels have been recognized by the Securities and Exchange Commission (SEC) as channels for financial disclosure. These channels are more widely available to investors and allow for interaction between users via postings and comments [26]. Social media, as a result, is a unique platform that provides proactive and interactive two-way communication between the firm and its stakeholders [27].

Therefore, this study aims to explore the link between two fundamental areas of corporate governance, board of directors' composition and social media disclosure practices. The paper was conducted to mainly address the following research questions. First, to what extent do the United Arab Emirates listed firms use the various social media platforms? What is the level of social media disclosure (SMFD) practiced by firms listed in UAE? Additionally, what relationship exists between corporate governance variables (board of directors' composition: board diversity, and board independence) and the level of financial information disclosure on social media in UAE?

The study used the case of United Arab Emirates as a context for numerous reasons. Based on the CIA [29], the UAE has a population of approximately 6.1 million people who enjoy a per capita GDP of $68,200 due to a diversifying economy that includes Dubai and the world's tallest building, the Burj Khalifa - Dubai. [24] claim that since the seventies, the government of UAE has invested heavily in the construction of infrastructures to create a favorable environment for foreign capital investments. The economy of UAE is transforming from being an undiversified oil dependent to a diversified non-oil dependent. Thus, the United Arab Emirates is viewed as the region financial hub with an ideal legal system and a suitable investment environment. According to Telecommunications Regularity Authority in UAE [49] in its fifth sector annual review published in 2014, it stated that 85% of the population in UAE uses the internet regularly, and basically, everybody owns a mobile phone, which could be applied to social media usage. Further, the studies that have examined the level of SMFD in the UAE are still negligible to the best of researchers' knowledge; hence, this study contributes knowledge to literature covers BOD diversity, disclosure, and social media in the UAE.

Despite the significance of this study to the business environment in the United Arab Emirates, it can also provide contribution to the Middle East countries, mainly to the Arab States of the Gulf, i.e., the GCC countries, as these countries have comparable social, political, and economic environments. This study is perceived to be of interest to policy makers, regulators, government, and the international investment public at the level of UAE, the Gulf Cooperation Council, and Middle East as well. It also covers the implications for companies regarding women nominations for board membership.

2 Literature Review

The agency theory by [40] attempts to underlie the disclosure practices of companies in relation to the corporate governance processes, mechanisms, and relations applied. It takes into consideration the conflict of interest caused by the separation between the shareholders (the principal) and the management (agents) of corporations. Based on several studies, there are arguments stated that managers have some incentives to make decisions that reduce the chance of losing their positions and/ or to increase their rewards [9] and [46].

Thus, the existence of conflict between the two parties (managers and shareholders) is behind the raise of agency costs that are related to costs of monitoring and costs of some activities used by the agent to assure shareholders that their interest will not be harmed [13]. In addition, such conflicts result in information asymmetries when managers may withhold information from shareholders that might not be in their interest. [1] agree that information asymmetries effect company financing and investment decisions , consequently influencing the cost of capital as well as asset pricing. As suggested by [34], the role of the board of directors in an agency framework is to solve agency issues between managers and shareholders by fixing compensation and changing managers that do not create value for the shareholders [28].

Despite no perfect recipe for a good corporate governance system, the below items are considered as basic principles to build what can be considered as a good corporate governance: clearly defined responsibilities and expectations for board members, disclosure and transparency requirements, well-defined shareholder rights, and mechanisms, both internal and external, to ensure compliance and hold board members accountable [13, 15, 21].

In other words, creating and maintaining a good corporate governance system requires participation from regulators, policy makers, and firms. From this perspective, and as mentioned in the introduction section of this paper, board composition (diversity and independence) is regarded as a significant issue faced by firms because Board of Directors proved to be a good internal governance mechanism that reduces agency conflicts and costs.

Based on [29], gender is the most debated diversity dimension in many areas including board diversity, politics and the society. The dimension of gender impact on the board is produced by the institutional context in which they occur, that is corporate governance and culture [2, 4, 23]. [28] list the benefits of board diversity as promotion of a better understanding of the marketplace, increased creativity and innovation, and effective problem solving. Hence, enhancing the independence of boards, and the frequency and quality of disclosure. Furthermore, it is argued that women have higher ethical standards in taking decisions and are more risk-averse than men; leading women to demand better and more financial disclosure and reducing the likelihood of earnings and disclosure manipulations [1]. On the other hand, some research provides evidence of female directors being more likely to be considered as tokens on boards, with a significant number of female directors designated with the aim of matching the demographic characteristics of the employees or meeting social or legal expectations [1, 44].

A study conducted by [3] propose that there is a higher probability that women will join committees that have monitoring roles, such as CG and audit committees, that are directly involved in increasing transparency. Moreover, they tend to hold management accountable for performance more strictly than men [51]. Along similar lines, [1] cite [43] claiming that the participation of women on the board encourages more effective communication to investors and increases the dissemination as well as the quality of information [37]. On the other hand, [33] found an inverse relationship between gender board diversity and information disclosure.

In GCC especially in Bahrain and UAE the increase in women leadership positions is recognized and supported. One main reason for this increase is a better and wider access to education for women through traditional methods and E-education which has led to a generation of more qualified, skilled, and confident women [18, 48]. Additionally, HH Sheikha Fatima Bint Mubarak [52] stated women hold 66% of public positions, 30% of them being top decision-making positions. To build on this knowledge, this study aims to measure the percentage of women on the board of directors in the UAE.

With regards to board independence, more than forty years ago [16, 17, 30] stated that there is no need to justify the importance of independent and outside board members (directors not involved in the direct operations of the firm), as the benefit of such a logical composition is a given. The non-executive directors play an important role in monitoring the actions of the CEO and executive directors to ensure that the shareholders' interests are maintained and to add to the diversity of skills and expertise of the directors [41]. [34] suggest that boards that are composed of a higher proportion of outside directors have greater monitoring ability over management. Yet, according to [46] independent directors are only employed on a part-time basis, so they are more likely to have other work commitments, they may lack the expertise necessary for understanding highly technical business issues and may have insufficient information when required to make key decisions.

The improvement in broadband technology has a significant influence on increasing the accessibility to the internet. Using the internet became cheaper, much easier, and faster for the users to communicate locally and internationally. So, the innovative financial reporting became more effective compared to traditional channels of financial reporting. [5] indicated that using the electronic reporting overcomes the restrictions faced under the traditional reporting system.

Online financial disclosure involves the publication of annual reports and financial information on the websites of firms. Several studies have investigated the level of online financial disclosure [7, 32]. However, a step forward in firms' disclosure practices is social media financial disclosure which offers a two-way communication between managers and investors. Social networking channels have provided a new way for managers to establish, re-establish, and maintain relationships through online interactions with stakeholders. While adoption of this disclosure method has been slow, companies are beginning to take advantage of social media as a disclosure channel [26]. A study by [43] found a positive and significant relationship between corporate governance mechanisms and social media (Twitter) financial disclosure as a complementary corporate disclosure channel. Their study mainly found that board effectiveness played an important role in reducing the information gap between management and investors through information dissemination on social media. This study therefore hypothesizes that there is a relationship between board composition (independence and gender) and the level of social medial financial disclosure.

3 Methodology

To address the study questions, primarily the relationship between the composition of board of directors and the level of social media financial disclosure in the United Arab Emirates, the data of this study is collected from a sample of 103 firms listed in the UAE financial markets using the period 2018–2019. Consequently, the study applies a two-stage process to measure (1) the level of usage of Social Media (SM) in UAE firms as well as (2) the level of Social Media Financial Disclosure of those firms. Therefore, some firms might use social media, but not for reporting financial information.

The first stage involved measuring the social media usage by UAE listed firms. In order to measure the percentage of usage of SM, the study use the binary data method, i.e., if a firm uses any type of social media platform (Facebook, Twitter, Instagram, Snapchat, YouTube, LinkedIn, others) it received a score of 1. However, if the firm did not use any form of SM platform it received a score of 0. Similarly, the second stage involved measuring the level of SMFD for firms using SM in stage 1. So, if a firm used SM to report financial information it received a score of 1, and if a firm did not use its SM to report financial information it received a score of 0.

The regression model below was developed to test the hypothesis, considering the SMFD as a dependent variable, board composition comprising independence and gender as independent variables. Additionally, the study used the level of SM usage, size of board of directors, firm age, firm size, and financial leverage as control variables.

Study Model:

$$SMFD_i = \beta_0 + \beta_1 BD\_ind + \beta_2 BD\_wom + \beta_3 BD\_size_i + \beta_4 SML_i + \beta_5 LF\_size_i + \beta_6 LVG_i + \beta_7 AGE_i + \varepsilon_i$$

Code	Variable Name	Operationalization
Dependent variable		
SMFD	Social media financial disclosure	This is a binary Wherein 1 means that the company discloses financial data and 0 otherwise
Independent Variables – Board compensation:		
BD_ind	Independent directors%	The percentage of independent members to the total board size
BD_wom	Women directors %	The percentage of women members to the total board size
Control Variables:		
BD_size	Board size	Natural logarithm of Total Assets
SML	Social media level	This is a binary Wherein 1 means that the company is has social media plat forms and 0 otherwise
LF_size	Firm size	Natural logarithm of Total Assets
LVG	Leverage	Total liabilities/ Total Assets
AGE	Firm Age	The difference between the establishing date of the firm and the report date
εi	Error	

4 Data Analysis

4.1 Descriptive Statistics

Table 1 provides the descriptive analysis of the model variables: independent, dependent, and control. The mean of board independence level was of 32% indicating a somewhat moderate of independence, with highest level of 88%, while the minimum level was 9%. With regards to gender diversity of the board of directors, UAE firms reported a maximum level of only 20%, with a mean of 2%. This figure shows that women are still under presented on the board of directors. One reason for this could be that UAE women do not apply for such positions as of yet. Based on [32], UAE women can be found in "departmental leadership" positions mainly in hospitality and services industry. The results, moreover, show that the overall level of firms that use social media was 84%, which is considered as a moderate level of usage of SM by UAE firms considering that 85% of the country's population regularly uses the Internet. Additionally, the results show that of the 84% of firms that use SM, 67% use SM for financial disclosure.

Moreover, Table 1 reports the control variables descriptive statistics. It shows that the mean of firm size measured by total Assets, was 72,425.99, with a minimum of 412,532.27 and a maximum 3,412,461.54, which indicates fairly large firms. The normality distributions of the total Assets were skewed, so natural logarithm was used in the regression analysis to reduce skewness and bring the distribution of the variables nearer to normality. Moreover, the mean leverage of the firms was approximately 35% showing firms with somewhat medium debts. Finally, firm age ranges from 2 to 50 with a mean of 26.56.

Table 1 Descriptive Statistics for continues variables

Variables	N	Minimum	Maximum	Mean	Std. deviation
BD_ind	206	0.09	0.88	0.3176	0.12215
BD_wom	206	0.00	0.20	0.0208	0.05010
BD_size	206	5	12	7.59	1.284
Firm Size	206	412,532.27	3,412,461.54	72,425.99	315,257.13251
LVG	206	0.23	0.56	0.3514	0.31632
AGE	206	2	50	26.56	14.094

Descriptive Statistics for discontinuous variables

Variables	Achieved		Not achieved	
	Number	Percentage	Number	Percentage
SMFD	69	67%	34	33%
SML	87	84%	16	16%

Table 2 Collinearity Statistics Test

Model	Tolerance	VIF
BD_ind	0.827	1.209
BD_wom	0.908	1.101
BD_size	0.828	1.207
SML	0.904	1.106
LVG	0.855	1.170
AGE	0.936	1.068
LF_size	0.934	1.071

Table 3 Autocorrelation test

Model	R	R^2	A. R^2	Std. error	Durbin-Watson
1	0.638	0.407	0.386	0.371	2.511

4.2 Validity

A Variance Inflation Factor (VIF) test was used to check the data for multicollinearity. This test was used as according to James et al. (2017) VIF calculates the severity of multicollinearity and measures how much the variance of an estimated regression coefficient is affected by collinearity. The VIF should be lower than 10 and tolerance should not be below 0.2 (Field, 2005). The VIF scores for each variable, both independent and dependent, are reported in Table 2. The results indicate that since no VIF score exceeded 10 for any variable in the model, while no Tolerance score was below 0.2. So, it was concluded that there is no threat of multicollinearity.

Additionally, the Durbin-Watson test was carried out to identify the presence of autocorrelation at lag 1 in the prediction errors or residuals from the regression analysis, the results are reported in Table 3. The Durbin Watson (D-W) value of the model was (2.511). Consequently, it is concluded that there is a negative autocorrelation founded in the model because the (D-W) value was more than 2.

4.3 Regression Results

Table 4 reports the findings of the regression analysis. The findings indicate that the model was reflecting the relationship between the variables in a statistically appropriate way. According to the table, the model has an adjusted R^2 of 0.407 which shows that the model explains approximately 40% of the variation in the SMFD amongst the UAE listed firms. Additionally, the probability of the F-statistic with a significance 0.000 means that the independent variables are significant in interpreting the SMFD level.

Table 4 Regression analysis

Variables	Beta	T. test	Sig
BD_ind	0.307	2.169	0.024**
BD_wom	0.519	3.331	0.001*
BD_size	0.145	2.418	0.017**
SML	0.632	10.986	0.000*
LVG	−0.177	−1.996	0.077***
AGE	−0.047	−0.829	0.408
F_size	−0.147	−2.597	0.010**
R^2		0.407	
F		19.394	
Prob. (F)		0.000	

* pv < 1%; ** pv < 5%; ***pv < 10%

The main hypothesis of the study states that there is a relationship between board composition (gender and independence) and SMFD by firms listed in the UAE markets. The result indicates that there is a significant and positive relationship between board composition and social media financial disclosure, i.e., the higher the number of independent members on the board and the higher the number of females on the board the higher the disclosure. This could be due to the fact that female leaders prefer monitoring positions where they are stricter with the managers than men. Thus, demanding more transparent information frequently. This is in line with [43] study which stated that the participation of women on the board encourages more effective communication to investors and increases the dissemination as well as the quality of information. However, it contradicts [33] who found an inverse relationship between gender board diversity and information disclosure.

With regards to board independence, this result is consistent with [31] who reported a positive association between the level of the independence of the board and the level of disclosure. According to [41], the independent directors play a critical role in monitoring the actions of the CEO and executive directors to ensure that the shareholders' interests are maintained and to add to the diversity of skills and expertise of the directors. This could be generalized to the usage of social media as a modern disclosure tool.

By analyzing the control variables, the study found a significant and positive relationship between the level of SMFD the level of usage of social media and board size. The positive relationship with board size could be that a larger sized board improve the directing and monitoring by bringing together a variation in the views, experiences and resources. This leads to greater transparency and reduction in the agency costs [38] Secondly, SMFD has a positive relationship with the social media usage because obviously the firms that use social media will use it as a disclosure tool to attract potential investors and satisfy current shareholder.

On the other hand, there was a significant and negative relationship with leverage and firm size. One reason for the negative relationship between leverage and the level of disclosure, is that firms with higher leverage tend to be more conservative and try not to disclose to stakeholders that they are indebted. Finally, the results show no relationship between age and the level of SMFD.

5 Conclusions

The study set out to link two fundamental topics of corporate governance: board composition and modern disclosure practices such as social media disclosure. The paper aimed to address several research questions mainly the association between corporate boards of directors' composition: board diversity, and board independence, and the level of social media financial disclosure in the UAE.

This study stated an overall level of social media usage in the UAE at 84% which is relatively high-level percent within the users of internet in the UAE, which is 85% of population. With regards to independence of board, the overall level of independent members on the board is 32% which is a considerably moderate level. Finally, the percentage of women on the board is on average 2% showing that women are still under presented on the board of directors in the UAE. As a result of testing the relationship between the dependent and independent variables, the study found a positive and significant relationship between the composition of the board and the level of social media financial disclosure.

From a practical perspective, this study provides contributions to UAE's government, policymakers and regulators with regards to two important issues, particularly gender diversity on the board as well as the modern disclosure tools that could be used by firms to reduce the agency problem. Policy makers and regulators can also make use of information from this research in setting new policies on social media financial disclosure. Furthermore, they have to pay more attention to the percentage of independent members and females on the board as they have great power in monitoring and influencing the decisions made by the directors as shown by the results of the study. The results of this study can be utilized in the GCC countries as they have political, economic, and social environments similar to the UAE. Additionally, from academic perspective, this study presents a unique point of views which would add to the literature examining corporate governance mechanisms, board diversity, and use of social media.

This paper suggests having a study that further investigates the relationship between social media disclosure and performance: financial, operational and stock. Future studies could also test the relationship between gender diversity and the level corporate social responsibility disclosure, or the level of corporate social responsibility disclosure on social media.

References

1. Abad D, Lucas-Perez M, Minguez V, Yague J (2017) Does gender diversity on corporate boards reduce information asymmetry in equity markets? BRQ Bus Res Quar 20(3):192–205
2. Abdullah SN, Ismail K, Nachum L (2016) Does having women on boards create value? The impact of societal perceptions and corporate governance in emerging markets. Strateg Manag J 37(3):466–476
3. Adams RB, Ferreira D (2016) Women in the boardroom and their impact on governance and performance. J. Financ. Econ. 94(2):291–309
4. Aguilera RV, Jackson G (2010) Comparative and international corporate governance. Acad Manag Ann 4:485–556
5. Almilia L (2009) Determining factors of internet financial reporting in Indonesia. Account Taxation 1(1):87–99
6. Al-Sartawi A (2016) Measuring the level of online financial disclosure in the Gulf Cooperation Council Countries. Corp Owner Control 14(1):547–558
7. Al-Sartawi A (2017) The effect of the electronic financial reporting on the market value added of the islamic banks in gulf cooperation council countries. In: 8th Global Islamic Marketing Conference, 4–6 May. International Islamic Marketing Association, Turkey
8. Al-Sartawi A (2020) Social media disclosure of intellectual capital and firm value. Int J Learn Intellect Capital 17(4):312–323
9. Al-Sartawi A (2020) Does it pay to be socially responsible? Empirical evidence from the GCC countries. Int J Law Manag 62(5):381–394
10. Al-Sartawi A (2020) Information technology governance and cybersecurity at the board level. Int J Crit Infrastruct 16(2):150–161
11. Al-Sartawi A (2019) Assessing the relationship between information transparency through social media disclosure and firm value. Manag Account Rev 18(2):1–20
12. Al-Sartawi A, Sanad Z (2019) Institutional ownership and corporate governance: evidence from Bahrain. Afro-Asian J Finance Account 9(1):101–115
13. Al-Sartawi A (2018) Ownership structure and intellectual capital: evidence from the GCC countries. Int J Learn Intellect Capital 15(3):277–291
14. Al-Sartawi A (2018) Institutional ownership, social responsibility, corporate governance and online financial disclosure. Int J Crit Account 10(3/4):241–255
15. Al-Sartawi A (2018) Corporate governance and intellectual capital: evidence from gulf cooperation council countries. Acad Account Financ Stud J 22(1):1–12
16. Sanad Z, Al-Sartawi A (2016) Investigating the relationship between corporate governance and internet financial reporting (IFR): eEvidence from Bahrain Bourse. Jordan J Bus Admin 12(1):239–269
17. Al-Sartawi A (2015) The effect of corporate governance on the performance of the listed companies in the gulf cooperation council Countries. Jordan J Bus Admin 11(3):705–725
18. Musleh Al-Sartawi AMA (2020) E-Learning Improves Accounting Education: Case of the Higher Education Sector of Bahrain. In: Themistocleous M, Papadaki M, Kamal MM (eds) Information Systems. EMCIS 2020. Lecture Notes in Business Information Processing, vol 402. Springer, Cham
19. Gupta M, Sikarwar TS (2020) Modelling credit risk management and bank's profitability. Int J Electron Bank 2(2):170–183
20. Gupta N (2019) Influence of demographic variables on synchronisation between customer satisfaction and retail banking channels for customers of public sector banks of India. Int J Electron Bank 1(3):206–219
21. Abdulrasool FE, Turnbull SI (2020) Exploring security, risk, and compliance driven IT governance model for universities: applied research based on the COBIT framework. Int J Electron Bank 2(3):237–265
22. Alhakimi W, Esmail J (2019) The factors influencing the adoption of internet banking in Yemen. Int J Electron Bank 2(2):97–117

23. Memdani L (2020) Demonetisation: a move towards cashless economy in India. Int J Electron Bank 2(3):205–211
24. Arafat W, Bing Z, Al-Mutawakel O (2018) Infrastructure Developing and Economic Growth in United Arab Emirates. Business and Economic Research. 8(1):95–114
25. Baker GP, Jensen MC, Murphy KJ (1988) Compensation and incentives: practice vs. theory. J Finance 43:593–615 (1988)
26. Trinkle BS, Crossler RE, France B (2015) Voluntary disclosures via social media and the role of comments. J Inf Syst Fall 29(3):101–121
27. Cade NL (2018) Corporate Social Media: How Two-Way Disclosure Channels Influence Investors Accounting, Organizations and Society, Forthcoming. SSRN: https://ssrn.com/abstract=2619249. https://doi.org/10.2139/ssrn.2619249
28. Carter DA, Simkins BJ, Simpson GW (2003) Corporate Governance, Board Diversity, and Firm Value, the Financial Review 38:33–53
29. Central Intelligence Agency' the World Factbook: Available at www.cia.gov Accessesed on 25 Mar 2018
30. Chandler M (1975) It's Time to Clean up the Boardroom. Harvard Business Review, September–October, pp 73–82
31. Chau G, Gray SJ (2010) Family ownership, board independence and voluntary disclosure: evidence from Hong Kong. J Int Account Audit Taxation 19(2):93–109
32. Ekramy SM (2017) Internet financial reporting determinants: a meta-analytic review . J Financ Report Account 15(1):116–154
33. Gul FA, Hutchinson M, Lai K (2013) Gender-diverse boards and properties of analyst earnings forecasts Account. Horizon 27(3):511–538
34. Fama EF, Jensen MC (1983) Separation of ownership and control. J Law Econ 24:301–325
35. Field A (2005) Discovering Statistics Using SPSS: (and Sex, Drugs and Rock "n" Roll). Sage, London
36. Grais W, Pellegrini, M.: Corporate Governance in Institutions Offering Islamic Financial Services: Issues and Options: World Bank Policy Research Working Paper No. 4052. Available at SSRN: https://ssrn.com/abstract=940709
37. Healy PM, Palepu KG (2001) Information asymmetry, corporate disclosure, and the capital markets: a review of the empirical disclosure literature. J Account Econ 31(1):405–440
38. Hidalgo RL, García-Meca E, Martínez I (2011) Corporate governance and intellectual capital disclosure. J Bus Ethics 100(3):483–495
39. James G, Witten D, Hastie T, Tibshirani R (2017) An Introduction to Statistical Learning, 8th ed. Springer, New York. ISBN 978-1-4614-7138-7
40. Jensen M, Meckling W (1976) The theory of the firm: managerial behaviour, agency costs, and ownership structure . J Financ Econ 3(4):305–360
41. Johl SK, Kaur S, Cooper B (2015) Board characteristics and firm performance: evidence from Malaysian public listed firms. J Econ Bus Manag 3(2):239–243
42. Joinson AN, Paine CB (2007) Self-disclosure, privacy and the internet. In: Joinson AN, McKenna KYA, Postmes T, Reips U (eds.) The Oxford handbook of Internet psychology, pp 237–252. Oxford University Press. Great Britain (2007)
43. Joy L (2008) Advancing women leaders: the connection between women board directors and women corporate officers. Catalyst, New York
44. Farrell KA, Hersch PL (2005) Additions to corporate boards: the effect of gender. J Corp Finance 11(1–2):85–106
45. Kang H, Cheng M, Gray S (2007) Corporate governance and board composition: diversity and independence of Australian boards. Corp Govern 15:195–207
46. Hussainey K, Aljifri K (2012) Corporate governance mechanisms and capital structure in UAE. J Appl Account Res 13(2), 145–160
47. Weir L, Laing D (2001) Governance structures, director independence and corporate performance in the U.K. Eur Bus Rev 13(2):86–94
48. Kemp LJ, Madsen SR, El-Saidi M (2013) The current state of female leadership in the United Arab Emirate. J Glob Respons 4(1):99–112

49. Telecommunication Regulatory Authority: UAE Telecommunications Sector Developments and Indicators, 5th Annual Sector Review (2014) https://www.tra.gov.ae/assets/q0Z5B3sI.pptx.aspx
50. TIAA-CREF (1997) TIAA-CREF Policy Statement on Corporate Governance (New York)
51. Triana M, Miller T, Trzebiatowski T (2014) The double-edged nature of board gender diversity: diversity, firm performance, and the power of women directors as predictors of strategic change. Organ Sci 25(2):609–632
52. UAE Interact: Women empowerment a concrete reality: Sheikha Fatima (2010) https://www.uaeinteract.com/docs/Women_empowerment_a_concrete_reality_Sheikha_Fatima_/43154.htm. Accessed 1 May 2018
53. Yang J, Liu S, Zhou D (2016) Voluntary Financial Disclosure on Social Media: Does Corporate Governance Matter? Available at SSRN https://ssrn.com/abstract=2836570

Exploring Organizational Strategies for Development of Digital Skills: A Case Study

Gianluca Prezioso and Emanuele Gabriel Margherita

Abstract Our contemporary society is increasingly based on globalization, technological progress and fast knowledge accumulation. Technological change pushed companies to change their structure, indeed, the fast integration of information and communication technologies (ICTs) is leading to a continuously evolving digital skills (DS) set necessary for employment and participation in society. This happens because employees' skills drive organizations' innovation capacity and competitiveness. In this study, we decided to analyse the manager's DS perception of a multinational firm present in 18 countries and operating in a manufacturing sector. Our results put in evidence that managers do not seem to have a clear and well-defined managers' perception about DS required in industries.

Keywords Digital skills · Manager's perception · Strategies · Digital transformation

1 Introduction

Our contemporary society is increasingly based on globalization, technological progress and fast knowledge accumulation [1]. Information, knowledge, and creativity are the new guideline for the modern economy [2, 3]. If on one hand the industrial economy has transformed into an economy based on information-oriented services [2], on the other hand these changes have reshaped the workplace and the work itself [4, 5]. Indeed, the fast integration of information and communication technologies (ICTs) is leading to a continuously evolving digital skills (DS) set necessary for employment and participation in society [2]. This happens because employees' skills drive organizations' innovation capacity and competitiveness [6].

G. Prezioso
University "G. d'Annunzio" of Chieti-Pescara, Viale Pindaro 42, Pescara, Italy

E. G. Margherita (✉)
Department of Economics Engineering Society and Organization – DEIM, University of Tuscia, Via del Paradiso, 47, 01100 Viterbo, Italy
e-mail: emargherita@unitus.it

A. M. A. Musleh Al-Sartawi (ed.), *The Big Data-Driven Digital Economy: Artificial and Computational Intelligence*, Studies in Computational Intelligence 974,
https://doi.org/10.1007/978-3-030-73057-4_19

Firms today are looking for highly skilled workers who have excellent technical preparation and at the same time have skills to face change [7, 8]. The Organization for Economic Co-operation and Development (OECD) (2017) and the European Commission too has emphasized that it is of fundamental importance to develop adequate skills to get the most from the digital economy and individual markets [9]. Today's high rate of change and the high influence of technology push companies to develop employees' DS in order to cope with this change [10, 11]. Moreover, several reports highlight how the acquisition of adequate digital skills is transformed into job opportunities [12], how prepare workers, carrying out certain activities, to learn or improve digital skills [13, 14] in order to prepare employees for the digital transformation and to "successfully navigate through an ever-changing, technology-rich work environment" [15]. For these reasons and to address the digital transformation correctly, it becomes important to recognizing a suitable set of DS [16].

In this study, we decided to adopt Van Laar et al. [10] twenty-first century digital skills classification in order to analyse the manager's DS perception of a multinational firm present in 18 countries and operating in a manufacturing sector. The aim of this research is to understand if manager's DS perception, related to the digital transformation, is in line or not with DS elaborated from the theory, what are the skills that managers perceive as needing to be implemented or improved. Moreover, we want to understand if and how manager's perception differs across different departments.

Van Laar et al. [10] conducted a literature review where they synthesized the most important academic literature related to 21st-century digital skills (DS). They elaborated a clear classification of the concept of skills needed in a digital environment. Specifically, they have identified which digital aspects should be integrated with the concepts of 21st-century skills [10]. These DS are today the reference point for innovation and competitiveness of workers and organizations [10] because are essential and useful to fulfil a wide range of occupational tasks [17]. A crucial aspect, for operational and managerial level, that organizations may fail and that shouldn't be underestimate is the process of defining the skills to be developed [18–20] or used for the personnel selection process [21]. Indeed, past research demonstrates that: managers neither have skill requirements top of mind [17, 22] and managers don't have a clear understanding of the role skill development plays in organizational management practices [22, 23]. Therefore, our study answers the following research question: *"What are the strategies, according to managers perception, to develop and enhance the digital skills within the organization?"*.

The paper is structured as follows: we report a literature review of previous evidence on the topic in Sect. 2. The method used to collect data is explained in Sect. 3. Data analysis are identified and explained in Sect. 4. Data are discussed in Sect. 5 In the last section we have discussed conclusion implications of our findings, limitations of the work and suggestions for future research analysis.

2 Literature Review

The globalization of markets led to more intense competition among firms, but also economic interdependence and broad collaboration through thousands of jobs have been lost due to relocation or automation [6, 24]. At the same time many typologies of jobs (e.g., engineering, management, education) where social interaction or creative skills are present, are less likely to suffer from job losses. This means that ICTs change the way of working day by day increasing efficiency of firms [25] and at the same time reshapes the required skills [26]. Indeed, the ever-increasing flexibility required in production and supply services has redesigned the workplace too [27]. More flexible working arrangements, broad information sharing, flatter decision-making structure, more distributed and decentralized decision making are just a few examples [27].

The added value that differentiates companies lies in the skills and abilities acquired over time. For this reason, firms are acquiring highly skilled workers who can adapt to the changing needs of work and deal with increasingly interactive and complex tasks [8]. Also, companies support and invest in those workers who are willing to update their knowledge and skills because the success of sustainable and high-performance organizations derives largely from these employees, representing the human capital [28]. Furthermore, the influence of technological change on demand for skills and employment should not be underestimated because the adoption of new technologies improve productivity and can move (displace) workers (from one place to another) [29].

Many employers are concerned about the technical gap between what students know and what they need [30] also they expect new graduates are able to understand how navigate in an digital and integrated environment in their new careers [31]. A comprehensive study of recent technological advances in demand and skill supply is provided by Brynjolfsson and McAfee [32]; Dachs and Peters studies [33] also highlight the positive correlation between job growth and workers' skills change. Moreover, the use of new technologies involves more skills and training in employees because these skills and training facilitate them in the implementation of new technologies [14, 34]. Mohnen and Roller studies confirm through statistical evidence that employee skills are an important innovation factor for companies due to the high rate of change [35]. Specifically, they observed and analysed the most important innovation obstacle in different industries and countries: the lack of skills.

Different researches, embedded Lewin and McNicol's [36] studies, contributed to the evolution of twenty-first century skills through ICT. Specifically, they talk about the importance of ICT in relation to the development of twenty-first century skills and how they lead to a successful career [36, 37]. Voogt and Roblin [38], in their studies, identified and explored some of these skills that go beyond the mere knowledge of specific software [7, 39] and concern: creativity, problem solving, critical thinking, productivity, digital literacy, collaboration, citizenship, and communication. Another great classification drafted by Claro et al. [39] puts in evidence the most important DS of the twenty-first century; in particular they elaborated four macro-classes of skills:

ability to solve cognitive tasks through the use of ICT, skills not related to technology (software), skills related to higher-order thought processes, cognitive abilities that promote continuous employee learning. Moreover, a wide range of classifications is present in literature. In order to comprehend and clarify different notions, Van Laar et al. [10] differentiate between technological skills, twenty-first century skills and twenty-first century digital skills. Talking about the technological skills, Hatlevik et al. [40] defined them as the skills that we need in order to use computers or Internet and to acquire DS of twenty-first century. Ferrari [9] analysed and elaborated a clear classification of what digital competencies are: evaluation and resolution of problems and technical operations, sharing, communication, creation of content and knowledge, collaboration, ethics and responsibility information management.

Going forward with the analysis of past studies, we find two of the most important "twenty-first century skills classification. The first, realized from the Partnership for the twenty-first century (P21) [27] differentiate three macro classes of skills, which are determined by sub-sets of skills: (i) learning skills (communication, creativity and innovation, critical thinking and problem solving), (ii) literacy skills (literacy, media and ICT information) and (iii) life skills (adaptability and flexibility, productivity and responsibility, leadership and responsibility initiative and self-direction, social and intercultural skills). The second, elaborated from the assessment and teaching of twenty-first century skills (ATC21S) and through the help of experts [41] identifies four macro-classes of skills, which are determined by sub-sets of skills: 1) ways of thinking (creativity and innovation, critical thinking, problem-solving and decision making, learning to learn and metacognition), 2) ways of working (communication, collaboration and group work), 3) tools for working (computer literacy, information technologies and communication literacy) and 4) living in the world (life and career, personal and social responsibility).

The most actual classification about the DS was elaborated by Van Laar et al. [10]. They, through a systematic literature review, synthesized the most important academic literature related to 21st-century digital skills (DS) elaborating the most actual classification of skills required in a digital environment. They specifically identified the digital aspects to integrate with the concept of twenty-first century skills or the human skills in the digital context that are necessary for individual employability [10]. Finally, a current studie that adopts the classification of Van Laar et al. [10] was developed by Prezioso et al. [42], which compare the Van Laar et al. classification [10] with the DS derived from interviews and job descriptions, concluding that DS are not in line with the theory.

2.1 Manager's Perception

The modern economy is increasingly based on knowledge and highly skilled human capital [43, 44] because they drive workforce organizations' competitiveness and innovation capacity [43, 45]. Indeed, the increasing complexity of modern society and the fast rate of change demands a high and flexible knowledgeable human capital base [46]. DS are strategically relevant for all employees [47], useful for solving complex business problems [48] and improve operational performance [49]. Yet, one of the problems, highlighted in several studies, concern the manager's perception of resources and competencies needed to obtain a flowless digital transformation [18–20].

Even though most managers recognize the importance of digital competencies, they have difficulties in the identification and assessment of those competencies when consensus from other manager's is not present [50]. This is because the management of the competences, itself requires specific skills, knowledge of the subjects and the appropriate techniques to increase them [51]. Indeed, as Mezias and Starback [52] affirmed: "manager's perception concept incorporates the whole that goes into managers' understanding of their work situations". Governance, in particular, must have all the necessary skills, so that everyone works in the same direction [51]. The manager's communication skills, for example, are related to the employee's perception of the workplace and understanding of the work itself [53–58].

Several past studies put in evidence the importance covered from manager's perception. Some analyse how information system (IS) manager's perception of key IS issues derive from the signals he or she receives inside or outside the organization [59]. In particular, the model elaborated from Watson [59] identifies three classes of contextual variables and characteristics that influence perception and behaviour of the IS manager: (i) environmental (characteristics of customers, markets, and competitors), (ii) organizational (the relationship between the IS manager and the CEO, the firm's culture, the perceived strategic value) and (iii) IS departmental (departmental resources, local problems, and the quality of personnel). In addition, he discovered individual variables (related to the manager) that could influence both perception of key issues and scanning [59]. In addition, Lynn Crawford's studies explored senior managers perception about the relationship between performance against standards and the effectiveness of project management performance in the workplace [60]. In this case results suggested that competence and perceptions are influenced by factors including the nature of the context and the types of project, the personality and behavioural characteristics of both the project supervisors and personnel [60]. Another interesting study discusses about manager's perception key factor underlining how middle manager perceptions about the firm internal environment is essential to every entrepreneurial process [18].

Few researchers studied the accuracy of managerial perceptions indicating their inaccuracy, others identified errors and corrective action [52]. Specifically, Mezias et al., [52] to solve these problems suggested to train and educate managers to correct misperceptions and build robust organizations. Moreover, they identified many variables that affect manager's perception and corrective action in order to tolerate misperception [52]. One more interesting descriptive and exploratory study, from Malaysian perspective, analyse managers' perception and managers' information requirements on information management and on knowledge managers [19]. This study put in evidence that managers' perception of the skills and qualifications of information managers may depart beyond the role of an information mediator increasing organizational performance and improving decision-making [19]. Moreover, Cheryl et al. studies analyse the different perception, between IT managers in industry and faculty in academia, about the importance of various skills for entry-level IT workers [61]. Especially considering individual skills they discovered that IT managers perception gave more importance than teachers to leadership skills, risk characteristics, entrepreneurial, hardware concepts and operating systems [61].

A recent and very interesting study was elaborated from Molla et al. in 2015 [62]. In order to understand the IT managers' perceptions, they used situational awareness theory while to identify IT managers' responses to digital disruption, they used disruptive innovation theory [62]. Situational awareness concern "the perception of the elements in the environment within a volume of time and space, the comprehension of their meaning, and the projection of their status in the near future" [63]. They put in evidence how perception, comprehension and projection, the three key concepts of situational awareness, are vital to response to the digital transformation in a proper way and useful for supporting the decision making process [62]. Another actual research, that adopted situational awareness theory, was elaborated by Prezioso et al. [23]. This study put in evidence how managers aren't often not properly aware of the DS required for a digital transformation process. In addition, they have difficulties to identify the DS present in the organization and that need to acquire to support digital transformation [23]. Specifically, managers should be prepared to take decision about the development of training to develop DS internally and acquiring DS externally [23].

Another actual research, that adopted situational awareness theory, was elaborated by Prezioso et al. [22]. This study put in evidence how managers aren't often not properly aware of the DS required for a digital transformation process. In addition, they have difficulties to identify the DS present in the organization and that need to acquire to support digital transformation [22]. Specifically, managers should be prepared to take decision about the development of training to develop DS internally and acquiring DS externally [22]. Another research, elaborated in 2019 from Van Laar et al. [26], put in evidence how perceived ease of use of ICT is considered one of determinants to the development of 21st-century digital skills among knowledge workers. Perceived easy involves "the perception of the difficulty of learning to use a particular technology" [26]. They concluded that perceived ease of use not only predicts basic technical skills but also content-related digital skills [26]. Indeed, other studies indicated that perceived ease-use is present in higher levels of ICT skills workers [64, 65] and cover an important role in the willingness to develop new skills [66].

Indeed, further research demonstrates that managers neither have skill requirements top of mind, nor have a clear understanding of the role skill development plays in organizational management practices [17]. This leads to digital skill insufficiencies that affect negatively task performance [23]. For these reasons it is of primary importance to properly plan and guide the digital transformation process through the evaluation and recognition of existing digital skills and the identification of missing skills [67]. But to do this properly, managers should be in possess of the right competences and have a clear idea about which resources and competences they need in their team [51].

3 Method

The objective of our research questions is to understand what the organizational strategies are, according to managers perception, to develop and enhance digital skills.

First, the review of the literature allowed us to identify, evaluate and synthesize the most relevant academic literature concerning DS and manager's perception in order to understand the phenomenon and the literature gaps. In addition, we used the interview method to test the relevance of manager's perception on DS framework in relation to the manufacturing labour market. We linked theory and practice in order to understand if manager's DS perception is in line or not with the theory. In order to investigate these issues we decided to adopt a case-study methodology because this method is specifically appropriate for investigating how a phenomenon emerges and generate contextualized explanations [68]. Moreover, when the boundaries between phenomenon and context are blurred, as in this case, this methodology is the preferred one [68] because allow the understanding and identification of different phenomenon dimensions [69].

Thanks to this qualitative research method managers and senior executives, working in a multinational company operating in a manufacturing sector, can express their individual perceptions on DS providing a-depth understanding on the organizational strategies to develop and enhance digital skills. In the majority of existing skill measurements, people are asked to evaluate how well they perform some skills presenting them a list of those skills [70]. This is the reason why these types of measurements typically collect data based on people's own estimations or perceptions of their digital skills [71].

3.1 Sample Selection

In this paper, we analyse a multinational firm operating in a manufacturing sector and present in 18 countries: Argentina, Brazil, China, Colombia, France, Germany, India, Indonesia, Ireland, Italy, Mexico, United Kingdom, Czech Republic, Russia, Spain, United States, Switzerland and Thailand. The firm is divided into 12 different departments each of which has different roles, objective and tasks. We called this firm "Digital" because we are not allowed to disclose the name. Digital's first investments in robotics, automation, training as well as educational courses for the employees date back to more than 35 ys ago. Thanks to these constant investments over the years, it has achieved extraordinary production results, developed a fully automated warehouse where people and robots work together and become the market leader. What 30 ys ago was produced in a year is now produced in a single day. Only last year, Digital invested $75.3 million in 2018, $68.2 million in 2017 and $66.2 million in 2016 in R&D; precisely the 7% (Company's Annual Report, 2019) of it profits. The aspects that convinced us to choose this company for our study are precisely the continuous investment in R&D over the years and the increasing number of specialized personnel engaged by this firm.

3.2 Data Collection

Data collection strategy is divided into three different steps; we collected multiple sources of data to establish construct validity [68]. First, in order to obtain background information about the firm, sector to which it belongs and process of designing and deploying skills, we collected information from journal articles, firm internal document, firm annual reports, databases, firms' official websites. Second, we gathered all firms' job descriptions (internal document) which are analytical written descriptions present for each organizational position containing the main characteristics of the position itself. These characteristics include name of the position, position in the organization chart, roles, main purposes, tasks assigned, skills required and relationships with other organizational positions. The job description is helpful for different reasons. It's, a practical tool useful for the recruiting process and for the job interview to assess the matching between the ideal role that line managers have in mind and the candidate's profile, it's the base of any form of job posting, offers a clear image of the person who covers that position and present a clear image to candidates. In addition, the job descriptions were particularly useful because they allowed us to efficiently and effectively identify the most suitable sample for our case study. Thirdly, as our principal source of data we conducted open-ended interviews with managers and senior executives of all departments of a multinational firm to ensure that we obtained multiple perspectives. Our goal is to develop a deeper understanding of the importance of the various digital skills of the twenty-first century according to managers' perceptions.

3.3 Semi-Structured Interviews Procedure

Interviews have been collected in 5 different production sites, 2 in Germany, 2 in Italy, and 1 in Switzerland. First, the company managers were informed of our arrival and the need to conduct interviews by the human resources department, with whom we had been working for a year. After that, the human resources department indicated the most experienced managers to contact for each department, to whom an email presentation was sent with a request for an interview. Once a positive response was received, a meeting date was set. We carried out 57 interviews, 53 interviews were face-to-face while 4 through Skype call. Interviews were conducted 37 in Italy, 18 in Germany and 2 in Switzerland between October 2018 and September 2019. The length of the interviews was between 13 and 75 min and the questionnaire was related to specific topics for the employee's DS. We adopted an inductive approach to determine what the participants brought to the study. Specifically, the interviews followed a semi-structured protocol that is structured in three part. In the first part we ask generical question about previous work and educational background, in the second one asks specific questions about firm's digitalization strategies and employee DS for the digitalization process and in the last part we ask to identify the most important DS, those missing and those to be developed necessary for the digitization process. Specifically, in this research we focus on the last part of the questionnaire, examining what are the most important organizational strategies perceived from interviewee. We used as a guideline the DS framework elaborated from Van Laar et al. [10] that was presented by presenting a card with each skill written on it with a short description. Specifically, we asked to the interviewees to identify the organizational strategies that they perceive as important, to develop and enhance the digital skills. Before each interview, it was clearly communicated to the interviewees that the data and results of the interviews would be treated confidentially and the informed consent of the participants to be interviewed via audio recorder was also requested. When we didn't allow to record the interview, we fully transcribed every detail during the interview, in order to preserve the veracity of the interview reports and conversations. Moreover, the transcriptions were submitted to interviewees for verification. The questionnaire is available upon request. Sample selection was carried out in order to ensure theoretical replicability [68] also the responses received during the interviews reflect the particular point of view of the respondents.

4 Data Analysis Procedure

In this study we analysed the manager's perception of organizational strategies to develop and enhance DS. We used a deductive to determine how participants perceive the theoretical construct approach, which we based on the literature. The theoretical constructs are the most used strategies, namely, training, talent and map of competencies, skills gap analysis, recruiting, long term strategy, job description analysis, coaching, job rotation, pilot project, and employer branding.

Moreover, during the interviews with managers was presented a card list, with a short description written of each digital skill, asking them if they recognized the skills as relevant and what the right the strategies are to develop and enhance these skills. On the base of their responses, we elaborated Table 1, dividing the data according to the pieces of evidence collected thought the interviews. We used Table 1 as a starting point of our data analysis and discussion. Table 1 is divided in 2 columns, on the left part we find the strategies that interviewees declared as important, to develop and enhance DS, during the interviews. While on the right part of the table, it is possible to observe the frequency of respondents itself divided into 2 columns: number of employees interviewed and the percentage of response for each strategy.

5 Findings and Discussion

Table 1 illustrates the results of our case study. Managers consider the development of digital skills in organization as long-term strategy (n = 7). Managers consider the training (n = 25) as the most useful strategies to enrich digital skills within the organizations.

Training are often organized by the management and technology developer during the adoption process of technology. Training aims at providing workers the proper digital skills to work with novel technologies [72, 73], and training can be proposed as refresher courses when digital technologies are retrofitted and workers need new digital skill to manage them. Coaching are also a popular strategy to develop digital skills to workers. Indeed, coaching are often used together with training. Coaching (n = 11) involves training-on-the-job or mentorship, where skilled workers help workers to learn digital skills to work novel technologies. The coaching can be proposed again when workers perform below standard for certain activities in order to increase their competencies to conduct the activities and work with the technologies [74]. Job rotation and pilot projects are few used to build digital skills in organization, and this indicates that digital skills are mandatory for each activity within the organizations rather than to be useful to certain tasks.

Moreover, talent and competence maps, skills gap analysis and job description analysis are not prominent strategies to build digital skills for the manager's perception. The first two strategies are as a purpose to check skills and competencies of the organization in order to find possible gaps of competencies of workers. Instead, the

Table 1 Results of the multiple case study

Strategy	Frequency	
	n	%
Training	25	44,6
Talent and map of competencies	2	3,6
Skills gap analysis	2	3,6
Recruiting	6	10,7
Long term strategy	7	12,5
Job Description analysis	1	1,8
Coaching	11	19,6
Job rotation	1	1,8
Pilot project	1	1,8
Employer branding	0	0
Total	56	

latter aims at detecting the needed digital skills through the analysis of the activities and tasks related to each job. This evidence confirms the extant literature which shows the difficulties of managers to plan the development of digital skills [50]. Indeed, the results show that strategy to acquire digital skills happen during the adoption of novel technologies by training and coaching rather than planning their acquisition a priori.

Furthermore, the recruiting strategy is useful for organization to acquire novel digital skills hiring novel workforce. The findings show that this strategy is less preferred compared to the training and coaching. This means that the organization, which we analysed, has a propensity to build digital skills to workers which have hired and possessing technical skills. Thus, the acquisition of adequate digital skills is transformed into job opportunities [12] because they have both technical skills and skills to manage novel technologies which make the workers ready to change [33, 35]. In line with this, managers do not consider the employer branding, which is, long-term strategy to manage the awareness and perceptions of employees, potential employees, and related stakeholders of a particular firm [75]. In modern economy employer branding is the promotion of the brand within and outside the organization and provides a competitive advantage to externally acquire, nurture and retain talent internally [76]. Promoting this strategy will lead internally to increased employee loyalty, while externally it will make the organization attractive to potential employees with specific skills [76]. For this strategy further research is needed to explore how this strategy are an effective way to attract DS talents and what actions an organization can develop to facilitate it.

6 Conclusion, Limitation and Future Studies

In this paper, we analyse manager's DS perception of a multinational firm facing digital transformation and operating in the manufacturing sector. The aim of the paper is to identify, according to manager's perceptions, the most relevant organizational strategies to develop and enhance DS within organizations. Integrating results derived from data and Table 1 is possible to observe how manager's perception organizational strategies to develop and enhance DS is focus on training, recruiting, long term strategy and coaching. Specifically, 44,6% of the interviewee, declared that training is the most important strategy to adopt in order to develop DS. Moreover, analysing the responses received during the interviews is evident that is present a lack of the right organizational strategies perception to develop DS. Expressly, no one declared the employer branding strategy as important. This strategy was appointed during the interviews but was not identified as important to develop DS. In addition, only a small part of the interviewee named talent and map of competencies, skills gap analysis, job Description analysis, job rotation and pilot project as important.

Most studies, in the industries context, analyse user behaviour, new business models or new technological developments neglecting the fundamental soft factor of the employees' skill level, necessary for the development of innovation capacity as well as for the adoption of technological and organizational changes [77]. Our results confirm past studies where managers do not seem to have a clear and well-defined managers' perception about which skill education and training in industries require [17] and which competences, resources and strategies they need in their team for undertaking digital changes [23]. This is worrying because it is the responsibility of managers to guide their employees appropriately, identifying specific knowledge requirements, capabilities and personal attributes for each role so that each employee effectively contributes to the job. Our recommendation to managers and organizations is to spend more time and attention to understand and adopt the right strategies, observing and analysing skill insufficiencies, developing digital skill profiles for each job function in order to succeed and stay competitive and develop day by day DS. Specifically, managers need to realize that DS must be continuously improved as a strategic factor [78] and it is of fundamental and strategic importance for organizations to priorities DS development strategies [17]. Organizations must recognize that the management of the competences, itself requires specific skills, knowledge [51]. For these reasons, it's of fundamental importance put in action corrective action through education and training. In addition, it's essential inform managers about every organizational and environmental properties and novelty in order to detect misperception or problems and build a strong organization [52].

6.1 Implications for Practitioners

This study is relevant for practitioners because findings indicate that managers and organizations can use the presented framework for evaluate, detect and verify digital strategies in order to obtain a flowless digital transformation. Human resources and managers of each department can benefit from a more explicit description or from the creation of a digital profile for each job function in order to fill the gaps of the skills that are needed by their employees, improve personnel selection, realize an appropriate workforce planning and reflect the real need of the firm. Moreover, considering that managers have difficulties in finding the right people with the right skills [79], the identification of strategies to develop the right DS for each department allow HR managers and recruiters to acquire the most talented workers, to retain the personnel already acquired and create the right workforce team. In addition, this framework gives more information and clarity to organizations and their employees about the content of ICT-related job aspects. This is important, particularly for senior workers, because if people don't understand why change is important, they will be reluctant to change [80, 81] and adapt their skills to the digital environment.

6.2 *Implications for Researchers*

The study has certain implications for researchers. Further researchers should address the lack of evidence related to manager's perception about the employer branding strategy. We encourage researchers to conduct qualitative studies – especially multiple case studies – in order to contribute to the knowledge accumulation of this strategy. Another fascinating research avenue is to investigate how organizations combine these digital skill strategies in order to develop a long-term strategy, and therefore, to investigate the value creation of these strategies. Our study is a qualitative, embedded single-case study of a multinational located in Europe. Thus, the results are mainly generalizable for European organizations. We encouraged researchers to conduct a similar study in different contexts, such as Asian, American companies in Germany and in organizations with different sizes In addition, future research can explore these limitations in order to understand if in a different business sector or in another manufacturing industry the manager's DS perception is in line with our study or not. Could be useful to understand if any correlations exist between manager's strategies perception and cultural aspect of the organization. Also, future research could investigate manager's strategies perception in small and medium enterprises respect big firms, the nationality, the culture and the age. This research is subject to some limitation, due to its exploratory nature. First, the small sample size, we collected data from a single multinational firm operating in a manufacturing sector and it's the reason why it's difficult to generalize our findings to all manufacturing industries or to different industries o countries. Second, this study gives a static view of the phenomenon because it doesn't take in consideration the possible cyclical variation of strategies and due it's cross-sectional rather than a longitudinal nature of managers in order to understand if and how these variables affect manager's DS perception. Also, further future studies could derive from the analyses of similarities and differences of different subsectors [82].

References

1. van de Oudeweetering K, Voogt J (2018) Teachers' conceptualization and enactment of twenty-first century competences: exploring dimensions for new curricula. Curric J. https://doi.org/10.1080/09585176.2017.1369136
2. van Laar E, van Deursen AJAM, van Dijk JAGM, de Haan J (2018) 21st-century digital skills instrument aimed at working professionals: conceptual development and empirical validation. Telemat Inform
3. Margherita EG, Esposito G, Escobar SD, Crutzen N (2020) Exploring the smart city adoption process: evidence from the belgian urban context. In Proceedings of the 6th International Workshop on Socio-Technical Perspective in IS Development (STPIS 2020), June 8–9, 2020. CEUR Workshop Proceedings (CEUR-WS.org)
4. Soulé H, Warrick T (2015) Defining 21st century readiness for all students: what we know and how to get there. Psychol Aesthetics Creat Arts ì
5. Gomber P, Koch JA, Siering M (2017) Digital finance and fintech: current research and future research directions. J Bus Econ

6. Anderson RE (2008) Implications of the information and knowledge society for education. In international handbook of information technology in primary and secondary education
7. Ahmad M, Karim AA, Din R, Albakri ISMA (2013) Assessing ICT competencies among postgraduate students based on the 21st Century ICT Competency Model. Asian Soc Sci
8. Carnevale AP, Smith N (2013) Workplace basics: the skills employees need and employers want. Hum Resour Dev Int
9. Ferrari A (2012) 21st Century Learning for 21st Century Skills. Springer, Berlin Heidelberg, Berlin, Heidelberg
10. van Laar E, van Deursen AJAM, van Dijk JAGM, de Haan J (2017) The relation between 21st-century skills and digital skills: a systematic literature review. Comput Human Behav 72:577–588
11. Margherita EG, Braccini AM (2020) Exploring the socio-technical interplay of Industry 4.0: a single case study of an Italian manufacturing organisation. In proceedings of the 6th international workshop on socio-technical perspective in IS development (STPIS 2020), June 8–9, 2020. CEUR Workshop Proceedings (CEUR-WS.org)
12. Curtarelli M, Gualtieri V, Jannati MS, Donlevy V (2016) ICT for work: digital skills in the workplace. Digital Single Market
13. Osman K, Hasnan AHS, Arba'at H (2009) Standard setting: inserting domain of the 21st century thinking skills into the existing science curriculum in Malaysia. Procedia Soc Behav Sci
14. Stone M (2014) Building digital skills through training
15. Grundke R, Marcolin L, Nguyen TLB, Squicciarini M (2018) Which skills for the digital era? OECD Sci Technol Ind Work Pap
16. Benaroch M, Chernobai A (2017) Operational IT failures, IT value destruction, and board-level IT governance changes. MIS Q 41:729–762
17. van Laar E, van Deursen AJAM, van Dijk JAGM, de Haan J (2019) Twenty-first century digital skills for the creative industries workforce: perspectives from industry experts. First Monday
18. Mcarthur R, Cullen FT, Wilcox P (2012) Middle managers' perception of the internal environment for corporate entrepreneurship: assessing a measurement scale. J Bus Ventur 9:2012
19. Karim NSA, Hussein R (2008) Managers' perception of information management and the role of information and knowledge managers: the Malaysian perspectives. Int J Inf Manage 28:114–127
20. Moeini M, Rivard S (2019) Responding—or not—to information technology project risks: An integrative model. MIS Q Manag Inf Syst 43:475–500. https://doi.org/10.25300/MISQ/2019/14505
21. Fitzgerald M, Kruschwitz N, Bonnet D, Welch M (2013) Embracing digital technology: a new strategic imperative. MIT Sloan Manag Rev 55
22. Prezioso G, Ceci F, Za S (2020) Employee skills and digital transformation: preliminary insights from a case study. ImpresaProgetto - Electron J
23. Van Deursen A, Van Dijk J (2014) Loss of labor time due to malfunctioning ICTs and ICT skill insufficiencies. Int J Manpow
24. Levy F, Murnane RJ (2015) The new division of labor
25. Margherita EG, Braccini AM (2020) IS in the cloud and organizational benefits: an exploratory study. In Lazazzara A, Ricciardi F, Za S (eds) Exploring Digital Ecosystems. Lecture Notes in Information Systems and Organisation, vol 33 Springer, Cham
26. van Laar E, van Deursen AJAM, van Dijk JAGM, de Haan J (2019) Determinants of 21st-century digital skills: a large-scale survey among working professionals. Comput Human Behav 100:93–104
27. Skills P (2008) for 21St C.: 21st Century Skills, Education & Competitiveness. a Resour. Policy Guid
28. Araújo J, Pestana G (2017) A framework for social well-being and skills management at the workplace. Int J Inf Manage

29. Castro Silva H, Lima F (2017) Technology, employment and skills: a look into job duration. Res Policy. https://doi.org/10.1016/j.respol.2017.07.007
30. Neill MS, Schauster E (2015) Gaps in advertising and public relations education: perspectives of agency leaders. J Advert Educ
31. Mishra KE, Mishra AK (2020) Innovations in teaching advertising: teaching digital on a shoestring budget. J Advert Educ
32. Brynjolfsson E, Mcafee A (2014) The digitization of just about everything. Second Mach. age Work. progress, Prosper. a time Brill. Technol
33. Dachs B, Peters B (2014) Innovation, employment growth, and foreign ownership of firms: a European perspective. Res Policy
34. Bartel AP, Lichtenberg FR (2006) The comparative advantage of educated workers in implementing new technology. Rev Econ Stat
35. Mohnen P, Röller LH (2005) Complementarities in innovation policy. Eur Econ Rev. https://doi.org/10.1016/j.euroecorev.2003.12.003
36. Lewin C, McNicol S (2015) The impact and potential of iTEC: evidence from large-scale validation in school classrooms. In Re-engineering the Uptake of ICT in Schools. pp 163–186. Springer International Publishing
37. Mishra KE, Wilder K, Mishra AK (2017) Digital literacy in the marketing curriculum: are female college students prepared for digital jobs? Ind High Educ. https://doi.org/10.1177/0950422217697838
38. Voogt J, Roblin NP (2012) A comparative analysis of international frameworks for 21st century competences: implications for national curriculum policies. J Curric Stud. https://doi.org/10.1080/00220272.2012.668938
39. Claro M, Preiss DD, San Martín E, Jara I, Hinostroza JE, Valenzuela S, Cortes F, Nussbaum M (2012) Assessment of 21st century ICT skills in chile: test design and results from high school level students. Comput Educ
40. Hatlevik OE, Ottestad G, Throndsen I (2015) Predictors of digital competence in 7th grade: a multilevel analysis. J Comput Assist Learn
41. Binkley M, Erstad O, Herman J, Raizen S, Ripley M, Miller-Ricci M, Rumble M (2012) Defining twenty-first century skills. In: assessment and teaching of 21st century skills. pp 17–66. Springer Netherlands, Dordrecht. https://doi.org/10.1007/978-94-007-2324-5_2
42. Prezioso G, Ceci F, Za S (2020) Is this what you want? looking for the appropriate digital skills set. In Digital Transformation and Human Behavior. pp 69–86. Springer, Cham
43. How to motivate and retain knowledge workers in organizations: a review of the literature. Int J Manag
44. Jara I, Claro M, Hinostroza JE, San Martín E, Rodríguez P, Cabello T, Ibieta A, Labbé C (2015) Understanding factors related to Chilean students' digital skills: a mixed methods analysis. Comput Educ
45. Picatoste J, Pérez-Ortiz L, Ruesga-Benito SM (2018) A new educational pattern in response to new technologies and sustainable development. Enlightening ICT skills for youth employability in the European Union Telemat Informatics. https://doi.org/10.1016/j.tele.2017.09.014
46. Kefela GT (2010) Knowledge-based economy and society has become a vital commodity to countries. Int NGO J
47. Vial G (2019) Understanding digital transformation: a review and a research agenda. J Strateg Inf Syst. https://doi.org/10.1016/j.jsis.2019.01.003
48. Dremel C, Herterich MM, Wulf J, Waizmann JC, Brenner W (2017) How AUDI AG established big data analytics in its digital transformation. MIS Q Exec
49. Anthony Byrd T, Lewis BR, Bryan RW, Byrd TA (2006) The leveraging influence of strategic alignment on IT investment: an empirical examination. Inf Manag 43:308–321. https://doi.org/10.1016/j.im.2005.07.002
50. King AW (2001) Managing organizational competencies for competitive advantage: the middle-management edge. Acad Manag Exec
51. Amidei R (2009) Governance ICT e competenze professionali. Impresa Progett J Manag

52. Mezias JM, Starbuck WH (2003) Studying the accuracy of managers' perceptions: a research odyssey. https://doi.org/10.1111/1467-8551.00259
53. Bailey DE, Kurland NB (2002) A review of telework research: findings, new directions, and lessons for the study of modern work
54. Desouza KC (2005) The internet in the workplace: how new technology is transforming work. Acad Manag Rev
55. Golden TD (2006) The role of relationships in understanding telecommuter satisfaction. J Organ Behav. https://doi.org/10.1002/job.369
56. Johlke MC, Duhan DF, Howell RD, Wilkes RW (2000) An integrated model of sales managers' communication practices. J Acad Mark Sci
57. Latapie HM, Tran VN (2007) Subculture formation, evolution, and conflict between regional teams in virtual organizations - lessons learned and recommendations. Bus. Rev, Cambridge
58. Martins N (2002) A model for managing trust. Int J Manpow
59. Watson RT (1990) Influences on the IS manager's perceptions of key issues: Information scanning and the relationship with the CEO. MIS Q Manag Inf Syst 14:217–230. https://doi.org/10.2307/248780
60. Crawford L (2005) Senior management perceptions of project management competence. Int J Proj Manag
61. Aasheim CL, Li L, Williams S (2009) Knowledge and skill requirements for entry-level information technology workers: a comparison of industry and academia. J Inf Syst Educ
62. Molla A, Cooper V, Karpathiou V (2015) IT managers' perception and response to digital disruption: an exploratory study. In ACIS 2015 Proceedings - 26th Australasian Conference on Information Systems
63. Endsley MR (1995) Toward a theory of situation awareness in dynamic systems. Hum Factors J Hum Factors Ergon Soc 37:32–64
64. Heerwegh D, De Wit K, Verhoeven JC (2016) Exploring the self-reported ICT skill levels of undergraduate science students. J Inf Technol Educ Res. https://doi.org/10.28945/2334
65. Verhoeven JC, Heerwegh D, De Wit K (2016) ICT learning experience and research orientation as predictors of ICT skills and the ICT use of university students. Educ Inf Technol
66. Edmunds R, Thorpe M, Conole G (2012) Student attitudes towards and use of ICT in course study, work and social activity: a technology acceptance model approach. Br J Educ Technol
67. Hess T, Benlian A, Matt C, Wiesböck F (2016) Options for formulating a digital transformation strategy. MIS Q Exec 15:123–139
68. Yin RK (2018) Case study research and applications: design and methods
69. Eisenhardt KM, Graebner ME (2007) Theory Building from cases: opportunities and challenges. Acad Manag J 50:25–32. https://doi.org/10.1002/job
70. van Laar E, van Deursen AJAM, van Dijk JAGM, de Haan J (2020) Measuring the levels of 21st-century digital skills among professionals working within the creative industries: a performance-based approach. Poetics
71. Kuhlemeier H, Hemker B (2007) The impact of computer use at home on students' Internet skills. Comput Educ
72. Margherita EG, Braccini AM (2020) Organizational impacts on sustainability of industry 4.0: a systematic literature review from empirical case studies. In Agrifoglio R, Lamboglia R, Mancini D, RF (eds) Digital Business Transformation. Lecture Notes in Information Systems and Organisation, Springer, Cham
73. Braccini AM, Margherita EG (2019) Exploring organizational sustainability of industry 4.0 under the triple bottom line: the case of a manufacturing company. Sustainability 11:36. https://doi.org/10.3390/su11010036
74. Margherita EG, Braccini AM (2020) Industry 4.0 technologies in flexible manufacturing for sustainable organizational value: reflections from a multiple case study of Italian manufacturers. Inf Syst Front
75. Backhaus K, Tikoo S (2004) Conceptualizing and researching employer branding. Career Dev Int 9:501–517

76. Minchington B, Kaye T (2007) Measuring the effectiveness of your employer brand. Hum Resour Mag 12
77. Kamprath M, Mietzner D (2015) The impact of sectoral changes on individual competences: a reflective scenario-based approach in the creative industries. Technol Forecast Soc Change
78. Manuti A, Pastore S, Scardigno AF, Giancaspro ML, Morciano D (2015) Formal and informal learning in the workplace: A research review. Int J Train Dev. https://doi.org/10.1111/ijtd.12044
79. Haukka S (2011) Education-to-work transitions of aspiring creatives. Cult Trends. https://doi.org/10.1080/09548963.2011.540813
80. Lazazzara A, Za S (2016) How subjective age and age similarity foster organizational knowledge sharing: a conceptual framework. In Lecture Notes in Information Systems and Organisation
81. Lazazzara A, Za S (2019) The effect of subjective age on knowledge sharing in the public sector. Pers Rev. https://doi.org/10.1108/PR-07-2018-0248
82. Hennekam S, Bennett D (2017) Creative industries work across multiple contexts: common themes and challenges. Pers. Rev

Achieving Sustainable Outcomes Through Citizen Science: Recommendations for an Effective Citizen Participation

Emanuele Gabriel Margherita

Abstract Citizen Science refers to a scientific project, managed by researchers, where volunteers are involved in studying and acquiring data related to a natural phenomenon. To this end, researchers design a platform to support this process. A successful Citizen Science project delivers important sustainable outcomes because it contributes to understanding and addressing environmental issues. For an effective Citizen Science project, researchers manage crucial actions: the management of volunteers' motivation and engagement and the development of a Citizen Science platform. With this study, I contribute to the literature discussing these three topics proposing recommendations to address these actions, based on a literature review. The study has certain implications for practitioners and researchers.

Keywords Citizen science · Literature review · Participation of volunteers · Citizen Science project design

1 Introduction

There is in an increasing interest of scholars and practitioners in studying how advanced information and communication technologies (ICT) contribute to the development of a more sustainable society [1]. To this end, ICT support the development of a sustainable society because they help "meet the needs of the present without compromising the ability of future generations to meet their own needs" [2 pg. 43]. ICT allows novel forms of interactions [3] and novel opportunities to exploit them in order to alleviate the impact of society on the environment [4]. Within this landscape, various studies investigated how these ICTs deliver sustainable value in the organizational setting and smart city [5–7], while others started exploring the role of Citizen Science for a sustainable society [8]. Citizen Science refers to a scientific project which is led by researchers and where citizens participate in the collection of

E. G. Margherita (✉)
Department of Economics Engineering Society and Organization – DEIM, University of Tuscia, Via del Paradiso, 47, 01100 Viterbo, Italy
e-mail: emargherita@unitus.it

A. M. A. Musleh Al-Sartawi (ed.), *The Big Data-Driven Digital Economy: Artificial and Computational Intelligence*, Studies in Computational Intelligence 974,
https://doi.org/10.1007/978-3-030-73057-4_20

data [9]. To this end, researchers make use of a platform which is used to collect data by citizens regarding various phenomena in natural contexts, such as bird watching [10, 11].

Citizen Science project delivers sustainable outcomes. It contributes to creating awareness over an environmental issue related to the disappearing of species, pollution and on social issues like diseases [12]. Also, citizen science help mitigating these issues by means of the collection of data which are then analyzed by experts in order to develop a solution [11]. Moreover, citizen science platform can help develop a community over a distinct issue that can community it to people who are not aware by word of mouth. In this way, the sustainable value of citizen science project can be amplified. In order that citizen science is successful, researchers should manage the motivation and the engagement of volunteers, as well as the development of a suitable platform. With this study, I want to contribute to the literature illustrating these three topics deeply. I conduct a review of studies [13] in order to provide a recommendation for successful citizens science projects.

In the next section, I present the theoretical framing of citizen science. In Sect. 3, I illustrate the research method which I conducted. In Sect. 4, I portray the results of the literature review. The study concludes in Sect. 5.

2 Citizen Science

In literature, there are various definitions regarding citizen science: "*Citizen science enlists the public in collecting large quantities of data across an array of habitats and locations over long spans of time*" [9]; "*Engaging non-scientists to survey ecosystems, a process known as citizen science has been adopted worldwide*" [14].

Summarizing, citizen science is a scientific project, in which researchers or scientists engage citizens or "non-scientists" to collect data regarding a precise phenomenon through an online platform (or mobile application) [9]. The project purposes are twofold to study a phenomenon and increase citizen knowledge about it [11].

It is important to clarify which Citizen Science is not new. Scientists and citizens have already collaborated: "*Charles Darwin (1809–1888) sailed on the Beagle as an unpaid companion to Captain Robert FitzRoy, not as a professional naturalist* [12, pg 1]". What is new is the employment of information and communication technologies. Indeed, afterwards, the choice of the scientific question, new socio-techincal actors intervene in the projects: the volunteers and the platform. Scientists (or researchers), the managers of the projects, should handle these two actors because the success of the project depends on them, as well [11, 15]. The volunteer should be found, and they should be involved in the project. Thus, scientists should motivate these people in entering in the citizen science projects and maintaining a high level of engagement [16]. Also, the platform should be managed in order to make it easy to use and ready to acquire data [16].

Table 1 Model for the development of a citizen science project

Model for developing a citizen science project	Subjects	Critical actions for
1. Choose a scientific question	Scientists	
2. Form a scientific an evaluator team	Scientists, platform	Platform
3. Develop, test, and refine protocols, data forms, and educational support materials	Platform	Platform, volunteers engagement
4. Recruit participants	Volunteers	Volunteers engagement, volunteers motivation
5. Train participants	Volunteers	Volunteers motivation, volunteers platform
6. Accept, edit, and display data	Platform, scientists	Volunteers motivation, volunteers platform
7. Analyze and interpret data	Scientists	Platform
8. Disseminate results	Scientists	Platform
9. Measure outcomes	Scientists, volunteers	Platform

In Table 1, I summarized the phases to develop a citizen science project, based on the work by Bonney et al., which are enriched by the actors involved in the phase and the actions that researchers manage for each phase. The project is designed and tested in the first three phases, together with the platform. In phase 4 and 5, volunteers are recruited and trained to use the platform. Data are then collected. In the last three phases, data are analyzed and interpreted. Results are disseminated, and the project is eventually evaluated [11].

The extant literature revealed several application fields of Citizen Science projects. Mainly, citizen science project aims at investigating phenomena involved in the natural context where particular species of animals are the main subject. Bird-watching was the first "hobby" which is transformed into a citizen science project (ebird) [11, 16]. Still, I found other topics of citizen science project having as the main aim the understanding of illness (malaria) or "soft" theme like the implication on traffic after a drive-ability law reform in Finland and the meteorology [17].

3 Research Method

I conducted a literature review [13] over Scopus in January 2019 with the following query searches: Citizen Science AND Motivation; Citizen Science AND "Engag*" OR "Engaging"; Citizen Science AND "Platform" OR "Proces*". In order to increase the validity and reliability of the research [18], I refined the search by choosing articles in English and only from journals. I merged all the articles which are extracted from the query searches in a single database. The database contains 450 articles. I read all the title and abstracts, and I selected the most pertinent papers for the purpose of the research. Afterwards, I read the entire papers and I select the more relevant to my research. I was guided by investigating the topic of the motivation and the engagement of volunteers and the development of a citizen science platform. The final database contains 17 papers.

4 Findings and Discussion

In this section, I present and illustrate the findings for the topic of motivation and engagement of volunteers as well as the platform of citizen science projects.

4.1 The Motivation of Volunteers for a Citizen Science Project

The motivation of volunteers refers to the reasons for which volunteers undertake and pursue a citizen science project.

A citizen science project is often well-defined and aims at studying a particular phenomenon. This means that volunteers undertake the citizen science project because they already know something regarding the phenomenon, and they are enthusiastic in knowing more. Accordingly, the project design should plan actions to maintain alive this enthusiasm and interest.

From the proposed model for the development of a citizen science project in Table 1, an important task, in phase 3 (the design part), is to create educational materials, because these materials can be offered to support participant understanding and satisfactory completion of project protocols [9]. Still, according to Dem et al. volunteers are keen on taking part in a scientific project and interested in knowing about the results [19]. Indeed, according to Landzandra volunteers wanted and reported a gain in knowledge about citizen science and the topics of the project [20].

The training phase is a further important action to manage. Indeed, in this phase, volunteers, previously engaged, are motivated to undertake the project, and they have to learn how to use the application. If the application is not clear and easy to use motivation tends to decrease, producing a negative impact on the project.

Motivating volunteers also include the development of reward actions for them. Being the participation free, volunteers do not aim at a monetary reward but a recognition [21]. Studies posited that volunteers want a recognition which has social and environmental value. Environmental reward refers to taking part in the scientific process and acquiring knowledge (or educational learning) through the data collections and interactions with scientists and the fulfilment of the project. The social reward refers to the knowledge that can be put into practice in the real-life and useful for the community [22].

These rewards should be designed prior to the beginning of the project: "*The majority of the citizen science programs focused on scientific outcomes, whereas the educational and social benefits of program participation, variables that are consistently ranked as important for volunteer engagement and retention, were incidental. Evaluators indicated usability, across most of the citizen science program sites, was higher and less variable than the ratings for participant engagement and retention [20, p. 568]*". Accordingly, scientists should design phase 8 – dissemination results both to present the results to other scientists (conferences, papers) both to the citizen in order to make the results tangible for the community.

In conclusion, to maintain the motivation of volunteers, managers should organize events to facilitate face-to-face interaction of volunteers along the process. Thus, it is important to provide users with tools to communicate in order to support social learning, community building and sharing [24]. This has implications for the design and management of online citizen science projects. Furthermore according to Cappa, beyond the enrichment of the task is to provide supports to the citizen through face-to-face interaction between scientists and volunteers, since information technologies (App or website) can isolate volunteers from the rest of the project [25].

4.2 The Engagement of Volunteers for a Citizen Science Project

The term *engagement* refers to the actions to communicate project and recruit volunteers. Despite, the query produced several results, very few papers focus on this topic, which is represented by *Phase 4 - Recruit participants* of our main schema. I proposed some extracts:

> "*During February 2006, 2007, and 2008, we recruited 58,35, and 26 volunteers respectively from the New York-New Jersey Trail Conference (NYNJTC); a recreational hiking association with a membership of about 10,000 individuals and about 100 clubs. These volunteers were recruited via an email flyer sent to the entire membership. We offered no material incentives. If volunteers were able to undertake the hiking and could attend the training sessions, they were accepted* [18, pg. 427]."

> "*Potential participants were informed about the project through (a) the website of the project; (b) an information desk in the Cretaquarium; (c) posters and leaflets which were distributed in the participating diving clubs and the tourist information offices. Often, divers were approached directly before their dives in the diving centres and usually expressed interest in participation*[1]".

> "*Aided by a social media campaign that raised awareness of the project in the two jurisdictions, citizen scientists were recruited in-person from community centers located in different areas in each city and the universities of Regina and Saskatchewan to ensure recruitment of a representative sample*[2]".

---

[1] https://www.ncbi.nlm.nih.gov/pmc/articles/PMC5893892/

[2] https://www.ncbi.nlm.nih.gov/pmc/articles/PMC5893892/

From these examples, I can state that to engage volunteers, the management of citizen science project should create a communication plan through different types of media. Indeed, Bonney et al. said. that participants can be recruited by a various techniques like press releases, listservs, direct mailings, advertisements, public service announcements, magazine and newspaper articles, brochures, flyers, and presentations, including posters and workshops at conferences of potential participants or their leaders [9].

Moreover, Pandya proposed a framework to engage diverse communities in citizen science in the U.S. starting to reflect on the reasons for the lack of participation (family resources: lack of transportation, access to natural areas as well as and scientific education) and concluding with a well-articulated framework [27]:

- Align research and education with community priorities. Here, Pandya et al. said that the research question is driven by a need of the studied community. In that way, it will be easier to find volunteers.
- Plan for co-management of the project. The citizen science management should propose a training course to make easier the data collection since inventible conflict should happen during the project.
- Engage the community at every step.
- Incorporate multiple kinds of knowledge. Successful participatory projects seek expertise from all participants and build processes and procedures (such as regular community meetings) to facilitate and validate that expertise.
- Disseminate results widely.

Concluding, these studies show that volunteers are challenging step without a strategy. There is a commitment to finding these volunteers. Firstly, the recommendation is to find volunteers within the community which is interested in the topic - the beauty of citizen science is that very specific and it is not difficult to find the main topic and their audience. Afterwards, the management of the project should encourage these volunteers to "convince" friend to help the project (word of mouth).

4.3 The Platform for a Citizen Science Project

Platform refers to the information technology (web site or mobile app) which is employed for two main tasks: to acquire, collect and analyze the information as well as allowing the volunteers to create a community and get their recognition (*in terms of scientific results*) [12]. Data are often stored in the cloud in order to reduce costs [28].

The platform is an actor intervening in several phases of the main framework. Indeed, the platform is designed in Phase 2 - Form a scientist/educator/technologist/evaluator team through a commitment of scientists and IT technicians. Thus, in Phase 3 - Develop, test, and refine protocols, data forms, and educational support materials, the platform is tested and evaluated. Here, the main goal is to make suitable the platform to collect the data in a scientific way and ensure data quality [9]. The data quality issue is crucial for citizen science since reliable data is not often accomplished. To address this issue, citizen science projects employ accurate technologies [10] or stringent protocol. Bear in mind that the employment of stringent protocol shall be presented with a user-friendly way. Otherwise, it causes reduction of volunteers [16]. Moreover, this protocol improves the reliability of data through the automation of location (GPS) and the species (or subjects of analysis) recognition as well as validation tools in the data gathering (If the volunteers fulfil the form entirely). Still, some works said that further development to improve the data quality is the creation of a data centre within different citizen science projects, where the data can be integrated [29, 30].

In phase 5 – Train participants, volunteers try and test the platform. Studies assert that the application should be user-friendly and presented as a means to engage the volunteers [9]. Phase 6 and 7 are the most important and crucial part of the project because all the data are gathered, analyzed and interpreted. These two phases are effective, whether the application has been well-designed [9].

Along with the entire project, the application should have "social skills". Indeed, the creation of a community is useful both to maintain high the moral/motivation of the volunteers both to have a trained public for future citizen science projects. Thus, the application should have functionalities for the communications among volunteers and scientists (a sort of social network of the project). Still, the application should contain functionalities to disseminate the results among the volunteers allowing to get their "recognitions".

Whether you are seeking practical advice to design a platform, Newman et al. reviewed how to develop a citizen science platform studying volunteer perceptions, and how to solve problems and improve the platform performance [16]. The framework highlights that a platform needs customizable data entry forms and allows to communicate volunteers among and other and with scientists.

5 Conclusion

The study illustrates three main actions that the developer of a citizen science project should manage: the motivation and the engagement of volunteers and the development of a platform. Based on a literature review, I summarize and propose discuss recommendations to address these actions. The study has certain implications for practitioners. I pinpoint various aspect that managers should consider during the design phase of the project and along with the project. I also highlight a model for the development of citizen science and actions to manage the volunteers and the development of the platform.

The study also has implication for researchers. The study illustrates a participative approach for citizen science, where the role of volunteers is crucial. The platform has two features, both related the data acquisition but also to maintain alive the commitment of volunteers. Further studies investigate this participative approach exploring further critical factors for an effective citizen science project, and what are the features to make easy-to-use a citizen science platform. Moreover, researchers should advance the knowledge related to the sustainable outcome of citizen science projects and when such project can be integrated into a Smart City architecture. The study has a limitation because I used the Scopus database for the literature review. Thus, I encourages researches to conduct a similar literature review employing different a database, such as Google Scholar and Web or Science.

References

1. Mikalef P, Pappas IO, Krogstie J, Pavlou PA (2020) Big data and business analytics: a research agenda for realizing business value. Inf Manag 57. https://doi.org/10.1016/j.im.2019.103237
2. WCED (1987) Our Common Future. Oxford University Press: Oxford
3. Margherita EG, Braccini AM (2020) Exploring the socio-technical interplay of Industry 4.0: a single case study of an Italian manufacturing organisation. In Proceedings of the 6th International Workshop on Socio-Technical Perspective in IS Development (STPIS 2020), June 8–9, 2020. CEUR Workshop Proceedings (CEUR-WS.org)
4. Margherita EG, Braccini AM (2020) Industry 4.0 technologies in flexible manufacturing for sustainable organizational value: reflections from a multiple case study of italian manufacturers. Inf Syst Front
5. Braccini AM, Margherita EG (2019) Exploring organizational sustainability of Industry 4.0 under the triple bottom line: the case of a manufacturing company. sustainability. 11:36. https://doi.org/10.3390/su11010036
6. Margherita EG, Braccini AM (2020) Organizational impacts on sustainability of industry 4.0: a systematic literature review from empirical case studies. In Agrifoglio R, Lamboglia R, Mancini D, RF (ed) Digital Business Transformation. Lecture Notes in Information Systems and Organisation. Springer, Cham. https://doi.org/10.1007/978-3-030-47355-6_12
7. Margherita EG, Esposito G, Escobar SD, Crutzen N (2020) Exploring the smart city adoption process: evidence from the Belgian urban context. In Proceedings of the 6th International Workshop on Socio-Technical Perspective in IS Development (STPIS 2020), June 8–9, 2020. CEUR Workshop Proceedings (CEUR-WS.org)
8. Sauermann H, Vohland K, Antoniou V, Balázs B, Göbel C, Karatzas K, Mooney P, Perelló J, Ponti M, Samson R, Winter S (2020) Citizen science and sustainability transitions. Res Policy 49:103978. https://doi.org/10.1016/j.respol.2020.103978
9. Bonney R, Cooper CB, Dickinson J, Kelling S, Phillips T, Rosenberg KV, Shirk J (2009) Citizen science: a developing tool for expanding science knowledge and scientific literacy. Bioscience 59:977–984. https://doi.org/10.1525/bio.2009.59.11.9
10. Bröring A, Remke A, Stasch C, Autermann C, Rieke M, Möllers J (2015) enviroCar: a citizen science platform for analyzing and mapping crowd-sourced car sensor data. Trans GIS 19:362–376. https://doi.org/10.1111/tgis.12155
11. Bonney R, Shirk JL, Phillips TB, Wiggins A, Ballard HL, Miller-Rushing AJ, Parrish JK (2014) Next steps for citizen science. Science 343(6178):1436–1437. https://doi.org/10.1126/science.1251554
12. Switzer A, Schwille K, Russell E, Edelson D (2012) National geographic fieldscope: a platform for community geography. Front Ecol Environ 10:334–335. https://doi.org/10.1890/110276
13. Webster J, Watson RT (2002) Analyzing the past to prepare for the future: writing a literature review. MIS Q 26
14. Bodilis P, Louisy P, Draman M, Arceo HO, Francour P (2014) Can citizen science survey non-indigenous fish species in the eastern mediterranean sea? Environ Manage 53:172–180. https://doi.org/10.1007/s00267-013-0171-0

15. Silvertown J (2009) A new dawn for citizen science. Trends Ecol Evol 24:467–471. https://doi.org/10.1016/j.tree.2009.03.017
16. Newman G, Zimmerman D, Crall A, Laituri M, Graham J, Stapel L (2010) User-friendly web mapping: lessons from a citizen science website. Int J Geogr Inf Sci 24:1851–1869. https://doi.org/10.1080/13658816.2010.490532
17. Sharples M, Aristeidou M, Villasclaras-Fernández E, Herodotou C, Scanlon E (2017) The sense-it app. Int J Mob Blended Learn 9:16–38. https://doi.org/10.4018/IJMBL.2017040102
18. Vom Brocke J, Simons A, Niehaves B, Niehaves B, Reimer K, Brocke J, Vom Simons A, Niehaves B, Niehaves B, Reimer K, Plattfaut R, Cleven A (2009) Reconstructing the giant: on the importance of rigour in documenting the literature search process. Eur Conf Inf Syst
19. Dem ES, Rodríguez-Labajos B, Wiemers M, Ott J, Hirneisen N, Bustamante JV, Bustamante M, Settele J (2018) Understanding the relationship between volunteers' motivations and learning outcomes of Citizen Science in rice ecosystems in the Northern Philippines. Paddy Water Environ 16:725–735. https://doi.org/10.1007/s10333-018-0664-9
20. Land-Zandstra AM, Devilee JLA, Snik F, Buurmeijer F, Van Den Broek JM (2016) Citizen science on a smartphone: participants' motivations and learning. Public Underst Sci 25:45–60. https://doi.org/10.1177/0963662515602406
21. Alender B (2016) Understanding volunteer motivations to participate in citizen science projects: a Deeper look at water quality monitoring. J Sci Commun 15:1–19
22. Hobbs SJ, White PCL (2015) Achieving positive social outcomes through participatory urban wildlife conservation projects. Wildl Res 42:607–617. https://doi.org/10.1071/WR14184
23. Wald DM, Longo J, Dobell AR (2016) Design principles for engaging and retaining virtual citizen scientists. Conserv Biol 30:562–570. https://doi.org/10.1111/cobi.12627
24. Jennett C, Kloetzer L, Schneider D, Iacovides I, Cox AL, Gold M, Fuchs B, Eveleigh A, Mathieu K, Ajani Z, Talsi Y (2016) Motivations, learning and creativity in online citizen science. J Sci Commun 15:1–23
25. Cappa F, Laut J, Nov O, Giustiniano L, Porfiri M (2016) Activating social strategies: face-to-face interaction in technology-mediated citizen science. J Environ Manage 182:374–384. https://doi.org/10.1016/j.jenvman.2016.07.092
26. Jordan RC, Brooks WR, Howe DV, Ehrenfeld JG (2012) Evaluating the performance of volunteers in mapping invasive plants in public conservation lands. Environ Manage 49:425–434. https://doi.org/10.1007/s00267-011-9789-y
27. Pandya RE (2012) A framework for engaging diverse communities in Citizen science in the US. Front Ecol Environ 10:314–317. https://doi.org/10.1890/120007
28. Margherita EG, Braccini AM (2020) IS in the cloud and organizational benefits: an exploratory study. In Lazazzara A, Ricciardi F, Za S (eds) Exploring Digital Ecosystems. Lecture Notes in Information Systems and Organisation, vol 33. Springer, Cham
29. August T, Harvey M, Lightfoot P, Kilbey D, Papadopoulos T, Jepson P (2015) Emerging technologies for biological recording. Biol J Linn Soc 115:731–749. https://doi.org/10.1111/bij.12534
30. Vitos M, Stevens M, Lewis J, Haklay M (2012) Community mapping by non-literate citizen scientists in the rainforest michalis vitos. Matthias Stevens, Jerome Lewis and Muki Haklay. 46:3–11

Outcomes of Smart Tourism Applications On-site for a Sustainable Tourism: Evidence from Empirical Studies

Stefania Denise Escobar and Emanuele Gabriel Margherita

Abstract In this study, we explore the outcomes of Smart Tourism on-site applications for Sustainable Tourism. Smart Tourism describes a plethora of advanced information and communication technologies applied in the tourism industry. Although various applications of Smart Tourism are described in the literature, the outcomes of ST for a Sustainable Tourism is still under research. To address this gap, we conduct a literature review over Scopus by seeking empirical Smart Tourism applications on-site. We assess the sustainability outcomes by using the framework of the three pillars of sustainability, composed of the economic, environmental and socio-economic pillars. The results show that Smart Tourism applications on-site support the three pillars of sustainability, but not simultaneously. The study has certain implications for practitioners and researchers.

Keywords Smart tourism · Sustainability framework · ICT in tourism development

1 Introduction

Tourism is defined by [1] as a socio-cultural and economic phenomenon entailing movement of people to a destination for personal or business purposes. Tourism is one of the world's largest economic industries contributing to 10.3% of the global GDP [2]. As one of the main sources of economic development, tourism has been widely recognized [3–11]. According to UNWTO [12] by 2030, 1.8 billion people will engage in the tourism industry.

The current tourism development strategies are unsustainable for destinations because they are resource-demanding in terms of capital, social value and natural resources [13]. Tourism mobility represents an issue for its negative environmental

S. D. Escobar (✉)
Free University Bozen, Piazza Università. 1, 39100 Bolzano, Italy
e-mail: Stefania.Escobar@economics.unibz.it

E. G. Margherita
Tuscia University, Via del paradiso, 47, 01100 Viterbo, Italy

A. M. A. Musleh Al-Sartawi (ed.), *The Big Data-Driven Digital Economy: Artificial and Computational Intelligence*, Studies in Computational Intelligence 974,
https://doi.org/10.1007/978-3-030-73057-4_21

impact related to the increase of CO2 emissions pollution [14]. Furthermore, the phenomenon of Overtourism is the negative impact of tourism on the quality of life of residents. It negatively impacts the quality of the visitors' experience and destination resources [15]. According to WTTC and OECD [2, 4], the industry is forecasted to continue growing steadily after the pandemic, meaning that the physical, ecological, social, economic, psychological and/or political capacity of a destination are and will be exceeded [7, 16, 17].

A way to enhance the sustainability of the tourism industry, and reach sustainable tourism, is the adoption of information and communication technologies (ICTs). They have proliferated recently [18] and is used extensively from operational and business perspectives [19–21] as well as in other settings representing key actions for sustainability [22–24]. Moreover, the constant progress in the development of ICTs has also induced significant transformations in tourist behaviour, leading to tourists to an easy acceptance of them. Using ICTs, the tourists are informed, empowered and wishes to obtain more personalized and better-designed experiences leading the industry to adapt tourism products and experiences [25–27].

Smart tourism (hereinafter ST) is the sum of all integrated efforts at a destination that takes advantage of smart technology to achieve environmental, social, economic sustainability, innovation and competitiveness [20]. ST considers the technologies which are used in the pre and post-travel experience and on-site travel experience. The most significant technologies of ST are cloud services, Internet of Things (IoT), end-user internet service system, social media platforms, user modelling, big data, open data and ubiquitous connectiveness [28, 29].

Although various ST applications are described in the literature, the sustainability outcomes of ST for sustainable tourism are in an early stage of research [20].

With this study, we address this gap investigating the sustainability outcomes of ST applications used on-site for sustainable tourism. To this end, we conduct a literature review over Scopus by seeking for empirical applications of ST. To assess the sustainability of these ST applications, we employ the framework of the three pillars of sustainability, that includes the economic, environmental and socio-cultural pillars [30]. It extends the previous sustainability perspective in tourism. That encourages only tourism strategy to safeguard the environment by adding tourism practices to increase the socio-economic conditions of people within a destination. Our study mainly answer the following research question: "*What are the outcomes of Smart Tourism applications on-site for a Sustainable Tourism?*".

Therefore, we contribute to the literature by illustrating the sustainability outcomes of ST applications on-site for sustainable tourism under the framework of the three pillars of sustainability.

2 Theoretical Background

Our study is framed in two different literature streams: smart tourism and the concept of sustainability in tourism. These literature streams are illustrated in this section.

2.1 The Smart Tourism Paradigm

Generally, the term "smart" has become a popular term to describe technological, economic and social developments fueled by smart technologies that rely on sensors, open data, big data, open API, new ways of connectivity between humans and machines and multi-device, networked exchange of information [20].

The term "smart" applied to the tourism industry brought the concept of ST which refers to the use of technologies (e.g. internet, mobile communication and augmented reality) to collect enormous amounts of data to provide real-time support to all stakeholders in the destination and context-aware as well as personalized experiences to tourists [20, 31, 32].

Gretzel et al. [20] explain the origins of ST as derived from traditional tourism and more recently e-tourism. E- tourism differs from ST in the travel sphere considered, the core technology used, the travel phase included. In essence, although in both concepts there is a predominant presence of ICT technologies to enhance the tourist experience, e-tourism merely handles the digital part of the experience during the pre and post-travel phases of a trip and websites are the core technology. Instead, ST aims at being a bridge between the digital and physical spheres of the travel experience and also includes the on-site phase of a trip with Big Data as the core technology.

To build successful strategies based on the ST concept, Gretzel et al. [20] describe ST as comprising five layers: (1) a physical layer (transportation, resources, and service infrastructures); (2) a smart technology layer (business solutions and consumer applications); (3) a data layer (data storage, open data and data-mining applications); (4) a business layer (innovation based on the available technologies and data sources); and (5) an experience layer (technology and data-enhanced experiences' consumption). The technical infrastructure integrated into the physical layer of a destination forms the basis of all other layers as ST most innovative feature entails the connection between the physical and digital world. The physical layer enables the technology layer to develop the data layer with a useful and varied kind of data. ST, through the data layer, combines different kinds of data to boost innovation and allow the creation of a business layer. Services and applications created by the business layer using ST data enable the development of improved touristic experiences and sustainability goals in a destination. Consequently, the experience layer considers tourists but also residents and hosts of a destination.

Smart technologies form the foundation of ST [21]. These technologies are described by [33] as those that sense (bringing awareness to everyday things), learn (using experience to improve performance), adapt (modifying behaviour to fit the environment), infer (drawing conclusions from rules and observations), predict (thinking and reasoning about next steps) and self-organize (self-generating and self-sustaining at technology level).

According to Zhang Ling-Yung et al. [29], the most significant technologies of ST are cloud services, Internet of Things (IoT), and end-user internet service system. However, it is worth also mentioning Social Media platforms, User modelling, Big Data, Open Data and ubiquitous connectiveness [28].

Cloud services enable the creation of multiple tourism applications and share information by providing scalable access [34].

IoT helps with information analysis, data collection, and system automation with the introduction of a host of sensors, chips, and actuators integrated extensively in the physical infrastructure of a destination. Finally, the end-user internet service system supports cloud services and IoT through applications and devices. The three ST technologies have a high value in the interaction at all levels of stakeholders in the tourism context. Indeed, these technologies can boost the achievement of relationships among various tourism and non-tourism stakeholders [18, 20, 35, 36]). These technologies can be adopted on existing infrastructure to increase the sustainability value of IT [37, 38].

Examples of ST technologies are SHCity, a mobile application that integrates real-time data and routing algorithms to enrich the tourism experience. SNBSOFT, a tourism business that uses open geospatial data to provide a mobile application with multilingual maps of the city for visitors [21].

Special mention needs to be made for Open Data and ubiquitous connectedness in a destination. These two elements are important in the development of an ST destination. Ubiquitous connectiveness under the form of Wi-Fi networks allows smart technologies to communicate with each other and facilitates interactions with ST end-users, who are the ones more in need of access points for ST content [21]. Making a Smart tourist rely on limited connection due to restrictive data plans or high roaming rates can damage the quality of the experience and harm destination competitiveness.

A good example is Seoul (South Korea) which was awarded as the world's most wired city offering also the fastest Internet speeds [39, 40]. Another example is El Hierro (Spain), the first island in the world to have Wi-Fi networks and WiMAX that covers the entire territory with 26 free points of access among touristic attractions and open public spaces so that benefits both residents and visitors. Plus, solar panels have been installed that provide power to the connection points to reduce the impact on the environment [41].

Open data, which is also seen as an essential feature for smart development [42], is defined as data freely accessed, used, modified and shared by anyone for any purpose [...] [43]. Indeed, an ST destination differs from a traditional one because it is a knowledge-based destination in which information is available to all stakeholders in a systematic and efficient way [44]. The city of Marbella (Spain) with the project "Open Data Marbella" offers an example of this concept with open data service open to all citizens [41]. Seoul (South Korea) is once again an example of good practice having national, metropolitan and tourism-specific open data that stimulates the development of tourism apps and consequently tourism businesses in the destination [21].

The ST scope is three-fold: on the one hand, the data transformed by advanced technology helps to create on-site personalized smart experiences for the tourist. On the other hand, it helps create smart business value-propositions with a clear focus on efficiency, sustainability and experience. Finally, ST helps develop tourist destination management strategies that guarantee the sustainable development of tourist areas, accessible to everyone that increases the quality of the experience at the destination as well as residents' quality of life [20, 41]. According to Rocha [65], computing and ICT coupled with big data and IoT enable governments and businesses to improve products and service delivery to citizens and tourists by analyzing the general and most common preferences, as well as being able to offer personalized services to each individual.

Therefore, several researchers acknowledge the potential of ST to enhance the experience of the tourist, the livelihood of residents and the efficient use of resources in a destination [20, 45].

ST allows the development of smart destinations capable of creating smart business ecosystems that ultimately develop smart experiences for tourists enhancing the competitiveness of a destination while making sustainable use of resources and enhancing the quality of life for residents.

2.2 *Sustainability Framework for Smart Tourism Development*

Historically, sustainability in Tourism is often focused on tourism actions or strategies directed towards the safeguard of the environment [17, 46, 47]. However, UNWTO [30] recommends extending this perspective by using the three pillars of sustainability to assess sustainable tourism. The three pillars of sustainability are the environmental, economic and socio-cultural aspects of tourism development.

The environmental pillar refers to the optimal use of environmental by tourism stakeholders that constitute a key element in tourism development, maintaining essential ecological processes and helping to conserve natural heritage and biodiversity.

The socio-cultural pillar refers to the consideration and respect of the socio-cultural authenticity of host communities, conservation of their cultural heritage and traditional values as well as contribution to intercultural understanding and tolerance.

Finally, the economic pillar refers to the ability of tourism businesses to ensure viable, long-term economic growth by providing socio-economic benefits distributed among all stakeholders fairly. This includes secure employment, income-earning opportunities, as well as a contribution to poverty alleviation. Besides, it refers to the ability to maintain a high level of tourist satisfaction by ensuring a meaningful experience to the tourist [30].

Although sustainability lies at the root of the smart concept [17], and some literature has addressed the relevance of sustainability in ST [46, 48, 49] there is a lack of studies that address all three pillars of sustainability simultaneously. Studies addressing social sustainability (poverty reduction and social inclusion) are still underrepresented. Both industry and research are more focused on technology implementation, showing little interest or concern for the environmental or social implications that technologies have in a destination [17, 45].

3 Methodology

We conducted a literature review over Scopus with the following keyword search: "Smart Tourism" OR "Smart Destination" AND "Sustainability" OR "Sustainable" [50]. We add in the keyword search Smart Destination because it is a component of the ST concept. We refined the research selecting only papers in English. We conducted the review in December 2020. The keyword research revealed 89 results. We read all the title and abstract of the papers, and we chose the more appropriate for our research [51]. We aimed at exploring empirical applications of Smart Tourism and the consequent sustainability outcomes. After reviewing the reference from the chosen papers, we reached a database composed of 3 studies illustrating 9 ST applications on-site, which we used for the literature review. To analyze the papers, we followed a qualitative coding technique [52].

We considered three criteria when choosing destinations: I) we considered those destinations that are considered "smart tourism destinations"; II) we then considered those smart tourism destinations that supported their strategies at least one pillar of sustainability. III) We checked that the sustainable strategies were the result of ST applications on site. Consequently, we excluded from our analysis of those destinations that have great sustainable based strategies but not based on smart tourism technologies.

4 Findings

Table 1 summarizes the outcomes of St applications on-site for a Sustainable Tourism. Several studies explore ST's impact on destination competitiveness [41, 53, 54].

Table 1 Outcomes of Smart Tourism Applications on the three pillars of sustainability (✓ represents the pillar supported by Smart Tourism application)

Smart Tourism Applications on-site	Sustainability pillar		
	Economic	Environmental	Socio-cultural
Aix-en-Provence (France)	✓	✓	
Copenhagen (Denmark)		✓	✓
Helsingborg (Sweden)		✓	
Helsinki (Finland)		✓	
Las Palmas (Spain)	✓		
Madrid (Spain)	✓		
San Sebastian (Spain)		✓	✓
Venice (Italy)		✓	✓
Zagreb (Croatia)		✓	

Through big data, AI, IoT and other state-of-the-art technologies, tourism destinations and businesses offer more personalized and varied experiences to tourists. They support the economic development and competitiveness of destinations [6, 55–58]. In the case of the city of Madrid, the destination is seeking to redirect the flow of tourists in the city with "Madrid 21 Destinos". This project makes use of smart technologies such as sensors and beacons for the monitoring of tourists throughout the city to promote different less visited areas of Madrid with the intent of redirecting the flow of tourists and enhancing destination competitiveness [54].

In Las Palmas (Spain), IoT sensors such as beacon trackers were located in big open shopping areas to stimulate the shopping tourism segment. That is a form of tourism carried out by individuals for whom the acquisition of goods (outside their place of residence) is a determining factor in their decision to travel [1]. This initiative wants to enhance the competitiveness of the destination. It helps local businesses to reach tourists through their mobile phones while visiting the city [41].

Zagreb (Croatia) instead represents an example of ST supporting the environmental pillar: Freewa Project has the goal of preserving drinking water sources. The project runs through a web platform (Freewa) that has a mobile app for finding free drinking water locations and reusable water bottles across the city. Zagreb has good water quality, and through this mobile application, not only tourists but also locals can enjoy it while contributing to the preservation of drinking water sources in the city [53, 59].

Another example of environmental sustainability is Aix-en-Provence with The project 'Aix Living Places'. The plan is to install hundreds of sensors in the streets of Aix-en-Provence. These sensors will collect data that will be analyzed by local startups to develop solutions that improve the city center for tourists and locals. Among these solutions, tracking pedestrians to ease the flow of people walking through the city, and the measurement of air quality [53, 60].

The city of Helsingborg is an example of environmental sustainability. Through the city's interactive online map, visitors can access a highly developed cycling infrastructure stimulating green tourism mobility across the city [61].

The 'Miljokajakken' initiative (The Green Kayak) in Copenhagen represents an example of environmental and social sustainability. Through this initiative, visitors can rent a kayak for free if they share on social media all good actions they made while visiting the city of Copenhagen and commit to collect rubbish while kayaking. Through social media, they can increase the visibility of the attraction but also of the good practices showing tourist responsible behaviour [53, 62].

Helsinki represents an example of environmental sustainability having the ambitious target of becoming a carbon–neutral city by 2030. For this goal, several hotels are offering to tourists digital carbon footprint calculators so that visitors are aware of their impact on the destination [63].

Concerning social sustainability, Venice (Italy) launched a campaign "#EnjoyRespectVenezia to convince visitors to behave responsibly in the city and to respect the environment, landscapes, art and Venice's locals. The campaign wants to raise awareness about tourism's impact firmly believing in responsible tourist behaviour as a key element for sustainable tourism development. Sensors are present all over the city to monitor the number of tourists. And through a website visitors have day-to-day forecasts of the number of expected visitors to the town and choose in which day to visit the destination. The website also offers tips on less visited areas of Venice worth seeing with the intent of redirecting the flow of tourists in the destination [53, 64].

Finally, San Sebastian (Spain) with the 'Live San Sebastian, Love Donostia' campaign aims to involve visitors in the city's sustainability. The destination created a list of good practices available for visitors in several languages and through various online devices. In this list, visitors are encouraged to public places clean, respect the resident's need for quiet at night, use public transport and to support local businesses. San Sebastian represents a good example of social sustainability by giving visitors the tools to visit the city with sustainable behaviour. The purpose is to allow tourists to enjoy the city while still respecting the area and those who live in it [63].

5 Discussion

According to the literature, ST applications on-site are mainly used to boost tourist experience and destination competitiveness, that is, economic sustainability. And to sustain and safeguard the environment. Our study adds a "sustainable piece" in the tourism literature, illustrating applications that support the socio-economic pillars. Based on our results, the socio-cultural pillar is supported when ST applications on-site support: the outcome of ST initiatives support on the socio-cultural pillar is a responsible behaviour of tourist during their stay at a destination in terms of their impact on attractions and residents (e.g., San Sebastian, Venice).

The support to the economic pillar includes initiatives promoting local businesses and destination competitiveness (e.g., Las Palmas, Madrid). Also, the environmental pillar is supported by limiting the use of resources by tourists for example by promoting the use of tab water (e.g., Zagreb) and promoting the use of green tourism mobility (e.g., Helsingborg).

Also, we found a lack of studies of ST applications on-site that supports the three pillars. This is a prominent research area because whether ST applications support the three pillars simultaneously boosts a Sustainable Tourism. Further studies are needed to explore the ST applications and how the three pillars are supported simultaneously.

Moreover, our results show a large variety of technologies used in the ST paradigm, which confirmed the literature in defining ST strategies. Indeed, many strategies considered ST-based are instead smart city strategies (e.g. Eolic plants whose purpose is to enhance the quality of life of the city yet the direct link with tourism is missing). Therefore, it may seem like part of a destination marketing strategy to promote these actions as smart tourism initiative which contributes to the terminology confusion that orbits around the ST concept. We looked at those destinations that employed smart tourism technologies to enhance the impact of tourism destinations. By doing so, we excluded those destinations which technologies were not smart tourism-related or those which did not employ smart tourism technologies in their sustainability strategies. The main point of this paper is to study how to reduce the impact of tourism in terms of economic, social and environmental sustainability in destinations. Therefore, with this paper, we go as far as to tell that aside for awards for being smart tourism capitals, there should be less confusion between a smart destination and a sustainable destination. Although a destination can be sustainable and smart, it is clear that there is still some confusion about what is sustainable and what is smart. Smart tourism is a tool to boost sustainability, yet it is not the only tool. Indeed, several sustainable strategies are not based on smart tourism. A destination can be sustainable without being smart. The fact that some initiatives are considered as smart tourism can be related to the confusion around the concept of smart tourism (which this paper tries to clarify) and the desire of increasing the visibility of the destination by promoting it differently.

5.1 *Implications for Practitioners and Researchers*

Through the combination of theory and examples showed, this paper has certain implications for academics and practitioners. The development of ST based experiences will alter the tourism industry in terms of destination development, businesses and experiences. It will offer a better future for tourism-based economies and sustainable tourism. Destination management organizations can use this study as a guideline of ST application in order to achieve sustainability.

This study has implication for researchers because it wants to make some clarity about the origin and definition of ST believing that a better understanding of the ST concept will lead to clearer ST strategies in destinations and businesses. In addition, by explaining the impact of technologies and potential benefits for all tourism stakeholders, this review pinpoints the positive relationship between ST and sustainable development. Still, by acknowledging the lack of research on sustainability and ST, this paper wants to encourage future studies on this still underexplored field to advance knowledge and possible solutions. We also invite scholars to study outcomes of ST applications, both on-site and in the pre and post-travel stages to explore the entire spectrum of ST technologies.

The study also has a limitation because we used Scopus database. Thus, a prominent research avenue is to extend our investigation employing further search databases, such as Web of Science and Google Scholar.

6 Conclusion

The study is motivated by a lack of studies which explore the out-comes of ST application for a Sustainable Tourism. To address this gap, we conduct a literature review over Scopus by seeking for empirical applications of ST and their sustainability outcomes We assess the sustainability we employ the framework of the three pillars of sustainability, that includes the economic, environmental and socio-cultural pillars, which is barely used in the tourism research. The results show that the outcomes ST applications are on the three pillars. Some studies support the economic pillar by promoting local businesses and enhancing destination competitiveness, some others the social pillar by enhancing the responsible tourist behaviour and finally the environmental pillar by limiting the use of resources from tourists in the destination. Moreover, we found the risk of the ST concept to become a tool for destination marketing to promote the destination is far from being eradicated due to a lack of clarity on the scope of ST.

The outcomes of ST applications on-site on the environmental pillars are limited use of natural resources as well as the reduction of pollution of tourism activities.

Moreover, the outcomes of ST applications on-site on the socio-economic pillars are initiatives that are increasing the responsible behaviour of the tourist at a destination. These initiatives also raise awareness of the negative impact of tourism activities. This pillar is barely investigated in the tourism literature and requires further studies. Lastly, to study ST strategies that reach sustainable tourism development, further research should explore those ST initiatives that support all three pillars at the same time.

References

1. UNWTO (2014) Global Report on Shopping Tourism. https://affiliatemembers.unwto.org/publication/global-report-shopping-tourism. Accessed 10 Dec 2020
2. WTTC (2020) Economic Impact Report. https://wttc.org/Research/Economic-Impact. Accessed 15 Dec 2020
3. Brandão M, Joia LA, Do Canto Cavalheiro GM (2020) Towards a smart tourism destination development model: promoting environmental, economic, socio-cultural and political values. Tour Plan Dev 17:237–259. https://doi.org/10.1080/21568316.2019.1597763
4. OECD (2020) OECD Tourism Trends and Policies 2020
5. Moll-de-Alba J, Prats L, Coromina L (2016) The need to adapt to travel expenditure patterns. A study comparing business and leisure tourists in Barcelona. Eurasian Bus Rev 6:253–267. https://doi.org/10.1007/s40821-016-0046-4
6. Sigalat-Signes E, Calvo-Palomares R, Roig-Merino B, García-Adán I (2020) Transition towards a tourist innovation model: the smart tourism destination: reality or territorial marketing? J Innov Knowl 5:96–104. https://doi.org/10.1016/j.jik.2019.06.002
7. Sigala M (2020) Journal Pre-proofs
8. Lee CC, Chang CP (2008) Tourism development and economic growth: a closer look at panels. Tour Manag 29:180–192. https://doi.org/10.1016/j.tourman.2007.02.013
9. Balaguer J, Cantavella-Jordá M (2002) Tourism as a long-run economic growth factor: the Spanish case. Appl Econ 34:877–884. https://doi.org/10.1080/00036840110058923
10. Brida JG, Risso WA (2008) The contribution of tourism to economic growth: an empirical analysis for the case of Chile. Eur J Tour Res 2:178–185
11. Kreishan FMM (2010) Tourism and economic growth: the case of Jordan. Eur J Soc Sci 15:63–68
12. World Tourism Organization (2017) UNWTO Tourism Highlights 2017 Edition. UNWTO Tour. Highlights. 10. https://www.e-unwto.org/doi/pdf/10.18111/9789284419029. Accessed 6 Dec 2020

13. Glyptou K, Paravantis JA, Papatheodorou A, Spilanis I (2014) Tourism sustainability methodologies: a critical assessment. In: IISA 2014 - 5th international conference on information, intelligence, systems and applications, pp 182–187. https://doi.org/10.1109/IISA.2014.6878832
14. Dubois G, Peeters P, Ceron JP, Gössling S (2011) The future tourism mobility of the world population: emission growth versus climate policy. Transp Res Part A Policy Pract 45:1031–1042. https://doi.org/10.1016/j.tra.2009.11.004
15. Goodwin H (2017) The challenge of overtourism. Responsible Tour Partnersh 4:1–19
16. Peeters P, Gössling S, Klijs J, Milano C, Novelli M, Dijkmans C, Eijgelaar E, Hartman S, Heslinga J, Isaac R, Mitas O, Moretti S, Nawijn J, Papp B, Postma A (2018) Research for TRAN committee-overtourism: impact and possible policy responses. Res. TRAN Committee-Overtourism impact possible policy responses, pp 1–255
17. Gretzel U, Jamal TB (2020) Guiding principles for good governance of the smart destination. Travel Tour Res Assoc Adv Tour Res Glob (2020)
18. Koo C, Shin S, Gretzel U, Hunter WC, Chung N (2016) Conceptualization of smart tourism destination competitiveness. Asia Pacific J Inf Syst 26:561–576. https://doi.org/10.14329/apjis.2016.26.4.561
19. Dorcic J, Komsic J, Markovic S (2019) Mobile technologies and applications towards smart tourism – state of the art. Tour Rev 74:82–103. https://doi.org/10.1108/TR-07-2017-0121
20. Gretzel U, Sigala M, Xiang Z, Koo C (2015) Smart tourism: foundations and developments. Electron Mark 25:179–188. https://doi.org/10.1007/s12525-015-0196-8
21. Gretzel U, Ham J, Koo C (2018) Creating the city destination of the future: the case of smart Seoul, pp 199–214. https://doi.org/10.1007/978-981-10-8426-3_12
22. Margherita EG, Braccini AM (2020) Organizational impacts on sustainability of industry 4.0: a systematic literature review from empirical case studies. In: Agrifoglio R, Lamboglia R, Mancini DRF (ed) Digital business transformation. Lecture notes in information systems and organisation. Springer, Cham. https://doi.org/10.1007/978-3-030-47355-6_12
23. Margherita EG, Braccini AM (2020) Industry 4.0 technologies in flexible manufacturing for sustainable organizational value: reflections from a multiple case study of italian manufacturers. Inf Syst Front (2020)
24. Margherita EG, Esposito G, Escobar SD, Crutzen N (2020) Exploring the smart city adoption process: evidence from the Belgian urban context. In: Proceedings of the 6th international workshop on socio-technical perspective in IS development (STPIS 2020), 8–9 June 2020. CEUR Workshop Proceedings (CEUR-WS.org)
25. Buhalis D, Foerste M (2015) SoCoMo marketing for travel and tourism: empowering co-creation of value. J Destin Mark Manag 4:151–161
26. Buhalis D, Law R (2008) Progress in information technology and tourism management: 20 years on and 10 years after the Internet—the state of eTourism research. Tour Manag (2008). https://doi.org/10.1016/j.tourman.2008.01.005
27. Gretzel U, Fesenmaier DR, Leary JTO (2006) Behaviour. Tour Bus Front Consum Prod Ind 9–18
28. Kontogianni A, Alepis E (2020) Smart tourism: state of the art and literature review for the last six years. Array 6:100020. https://doi.org/10.1016/j.array.2020.100020
29. Ling-Yung Z, Li N, Min L (2012) On the basic concept of smarter tourism and its theoretical system. Tour Trib Lvyou Xuekan 27:66–73
30. UNWTO (2004) Tourism Highlights Edition 2004
31. Hunter WC, Chung N, Gretzel U, Koo C (2015) Constructivist research in smart tourism. Asia Pacific J Inf Syst 25:105–120. https://doi.org/10.14329/apjis.2015.25.1.105
32. Tu Q, Liu A (2014) Framework of smart tourism research and related progress in China. In: Proceedings of the international conference on management and engineering (CME 2014), pp 140–146
33. Derzko W (2006) Smart technologies in the new smart economy. In: 1st technology futures forum (TFF) VTT Valimo (Metallimiehenkuja 2). Otaniemi, Espoo, Finland

34. Margherita EG, Braccini AM (2020) IS in the cloud and organizational benefits: an exploratory study. In: Lazazzara A, Ricciardi F, Za S (eds) Exploring digital ecosystems. Lecture notes in information systems and organisation, vol 33. Springer, Cham
35. Atzori L, Iera A, Morabito G (2010) The internet of things: a survey. Comput Netw 54:2787–2805. https://doi.org/10.1016/j.comnet.2010.05.010
36. Dikaiakos MD, Katsaros D, Mehra P, Pallis G, Vakali A (2009) Cloud computing: distributed internet computing for IT and scientific research. IEEE Internet Comput 13:10–11. https://doi.org/10.1109/MIC.2009.103
37. Lamsfus C, Martín D, Alzua-Sorzabal A, Torres-Manzanera E (2015) Smart tourism destinations: an extended conception of smart cities focusing on human mobility. Inf Commun Technol Tour 2015:363–375. https://doi.org/10.1007/978-3-319-14343-9_27
38. Lopez de Avila A (2015) Smart destinations: XXI century tourism. In: ENTER 2015 conference on information and communication technologies in tourism, Lugano, Switzerland, pp 4–6
39. Wired Magazine (2002) The bandwidth capital of the world. https://www.wired.com/2002/08/korea/
40. TechRepublic (2015) 20 of the world's most connected, innovative cities. https://www.techrepublic.com/pictures/photos-20-of-the-worlds-most-connected-innovative-cities/
41. SEGITTUR: Informe destinos turísticos inteligentes: construyendo el futuro
42. Meijer A, Bolívar MPR (2016) Governing the smart city: a review of the literature on smart urban governance. Int Rev Adm Sci 82:392–408. https://doi.org/10.1177/0020852314564308
43. Open Knowledge International: Open data
44. Del Chiappa G, Baggio R (2015) Knowledge transfer in smart tourism destinations: analyzing the effects of a network structure. J Destin Mark Manag 4:145–150. https://doi.org/10.1016/j.jdmm.2015.02.001
45. Gretzel U, Collier de Mendonça M (2019) Smart destination brands: semiotic analysis of visual and verbal signs. Int J Tour Cities 5:560–580. https://doi.org/10.1108/IJTC-09-2019-0159
46. Shafiee S, Rajabzadeh Ghatari A, Hasanzadeh A, Jahanyan S (2019) Developing a model for sustainable smart tourism destinations: a systematic review. Tour Manag Perspect 31:287–300. https://doi.org/10.1016/j.tmp.2019.06.002
47. Mehraliyev F, Chan ICC, Choi Y, Koseoglu MA, Law R (2020) A state-of-the-art review of smart tourism research. J Travel Tour Mark 37:78–91. https://doi.org/10.1080/10548408.2020.1712309
48. Perles Ribes JF, Ivars Baidal J (2019) Smart sustainability: a new perspective in the sustainable tourism debate *. Investig Reg 2019:151–170
49. González-Reverté F (2019) Building sustainable smart destinations: an approach based on the development of Spanish smart tourism plans. Sustainability 11:6874. https://doi.org/10.3390/su11236874
50. Webster J, Watson RT (2002) Analyzing the past to prepare for the future: writing a literature review. MIS Q 26
51. Vom Brocke J, Simons A, Niehaves B, Niehaves B, Reimer K, Vom Brocke J, Simons A, Niehaves B, Niehaves B, Reimer K, Plattfaut R, Cleven A (2009) Reconstructing the giant: on the importance of rigour in documenting the literature search process. Eur Conf Inf Syst (2009)
52. Corbin J, Strauss A (2015) Basics of qualitative research. Techniques and procedures for developing grounded theory. SAGE Publications Inc., Thousand Oaks (2015)
53. Commission E (2020) Compendium of best practices "2019 & 2020 European Capital of Smart Tourism competitions". https://smarttourismcapital.eu/wpcontent/uploads/2019/07/Compendium_2019_FINAL.pdf. Accessed 15 Nov 2020
54. Ivars-baidal JA, Maria HG, de Sofia MM (2019) Integrating overtourism in the smart tourism cities agenda. e-Review Tour Res 17:122–139
55. Cimbaljević M, Stankov U, Pavluković V (2019) Going beyond the traditional destination competitiveness–reflections on a smart destination in the current research. Curr Issues Tour 22:2472–2477. https://doi.org/10.1080/13683500.2018.1529149
56. Feng J, Yang Y, Shen H, Cai Z (2017) Development of lighting control system for smart hotel rooms. Int J Performability Eng 13(6):913–921. https://doi.org/10.23940/ijpe.17.06.p12.913921

57. Straker K, Wrigley C (2018) Engaging passengers across digital channels: an international study of 100 airports. J Hosp Tour Manag 34:82–92. https://doi.org/10.1016/j.jhtm.2018.01.001
58. Park JH, Lee C, Yoo C, Nam Y (2016) An analysis of the utilization of Facebook by local Korean governments for tourism development and the network of smart tourism ecosystem. Int J Inf Manag 36:1320–1327. https://doi.org/10.1016/j.ijinfomgt.2016.05.027
59. FREEWA (2019) FREEWA: FREE WATER PROJECT. https://freewa.org/. Accessed 20 Dec 2020
60. thecamp (2019) Transform your organization to face today's biggest challenges. https://thecamp.fr/en/support-positive-transformation. Accessed 20 Dec 2020
61. Kattegattleden: Kattegattleden - a cycle path along Sweden's west coast
62. Kayak Republic (2019) GreenKayak. https://kayakrepublic.dk/en/diverse/greenkayak/. Accessed 20 Dec 2020
63. Ilmasto H (2019) Helsinki's climate actions. https://www.stadinilmasto.fi/en/. Accessed 28 Dec 2020
64. Città di Venezia (2019) #EnjoyRespectVenezia. https://www.comune.venezia.it/it/enjoyrespectvenezia. Accessed 28 Dec 2020
65. Rocha J (2021) Smart tourism and smart destinations for a sustainable future. Encycl UN Sustain Dev Goals Decent Work Econ Growth 871–880. https://doi.org/10.1007/978-3-319-95867-5_88

A Conceptual Framework of Financial Inclusion: The Links with Individuals, SMEs, and Banks

Fadi Shihadeh

Abstract This chapter presents the conceptual framework of financial inclusion with linking to individuals, SMEs, and banks as these entities considering as demand and supply sides of financial inclusion. This paper address the definition and main factors of financial inclusion. Also, the importance and influence of financial inclusion in economic development and sustainability. Furthermore, how enhancing financial inclusion could influence people life and SMEs business, through access and using the formal financial services. Beside individuals and SMEs, this study drowns how banks -as supply side- could develop their business and positively enhancing their performances. The recommendations focusing on formal institutions to develop the laws and enhancing the infrastructure, also motivating the financial services providers to develop more services according to people needing.

Keywords Financial inclusion · Banks services · Individuals · SMEs

1 Introduction

Global organizations identify financial inclusion as one of the main factors of poverty alleviation in less-developed countries [1]. According to the Consultant Group to Assist the Poor (CGAP), "financial inclusion means that households and businesses have access to and can effectively use appropriate financial services. Such services must be provided responsibly and sustainably, in a well-regulated environment" [2]. Where the World Bank defines financial inclusion as "access to and use of services provided responsibly and sustainably, and the delivery of financial services at affordable costs to disadvantaged and low-income segments of society." The Palestine Monetary Authority (PMA) defines financial inclusion as enhancing access to financial services for all groups in society, at fair, transparent, and affordable costs [3]. Where, [4] define financial inclusion as the provision of financial services that benefit all members of society, without discrimination, at an affordable price, and in a suitable

F. Shihadeh (✉)
Palestine Technical University-Kadoorie, Tulkarm, Palestine
e-mail: f.shehadeh@ptuk.edu.ps

A. M. A. Musleh Al-Sartawi (ed.), *The Big Data-Driven Digital Economy: Artificial and Computational Intelligence*, Studies in Computational Intelligence 974,
https://doi.org/10.1007/978-3-030-73057-4_22

place, time, and form. The central point is that households (all members) and businesses should have access to financial services, especially those that face obstacles to financial inclusion, such as females, poors, youth, and small- and medium-sized enterprises (SMEs). Further, financial services should be designed to cover the needs of community members at affordable prices and continuously.

This study addresses the main points of financial inclusion. First, financial inclusion starts from "access," which is considered the main step toward the use of financial services. This means having an account in a formal financial-services institution or having the ability to open an account and access financial services without facing obstacles. Access is a supply-side factor informal financial service, such as banks, post offices, Schmukler of access means that financial-services providers are responsible for increasing their penetration levels within a country, especially in rural areas. Further, government institutions, in collaboration with financial-services providers, should focus on financial awareness and encourage society's members to access and use financial services, [5]. Presently, more than half of the world's adult population does not have a formal bank account, [6]. This high percentage indicates that global and official country-level institutions should work to enhance access levels, worldwide.

The second main factor is the "usage" of financial services. Having access to financial services does not necessarily translate to financial inclusion. For financial inclusion to lead to economic development, individuals and firms must be motivated to use financial services. This means developing services that meet their needs. For the disadvantaged, in particular, a lack of money is the primary factor behind their lower education levels, poor health outcomes, lower investment in job-creation activities, and fewer savings for the future.

The third main factor is that financial services providers provided responsibly and transparently for all individuals, groups, and firms in a country. Finally, these services should be offered to all at affordable and reasonable prices and continuously.

2 Why Financial Inclusion?

When members of a society are not included in its financial system (for example, the poor and SMEs), they need to rely on their limited resources to meet their financial needs and pursue promising growth opportunities. This lack of financial inclusion may hamper a country's economic growth and development. "Financial inclusion can be a key driver of economic growth and poverty alleviation, as access to finance can boost job creation, reduce vulnerability to shocks and increase investments in human capital" [7]. Several types of research have been conducted to empirically test the effect of financial inclusion on economic growth, income inequality, and unemployment in several countries and regions [1, 8–10].

Global organizations seek to achieve sustainable development in less-developed economies through achieving several goals: poverty reduction, improved health outcomes, higher education levels, gender equality, increased employment, economic growth, and sustainable development, where the financial inclusion addressed as an enabler of sustainable development goals [11]. Having access to financial sources encourages disadvantaged to save, invest, innovate, and create their businesses and, thus, create more jobs and decrease unemployment [1].

Having access to and use of formal financial sources helps the disadvantaged cover their needs, and some of these needs are urgent, such as medical care or meeting daily needs. Medical care and education are considered the leading indicators of and targets for sustainable development goals. Thus, enhancing the use of formal borrowing can enhance the sustainable development process. Youth, poor, and females, in particular, have less opportunity to earn money [2]; therefore, financial inclusion would allow them to borrow to invest in their future through education, by creating their businesses and by innovating through the use fast-developing technologies. Moreover, the financial inclusion of this disadvantaged group could positively reflect on the economy in several ways, including more money turning over in the economy, more job opportunities, more investments, and greater overall well-being for community members.

3 The Mechanism of Financial Inclusion

Researchers see financial inclusion as a key to financial development and economic sustainability. As most of the financially excluded are the poor, females, the illiterate, and youth who mainly live in rural areas and who make up society's most disadvantaged, [2, 12], researchers have focused on how these individuals deal with financial inclusion. Enhancing their level of access to financial sources is a key factor in improving their lives through employment, education, and healthcare. Enhancing financial inclusion means addressing both supply and demand sides of this vital factor, [13].

Moreover, It is not enough to encourage individuals and SMEs to access and use formal financial sources, while financial services providers fail to offer suitable and affordable services for the community members. Formal institutions, such as central banks, other financial regulatory authorities, ministries of finance, and other government institutions with an interest in financial inclusion and financial services, should work together to encourage the financial services industry to develop products and services that fit the real needs of community members. Banks as leading financial services providers can play an essential role [14]. They have the resources, technologies, tools, and manpower to develop and innovate the types of new products and services that can cover the needs of the disadvantaged, thereby helping them to become employable or to create and develop their businesses. This all leads to increased employment opportunities, poverty alleviation, and reduced unemployment, all of which lead to sustainable economic development. Meanwhile, an analysis of the individual characteristics that influence financial inclusion could lead to policies, regulations, and strategies that help shape a financial system that enhances the condition of people who have been excluded from the formal system. Figure 1, presents the mechanism of financial inclusion on banks, individuals, and SMEs.

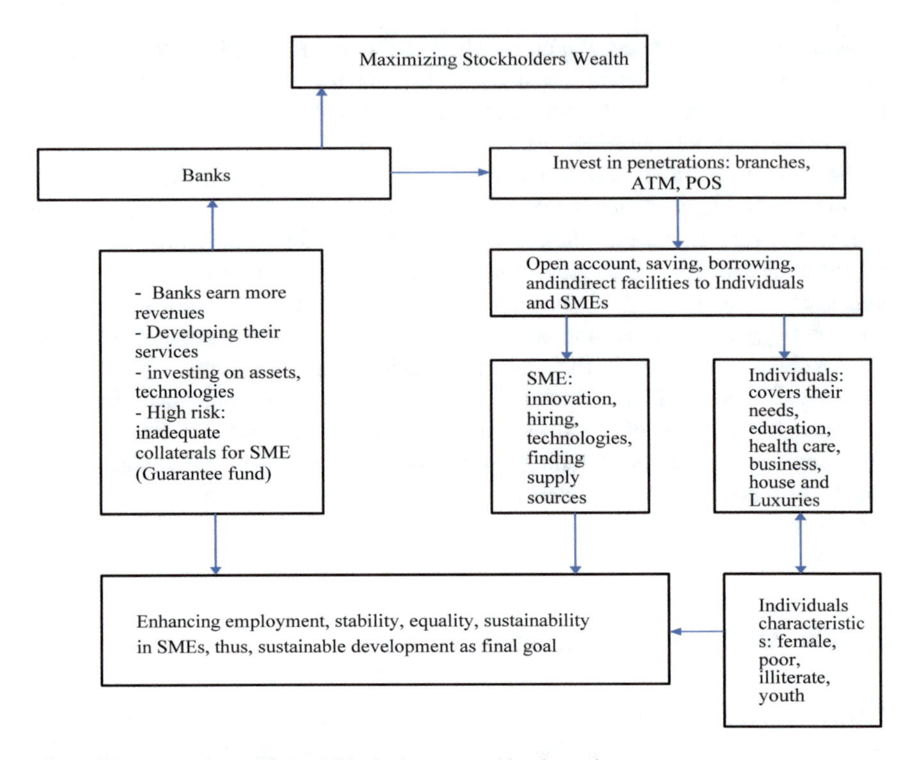

Fig. 1 The mechanism of financial inclusion. created by the author

4 How Financial Inclusion Influence the Demand and Supply Entities

As mentioned in the above sections, the philosophy of financial inclusion means that the formal financial system includes all citizens and entities to use the financial services. This section addresses the influence of financial inclusion on individuals and SMEs.

4.1 The Influence on Individuals

In their efforts to achieve sustainable development, policy-makers in developing countries continually face the triple challenge of poverty, inequality, and unstable economic growth [15]. Access to and use of formal financial sources for business development, education, healthcare, housing, and other needs can change individuals' lives for the better and, thus, enhance economic growth and sustainable development.

Financial inclusion is not the final target. Rather, it is a key factor in achieving development goals of economic growth, poverty reduction, and gender equality in education, health care, and work opportunities and, thus, in having the necessary resources for sustainable development [1]. Governments are working to ensure that all citizens are involved in the development process. Strategies aim to include them in the formal financial system, where they can open an account, make deposits, save money and use the formal financial services so they can cover their needs and participate in the development process. Poverty and inequality naturally occur in society. The question is to what minimum level these factors can be reduced without hindering the process of economic growth and sustainability.

In the past decade, several studies have examined the influence of financial access, as a financial inclusion indicator, on economic growth and development [e.g. 16, 10, 17]. The results show that enhancing financial inclusion leads to a decrease in inequality; however, its influence on poverty alleviation is not evident. Enhancing financial inclusion positively reflects on economic stability by attracting deposits to the banking system [10]. Further, Jin [18] points out that, in African and Latin American countries, financial inclusion can reduce the gap between the rich and poor. Where the recommendation is that governments promote financial inclusion by restructuring financial regulations to enhance the level of financial access to the poor and middle-income members of society.

Having a formal account consider as primary driver toward the use of formal financial services. Global institutions such as the World Bank, IMF, and other organizations have been working with central banks and other formal financial institutions to formulate policies and regulations that aim to increase the percentage of citizens with formal accounts. However, some barriers remain; these are addressed in World Bank surveys that measure financial inclusion, worldwide, [19]. In the survey, respondents noted their reasons for not having a bank account as being lack of money, too far away, too expensive, lack of trust, lack of documentation, religious reasons, or if a family member has an account [17, 20]. For "lack of money", the reason given is that some banks charge a fee for opening an account and another fee for managing an account. Another issue is that the disadvantaged do not have enough money to save at the bank. Thus, they do not need to open an account just for the sake of having one. For "too expensive", people were deterred by the fee charged for opening an account for the first time or for the monthly fee for maintaining an account [21].

"Lack of documentation" means that people do not have identity cards to use at a bank. For "lack of trust", people reported that they do not trust formal financial institutions with their savings nor do they trust their services. For "religious reasons", such as the Islamic religion, there are rules associated with dealing with financial services institutions, especially with banks. These rules influence the access level to formal financial institutions [22, 23]. Furthermore, in some economies and countries, if someone in the family has a formal account, then this is enough for other members to not have one. These reasons addressed in the World Bank survey and considered as barriers to not have a formal account.

Some of these barriers are related to the demand side of financial services, such as lack of money, if a family member has a formal account, lack of documentation, and religious reasons; whereas other barriers are related to the supply side, such as too expensive, too far away, and lack of trust [24]. Governments can devise rules and regulations that encourage banks and other financial services providers to enhance citizens' abilities to obtain formal accounts and, thus, to use the formal financial system. For example, governments and formal institutions can work together to enhance financial awareness among citizens so they can see the importance of having an identity card. They can also work together to develop services that are sensitive to religious issues and that can be offered through specialized financial institutions [22, 23, 25–29]. Further, on the supply side, financial services providers can either reduce their fees or offer to open accounts free of charge; they can also increase banking penetration in rural areas to reach all citizens. More effort can be made to enhance trust in the financial system and financial services providers, such as the banks.

4.2 The Influence on SMEs

SMEs are important to an economy due to their percentage of the total number of firms and, thus, their contribution to GDP, in creating jobs and adding value to economic productivity [30]. In emerging economies, SMEs contribute to approximately 40% of GDP and around 60% of total employment [31].

Due to the importance of SMEs to an economy, researchers, policy-makers, and economists have examined the factors that influence SMEs' performance and, thus, their contribution to employment and growth. They found that access to financial sources is the main obstacle to SMEs' growth [30, 32, 34–36]. The nature of SME firms limits their access to resources, mainly financial, technological, and labor. SMEs in developing countries suffer from low credit ratings. In the Middle East, North Africa, Afghanistan, And Pakistan (MENAP), less than 10% of SMEs have access to the financial resource, through lines of credit or other resources because they do not meet credit-rating requirements. Most studies found several issues, including lack of collateral, weakness in the management of financial systems, and unstable profitability [37, 38]. These limitations continuously reflect on their ability to innovate and grow. To ensure that SMEs fulfill their role in innovation, job creation, and economic stability and growth, government institutions and formal financial sources should work to assist SMEs in overcoming the obstacles they face. The global agenda should aim to enhance financial inclusion through reforming rules and regulations and the financial infrastructure to provide a suitable environment for SME growth, [12].

Reforming the infrastructure and innovating services to address the needs of SMEs will enable banks to lend more to these entities, at lower risk, thereby, increasing profits in the long term [33]. When banks attract SMEs' deposits and they improve their liquidity by improving the lending environment. However, enhancing financial inclusion for SMEs depends on the initiatives taken on two sides. On the one hand, for governments and official institutions, reforming the financial infrastructure would mean developing rules and regulations related to collateral issues and the prices of products and services. Financial-services should develop product services that suit the needs of SMEs in terms of both the type and price [30]. On the other hand, SMEs can aim for efficient management and financial systems to enhance their performance and decrease their risk.

Financial inclusion for SMEs means providing access and use of a diversity of formal financial services and funding sources continuously. It also means making them aware that some banks and monetary financial institutions offer analysis and technical consultancy services. These services and funding sources can help SMEs innovate, develop their products and distribution channels, reach new markets, and invest more in technologies to enhance their performance and grow their businesses. These develoments could positively reflect on job creation, reduction in poverty and inequality, and increased innovation and improved living standards and, thus, economic growth. Therefore, achieving sustainability development goals (SDG).

4.3 The Influence on Banks

Banks are the main players in the financial sector. They enhance financial inclusion and their performance, profits, and revenues through their variety of services and innovation, including online banking, digital channels, through penetration, and investments.

4.3.1 Banking Services

Banks can benefit from the developing services that meet clients' daily needs [13, 14]. One central principle for firms is continuity. Banks, as firms that work under this principle, offer services and facilities to individuals and SMEs to earn more revenues and profits. Community members increasingly seek new services to cover their needs, according to the developments of their daily lives. Therefore, to efficiently maintain the continuity of their products and services, banks should ensure that they keep pace with the needs of their clients. Growing their operations though enhancing financial inclusion means continually developing banking services.

4.3.2 Banking Penetration

Banks use multiple channels through which to supply their products and services to their customers (i.e., branches, mobile banking, ATMs, internet banking, and telephone banking), [39]. To earn more profits, banks expand their networks through POS, ATMs, branches, and other electronic terminals [40, 41]. They aim to attract deposits, reach more customers, and offer direct and indirect credit and other services. Opening new branches allow banks to invest in technology and equipment and hire more staff. This benefit the local economy as it allows individuals to access formal financial services, especially in disadvantaged areas [42]. Through branches, banks can offer their services and receive feedback from their customers [14, 29]. This enables them to innovate services that are directed to youth, women, craftsmen, and farmers, population groups that are typically marginalized when it comes to access to credit. In offering these services through their branches, banks can earn more profits, especially if this expansion comes with innovative services and suitable conditions for credit, such as reasonable collateral requirements and costs, [13, 43].

When banks invest generate new services to address customers' needs, they achieve suitable returns, which they use to enhance their performance. Taking this approach means an increase in capital expenditures and profits might decrease in the short run, due to the cost increment. As most banks cannot achieve earnings from this type of short-term development, they aim to enhance their earnings over the long term.

There are also costs related to ATMs and ATM services. These include the purchase or lease and maintenance of the machines and the cost of new technology and security. These conditions apply to services at the point of sale. As these are deemed essential services and as essential services are usually offered free of charge, they do not provide direct revenues for banks. However, banks that provide these services can generate earnings by attracting new customers and their deposits. This approach enhances its ability to lend money to (demand-side) clients. ATMs also enhance customer loyalty [44]. Increasing the number of ATMs could also lead to new customers opening up accounts and, thus, increased bank deposits [45]. Some banks use ATMs to lend money to customers, offer new services, or allow customers to pay their bills (school fees, university tuition, telephone charges, electricity and water charges, and taxes) [46, 47].

4.3.3 Services Innovation

[45] linked service innovation with improvements in having a formal account and payment tools, such as electronic cards. Enhancing these tools improves financial inclusion level. However, the associated costs include risk, market research, advertising, training, and delivery channels. These types of costs have no direct effect on profits until customers use the new services, but banks can earn profits, from new services, in the long term, [14]. However, creating a new service also poses challenges for banks and the financial system because customers may not find the new service to be useful and may not utilize it [26].

Innovation as an essential issue for any sector of the economy. Therefore, Banks innovates their services to enhance their competitive advantage [48, 49]. Several empirical studies address the significant impact of financial innovation on banks' performance [50, 52, 52–55]. On the one hand, these innovations must take into account the needs of individuals and SMEs, thus, how these services can enhance an individual's daily lives and SMEs' performance. On the other hand, researchers argue about the causality of financial innovation on financial institutions' performance, especially that of the banks.

There are two approaches to service innovation. On one side, banks can innovate their services to cover their customers' needs. They can use their networks (i.e., branches, ATMs, offices, online channels) to find out what customers are looking for, and at what price, through feedback and inquiries. Thus, banks can find gaps in the services offered in the community and, then, develop these services or introduce new ones and attract more deposits and new clients. On the other side, banks can work on market research to find out what new services can be developed. They can motivate their customers to see a new need, based on the level to which the economy is developing and the level of technology use and business entrepreneurship. This kind of innovation could help banks attract more clients, deposits, and decrease their risks through the diversification of services, thereby, increasing revenues and profits.

4.3.4 Online Banking and Digital Channels

Credit cards, debit cards, mobile banking, and internet services are used to deal with financial services, both locally and internationally. Banks use these channels to offer and promote their services and, thus, enhance their investment and profits. They also reduce their risk. For example, before extending a line of credit, banks require customers to guarantee that both the credit used and the interest owed will be covered. Banks have also addressed the issue of increased technological risk and the possibility of fraud and have reduced credit and liquidity risks [44]. Banks earn more profits from credit cards, but they also suffer from the additional risks associated with extending credit [56].

Increasing the use of online tools allows banks to invest more in these tools. Offering online banking as a service decreases the number of visits to branches and reduces the associated costs [58], for example, in the use of paper and employees' time spent on delivering services on a person-to-person basis at bank branches. Through online banking, customers can easily access financing, transfer money, and pay utility bills, thereby, increasing monetary turnover, business innovation, and work opportunities in the economy [45]. Moreover, as online banking is less costly, banks can earn more revenues from individual transactions. Some of these revenues do not directly come from online services. For example, Banks usually offer online banking to new users free of charge; this encourages existing customers to use this service, and it could attract more customers to open an account, thus leading to the increased use of other banking services, [59].

Online banking can help banks reduce operating costs and increase operating revenues, thereby, achieving efficiencies [59]. Further, enhanced customer use of online banking will reflect on banks' performance [60] When banks encourage customers to open accounts through online banking, where they can transfer money or make deposits at any time, these banks achieve more efficiency and profits than other banks [16]. However, the efficiencies gained and profits earned depend on several issues, such as internet access and cost, and prospective clients' financial awareness [62].

4.3.5 Banking Investment

Banks base their targets on maximizing profits and revenues, thus, maximizing stockholder value [13, 63]. They invest their capital in opening new branches, offices, and ATMs, acquiring new technologies and equipment, in market research, hiring and training staff, and innovations and developing new services. These types of investments have helped banks achieve and enhance their targets, including increasing financial inclusion.

The philosophy of financial inclusion means including more citizens in the formal financial system of government institutions and financial-services providers. The banks are the leading financial-services providers in terms of attracting deposits and offering credit and other services. To provide these services, they seek to earn more revenue and enhance their performance indicators. They also focus on decreasing their costs and risks by balancing their loan portfolios; that is, they balance their loans to individuals and SMEs with other kinds of investments, such deposits in other banks and financial institutions.

Whereas banks seek to enhance their performance indicators, citizens want to cover their daily needs, enhance their living standards, and establish their businesses through access to financial sources. Thus, financial inclusion means matching banks' targets and people's needs. When banks enhance financial inclusion through banking penetration and the provision of suitable services at affordable prices, this can increase their profits and performance indicators.

4.3.6 Banking Competition

Banks compete within economies, countries, and regions. This competition could be enhanced through increased banking penetration (i.e., branches, offices, ATMs, POSs); services innovation; developing mechanisms to offer services through affordable pricing, collateral requirements, and other procedures; and activating and enhancing online services [13]. This competition motivates banks to develop their business and services, thus, attracting more clients and deposits, which are considered the primary revenue sources for commercial banks. These developments in financial services provide an appropriate environment for using formal financial services to develop businesses, innovate, and encourage entrepreneurship among youth, females, and small firms, thereby creating more jobs, decreasing unemployment, and achieving economic growth and sustainable development.

4.4 Financial Development, Economic Growth, and Stability

One key to enhancing economic development is enhancing financial inclusion [64, 65]. One main factor in enhancing financial inclusion is to reform the financial infrastructure through laws and regulations, payment systems, and collateral requirements.

A considerable amount of research points to the need for countries and economies to direct more effort toward reforming the infrastructure of their financial systems [14]. This reform would include financial intermediation for all types of financial institutions (i.e., banks, money exchangers, stock markets, brokerage companies, and insurance companies, microfinance institutions, leasing companies). These institutions provide several services to society's individuals and firms. Therefore, reforming the regulations that link these institutions and their customers will reflect on the financial-development indicators and the access to and usage of financial services [66–68].

Developments in financial intermediation should reflect on the level of trust in these institutions and, thus, their ability to offer their services through savings and deposits. Many companies offer their services through banks as they play a critical role in the surplus and deficits of business units. Some of these services include insurance and stock purchases. Therefore, through a policy of financial inclusion whose goal is to include more people in the financial system, banks can attract deposits, offer loans, and other services.

Reforming the financial system reflects on both financial inclusion (individuals and firms at the micro-level) and financial development (stock markets, private credit as a percentage of GDP at the macro level). In some empirical studies, no relationship is made between financial development and economic growth as targets for countries [69]. Other studies point out the general relationship between these two factors (Levine, [70–75]. Where, other studies found that there is a definite relationship between the ability of the financial system to ensure financial inclusion and how a country develops, [76]. They also point out that there is a definite significant relationship between financial-inclusion indicators and economic growth. Enhancing the level of credit in an economy could reflect on its growth and development [77].

Furthermore, reforming the financial system to achieve financial inclusion will mean that more individuals and firms will become engaged in the financial system, and this can enhance the financial stability of the economy [66, 78]. In addition, empowering households and SMEs will enable them to face financial and economic shocks [79, 80]. Indirectly, a financial inclusion that enhances access to and usage of financial services could enhance productivity and innovation, both of which lead to economic growth [74, 75, 81].

5 Conclusion and Policy Recommendations

Financial inclusion considered a vital topic in economics and financial studies. It contributes to improved economic and social indicators, particularly among the disadvantaged members of societies, and it significantly contributes to economic growth and sustainable development. This study presents the mechanism and theoretical framework of financial inclusion. This framework shows the definition and dimensions of financial inclusion and its importance, according to the relevant global institutions. It discusses the influence of enhancing financial inclusion on both demands-

(individuals and SMEs) and supply-side entities (mainly banks). It also presents how financial inclusion influences banks' activities, such as in investment in products, services, innovation, and network expansion, and bank competition.

The main points show the vital role financial inclusions play in social and economic development. The finance literature has widely examined financial inclusion, its promotion and enhancement, and its impact on individuals and SMEs. Banks use their resources and technologies to develop products and services at affordable prices, under fair conditions, and extend their reach to the disadvantaged members of society. In this way, they enhance their performance, reduce their risk, and earn profits, while contributing to increased standards of living and overall economic development. When individuals have access to formal financial services, they can enhance their ability to increase their level of education and access to health care and invest in their businesses. This results in poverty alleviation and the achievement of equality between males and females in opportunities for work. Further, SMEs can use financial sources to develop their businesses, find new markets, and improve their growth and sustainability.

This study presented a theoretical framework of the connection between banks' targets and citizens' needs through financial inclusion. The connection between formal financial institutions, particularly the banks, and citizens and SMEs, directly contributes to long-term economic growth and sustainability. Moreover the current study recommend that future research could focus on online banking services according to the COVID 19 pandemic and how using online services will influence living for individuals and firms thus the influence on banks performance and risk.

Reference

1. Consultative group to assist the poor, CGAP (2016) Achieving the sustainable development goals, the role of financial inclusion. https://www.cgap. org/publiccations/ achieving-sustainable-development- goals.
2. Consultative group to assist the poor, CGAP (2012) Annual report, advancing financial access for the world's poor. www.cgap.org/sites/de- fault/files/CGAP-Annual-Report-Dec-2012.pdf
3. Palestine monetary authority (2018) "www.pma.ps"
4. Aduda J, Kalunda E (2012) Financial inclusion and financial sector stability with reference to kenya: a review of literature. J Appl Financ Bank SCIENPRESS Ltd 2(6):1–8
5. Wang X, h., & Shihadeh, F. h. (2015) Financial inclusion: policies, status, and challenges in palestine. Int J Econ Financ 7(8):196–207. https://doi.org/10.5539/ijef.v7n8p196
6. World Bank (2017) Small and medium enterprises (SMEs) finance. https://www.worldbank.org/en/topic/financialsec- tor/brief/smes-finance
7. Ben Naceur S (2014) ACCESS to finance for small and medium-sized enterprises in the MENAP and CCA regions. International Monetary Fund. https://www.imf.org/external/pubs/ft/reo/2014/mcd/eng/
8. Turegano DM, Herrero AG (2018) Financial inclusion, rather than size, is the key to tackling income inequality. Singapore Econ Rev 63(01):167–184
9. Claessens S (2006) Access to financial services: a review of the issues and public policy objectives. World Bank Res Observer 2 21(2): 207–40. https://doi.org/10.1093/wbro/lkl004

10. Neaime S, Gaysset I (2018) Financial inclusion and stability in MENA: evidence from poverty and inequality. Financ Res Lett (24):230–237. https://doi.org/1 0.1016/j.frl.201 7.09.007
11. World bank (2020). https://www.worldbank.org/en/topic/financialinclu- sion/overview
12. Angori G, Aristei D, Gallo M (2019). Lending technologies, banking relationships, and firms' access to credit in Italy: the role of firm size. Appl Econ 51(58): 6139–6170. https://doi.org/10.1080/00036846.2019.1613503
13. Shihadeh F, Liu, B (2019) Does financial inclusion influence the banks risk and performance? Evidence from global prospects. Acad Acc Financ Stud J 23(3):1–12. https://www.abacademies.org/articles/DoesFinancial-Inclusion- Influence-the-Banks-Risk-and-Performance-1528–2635–23–3–403.pdf
14. Shihadeh FH, Hannon A, Guan J, Ul Haq I, Wang X (2018) Does financial inclusion improve the .banks' performance? Evidence from Jordan. In: John W. Kensinger (ed.) Global tensions in financial markets (research in finance, volume 34) emerald publishing limited pp 117–138. https://doi.org/10.1108/S0196-382120170000034005
15. Novignon J, Nonvignon J, Mussa R (2018) The poverty and inequality nexus in Ghana: a decomposition analysis of household expenditure components. Int J Soc Econ 45(2):246–258. https://doi.org/10.1108/IJSE-11-2016-0333
16. Honohan P (2004) Financial development, growth and poverty: how close are the links. In: Goodhart, C. (Ed.), financial development and economic growth: explaining the links. Palgrave, London, UK
17. Demirgüç-Kunt A, Klapper LF (2012) Measuring financial inclusion: the global findex database, World bank policy research working paper 6025, World Bank, Washington, DC
18. Jin D (2017) The inclusive finance have effects on alleviating poverty. Open J Soc Sci 5:233–242. https://doi.org/10.4236/jss.2017.53021
19. Demirguc-Kunt A, KlapperL, Singer D, Ansar S, HessJ (2018) Global findex database 2017 measuring financial inclusion and the fintech revolution. Washington, DC: World Bank. © World Bank. https://openknowledge.worldbank.org/handle/10986/29510 License: CC BY 3.0 IGO
20. Demirgüç-Kunt A, Klapper L, Randall D (2014) Islamic finance and financial inclusion:measuring use of and demand for formal financial services among Muslim adults. Rev Middle East Econ Financ 10(2):177–218
21. Allen F, Demirgüç-Kunt A, Klapper L, Peria MSM (2016) The foundations of financial inclusion: understanding ownership and use 30. http://dx.doi.org/https://doi.org/10.1016/j.jfi.2015.12.003
22. Demirgüç-Kunt A, Klapper LF, Singer D (2013) Financial inclusion and legal discrimination against women: evidence from developing countries, Policy research working paper 4616, The World Bank
23. Ben Naceur S. (2014), "Access to finance for small and medium-sized enterprises in the MENAP and CCA regions, International Monetary Fund. www.imf.org/external/pubs/ft/reo/2014/mcd/eng/
24. Zins A, Weill L (2016) The determinants of financial inclusion inAfrica. Rev Dev Financ 6(1):46–57
25. Zhang Q, Chen R (2015) Financial development and income inequality in China: an application of SVAR approach. Procedia Comput Sci 55:774–81. 10.1016j.procs.2015.07.159
26. Beck T, Senbet L, Simbanegavi W (2015). Financial inclusion and innovation in Africa: an overview. J African Econ, 24(sup1): i3–i11. doi: https://doi.org/10.1093/jae/eju031
27. Beck T, Demirgüc-Kunt A, Honohan P (2009). Access to financial services: measurement, impact, and policies. World Bank Res Observer, 24(1):119–145. https://doi.org/https://doi.org/10.1093/wbro/lkn008
28. Shihadeh FH (2018) How individual's characteristics influence financial inclusion: evidence from MENAP. Int J Islamic Middle East Financ Manage 11(4):553. https://doi.org/10.1108/IMEFM-06-2017-0153
29. Shihadeh FH (2019) Individual's behavior and access to finance: evidence from Palestine. Singapore Econ Rev. https://doi.org/10.1142/S0217590819420025

30. Gozzi JC, Schmukler SL (2015) Public credit guarantees and access to finance. Euro Econ Banks Regul Real Sect 2:101–117
31. World Bank (2017). Small and Medium Enterprises (SMEs) finance. https://www.worldbank.org/en/topic/financialsec-tor/brief/smes-finance
32. International Monetary Fund (2014). Access to finance for small and medium-sized enterprises in the MENAP and CCA regions. https://www.imf.org/external/pubs/ft/reo/2014/ mcd/eng/
33. Wellalage N, Locke S (2017). Access to credit by SMEs in South Asia: do women entrepreneurs face discrimination. Res Int Bus Financ, 41:336–346. http://dx.doi.org/https://doi.org/10.1016/j.ribaf.2017.04.053
34. Shinozaki S (2012). A new regime of SME finance in emerging Asia: empowering growth-oriented SMEs to build resilient national economies. ADB working paper series on regional economic integration 104
35. De la Torre A, Martnez Pería MS, Schmukler SL (2010) Bank involvement with SMEs: beyond relationship lending. J Bank Financ 34(9):2280–2293
36. Petersen MA, Rajan RAG (1994) The benefits of lending relationships: evidence from small business data. J. Fin. 49 (1):3–37
37. Beck T, Demirguc-Kunt A, Maksimovic V (2008) Finacing patterns around the world: the role of institutions. J. Fin. Econ. 90:467–487
38. Hoehle H, Huff S (2012) Advancing task-technology fit theory: a formative measurement approach to determining task-channel fit for electronic banking channels. In: Hart DN, Gregor SD (eds) Information systems foundations: theory building in information systems. ANU E Press, Canberra, pp 133–169
39. Berger AN, Leusner JH, Mingo JJ (1997) The efficiency of bank branches. J Monetary Econ 40(1):141–162
40. Hensel ND (2003) Strategic management of cost efficiencies in networks: cross-country evidence on European branch banking. Euro Financ Manag 9(3):333–360
41. Nguyen HLQ (2015). Do bank branches still matter? the effect of closings on local economic outcomes. Department of economics, Massachusetts Institute of Technology, Cambridge, MA
42. Shihadeh F (2020) The influence of financial inclusion on banks' performance and risk: new evidence from MENAP. Banks Bank Syst 15(1):59–71. https://doi.org/10.21511/bbs.15(1).2020.07
43. Monyoncho, l (2015) relationship between banking technologies and financial performance of commercial banks in kenya. Int J Econ Comm Manag 3(11):784–814
44. Frame WS, White lJ (2012) Technological change, financial innovation, and diffusion in banking. in The Oxford handbook of banking, chapter 19, Oxford University Press
45. Martins C, Oliveira T, Popovič A (2014) Understanding the internet banking adoption: a unified theory of acceptance and use of technology and perceived risk application. Int J Inf Manag 34(1):1–13
46. Lee KC, Chung N (2009) Understanding factors affecting trust inand satisfaction with mobile banking in Korea: a modified DeLone and McLean's model perspective. Interact Comput 21(5–6):385–392
47. Frame WS, White lJ.(2004). Empirical studies of financial innovation: lots of talk, little action? J Econ Lit 42(1):116–144
48. Batiz-Lazo B, Woldesenbet K (2006) The dynamics of product and process innovation in UK banking. Int J Fin Serv Manag 1(4):400–421
49. Ngari JMK, Muiruri JK (2014) Effects of financial innovations on the financial performance of commercial banks in Kenya. 4(7): 51–57. www.ijhssnet.com
50. Akhisar I, Tunay B, Tunay N (2015). The effects of innovations on bank performance: the case of electronic banking services. In: World conference on technology, innovation and entrepreneurship, procedia – social and behavioral sciences, vol 195, pp 369–375
51. Sansone M, Formisano V (2016). marketing innovation and key performance indicator in banking. Int J Mark Stud 8(1): 44–56. https://doi.org/10.5539/ijms.v8n1p44
52. Mabrouk A, Mamoghli C (2010) Dynamic of financial innovation and performance of banking firms: context of an emerging banking industry. Int Res J Financ Econ 5:2010

53. Akram JK, Allam MH (2010). The impact of information technology on improving banking performance matrix: jordanian banks as case study. In: European, mediterranean & middle eastern conference on information systems 2010. 12–13 April 2010, Abu Dhabi, UAE.
54. Hasan I, Schmiedel H, Song L (2010). Return from retail banking and payments. Bank of Finland research, discussion papers no 3
55. Sinkey JF, Nash RC (1993). Assessing the riskiness and profitability of credit-card banks. J Finan Serv Res 7:127–150. https://doi.org/https://doi.org/10.1007/BF01046902
56. Humphrey DB (1994) Delivering deposits services: ATMs versus branches. Econ Q Fed Reserve Bank Richmond 80:59–81
57. Shihadeh F (2020). Online payment services and individuals behavior: new evidence from the MENAP. Int J Electr Bank. https://www.inderscience.com/info/ingeneral/forthcoming.php?jcode=ijebank.
58. Simpson J (2002) The impact of the internet in banking: observations and evidence from developed and emerging markets. Telematics Inform 19:315–330
59. Stoica O, Mehdian S, Sargu A (2015) The impact of internet banking on the performance of romanian banks: DEA and PCA approach. Procedia Econ Financ. 20:610–622. https://doi.org/10.1016/S2212-5671(15)00115-X
60. Hernando I, Nieto M (2007) Is the Internet delivery channel changing banks' performance? the case of Spanish banks. J Bank Financ 31(4): 1083–1099. https://doi.org/https://doi.org/10.1016/j.jbank- fin.2006.10.011
61. Atay E (2008) Macroeconomic determinants of radical innovations and internet banking in Europe. Ann Univ Apulensis Ser Oeconomica 2:10
62. Pilloff SJ (1996) Performance changes and shareholder wealth creation associated with mergers of publicly traded banking institutions. J Money Credit Bank 28(3):294–310. https://doi.org/10.2307/2077976
63. Chibba M (2009) Financial inclusion, poverty reduction and the millennium development goals. Eur J Dev Res 21:213–230
64. Pouw NG, Gupta (2017) Inclusive development: a multi-disciplinary approach. Curr Opin Environ Sustain 24:104–108. https://doi.org/10.1016/j.cosust.2016.11.013
65. World Bank (2013). Financial Inclusion in Brazil: Building on Success 8314
66. Prasad E 2010 Financial sector regulation and reforms in emerging markets: an overview. NBER Working Paper 16428, Cambridge, MA
67. Cull R, Demirgüç-Kunt A, Lyman T (2012) financial inclusion and stability: whatdoes research show? CGAP, Washington DC
68. Hassan MK, Sanchez B, Yu JS (2011) Financial development and economic growth: New evidence from panel data. Q Rev Econ Financ 51(1):88–104. https://doi.org/10.1016/j.qref.2010.09.001
69. Levine R (1997) Financial development and economic growth: views and agenda. J Econ Lit XXXV, 688–726. https://www.jstor.org/stable/2729790
70. Levine R (2002) Bank-based or market-based financial systems: which is better? J Financ Intermediation 11(4):398–428
71. Levine R (2005) Finance and growth: theory and evidence. Handbook Econ Growth 1:865–934
72. Beck T, Levine R (2004) Stock markets, banks, and growth: panel evidence. J Bank Financ 28(3):423–442
73. Levine R, Zervos S (1998). Stock markets, banks, and economic growth. Am Econ Rev, 537–558
74. Levine R, Loayza N, Beck T (2000) Financial intermediation and growth: causality and causes. J Monetary Econ 46:31–77. https://doi.org/10.1016/S0304-3932(00)00017-9
75. Agyekum F, Locke S, Hewa Wellalage N (2016) Does financial accessibility and inclusion promote economic growth in low income countries (LICs). In: Elaine M (ed) Financial performance analysis, measures and impact on economic growth. Nova Science Publishers Inc., New York, pp 158–186
76. Corrado G, Corrado L (2015) The geography of financial inclusion across Europe during the global crisis. J Econ Geo 15:1055–1083

77. Rusu V (2017). Financial inclusion and stability: linkages among financial development and economic growth. J Fin Innov https://doi.org/https://doi.org/10.15194/jofi_2017.v0.i0.57
78. Raddatz C (2006) Liquidity needs and vulnerability to financial under- development. J Finan Econ 80:677–722
79. Beck T, Lundberg M, Majnoni G (2006) Financial intermediary development and growth volatility: do intermediaries dampen or magnify shocks? J Int. Money Finan 25:1146–1167
80. Levine R (1999) Law, finance, and economic growth. J Fin Intermediation 8:36–67
81. Ayyagari M, Demirguc-Kunt A, Maksimovic V (2011) Small vs. young firms across the world : contribution to employment, job creation, and growth Policy Research Working Paper Series 5631, The World Bank.

Sustainable Development: The Case of the Smart City

Nadia Mansour and Nabila Smaili

Abstract The economic and environmental crises are having a major impact on several countries, particularly urban territories, as a result; Effective treatment of urbanization problems has become a global priority today. For several years, urban development and related issues have been the subject of numerous international conferences. In this context, making cities smart and sustainable requires an in-depth rethinking of resource access models, transportation, waste management, building cooling, and especially energy management (generation and transmission).

Keywords Smart city · Sustainable development · New information · Communications technologies · Urban development

1 Introduction

The United Nations organization announces that two out of three people will live in cities in 2030 and that cities today occupy 2% of the world's surface, are home to 50% of the world's population, consume 75% of the energy produced and are the source of 80% of CO2 emissions.

The economic and environmental crises are therefore having a major impact on urban territories; effectively addressing urbanization problems has become a global priority today. Urban development and related issues have thus been the subject of numerous international conferences for several years.

The 41st World Expo, which took place in Shanghai in 2010, with the theme "Better City, Better Life", showed that the city is the source of new opportunities for individuals, but also of new challenges (fight against greenhouse gases and air

N. Mansour (✉)
University of Sousse-Tunisia and University of Salamanca, Salamanca, Spain
e-mail: Mansour.nadia@usal.es

N. Smaili
University of Mouloud Mammeri, Tizi-Ouzou, Algeria

A. M. A. Musleh Al-Sartawi (ed.), *The Big Data-Driven Digital Economy: Artificial and Computational Intelligence*, Studies in Computational Intelligence 974,
https://doi.org/10.1007/978-3-030-73057-4_23

pollution, The challenge of the development of new transport modes (e.g., questioning of certain modes of transport due to the scarcity of fossil fuels, problems posed by industrial disasters and insecurity, excessive waste production, increasing energy consumption), which are not properly taken into account, can lead to further environmental degradation, poverty, and exclusion.

According to [17]: "Nowadays, the popular term "smart city" implies the intertwining of several parallel social and technical processes in modern society. Firstly, it incorporates the process of scientific and technological development and the constant diffusion of technology as well as their implementation into all spheres of human life. The second process is the desire to improve the quality of life and create comfortable living conditions, which characterizes the modern development of the urban environment. The third process reveals the transformation of a territorial management system, the use of innovative methods to the allocation of resources as well as the setting of tasks, and the coordination of their implementation. It is assumed that synergy across three processes should lead to the creation of new social values and urban sustainable development".

In this context, making cities smart and sustainable means trying to reduce the impact of harmful effects on the environment, but also trying to rethink in depth the models of access to resources, transport, waste management, air-conditioning of buildings and above all energy management (production, transport).

This reflection started from a simple observation of these observations and led to the following questions: what is meant by a smart city? What are the salient features of the smart city and its various roles; and which actors are candidates for this role? In other words, it is a question of knowing:

How can we prepare the transformation of cities for these new challenges to guarantee a better living environment for future generations? How can an intelligent city be considered as a vector of sustainable development?

2 The Objectives Pursued

In this context, the objectives of this study are as follows:

- Elucidate what is meant by the intelligent city and to propose a grid for analyzing the dynamics of innovation in this field.
- The various vital functions concerning mobility, water and energy supply, transport networks, and waste management are reviewed to respond to new ecological issues and the comfort requirements of city dwellers.

To answer the fundamental question mentioned above, a theoretical framework has been chosen and a case study will be presented on the importance of the Dubai Smart City strategy.

The first section is devoted to a review of targeted literature, the second section discusses the components and actors of a smart city, the third section focuses on the strengths and limitations of the smart city, the fourth section analyzes the links between the smart city and sustainable development, and the last section presents the case study on the Smart City of Dubai.

2.1 Smart City: Literature Review

While the expressions "future cities", "eco-city", "intelligent cities", "compact cities", "innovative cities", "green cities", are used in a stable way to characterize the cities of tomorrow, the expression of Anglo-Saxon origin "smart cities" is increasingly used [20].

The expression was born in the 1990s. Three phenomena are often identified to explain its origin and popularity:

- The popularity of the term "smart city" is the product of an advertising campaign by a private firm in the context of profit-seeking: this expression is first of all the result of a market re-conquest strategy put in place by the IBM firm. Wishing to increase its profits in a period of recession, the firm identified cities as a huge potential market, by associating it with information and communication technologies [25].
- An expression that is in line with other terms to capture the emergence of new technologies in urban spaces.
- To think about the city of the future, several urban thinkers have taken an interest in the city of tomorrow, a city where technology still plays a role. The "intelligent city" remains, in a way, the culmination of these different thoughts. The success of this name is also due to its appearance in a, particularly favorable context. Cities would indeed be faced with four major phenomena, requiring a series of actions to be put in place:

1) Growing urbanization
2) Climate change and awareness of resource scarcity,
3) The reduction of budgets
4) And the competition between cities.

Faced with these phenomena, the "intelligent city" appears, thanks to technology, as a possible response. Indeed, the "link with technology is perfect, even if it can refer to a great diversity of usages and levels of appropriation" [8].

Also, "Smart cities have found increasing popularity over the last decade in public policy circles and academia. However, the definitions and characteristic elements of the concept have remained somewhat inconsistent. In literature a variety of issues, including Information and Communication Technology (ICT), universal connectivity, knowledge and creativity, big data and open data [29], social capital, business and entrepreneurship, smart community, ecological sustainability… were used to characterize the smart city discourse, bewildering the meaning of the concept" [23].

In this context, the smart city can refer to the aggregation of aspects of what can make a city economically competitive, efficiently managed, and pleasant to live in.

Other factors come into play: the city will be said to be smart if it promotes innovation (smart economy), invests in training (smart people), is well-governed (smart governance), has a good quality of life (smart living) and good environmental performance (smart environment) in addition to sustainable mobility (smart mobility), [1].

The intelligent city is simply the one that responds to the normative vision that the current discourse proposes: a competitive, environmentally virtuous, democratically participatory, energy-efficient city, concerned about the quality of life of its inhabitants, [1].

There is no univocal and consensual definition of the concept of "intelligent city". The operationalization and application of the concept vary according to the country, the territory, the context, and the territorial stakes.

According to [23] and based on the information and communication technology, to define a smart city, they represented different domain emphasis:

- Information and communication technology (ICT): A smart city is an instrumented, interconnected, and intelligent city that uses ICT to sense, analyze, and integrate critical information on core systems in running cities [14].

A smart city refers to a local entity—a district, city, region, or small country—that takes a holistic approach to employ information technologies with a real-time analysis that encourages sustainable economic development [26].

- Business-led urban development, entrepreneurship, and creative industries: A key characteristic of smart cities is their underlying emphasis on business-led urban development, the domination of neoliberal urban spaces, and a subtle shift in urban governance from managerial to entrepreneurial forms; and cities being shaped increasingly by big-business and/or corporations [13]
- Community development and social capital: A smart city will be a city whose community has learned to learn, adapt, and innovate within the emerging technological age [6]A smart community is a community that has made a conscious effort to use information technology to transform life and work within its region in significant and fundamental rather than incremental ways) [5].
- Learning and knowledge-based development: Smart cities are territories with high capacity for learning and innovation, which is built in the creativity of their population, their institutions of knowledge creation, and their digital infrastructure for communication and knowledge management [16]
- Sustainable development: A city is smart when investments in traditional infrastructure, social development and modern (ICT) communication infrastructure fuel sustainable growth and high quality of life, with a wise management of natural resources [31]

Besides, according to [9] a city can be intelligent in several areas of its management: economy, mobility, environment, inhabitants, lifestyle, and administration are the six areas of city intelligence that must be improved by cities using the latest technologies and innovative thinking.

The theoretical aspects and practical manifestations of the intelligent city vary enormously from one country to another, but the integration of the latest information and communication technologies is the common denominator in all aspects.

2.2 *Components and Actors of an Intelligent City*

The work of [9] is most often used to demonstrate the six components of the smart city: smart governance, smart economy, smart environment, smart living, smart mobility and smart people.

The interaction and cooperation between several actors can lead to the emergence of smart cities at several levels:

- The intelligence of tools through innovation and digital technology;
- An intelligence of the organization of spaces thanks to the engineering of urban planning and the emergence of new functionalities in cities;
- The intelligence of the citizen, who accesses relevant information and can thus make better decisions, creating a virtuous circle.

The smart city is then a community that understands its issues and is proactive in developing solutions to these issues.

Intelligent Governance

Governance is a fundamental pillar when it comes to the implementation of the smart city since this concept requires specific governance. This governance, which will have to be more open and transparent, will work in conjunction with the various city departments.

The State and local authorities are often seen as the main actors in transition and development. Many movements are calling for a more secure city, for a more pleasant living environment, for a more participatory democracy, which depends to a large extent on the developments carried out by public actors.

Intelligent governance means modernizing the administration to meet the new governance challenges. Modernization of public action to better organize local power through optimization of public policies and an analysis of their effects, finally, participatory management and close collaboration between the different actors and citizens where new information and communication technologies are used as a lever between decision-makers, public actors and citizens; where information is transparent.

The three pillars of smart governance are transparency, cooperation, and participation. Firstly, transparency is a smart city concept is a social practice that takes its meaning above all in the accessibility of data, the demonstration of the steps taken and the explanation in a pictorial way of what is done by the administration. Integrated governance of cooperation consists of breaking down the silos and working transversally between the different sectors of activity and services of the city.

Secondly, governance within an intelligent city also requires participation and cooperation with organizations and companies. When we speak of integrated governance, we are referring to governance that makes decisions that go hand in hand with the different city services, all of which are facilitated by the use of digital and technological tools.

Thirdly, the participation of citizens or organizations is essential to build a project that corresponds to the needs of the city's users, whether they are residents or workers.

Smart Economics
A smart economy is a basic foundation for sustainable development [32]. It is based on the spirit of innovation and entrepreneurship and the creation of decent jobs; on productivity and market flexibility by using local wealth and promoting the attractiveness of the territory. One of the main motivations for becoming smart is the power to become an attractive city on the international scene but above all a desire for economic development [12].

Intelligent Mobility
It involves integrating different modes of transportation- rail, car, cycling, and walking. This integration reduces the environmental footprint, optimizes the use of urban space, and offers city dwellers a wide range of mobility solutions to meet all their needs. Also, the city of tomorrow will need to implement the latest public transport and electric mobility technologies.

Mobility assistance services can focus on urban navigation, multimodality, congestion management, real-time knowledge of free parking spaces, and real-time regulation of vehicle energy consumption, renting or sharing means of transport, services to avoid air pollution.

Access to real-time transport data via electronic screens in stations, traffic conditions on the road network, waiting time at stops, and public transport stations, breakdowns, and in short, better management of urban flows.

Location of network access points, access to information sources, geolocation services, payment services. This will ensure a single system that is efficient, easily accessible, affordable, safe, and environmentally friendly.

Smart Environment
Water management, waste management, and energy management are at the heart of a city's environmental concerns. In a smart city, the various technological tools allow, among other things, for the protection and preservation of our natural resources and natural environments.

In the field of energy, intelligent electricity distribution network computer technology can optimize the production and distribution of electricity while adjusting to demand: cities will have to strengthen their activities in terms of energy efficiency (development of low consumption street lighting or variable lighting according to periods of high energy consumption) and will have to set up local energy production systems (photovoltaic solar panels on the roofs of buildings, production of electricity from waste, etc.), and will have to develop a system of local energy production (photovoltaic solar panels on the roofs of buildings, production of electricity from waste, etc.).); saving energy through new technologies is also what an intelligent environment is all about.

Concerning waste, cities will have the mission to reduce, or even avoid, their waste production and to set up efficient waste recovery, recycling and reclamation systems (the process by which a material waste or useless product is transformed into a new material or product of higher quality or utility), finally, detectors to alert in case of leakage would be a way to save an important natural resource which is water.

Smart Housing

Intelligent housing can be applicable at different scales. It can be a safe, culturally rich, and culturally safe living environment that provides health and education services. Also, it can involve the development of green neighborhoods or eco-neighborhoods (because the model of urban sprawl - costly in terms of space, public amenities, and energy - which has been dominant until now is no longer possible). Urban forms must be reinvented which, at the same time, respect essential privacy, ensure sufficient sunlight, allow for change, and promote "living together". These may be green homes or even homes that are certified according to Leadership in Energy and Environmental Design (LEED)[1] or homes that include certain green components.

Faced with these challenges, various solutions can be implemented using intelligent elements. The energy efficiency of a building, both passive (use of insulating materials) and active (energy management, regulation) must be improved. For this purpose, self-consumption can be used.

Also, the circular economy can be a source of opportunities to avoid wastage of materials and promote recycling.

Intelligent Citizen

The citizen is an important stakeholder in the smart city [30]. Indeed, his participation is required, whether in the consultation phase upstream or during the implementation phase, as an actor for the protection of the environment, in economic matters, or the social aspect within his community. Moreover, the intelligent citizen is the one who will use the new technological tools, in particular, to participate in public debates and neighborhood life; this will give him access to training and distance education.

The design of a sustainable and intelligent city requires the coordination of the different stakeholders in charge of its development. Indeed, projects involve the integration of increasingly complex technologies and the association of actors specializing in various sectors, to develop the territory as a whole.

The citizen is an important stakeholder during the process and afterward, i.e. during implementation. Businesses, whether in the field of information technology,

---

[1] Leadership in Energy and Environmental Design (LEED) is a North American system for standardizing high environmental quality buildings created in 1998. The evaluation criteria include energy efficiency, water consumption efficiency, heating efficiency, use of locally sourced materials, and reuse of surplus materials.

technology or transport, universities, political or public actors, and the government are also important stakeholders. Leveraging our collective intelligence, the sum of knowledge, skills, and intellectual capacities of individuals, will foster the emergence of creative, evaluative, and problem-solving skills [11].

The project of becoming an intelligent city is a project that requires the expertise of many professionals: administrators, architects, engineers, and urban planners will be called upon to review the planning and development of the city.

3 The Strengths and Limitations of a Smart City

3.1 Strengths of a Smart City

3.1.1 An Efficient and Streamlined City

An intelligent city is synonymous with a bulwark against chaos; the city would become more efficient because streamlined by the supply of data that allows to supervise the city, make it more controllable and above all more receptive and reactive.

3.1.2 Improved Quality of Life

The intelligent city aims, among other things, at improving the quality of life, governance, urban policy development, etc. It is even envisaged that new information technologies will produce smarter citizens who will in turn adopt smarter behaviors. The smart city must attract smart people. Thus a smart city would be a means to foster social innovation, social justice, and civic engagement. As such, it is seen as a sustainable city that accumulates mental and behavioral changes.

However, while there are many positive aspects of the smart city, it also has some limitations that need to be detailed.

3.2 The Limits of the Smart City

- The Smart City is based more on supply than on-demand. This has two consequences: on the one hand, supply determines the quantity and price of products (management, operation, and maintenance costs can be very high) and on the other hand, supply determines the price of products (the cost of management, operation, and maintenance can be very high). The rapid evolution of technologies can lead to obsolescence, creating new needs and related costs); on the other hand, it leads to formulas that are disconnected from the social context in which they are developed [3].

[15] points out in this respect that private companies sell "solutions" to cities and these solutions ignore the historical, political, social, territorial, and cultural context of each community. This results in a mismatch between the product sold and the needs felt.

- Security and ethical failures in terms of managing the big data [21] generated by technological innovation will lead to the creation of new control processes that may eventually lead to forms of surveillance of individuals that erode privacy in various forms (insecurity, disclosure of information, exposure, appropriation, blackmail, distortion, intrusion).

Moreover, the digital system put in place is by definition vulnerable to attacks of all kinds (hacking, malfunctions, and accidents). Thus, securing the digital system requires updating the technology, which entails costs, regularly.

- Thirdly, social and territorial inequalities will be reinforced or even increased: for some, the smart city carries the risk of reinforcing or even creating new inequalities within the same territory, we are witnessing a development based only on certain very circumscribed neighborhoods and on certain specific sectors (energy and transport in particular).

4 The Links Between Smart Cities and Sustainable Development

The concept of the smart city is relatively recent and currently poorly framed. Contrary to sustainable development, where several approaches are possible when we talk about the smart city we are in the absence of an operational framework. However, the smart city is very much focused on the use of new technologies. In particular, it is the incorporation of these technologies into the majority of services, infrastructures, and equipment that make the smart city smart.

Sustainable development is the development that meets the needs of the present without compromising the ability of future generations to meet their own needs.

The smart city is seen as a tool for achieving sustainable development. Based on this UN definition, it is possible to make connections between the two concepts "sustainable development" and "smart city". Indeed, both concepts deal with economic, social, and environmental concepts. The notion of governance is also an inseparable dimension of both concepts. However, the operationalization and implementation of the first concept are broader than the second and applicable to different sectors.

Within this framework, and to become the world's smartest city by 2021, Dubai is pursuing one of the most ambitious information and communication technology (ICT) integration programs ever undertaken. Also, Dubai Smart City's major strategy is revolutionizing the way residents and visitors live, work, and play.

5 Case Study: Dubai Smart City's Important Strategy

Dubai's goal of becoming the world's smartest city is progressing rapidly with the implementation of a series of groundbreaking initiatives designed to transform the city into a world leader in innovation and a popular investment destination.

Government authorities have already announced plans to implement more than 100 initiatives to foster collaboration between the public and private sectors, covering six priority areas: transport, society, lifestyle, economy, government, and environment.

Also, according to the Dubai Smart City Strategy launched in 2014, the authorities are embarking on more than 545 planned initiatives and undertakings designed to redefine the way residents and visitors live in the city.

Major strategic goals include: transforming more than 1,100 essential government services into smart services, to be delivered primarily online; introducing autonomous vehicles and intelligent transport services; providing free, high-speed Wi-Fi connection throughout the emirate; and developing a data-driven economy that the authorities expect to generate an additional AED 10.4 billion (€2.36 billion) to GDP by 2021.

At the heart of the success of Dubai Smart City's strategy is the proliferation of smartphones in the emirate. According to Google's Global Smartphone Penetration Index, almost eight out of ten UAE residents have at least one smartphone, making the country the world's largest smartphone market. UAE residents are also the world's largest consumers of mobile data.

The collection of user-generated data will enable Dubai authorities to plan and implement a wide range of smart services using ICT, which would benefit the transport and health sectors, but not only.

The introduction of the Data Act in Dubai in 2015 has also helped to ensure that the private and public sectors share data in line with international best practice in terms of anonymization and standardization, in order to facilitate connectivity and access to services and information. This "will lead to better decision-making…and stimulate creativity and innovation in different sectors," said Younus Al Nasser, Deputy Director-General of the Dubai Smart City Office (2020).

Also, the success of Dubai's Smart City strategy is underlined by the volume of international investment that the city's high-tech industries have attracted over the past 24 months.

The UAE attracted 70% of all investment in the Middle East and North Africa start-ups in 2018, the majority of which was made in Dubai, according to the latest MENA Investment Report [18].

At the same time, foreign direct investment (FDI) increased by 41% in 2018 compared to the previous year, according to government data.

The emirate attracted 38.5 billion dirhams ($10.5 billion) of FDI in 2018, with the United States, India, Spain, China, and the United Kingdom accounting for 70% of the total.

A Smart City is a Happy City
The happiness of citizens and visitors to Dubai is seen as a key component in the emirate's Smart City strategy. A city-wide ICT program was introduced in 2016, giving people a chance to give their opinions, good or bad, on a range of issues from the quality of essential services to leisure activities.

A study conducted in the same year showed that 83% of Dubai residents are happy to live in the city. The Dubai Happiness Agenda aims to increase this figure to 95% by 2021.

Ultimately, this innovative approach will make Dubai a model for future Smart City developments around the world.

However, Dubai's intelligent transformation seems to be in progress… The challenge will be to ensure that the protection of residents' private data will also be smart. Also, while these applications could mean the end of red tape and paper, another battle of the "Smart City" is to get an intelligent supply of electricity.

6 Conclusion

Discourses on the smart city are gaining ground. They are currently mobilizing medium and small cities, which are also facing budgetary austerity. Faced with the deluge of data and the multiplication of so-called "intelligent" product vendors, their vulnerability is increased, but the need to mobilize all available resources to develop their territory remains.

It can be said that becoming an intelligent city means innovating while implementing a panoply of technological tools at the service of the city. Becoming a smart city will enable the latter to become more prominent on the international scene while promoting its power of attraction.

Current and future challenges cannot, therefore, be solved by technical solutions alone. Some require a social approach and a change of mentality, while others must incorporate technological advances without eliminating the social and political dimensions of their use. For example, the Smart Dubai agency has set up a charter of ethics and principles for the management of private data and the security of the city's inhabitants.

So, technology is never neutral, it has advantages but also disadvantages, and it is not yet possible to determine whether the "intelligent" solutions proposed will have a positive and lasting impact on the population over time.

References

1. Amel A, Alain R (2014) Le rôle des territoires dans le développement des systèmes trans-sectoriels d'innovation locaux: le cas des smart citie"s. Revue Innovation 2014/1(43)
2. Angelidou M (2014) Smart city policies: a spatial approach. Cities 41:3–11

3. Angelidou M (2015) Smart cities: a conjuncture of four forces. Cities 47:9–106
4. Barsoum J-F (2015) Devenir intelligent, oui mais pourquoi? La ville intelligente. Urbanité: 24–26
5. California Institute for Smart Communities (2001). Smart communities guide book. http://www.smartcommunities.org/guidebook.html
6. Coe A, Paquet G, Roy J (2001) E-Governance and smart communities: a social learning challenge. Soc Sci Comput Rev 19(1):80–93. https://doi.org/10.1177/089443930101900107
7. Cohen B (2011) Smart city wheel. Boyd Cohen smart cities. Urban and climate strategist Boyd Cohen. http://www.boydcohen.com/smartcities.html
8. Douay N, Henriot C (2016) La Chine à l'heure des villes intelligentes. L'information géographique 3: 89–102
9. Giffinger R (s.d.) (2018) The smart city model. In: European smart cities. http://www.smart-cities.eu/model.html
10. Gil-Garcia JR, Pardo TA, Nam T (2015) What makes a city smart? identifying core components and proposing an integrative and comprehensive conceptualization. Inf Polity 20: 61–87. https://doi.org/10.3233/IP-150354
11. Goulet F, Gravel F, Grondin H, Lessard M-J (2015) Les urbanistes, maîtres d'oeuvre du territoire numérique. La ville intelligente. Urbanité: 21–23
12. Harrison C, Donnelly, I (2011) A theory of smart city. J Technol 24–37. http://journals.isss.org/index.php/proceedings55th/article/viewFile/1703/572
13. Hollands R G (2008) Will the real smart city please stand up?. City, 12(3): 303–320. https://doi.org/10.1080/13604810802479126
14. IBM (2008) A smarter planet: the next leadership agenda. Council on Foreign Relations. https://www.cfr.org/event/smarter-planet-nextleadership-agenda
15. Kitchin R (2014) The real-time city? big data and smart urbanism. Geo J 79(1): 1–14
16. Komninos N (2002) Intelligent cities: Innovation, knowledge systems, and digital spaces. Routledge, London
17. Lyudmila V, Polina K, Felippe C (2017) Smart cities prospects from the results of the world practice expert benchmarking. Procedia Comput. Sci 119:269–277
18. Magnitt (2019). 2019 was a record-breaking year for MENA's startup ecosystem: exits and investments at an all-time high. 2019 MENA Venture: Investment Summary. https://magnitt.com/news/51386/magnitt-report-2019-record-breaking-year-mena-startup-ecosystem-exits-investments
19. Mahizhan A (1999) Smart cities. the Singapore case. Cities 16(1): 13–18
20. Mair E, Moonen T, Clark G (2014) What are future cities? origins, meanings and uses. catapult, futures cities. The business of cities for the foresight. https://www.gov.uk/government/uploads/system/uploads/attachment_data/file/337549/14-820-what-are-future-cities.pdf, consulté le 07 Novembre 2018
21. Mansour N, Salem SB (2020) Examination of big data analytics and customer segmentation in the banking sector: learning for BNP Paribas bank of France. In: Econder 2020 3rd international economics business and social sciences congress
22. Sandra B, Jérémy D (2017)"La ville intelligente: Origine, définitions, forces et limites d'une expression polysémique"..Rapport de l'Institut national de la recherche scientifique Centre-Urbanisation Culture Société. 10
23. Sarbeswar P, Hoon H (2019). Cutting through the clutter of smart city definitions: a reading into the smart city perceptions in India. City Cult Soc 18
24. Smart Grids: Rapport de la commission de régulation de l'énergie. (2018)
25. Söderström O, Paasche T, Klauser F (2014) Smart cities as corporate storytelling. City 18(3):307–320
26. Smart Nation Programme Office (2016). Smart nation Singapore. https://www.smartnation.sg/
27. Townsend AM (2014) Smart cities: Big data, civic hackers, and the quest for a new utopia. New-York, Norton & Compagny
28. Sueur, JP (2017) Villes du futur, futur des villes: Quel avenir pour les villes du monde?", Rapport d'information fait au nom de la Délégation sénatoriale à la prospective 594

29. Xiao X, Xie C (2021) Rational planning and urban governance based on smart cities and big data. Environmental Technology & Innovation
30. Ya-Ting C (2020) Building virtual cities, inspiring intelligent citizens: digital games for developing students' problem solving and learning motivation. Comput Educ 59(2):365–377
31. Yigitcanlar T, Kamruzzaman M, Foth M, Sabatini-Marques J, da Costa E, Ioppolo, G (2019) Can cities become smart without being sustainable? a systematic review of the literature. Sustain Cities Soc https://doi.org/10.1016/j.scs.2018.11.033
32. Yudo A, Mahesa R, Yudoko, G (2019) Dataset on the sustainable smart city development in Indonesia. J Brief 25. https://doi.org/10.1016/j.dib.2019.104098

The Artificial Intelligence in the Audit on Reliability of Accounting Information and Earnings Manipulation Detection

Mohammad Hussein Rahahleh, Ahmad Husni Bin Hamzah, and Norfadzilah Rashid

Abstract Providing reliable and relevant accounting information is the accounting profession key responsibility. The sound internal control system and management and the employees' ethical and integrity characteristics, is what depends on it in the reliability and relevance of accounting information and earning manipulation detection. This paper explains how artificial intelligence operates in the audit on the reliability of accounting information and allows managers to obtain high-quality accounting information and earnings manipulation detection by reducing information risk. There are many papers on the research suggested using Artificial Intelligence in auditing and accounting, but it is still not clear to show how to earnings manipulation detection and reduces information risk by using Artificial Intelligence. This paper proposes how to use the Artificial Intelligence practically to automate removing the audits weaknesses. This, in turn, leads to earnings manipulation detection and reduces detection risk, control risk and enhance audit quality by minimizing the risk accounting information and earnings manipulation detection.

Keywords Artificial intelligence · Audit · Accounting information · Earnings manipulation

1 Introduction

Artificial Intelligence is a mixture of equipment and software to replace human intelligence for solving complex business difficulties by using learning, reasoning, elucidating and recognizing patterns the same as human experts. Also, Artificial Intelligence utilizes an expert system as a substitute for an expert person and replaces

M. H. Rahahleh · A. H. B. Hamzah (✉) · N. Rashid
Faculty of Business and Management, Universiti Sultan Zainal Abidin, (UniSZA), Kuala Nerus, Terengganu, Malaysia
e-mail: ahmadhusni@unisza.edu.my

N. Rashid
e-mail: norfadzilah@unisza.edu.my

A. M. A. Musleh Al-Sartawi (ed.), *The Big Data-Driven Digital Economy: Artificial and Computational Intelligence*, Studies in Computational Intelligence 974,
https://doi.org/10.1007/978-3-030-73057-4_24

the human intelligence with the machine intelligence. Moreover, Artificial Intelligence helps managers greatly in terms of making decisions by presenting more reliable information, reducing repetitive decisions, fact processing data analysis, and simplifying complex decision factors [28].

The application of Information and Communications Technology was undertaken firstly on the basic accounting systems. It is argued that Accounting is the first business field in which the methods and strategies of Information and Communications Technology have been implemented [13]. Although financial modelling packages soon reported to be very advantageous in the accounting analytical aspects [14]. Moreover, due to the conservative approach of its practitioners, [7] considered the pace of Communications Technology and Information adoption accounting slow as a profession. By the 1990s, however, the profession had been forced to computerize its operations as a method of improving efficiency, reducing expenses, and resisting competition [25].

According to ICAEW in 2017, issued the article about "Artificial intelligence and the future of accountancy" and employing this method in the profession of accounting and auditing. The Institute viewed from several achievements in three perspectives: application for accounting, understand technology, and long-term visions. Establishing and developing knowledgeable organizations by means of artificial intelligence techniques is growing fast [23, 36]. The companies in the global economy and business deal with the issues of technology to survive where Artificial Intelligence work as an effective solution for this issue [20]. An Artificial Intelligence Survey (2017) showed that, compared to 2016, Forrester Research has projected a "greater than 300% rise in investment in artificial intelligence in 2017" This as a proof of the sector's fast global economic growth [33]. Banking and Manufacturing would benefit more from Artificial Intelligence. The application of Artificial Intelligence in accounting involves producing accurate, big data analysis, on-time authentic accounting information for the clients. Also, Artificial Intelligence has been known to have a great impact on auditing and accounting practices and on the structure of the internal control as well [10].

In a variety of tasks, Information and Communications Technology tools are now widely used from basic assignments like arithmetic calculations to more sophisticated ones like statistical analysis and flowcharting. These tools involve audit toolkits consisting of standard software packages and purpose-written software, audit inquiry programs, logit models, checklists (capable of implementing in-depth test of data), internal control templates, and integrated audit monitor modules expert systems widely utilized to identify a system's strengths and weaknesses. Deloitte's Visual Assurance and Price water house Coopers Risk Control Workbench are examples of such templates.

With the advancement in computer technology and as previously predicted by [1, 9], Information and Communications Technology devices like Electronic Data Interchange and image processing are steadily replacing traditional. Also, the generality of the accounting firms has incorporated the application of artificial intelligence as part of their integrated audit automation systems in making audit judgments. Thus, Audit trails are entirely altering the whole audit process which will affect the reliability of accounting information and earnings manipulation detection.

In the long-term vision, the accounting profession can benefit from Artificial Intelligence by the latter focusing on the accounting profession purpose for an organization utilize accounting information to take good decisions by the information users. In exploiting influential technologies, think radically and being adaptable. In studying technology, Artificial Intelligence enhances human decision-making, machine learning strengths, and the use of decision process of enterprise information management [5].

The aim of the profession in the accounting information system is to provide different users with accurate financial information to make correct decisions. To provide such accurate information, accounting systems often rely on the internal audit. Thus, examining the way Artificial Intelligence helps managers to limit the internal audit drawbacks to generate beneficial accounting information for the users and earnings manipulation detection would be questionable. Accounting systems are, therefore, moving rapidly towards being more incorporated with and intelligent logic through the application of Artificial Intelligence [11]. This study will demonstrate how Artificial Intelligence in internal audit will improve the reliability of accounting information and earnings manipulation detection.

Despite the metamorphosis the audit profession has experienced in the last 150 years, the auditing core theme continues to provide an expert opinion of an independent third-party on the truth and fairness of financial information provided by the management. Therefore, auditing is considered to encompass an information-intensive number of activities including gathering, organizing, processing, evaluating, and presenting data with a view to generating an accurate audit opinion. Therefore, the Artificial Intelligence benefits in accounting in the future including increasing produced reliable financial information and more accurate information and timeliness for decision-makers, and simplifying complicated cases in auditing and accounting [3].

As the decision aids which relies on Information and Communications Technology continue to cause challenges in the modern world of business at the same time with the growing burden on auditors to take a more active part in the corporate entities' control and governance. The purpose of this study is to investigate the way Artificial Intelligence can help firms to eliminate the drawbacks of internal control to generate authentic accounting information and earnings manipulation detection. Although there are many discussions on the important role of Artificial Intelligence in decision making in business, no studies have explained how Artificial Intelligence would enhance the accounting information quality by improving the internal audit. This gap explains the way Artificial Intelligence may work to decrease the risk of accounting information to promote the information user's confidence. The main aim of this paper is to address the gap in the Artificial Intelligence role in eliminating the risk of accounting information and earnings manipulation detection.

2 Literature Review

2.1 Artificial Intelligence and Auditing and Accounting Information

With the population of the world likely to increase and for the transaction's complexity, the audit procedures implementation would become highly software-dependent. Therefore, Expert Systems and Artificial Intelligence are valuable and, perhaps, unavoidable in the performance of today's audit [15]. There has been a continuous effort over the last two decades in developing the systems based on artificial intelligence to help auditors in making judgments [2]. These systems help auditors in making better decisions and increase the quality of accounting information and decrease the manipulation such as earning management by seriously considering the potential omissions and biases that could have normally happened in complete manual processes in making decisions. It is commonly thought that these systems need to be used as a mere assistance or input into the final determination of audit outcomes by the auditor because of the degree of sensitivity and versatility these judgments necessitate [2, 24].

Three basic iterative stages should be encompassed in a typical decision process. They are intelligence that includes collecting data, diagnosing problems, objectives, structuring problems, and validating data. also, a design that quantifying objectives, comprises manipulating data, assigning risks or values to alternatives, and producing alternatives. Moreover, the choice, which is simulating results of alternatives, encompasses providing statistics on alternatives, selecting among alternatives, and explaining the choice, clarifying alternatives [1]. Thus, Artificial intelligence is an essential section of the decision aids family which strives to be improved and used in the modern business managerial and technical operations with auditing included which is enhancing the quality of accounting information.

The gains have been described as accruable to the audit from using artificial intelligence by auditors for audits. They encompass: 1. the effectiveness and efficiency 2. the consistency 3. a structure for audit tasks 4. enhanced decision making and communication 5. improved staff training 6. expertise progress for beginners and shorter decision time, these benefits lead to increasing the reliability of accounting information [2, 16, 17, 29].

Nonetheless, the following weaknesses have been recognised as potential from adopting systems based on artificial intelligence. These include 1. prolonged decision processes as an outcome of seeking further alternatives 2. the high expense of building, updating, and sustaining systems 3. the novices' knowledge base inhibition 4. the professional judgment skills development inhibition 5. the risk of transferring tools to competitors and the possibility of using them in a court of law against the auditor for relying heavily on the decision aids evidence [23, 26, 37]. However, the implementation of systems based on artificial intelligence in reaching a judgment can be described as a double-edged sword. It is possible that the auditor will be responsible for not properly utilizing a modern decision aid in arriving at a judgment that is realized to be inaccurate just as he may be liable for depending in his judgment merely on an expert system to make a wrong judgment [4, 34].

2.2 *Modern and Traditional Audit Methods*

In prior decades, in providing a complete balance sheet containing no errors, conventional audit methods used to depend on accounts' detailed verification where no techniques for testing or sampling involved. As time goes on and economies grow, auditors evaluated the companies' financial statements Ruth and fairness based on the developed sampling techniques [21]. Furthermore, modern risk-based auditing was employed from the mid-1980s onwards. By doing this, further interest was shown to such fields that are more likely to have errors [31]. As a consequence, auditors turned to identifying, reporting, and assessing fraud explicitly in conformance to regulators' serious concern regarding corporate governance issues.

Regrettably, today, conventional manual auditing techniques are still being used by many auditors. These techniques which involve manual verifications, inventory counts, document vouching by sampling, and the use of ratios and/or basic statistics are considered conservative. However, they have little significance in business environment in the modern world because they are backward and slow [36]. Moreover, one of the traditional auditing techniques major drawbacks is that they cannot fully fulfil audit verification requirements as they are conducted manually with inadequate sample data [20].

However, the advancement of artificial intelligence has made the constant auditing process more efficient as it speeds up the process of auditing by automation. Furthermore, artificial intelligence helps auditors to expand the process in a timely way as well to cover the whole population. With innovations and continuous technology growth, auditors are expected to offer valuable services and advise management beside improving the financial statement credibility. In order to predict the expected outcomes of any transaction performance process today, Auditors have to use up-to-date analytical techniques and benefit from business IT automation and acceleration. [21].

2.3 *Artificial Intelligence and Internal Control Systems and Earning Manipulation*

The Artificial Intelligence integrates judgmental activities about the way deficiency in audit and internal controls can be eliminated. Much like a financial advisor, and offers the suitable solution depending on the reality of the system condition and data, automated analysis, and find the most robotic-assisted suitable solution will be implemented. The Artificial Intelligence is, thus, capable of searching for vast data sources, inferencing, and providing knowledge-based recommendations. Artificial Intelligence will then use automated data analysis and search to overcome the internal controls weakness and limit the manipulation such as earning management without the need for humans to intervene and judgment [5].

The previous study listed ten accounting topics subjects that can be incorporated with Artificial Intelligence. These subjects are detection of management fraud, financial and economic analysis, credit authorizing and screening, mortgage risk analysis, risk analysis of fixed income investment, regularities detection in security price movement, risk rating of exchange-traded, prediction of default and bankruptcy, the techniques of machine learning to identify characteristics of fraud and artificial intelligence in marketing automatically [25].

According to [5, 7] provide a practical model implementing Artificial Intelligence for providing quality accounting information by overcoming the audit and internal controls weaknesses in nearly every industry. The characterizes of efficiency model by more routine activities based on procedures, criteria, and rules that strengthen the audit and internal controls. The primary objective is to build such controls that meet quality performance, cost-benefit, and apply consistently in overcoming the weaknesses. With this Artificial Intelligence solution, people are engaged in accuracy monitoring and the way rules have to cope with the change of business conditions. The capabilities of machine learning need to be applied to these rules. For instance, internal controls monitor online data validity with a minimal interference by humans by employing codified knowledge and logic and making decisions about data reliability and accuracy.

At the model who was invented by [5], the Artificial Intelligence solution helps generating new control objectives, methods and methodology by recommendations based on the current control environment. An intelligent software that can analyse a control activity and give recommendations after that to raise the possibility of putting the control at risk is an instance of the ability of Artificial Intelligence in augmenting creative controls. Also, by analysing transactions and account balances, the model would impede those earning management techniques. Technology helps in finding alternatives and optimizing recommendation while humans make decision and perform [7].

2.4 Artificial Intelligence and Earnings Manipulation Detection

Academic research focusing on the fields of earnings manipulation has experienced a significant rise in the last few decades. Earnings management, as [28] stated, represents maintaining the accounting practices within legality limits in compliance with the accounting rules and standards established by the Generally Accepted Accounting Principles (GAAP). A lack of knowledge can be noticed regarding the management fraud characteristics and extensions. Moreover, the auditors have insufficient skills needed for detecting manipulated financial statements, and by using new techniques, managers are intentionally attempting to trick the auditors [14, 18, 30]. So, the detection of earning management by utilizing normal audit procedures turns out to be an extremely difficult task.

Further, the nature of accruals accounting provides managers with a great deal of flexibility in identifying the actual earnings that a company reports in any (such as research and development outlays or advertising expenses). In addition, they can somehow adjust the recognition timing of revenues and costs by, such as, delaying losses recognition by waiting to establish reserves of loss or advancing sales revenue recognition through credit sales [35]. Standard auditing procedures become insufficient for the managers who realize an audit's limitations. These limitations imply the need for more analytical techniques to effectively detect earnings management practices and go towards artificial intelligence to minimize and detect the manipulation throw earning management [33].

3 Conclusion

This paper shows the role and development of the process of artificially intelligent systems in auditing on reliability of accounting information and earnings manipulation detection in the light of their various advantages of artificially intelligent and some weaknesses recognized in the existing literature. It also addressed the importance of the implication of artificially intelligent systems by auditors in developing the audit process and the user trust in accounting information and limit the manipulation. Precisely, it investigated research efforts on employing expert systems and networks in auditing and their implications.

The studies conducted in this area were discussed in a way that highlighted a n important research gap that future studies could fill. These areas involve measuring the artificial intelligence impact on internal control systems' monitoring and design, meeting the gains of implementing these intelligent agents, and implications of employing these systems for audit firms to developing the performance in detection the earning management and enhancing the information of accounting. Moreover, this study has contributed to knowledge to describe the effect of implementing artificial intelligence in overcoming audit weaknesses to provide quality accounting information.

This paper focuses largely on the viability side of applying artificial intelligence in auditing, with limited discussion on the necessity side to the feasibility side of Artificial Intelligence application. It is recommended for more future research to be conducted on why it is valuable to implement artificial intelligence in auditing and what aspects cause auditing to improve in removing audit weakness to produce quality accounting information.

References

1. Abdolmohammadi MJ (1987) Decision support and expert systems in auditing: a review and research directions. Account Bus Res 17(66):173–185
2. Abdolmohammadi M, Usoff C (2001) A longitudinal study of applicable decision aids for detailed tasks in a financial audit. Int J Intell Syst Account Financ Manag 10:139–154
3. AlAli M, Almogren A, Hassan MM, Rassan AB (2018) Improving risk assessment model of cyber security using fuzzy logic inference system. Comput Secur 74:323–339
4. Ashton RH (1990) Pressure and performance in accounting decision settings: paradoxical effects of incentives, feedback and justification. J Account Res 28:148–186
5. Askary S, Abu-Ghazaleh N, Tahat YA 2018 Artificial intelligence and reliability of accounting information. In: Conference on e-business, e-services and e-society. Springer, Cham, pp 315–32
6. Barras R, Swann J (1984) The adoption and impact of IT in the UK accountancy profession. The Technology Change Centre, London
7. Bataller C, Harris J (2018) Turning artificial intelligence into business value. Accenture Emerging Technology Group, Today
8. Bell TB, Knechel WR, Payne JL, Willingham JJ (1998) An empirical investigation of the relationship between the computerisation of accounting system and the incidence and size of audit differences. Audit J Pract Theory 17(1): 13–26
9. Brown CE, Coakley J, Phllip ME (1995) Neural networks enter the world of management accounting. Manage Account 51–57
10. Brown CE, Murphy DS (1990) The use of auditing expert systems in public accounting. J Inf Syst 63–72 (Fall)
11. Canada J, Sutton SG, Kuhn JR (2009) The pervasive nature of IT controls. Int J Account Inf Manage 17(1):106–119
12. Carr JG (1985) IT and the accountant, summary and conclusions. Gower Publishing Company Ltd./ACCA, Aldershot
13. Clark F, Cooper J (1985) The chartered accountant in the IT age. Coopers & Lybrand and ICAEW, London
14. Coderre GD (1999) Fraud detection. using data analysis techniques to detect fraud. Global Audit Publications, Vancover
15. Dalal C (1999) Using an expert system in an audit: a case study of fraud detection. ITAUDIT 2 May 15
16. Eining M.M, Dorr PB (1991) The impact of expert system usage on experiential learning in an auditing setting. J Inf Syst 1–16
17. Elliott RK, Kielich JA (1985) Expert systems for accountants J Account 126–134
18. Fanning K, Cogger KO, Srivastava R (1995) Detection of management fraud: a neural network approach. In: Proceedings of the 11th conference on artificial intelligence for applications. ISBN:0-8186-7070-3, pp 220
19. Kahraman C, Kaya I, Çevikcan E (2011) Intelligence decision systems in enterprise information management. J Enterprise Inf Manag 24(4):360–379
20. Kuenkaikaew S, Vasarhelyi MA (2013) The predictive audit framework. Int J Digit Account Res 13:37–71
21. Lee TH, Ali A (2008) The evolution of auditing an analysis of the historical development. J Mod Account Audit 4(12):1–8
22. Lu H, Li Y, Chen M, Kim H, Serikawa S (2018) Brain intelligence: go beyond artificial intelligence. Mob Netw Appl 23(2):368–375
23. Mackay J, Barr S, Kletke M (1992) An empirical investigation of the effects of decision aids on problem-solving processes. Dec Sci 23:648–672
24. Manson S, McCartney S, Sherer M (2001) Audit automation as control within audit firms. Acc Audit Accountability J 14(1):109–130
25. Moudud-Ul-Huq S (2014) The role of artificial intelligence in the development of accounting systems: a review. IUP J Acc Res Audit Pract 13(2):7–19

26. Murphy D (1990) Expert systems use and the development of expertise in auditing: a preliminary investigation. J Inf Syst 18–35 (Fall)
27. Novac C (2000) Artificial intelligence system for decision -making process. Ovidius Univ Ann Constantza Ser Civil Eng 1(2): 261–266
28. Paolone F, Magazzino C (2014) Earnings manipulation among the main industrial sectors: evidence from Italy. Economia Aziendale Online 5(4):253–261
29. Pieptea DR, Anderson E (1987) Price and value of decision support systems. MIS Q 514–527
30. Porter B, Cameron A (1987) Company fraud – what price the auditor?. Acc J, 44–47
31. Salehi M (2008) Evolution of accounting and auditors in Iran. J Audit Pract 5(4):57–74
32. Segars S (2017) AI today, Ai tomorrow. In: Ltd NRP (ed) Global artificial intelligence survey. UK, Arm Northstar
33. Spathis C (2002) Detecting false financial statements using published data: some evidence from greecec. Manag Audit J 17(4):179–191
34. Sutton SG, Young R, McKenzie P (1994) An analysis of potential legal liability incurred through audit expert systems. Intell Syst Financ Manag 4:191–204
35. Todoroi D (2013) How to create adaptable ROBO-intelligences? Acad Econ Stud Econ. Inform 13(1):27–39
36. Verner J (2012) Why you need internal audit at the table. Business finance. Accessed 7 Sept 2016
37. Yuthas K, Dillard J (1996) An integrative model of audit expert systems development. Adv Accounting Inf Syst 4:55–79

The Role of Digital Transformation in Increasing the Efficiency of Banks' Performance to Enhance Competitive Advantage

Abdul Rahman Mohammed Suleiman Rashwan and Zainab Abd-Elhafiz Ahmed Kassem

Abstract The research aims to identify the role of digital transformation in increasing the efficiency of the performance of banks listed on the Palestine Stock Exchange, identifying the digital transformation in banks and the role of digital transformation in increasing the efficiency of bank performance to enhance competitive advantage and achieve a stable financial situation. The study recommended that banks review the quality mechanisms of digital services constantly and take action to improve these services, use technology to monitor the quality of electronic services, develop innovative and innovative models to employ emerging electronic technologies to monitor performance, commit to tasks and responsibilities in accordance with the governance framework and predict deviations before they occur, and to report proactively to the management of the system. Senior departments and departments concerned with the employment of emerging electronic technologies in monitoring performance indicators in order to enhance competitive advantage.

Keywords The privatization · Bank performance · Palestine Stock Exchange

1 Introduction

The rapid development and increase in the volume of information has complicated the process of controlling and benefiting from applications that have spread across the banking business and at all levels to achieve progress and business performance effectively and efficiently. In the past, the banking sector has not been able to meet these digital challenges in all elements and the need has become more urgent than ever to digitally transform financial institutions, and the implementation of this type of technology will also facilitate the interconnection between financial institutions and each other or between companies and the banking sector, which will have a positive

A. R. M. S. Rashwan
University College of Science and Technology, Jerusalem, Palestine

Z. A.-E. A. Kassem (✉)
Faculty of Commerce, Ain Shams University, Cairo, Egypt

A. M. A. Musleh Al-Sartawi (ed.), *The Big Data-Driven Digital Economy: Artificial and Computational Intelligence*, Studies in Computational Intelligence 974,
https://doi.org/10.1007/978-3-030-73057-4_25

and clear impact on the public as they are the main beneficiaries of these services as well as their quality and ensure a reliable and coherent source of information.

Digital transformation has become a necessity for all financial institutions seeking to develop, improve their service, facilitate access to beneficiaries and digital transformation not only the application of technology within banks, but also a comprehensive program that affects financial institutions, and the way and style of their work both internally and externally and also by providing services to the target audience to make services faster and easier. Digital transformation also contributes to connecting the banking sector to the public or private sectors so that joint work can be done flexibly and harmoniously. The necessity has become more urgent than ever for the digital transformation of financial institutions, mainly due to the rapid development in the use of information technology tools and tools in all aspects of life, whether related to transactions with the government sector or the private sector or for individuals, so there is clear pressure from all segments of society on financial institutions to improve their service to all channels.

Digital transformation also saves cost and effort significantly and improves operational efficiency and regulates it, and works to improve its quality and simplify procedures for obtaining services provided to the public and creates opportunities to provide innovative and creative services away from traditional methods of service delivery, which in turn will contribute to the creation of a state of satisfaction and acceptance from the public towards banking services, and mobile applications and e-commerce sites are one of these methods, and once this concept is implemented will be a huge amount of data and information that will in turn help decision makers in the banking sector listed on the Stock Exchange. Palestine to monitor performance and improve the quality of its services in addition to analyzing this data and information that will facilitate decision-making, setting goals and strategies, and attracting domestic and foreign investments.

2 Literature Review

- The study [4] (Asmara 2019) aimed to promote financial inclusion in Egypt because of its importance in expanding the circle of beneficiaries of financial services that will contribute to the empowerment of society as a whole and enhance the financial independence of individuals and achieve sustainable development in addition to the optimal use of resources and the transformation of the informal economy into a formal economy, and the results of the study reached several proposed approaches, the most important of which is the conversion of cash transactions to non-cash from During the official accounts and the mechanization of all financial and non-financial services and the increased use of electronic financial services, the most important recommendations of the study are to find new and unconventional solutions to help the Central Bank of Egypt to enhance the levels of financial inclusion by preparing a five-year timetable for the conversion of monetary activities to non-cash in coordination with the Egypt Plan 2030, electronic connectivity between banks and official agencies and the obligation of individuals and companies to serious controls to help activate financial coverage.

- The study [6] (Younis 2019) examined the concept and benefits of digital transformation and knowledge of the challenges it faces and know how important it is in Saudi banks, and the researchers used the field study on Saudi Al Rajhi Bank to learn about the importance of digital transformation and its impact on improving the quality of banking services and achieving customer satisfaction, which reflects on increasing the utilization of digital services and achieving digital leadership for Saudi banks. The results of the study reached: The importance of digital transformation in various sectors, especially the Saudi banking sector, with a relationship between the importance of digital transformation, improving the quality of digital banking services and increasing customer satisfaction, which reflects their increased interest in website services and achieving digital leadership for Saudi banks. One of the most important recommendations of the study is the need for a clear strategy for the digital transformation in all sectors of the country to accelerate the transformation of society into a digital information society, with the creation of a new job, namely digital employees with their representation as executives in the Board of Directors, with the need to recognize the dimensions of the quality of digital banking services, and to be careful to apply them by Saudi banks, and work to raise the skills of employees in providing digital banking services, while developing Their capabilities and training in the latest technologies in order to keep up with modern digital services, leading to increasing the competitiveness of banks, achieving digital leadership for Saudi banks, and also developing and launching new business models consistent with digital transformation.
- The study [1] (Filgueiras et al. 2019) analyzed the digital transformation of public services in the Brazilian Federal Government, and used the field study, which included 85 federal organizations, 1,740 public services are examined according to various factors explaining why a particular public service is digitized. The study also discusses the digital transformation of governments as an institutional change process in public institutions, taking into account the role of agents, the contexts of options and the factors that explain the decision to digitize public services. The use of technology encourages changes in the structure of government services, so that it can lead to unequal, inconsistent and incomplete processes that can promote the involvement or exclusion of citizens. Think about the process of digital transformation. It is necessary to design policies that may allow consistent, coordinated and homogeneous digitization of public services. The study's recommendations were that policy design, integration and institutional arrangements should be further strengthened through theoretical discussion for full and harmonious digital transformation.
- The study (The Liang L. et al. 2018). [2] study examined how entrepreneurs in small and medium-sized enterprises (SME) with inadequate capacity and limited resources are transforming their companies, a phenomenon that is still under study in existing literature. This model expands our understanding of both digital entrepreneurship and digital transformation.

- Reviewed the study (Loonam et al. 2018) [3] 10 case studies of literature, digital transformations and developed a conceptual framework to support researchers and practitioners and analyze the approach adopted by these organizations to successfully implement digital technologies. This article sought to obtain lessons from case writing in exploring digital transformation.
- A study [5] (Anna 2017) examined one of the major changes in the digital transformation industry that is undergoing a profound transformation of the banking system. The study also aims to identify the digital transformation in the banking sector, identify what fintech banks and fintech companies are developing in the market, and also point out that technology itself will not disrupt the banking industry, and the results of the study suggest that continued competition in the future will depend largely on the decisions that banks are developing in the market. Given that the problem of innovation is that it is unpredictable in terms of timing, size and consequences, the future of banks also expects the landscape to be strongly shaped by digital technology and competitors. Non-traditionalists.

Comment on Previous Studies

Most of the previous studies focused on the relationship between digital transformation and government establishment, companies or banks, but the variables did not meet in a single study, which distinguishes the current study, where it dealt with the relationship between digital transformation and the efficiency of the performance of banks listed on the Palestine Stock Exchange to enhance competitive advantage.

The scarcity of Arab research and its lack of study of the role of digital transformation in increasing the efficiency of banks' performance to enhance competitive advantage and not study the variables on banks listed on the Palestine Stock Exchange. Due to the importance of the topic, it had to be studied here, the research gap and the added achieved.

3 Research Methodology

Applied study: Where this aspect dealt with the following procedures:

First: The Curriculum

In this study, the researchers relied on the descriptive analytical method after ascertaining the sincerity and consistency of the study tool, namely the resolution.

Second: The Society and Sample of the Study

The study community consists of branch managers, accountants, internal auditors and financial controllers working in banks in the Gaza Strip, of which (310) employees were selected, and a random sample of (140) employees was selected, where a questionnaire was distributed to the sample of the study, and 128 questionnaires were recovered with a recovery rate (91%) It's almost.

Table 1 Correlation coefficient between the total score of each axis and the total score of the resolution

R.M.	Axis	Link coefficient	Probability value
1	There is a role for digital transformation in increasing the efficiency of bank performance to enhance competitive advantage	0.753	0.000
2	There is a role for digital transformation in attracting investment stake for banks and achieving a stable financial situation	0.7890	0.000

Fourth: The Validity of the Questionnaire

The validity of the questionnaire means making sure that it measures what is prepared to measure it, as it means "the inclusion of the resolution for all the elements that must be included in the analysis on the one hand, and the clarity of its paragraphs and vocabulary on the other hand", so that it is understandable to everyone who uses it, where the researcher has ascertained the validity of the questionnaire in two ways:

A—Virtual Honesty: Where the questionnaire was presented to a group of arbitrators, consisting of a number of faculty members in Palestinian universities in the Gaza Strip and specialists in accounting and statistics, the researchers responded to the opinions of the arbitrators by making the necessary deletion and modification in the light of the proposals submitted, where the questionnaire came out in its final form.

B—Constructive Honesty

The structural honesty of the resolution paragraphs was calculated on the single (140) research sample, calculating the correlation coefficient between the total score of each axis and the total score of the resolution, and Table 1 shows that all correlation coefficients in all resolution axes are at 0.05 as the probability value is less than 0.05.

Fifith: Stability

The stability of the resolution means ensuring that the answer will be approximately the same if repeatedly applied to the same persons at another time, and the Alpha Kronbach method has been used to measure the stability and credibility of the resolution, with Table 1 showing alpha kronbach coefficients for each resolution axis.

Table 2 shows that alpha kronbach transactions ranged from 0.749 to 0.766, which were high stability, while honesty transactions ranged from 0.831 to 0.854, indicating that the resolution was stable and honest.

Sixth: Testing the Hypotheses of the Study

Testing the First Hypothesis: There is a role for digital transformation in increasing the efficiency of the performance of banks listed on the Palestine Stock Exchange to enhance competitive advantage.

Table 2 Alpha Kronbach alpha coefficients for resolution stability

R.M.	Axis	Number of paragraphs	Kronbach Alpha coefficient (stability)	Honesty factor
1	There is a role for digital transformation in increasing the efficiency of bank performance to enhance competitive advantage	12	0.7660	0.854
2	There is a role for digital transformation in attracting investment stake for banks and achieving a stable financial situation	12	0.749	0.831
All axes are together			**0.7580**	**0.843**

Table 3 The role of digital transformation in increasing the efficiency of banks' performance to enhance competitive advantage

M	Paragraph	Arithmetic medium	Standard deviation	Relative weight	T test value	Probability value (.sig)	Order
1	Digital Transformation helps the Bank take steps to develop a strategic digital transformation plan in order to improve its performance	1.70	0.50	32.00	1.560	*0.000	12
2	Digital transformation helps the Bank implement initiatives and report periodically according to objectives and performance indicators	3.76	0.56	75.24	17.548	*0.000	6
3	The Bank monitors the performance indicators of electronic processes and procedures and conducts measurements and analysis through periodic reports with the aim of improving performance to enhance competitive advantage	3.77	0.85	75.56	11.891	*0.000	5

(continued)

Table 3 (continued)

M	Paragraph	Arithmetic medium	Standard deviation	Relative weight	T test value	Probability value (.sig)	Order
4	The Bank provides proactive reports to senior management and departments involved in the employment of emerging electronic technologies in monitoring performance indicators in order to enhance competitive advantage	3.86	0.78	77.36	14.378	*0.000	2
5	The Bank is working on a plan to develop and prepare employees and raise their level of knowledge of the digital transformation process	2.83	0.87	56.67	5.221	*0.000	10
6	The Bank measures the impact of training in digital transformation on upgrading its performance and the services it provides to enhance competitive advantage	3.43	1.02	68.62	5.490	*0.000	8
7	The Bank works by constantly educating its employees to develop their performance to help achieve its objectives related to digital transformation	3.70	0.90	74.02	10.038	*0.000	7
8	The digital transformation is leading to the development of human staff in all departments and branches of the Bank in a way that integrates the digital transformation process	3.86	0.98	56.69	4.539	*0.000	9

(continued)

Table 3 (continued)

M	Paragraph	Arithmetic medium	Standard deviation	Relative weight	T test value	Probability value (.sig)	Order
9	The Bank develops innovative and innovative models for employing emerging electronic technologies to monitor performance and commitment to tasks and responsibilities in accordance with the governance framework and to predict deviations before they occur	3.80	0.87	76.04	12.072	*0.000	4
10	The Bank adopts standards for the quality of digital services in cooperation with the relevant authorities to raise the level of performance and quality of the services provided	3.81	0.86	76.16	12.122	*0.000	3
11	The Bank is constantly reviewing the quality of digital services and taking the necessary measures to improve these services and employ technology in monitoring the quality of electronic services	3.90	0.76	77.55	14.387	*0.000	1
12	The Bank develops electronic applications in innovative and innovative ways that have contributed to its performance	2.09	0.860	41.78	5.837	*0.000	11
All sections of the field together		**3.30**	**0.82**	**65.68**	**8.510**	***0.000**	–

* Correlation D statistically at a level of $\alpha \leq 0.05$

- Paragraph 11 of the Bank constantly reviews the quality of digital services and takes the necessary measures to improve these services and use technology in monitoring the quality of electronic services has obtained the highest average account of (3.90) and relative weight (77.55%).
- Paragraph (1) helps the Bank by taking steps to develop a strategic digital transformation plan in order to improve its performance efficiency has got the lowest average account of (1.70) and relative weight (32.00%).

- The arithmetic average for all domain paragraphs was found to be equal to (3.30) i.e. relative weight (68.65%) It is greater than the neutral relative weight value (60%), the calculated (T) test value equal to (8.510) and is greater than the tablitic (T) value equal to (1.95), and the probability value (.sig) equals (0.000) It is less than (0.05), so the field is statistically indicative at the level of indication ($a \leq 0.05$), which indicates that the response level for this area has exceeded the average score of (3), which means that there is approval by the sample members for the paragraphs of this area.
- **The result of the hypothesis:** According to the previous table and the analysis of the data statistically and comment on them we find that (T) table less than (T) calculated, it is possible to accept the research hypothesis that states that "there is a role for digital transformation in raising the efficiency of the performance of banks to enhance competitive advantage."

The researchers believe that the approval of the individuals of the sample that the digital transformation has a role in raising the efficiency of the performance of banks to enhance competitive advantage, and may be due to the fact that banks operating in the Gaza Strip are aware of the need for digital transformation and what it achieves to raise the efficiency of their performance and enhance the competitive advantage by providing electronic services to customers and beneficiaries and the extent of satisfaction with those services provided.

(2019), (2019), [1–3], (Anna Omarini 2017).

4 Results and Recommendations

Results: The researchers reached the following conclusions:

1. Banks listed on the Palestine Stock Exchange are constantly reviewing the quality mechanism of digital services and the necessary measures are taken to improve these services and use technology in monitoring the quality of electronic services.
2. Banks listed on the Palestine Stock Exchange are developing innovative and innovative models to employ emerging electronic technologies to monitor performance and commitment to tasks and responsibilities in accordance with the governance framework and predict deviations before they occur.
3. Banks listed on the Palestine Stock Exchange report proactively to senior management and departments involved in the use of emerging electronic technologies in monitoring performance indicators in order to enhance competitive advantage.
4. Palestinian banks monitor the performance indicators of electronic processes and procedures and conduct measurements and analysis through periodic reports with the aim of improving performance to enhance competitive advantage.

5. Banks listed on the Palestine Stock Exchange are working on a plan to develop and prepare employees and raise their level of knowledge of the digital transformation process.
6. Banks listed on the Palestine Stock Exchange adopt standards for the quality of digital services in cooperation with the relevant authorities to raise the level of performance and quality of the services provided.
7. Banks listed on the Palestine Stock Exchange measure the impact of training in digital transformation to enhance its performance and the services it provides to enhance competitive advantage.
8. Banks listed on the Palestine Stock Exchange are constantly educating their employees to develop their performance to help achieve their digital transformation goals.
9. Banks listed on the Palestine Stock Exchange use electronic systems to analyze the behavior of beneficiaries and investors to measure their satisfaction with all its services and analyze the results and take the necessary actions.
10. The digital transformation can be used to provide the required information in a timely manner to all beneficiaries of the services of banks listed on the Palestine Stock Exchange.

Second: Recommendations: Based on the following results, the researchers recommend the following recommendations:

1. The need for banks listed on the Palestine Stock Exchange to make training on digital transformation one of the most important priorities that senior management should focus on and use as an effective tool to prepare the necessary competencies to perform electronic work by providing more opportunity to use modern technology.
2. The need for the senior management of the banks listed on the Palestine Stock Exchange to study, analyze and strategic plan to support the proper transition of digital transformation through the vision of far-reaching insight and a clear strategy based on the methods of modern change.
3. Digital transformation requires the availability of high skills not only in management skills but also needs to combine management and technological skills, which help the senior management of banks to respond quickly and act to remedy emergency problems.
4. Work to support the process of digital transformation in banks listed on the Palestine Stock Exchange as a priority, and the need to provide the potential to ensure its success.
5. Directing accounting, financial and administrative research to know and study the various aspects of digital transformation in order to delve deeper and learn about the real role it plays in order to raise the efficiency of the financial and administrative performance of banks.

The need to contain the accounting and administrative curricula in the faculties of economics and management sciences in Palestinian universities to study the digital transformation in all its aspects.

Reference

1. Filgueiras F, Cireno F, Palotti P (2019) Digital transformation and public service delivery in Brazil. Latin American Policy, vol 10, no 2, pp 195–219, 2019 Policy Studies Organization, Wiley Periodicals, Inc.
2. Li L, Su F, Zhang W, Mao J-Y (2018) Digital transformation by SME entrepreneurs: a capability perspective. 28:1129–1157. https://wileyonlinelibrary.com/journal/isj©
3. Loonam J, Eaves S, Kumar V, Parry G (2018) Towards digital transformation: lessons learned from traditional organizations 27(2):101–109. https://wileyonlinelibrary.com/journal/jsc©
4. Asmara AH (2019) A proposed model to activate financial inclusion through digital transformation to achieve Egypt Vision 2030. In: The 24th annual crisis research conference entitled managing the digital transformation of the Egypt vision 2030 application, Ain Shams University
5. Anna O (2018) The Digital Transformation in Banking and The Role of Fin Techs in the New Financial Intermediation Scenario. https://mpra.ub.unimuenchen.de/85228/, MPRA Paper No. 85228, UTC, Bocconi University- Department of Finance- Via Roentegen, Milano, Italy, pp 1:12.
6. Yons AS, Andys M (2019) The dimension of digital transformation in the banking sector - a debt study applied to Al Rajhi Bank in Saudi Arabia. In: 24th annual crisis research conference entitled: managing the digital transformation of the application of vision Egypt 2030, Ain Shams University. Second: Foreign references

Relationship Between Financial Technology and Financial Performance

Azzam Hannoon, Abdalmuttaleb M. A. Musleh Al-Sartawi, and Azam Abdelhakeem Khalid

Abstract The aim of current study is to investigate the impact of financial technology on the performance of Bahraini banks. The results indicates that the total level of applying financial technology by Bahraini banks was 70.51%, and it has a significant positive relationship with the financial performance. The study recommends that Bahraini banks to include more information about financial technology in the financial statements, and to indicate the level of resources invested in.

Keywords Financial technology · Performance · ROE · Banks · Bahrain

1 Introduction

Financial technology (Fintech) term is used to describe the adoption of new technologies that enhance the performance of financial services. Fintech can help different stockholders in managing and controlling their financial operations through different types of tools, devices, and software. Fintech started with the beginning of the twenty-first century, but now it is adopted in different sectors and industries such as education, retail banking, fundraising and nonprofit, and investment management, to name a few [5]. Fintech also includes cryptocurrencies as a new tool to trade and exchange money in an open uninstructed global market system [8]. After the financial crises, the world started looking for new tools to control its financial systems because of the disadvantages of the traditional controlling systems, which opens the road to the people and the governments to invest in the most cutting-edge technology to enhance the monitoring and the measurements of the compliance with the rules and regulations which will enhance the overall performance [7]. In the last 15 years,

A. Hannoon (✉)
American University in the Emirates, Dubai, UAE

A. M. A. M. Al-Sartawi
Ahlia University, Manama, Kingdom of Bahrain

A. A. Khalid
Universiti Pendidikan Sultan Idris, Perak, Malaysia

A. M. A. Musleh Al-Sartawi (ed.), *The Big Data-Driven Digital Economy: Artificial and Computational Intelligence*, Studies in Computational Intelligence 974,
https://doi.org/10.1007/978-3-030-73057-4_26

Bahrain invested a lot in technology and artificial intelligence as part of its vision toward the knowledge-based economy and the digital economy [5]. Accordingly, the current study is trying to answer the following questions: (1) what is the level of financial technology? And (2) what is the relationship between financial technology and financial performance? This paper is arranged as follows. Section 2 is about published literature review. Section 3 discusses the methods that were employed for data collection. Section 4 discusses the results, and finally, Sect. 5 provides the conclusion and recommendations.

2 Literature Review

In consideration of the fact that financial technology is increasingly evolving, there has been no consensus, about a single definition. As of now, most of the focus has been on companies that have high reliance on technological innovations to enhance their capacity to provide financial services [10]. FinTech can be described as technology-enabled financial solutions that cover up the whole commodities' scope conventionally offered by banks [5] and at the same time achieved the IT governance and applying the cybersecurity concepts [17, 19, 24, 28]. FinTech is also explained as a new form of monetary service trade that merges IT with monetary services such as remittances, payments, and management of assets as well [6, 12]. FinTech can be differentiated from a conventional financial business based on the degree to which they participate in financing, management of assets, and facilitation of payments among other aspects.

Under the broad category of financial technologies, there are four segments including Financing, management of assets, facilitation of payments, and other financial technologies. Financing is broken down to crowdfunding and credit factoring with activities such as investing and lending among groups of people [9]. Management of assets could include trading in social areas and personal management which is considered as important parts of intellectual capital [15, 20, 22]. Facilitating payments are made to officials to expedite an administrative process, payments are intended to ease the process of a given service that the client is legally entitled to [26, 27, 29, 30]. Finally, other financial technologies consist of insurance applications, search engine sites, and IT infrastructure for financing [1].

In terms of volume, it has been difficult to estimate the size, the speed of growth, and the impact that FinTech has on the conventional banking industry parameters. It can be noticed that in the year 2016, the value of funds invested in this industry was more than $25 Billion across at least 1000 agreements [2, 14], which indicates that the financial technology firms become a new opportunity for investments.

Another report has indicated that there has been an industry value of $14 billion emanating from close to 850 deals. By the end of the same year, there was at least $100 billion for around 9,000 firms. In terms of year-on-year growth, the industry funding grew by 75% between 2014 and 2015 and has been seen to keep increasing in the years after.

[4] argued that ICT has conveyed an absolute change of standards on the performance of financial institutions and on the delivery of services to the clients in the banking business. In a proposal to take up with worldwide growth, advance the delivery of customer services, as well as lessen the transaction costs, banks have ventured enormously in technology, moreover, banks have broadly accepted financial technology networks for delivering an extensive range of value-added commodities. Financial technology acts as a facilitator for improved production as well as monetary advance at the firm's intensity [11]. Financial technology makes commodities more accessible and inexpensive by lowering costs of for the bank's trading transactions [4, 13].

Bahrain as one of the first middle eastern countries that is investing in the technology, the Central Bank of Bahrain's efforts towards digital transformation have increased, The FinTech and Innovation Unit has been developed to ensure that services are offered through encouraging sufficient regulatory framework that fosters FinTech innovation in the financial sector [1, 3]. There are several initiatives that allow startups and the FinTech firms and licenses to offer their clients with innovative banking, as well as financial solutions in addition to regulations meant for both modern and sharia-compliant services. The advancement of FinTech solutions and the advantages associated play a significant role in the regulatory developments on both regional and international levels. Such regulatory developments are essential to guide the FinTech initiatives and all the stakeholders in the financial sector in Bahrain. Furthermore, using FinTech will lead enhance the way of disclosing information to achieve equality in sharing and using the information between the decision-makers such as using the internet or social media [16, 18, 21, 23]. On the other hand, using such technology will increase the need toward will skilled employees so the educational institutions will change their way of teaching and delivering their curriculum [25, 28].

3 Research Methodology

The data was collected from 12 banks working in Bahrain depending on the availability of the information needed to conduct the research. The FinTech level was calculated depending on a dummy variables checklist which is developed by the researchers based on literature review and distributed to each bank in the sample. Where the item will assign one (1) if it is achieved by the bank and (0) zero otherwise, then the average of the total items achieved by the banks was calculated to indicate the level. The research model was developed taking into consideration ROE as a dependent variable and FinTech level as independent variable along with some control variables that are firm size, age, and financial leverage. All the financial variables (dependent and control) were taken from Bloomberg database for the selected banks for the year 2019.

$$\boldsymbol{ROE_j = \beta_0 + \beta_1 FinTechL_i + \beta_2 BSi_i + \beta_3 BA_i + \beta_4 BFl_i + \varepsilon_j}$$

Where:
ROEi is Return On Equity in bank (i), CLi is FinTech level for bank (i), BSii is the bank size (i), BAi is the bank age (i), and BFli is the bank financial leverage (i).

4 Data Analysis

The descriptive statistics, Table 1, shows that overall average of the dependent variable, ROE was 0.22 in 2019. ROE result indicates that there is a good level of efficiency in controlling and managing the Bharani banks. On the other, the results show that the FinTech level by the banks was 70.51%, which is considered as a moderate level for the Bharani banks. According to Regarding the control variables, the total assets of the banks which were used as an indicator of bank size, shows that the mean size was 2.32E6 million, with a minimum of 53,226 million and a maximum 761,523 million. However, as the normality distributions of total assets are generally skewed, natural logarithm was used in the regression analysis to reduce skewness and bring the distribution of the variables nearer to normality. Moreover, the mean leverage for the banks was approximately 52.60% with a minimum 3%, indicating banks with somewhat high debts and a maximum of 95%, signifying very high debts. With regards to the age, it was ranged from 5 to 52 with a mean of 20.21 years old, which might lead to that Bharani banks had good experience and flexibility in applying FinTech which is interpretating the good level of FinTech achieved as mentioned above.

4.1 Validity and Reliability

To assess validity, several tests were conducted. The study checks for multicollinearity, as seen in Table 2, by conducting the variance inflation factor (VIF) indicating that no score exceeded 10 for any variable in the model. Similarly, the tolerance test, which is the inverse of the VIF, suggested that no score was below 0.2.

Table 1 Descriptive statistics for the variables

Variable	Min.	Max	Mean	S. D
FinTech	0.9059	0.302	0.7051	0.1267
ROE	−0.0.10	0.45.15	0.22	1.7214
Leverage	0.03	0.95	0.5260	0.2876
Age	5	52	20.21	11.077
Size*	53,226	761,523	2.32E6	1.118E5

*Millions

Table 2 Normal distribution and collinearity statistics tests

Model	VIF	Skewness	Kurtosis	Shapiro-Wilk Test	
FinTech	3.673	−2.419	2.154	0.234	0.3467
Lev	2.259	−0.976	−0.345	0.624	0.0905
Age	2.012	0.755	−2.268	0.757	0.0876
Size	2.274	3.334	2.311	0.825	0.0023

It was, therefore, concluded that no problems were found with regards to collinearity in the model.

Additionally, Table 2 reports the normality test, where the skewness test and the kurtosis test suggest that all the predictive variables are normally distributed except for firm size. Furthermore, the Shapiro–Wilk test was used to test the normality of the collected data for the variables. Table 2 reported a significance level of more than 0.05 for all the variables except for size; thus, it can be assumed that the data are normally distributed. Regarding the firm size, the study has considered the natural logarithm to transform the data to better fit the normal distribution before conducting the regression analysis. It should be noted that an autocorrelation test was not conducted in this research as the data used are cross-sectional.

4.2 Regression Analysis

Table 3 reports the empirical results of the regression analysis of the study model. It also shows the coefficients of determination, where the value of the F- statistic for the model was more than the F- scheduled at confidence level 95% which is 2.512, with a p-value less than 0.05, deeming the model as significant.

Table 3 Regression analysis results

Variables		Beta	t	Sig.
FinTech		0.711	3.212	0.002**
Lev		0.221	0.761	0.221
Age		0.901	2.118	0.032*
Size		0.711	2.601	0.021*
R	R2	Ad. R2	F-Stat	Prob. (F)
0.916	0.613	0.571	2.512	0.010

*P < 0.05 level
**P < 0.1 level

With regards to testing the hypothesis, the model reported a positive and significant relationship between the FinTech and ROE. This is in line with prior research which indicates that FinTech positively influences a financial institution's performance [3, 5] and contradict with [1] and [7], who found that FinTech had no effect on the financial performance. Moreover, the regression analysis shows a positive and significant relationship between size, age and ROE. This could be because older and larger firms have higher flexibility and they are investing more in FinTech tools because of their vast experience and availability of resources.

5 Conclusion and Recommendations

The aim of this study is to examine the impact of FinTech on financial performance of Bahraini Banks, the literature consisted of various research regarding the matter. Examples include the impacts on firms: post data breach, effect on shareholders wealth and impact on the firm's image/ reputation. Some of the previous literature concluded that there is no significant relationship between FinTech and the performance of financial institutions.

The current study collected data using a checklist developed by the researchers and distributed to the Bahraini banks to be filled out. The filled-out checklist was examined against the 2019 financial performance reports of the sample banks taken from Bloomberg database. The ROE ratio used to determine the performance of the banks. The results concluded that there is a significant relationship between FinTech, and financial performance of the Bahraini listed banks.

The recommendation of the current study was driven from the limitation has been faced, as the study depends on a checklist to calculate the FinTech, therefore, it is recommended that the financial technology to be included by the listed banks in their financial statements and to indicate the level of resources invested in. Future studies can consider using a larger sample size to avoid perfect multicollinearity in addition to unobserved heterogeneity. Also, future studies can add more control variables to be tested on the impact of FinTech and financial performance such as analyzing the impact of gender on FinTech adapting, practice and behavior.

References

1. Chishti S, Barberis J (2016) The FinTech Book: the financial technology handbook for investors, entrepreneurs and visionaries. Wiley, Hoboken
2. Bartlett R, Morse A, Stanton R, Wallace N (2018) Consumer-lending discrimination in the era of FinTech. Unpublished working paper. University of California, Berkeley. https://lendingtimes.com/wp-content/uploads/2018/11/discrim.pdf
3. Dhar V, Stein RM (2017) FinTech platforms and strategy. Commun ACM 60(10):32–35

4. Buckley RP, Arner DW, Veidt R, Zetzsche DA (2019) Building FinTech ecosystems: regulatory sandboxes, innovation hubs and beyond. UNSW Law Research Paper, pp 19–72. https://papers.ssrn.com/sol3/papers.cfm?abstract_id=3455872
5. Ali W, Muthaly S, Dada M (2018) Adoption of shariah compliant peer-to-business financing platforms by SMEs: a conceptual strategic framework for FinTech in bahrain. https://pdfs.semanticscholar.org/0c16/cfe3da4583db360eb546d3a63fa597aa6856.pdf
6. Dapp T, Slomka L, AG DB, Hoffmann R (2014) Fintech–the digital (r) evolution in the financial sector. Deutsche Bank Research", Frankfurt am Main. https://www.dbresearch.com/PROD/RPS_EN-PROD/PROD0000000000451941/Fintech_%E2%80%93_The_digital_%28r%29evolution_in_the_financia.pdf
7. Buchak G, Matvos G, Piskorski T, Seru A (2018) Fintech, regulatory arbitrage, and the rise of shadow banks. J Financ Econ 130(3):453–483
8. di Castri S, Plaitakis A (2018). Going beyond regulatory sandboxes to enable FinTech innovation in emerging markets. SSRN 3059309. https://papers.ssrn.com/sol3/papers.cfm?abstract_id=3059309
9. Lee I, Shin YJ (2018) Fintech: Ecosystem, business models, investment decisions, and challenges. Bus Horiz 61(1):35–46
10. Micu I, Micu A (2016) Financial technology (Fintech) and its implementation on the Romanian non-banking capital market. Practical Appl Sci IV(2(11)):379–384
11. Mugenda OM, Mugenda AG (2012) Research methods: quantitative and qualitative approaches. Acts Press, Nairobi
12. Nicoletti B, Nicoletti W, Weis (2017) Future of FinTech. Palgrave Macmillan, Basingstoke
13. Puschmann T (2017) Fintech. Bus Inf Syst Eng 59(1):69–76
14. Schweitzer ME, Barkley B (2017) Is 'Fintech' Good for Small Business Borrowers? Impacts on Firm Growth and Customer Satisfaction. https://ideas.repec.org/p/fip/fedcwp/1701.html
15. Al-Sartawi A (2020) Social media disclosure of intellectual capital and firm value. Int J Learn Intellect Capital 17(4):312–323
16. Al-Sartawi A (2020) Does it pay to be socially responsible? Empirical evidence from the GCC countries. Int J Law Manag 62(5):381–394
17. Al-Sartawi A (2020) Information technology governance and cybersecurity at the board level. Int J Crit Infrastruct 16(2):150–161
18. Al-Sartawi A (2019) Assessing the relationship between information transparency through social media disclosure and firm value. Manag Acc Rev 18(2):1–20
19. Al-Sartawi A, Sanad Z (2019) Institutional ownership and corporate governance: evidence from Bahrain. Afro-Asian J Finan Acc 9(1):101–115
20. Al-Sartawi A (2018) Ownership structure and intellectual capital: evidence from the GCC countries. Int J Learn Intellect Capital 15(3):277–291
21. Al-Sartawi A (2018) Institutional ownership, social responsibility, corporate governance and online financial disclosure. Int J Critical Acc 10(3/4):241–255
22. Al-Sartawi A (2018) Corporate governance and intellectual capital: evidence from gulf cooperation council countries. Acad Acc Financ Stud J 22(1):1–12
23. Sanad Z, Al-Sartawi A (2016) Investigating the relationship between corporate governance and internet financial reporting (IFR): evidence from bahrain bourse. Jordan J Bus Adm 12(1):239–269
24. Al-Sartawi A (2015) The effect of corporate governance on the performance of the listed companies in the gulf cooperation council countries. Jordan J Bus Adm 11(3):705–725
25. Musleh Al-Sartawi AMA (2020) E-learning improves accounting education: case of the higher education sector of bahrain. In: Themistocleous M, Papadaki M, Kamal MM (eds) Information systems. EMCIS 2020. Lecture notes in business information processing, vol 402. Springer, Cham
26. Gupta M, Sikarwar TS (2020) Modelling credit risk management and bank's profitability. Int J Electron Bank 2(2):170–183
27. Gupta N (2019) Influence of demographic variables on synchronisation between customer satisfaction and retail banking channels for customers' of public sector banks of India. Int J Electron Bank 1(3):206–219

28. Abdulrasool FE, Turnbull SI (2020) Exploring security, risk, and compliance driven IT governance model for universities: applied research based on the COBIT framework. Int J Electron Bank 2(3):237–265
29. Alhakimi W, Esmail J (2019) The factors influencing the adoption of internet banking in Yemen. Int J Electron Bank 2(2):97–117
30. Memdani L (2020) Demonetisation: a move towards cashless economy in India. Int J Electron Bank 2(3):205–211

Investigate the Effects of Behavioral Factors on Job Performance: A Conceptual Paper

Aya Naser Magableh, Khatijah Omar, and Jasem Taleb Al-Tarawneh

Abstract This is a conceptual paper on the effect of behavioral factors on job performance, and to investigate the mediating effect of employee engagement between behavioral factors and job performance in Jordanian commercial banks. This paper aims to present literature reviews related to investigating behavioral factors. Also, this paper proposed a model that enables the organization to achieve job performance by includes some factors which are talent management, quality of work-life, and organizational climate. However, employee performance refers to behaviors that are relevant to organizational goals and that are under the control of individual employees. Job performance is the feeling of the employee about his job. Likewise, all factors need to harmonize and align together to reach the desired progress, success, and sustainability. This paper suggests statistical techniques such as normality tests, multicollinearity test, and EFA using the SPSS. For the measurement model and hypothesis testing using smart PLS. Likewise, the conclusion that was drawn is that the effect of investigating Behavioral factors including talent management, quality of work-life, and organizational climate maybe help increase job performance. For future research, this paper recommends that all variables affecting the individual factors of employees should be included and applied to a service sector to confirm the results.

Keywords Behavioral factors · Talent management · Quality of work-life · Organizational climate · Job performance

A. N. Magableh (✉) · J. T. Al-Tarawneh
Faculty of Business, Universiti Malaysia Terengganu, 21030 Kuala Nerus, Terengganu, Malaysia

K. Omar
Faculty of Business, Economics and Social Development, Universiti Malaysia Terengganu, 21030 Kuala Nerus, Terengganu, Malaysia
e-mail: Khatijah@umt.edu.my

A. M. A. Musleh Al-Sartawi (ed.), *The Big Data-Driven Digital Economy: Artificial and Computational Intelligence*, Studies in Computational Intelligence 974,
https://doi.org/10.1007/978-3-030-73057-4_27

1 Introduction

Job performance refers to the degree of achievement and completion of the tasks that make up the job of an individual, and it reflects how the individual is achieved or satisfied with the job requirements. Besides, the issue of job performance has gained great importance to the person in charge of the institution as it represents one of the factors or determinants that are used in evaluating the organization that he heads or manages. Therefore, most officials attach great importance to the performance of workers in institutions. However, performance is a reflection of each individual's capabilities, as well as a reflection of the performance of these institutions and their degree of effectiveness and efficiency.

Increased job performance is an essential goal for businesses to sustain their market growth. As a result, the corporate initiative is geared towards optimizing performance outcomes, taking into account the organizational sense in which performance is generated [19]. Formulated in this sense, social considerations, such as societal expectations or the effects of emerging innovations, typical of all organizations, are part of success improvement processes and must be taken into account as a major research concern [22, 29]. However, enterprises all over the globe in both developed and developing markets are running in exceedingly and dynamic circumstances [3]. Accordingly, the growth and surviving within such a mutable set of surroundings are vastly necessitating high concentrating efforts to realize the organizational goals [81].

Each organization has it is own system in management and evaluation of performance, and whether it relates to the performance of individuals, or the performance of departments or teams, or the performance of the organization as a whole, it is through performance measurement, an organization can control the programs, systems, and thus on the achievement of its objectives. Performance evaluation is a means that drive managerial units to work vitality and activity, where make presidents follow the duties and responsibilities of their subordinates on an ongoing basis, and paid subordinates to work effectively [8], and the importance of this medium appeared when looking at the areas in which they have used the results of the performance evaluation, and most important: improving and developing employee performance, and the adoption of this evaluation as a means for determining bonuses and increments, a tool for the detection of training needs, and a means to judge the safety of selection, recruitment and training policies, and objective basis for drawing these policies [75].

In the last few decades, the banking industry in Jordan has gained much importance as it directly affects the economy of the country. There has been a change in quantity and quality service provision to the customers. Due to a large number of banks in Jordan, competition has started growing, and each bank needed to follow the strategic approach to get a broad market share and to earn the right profit margin. In Jordan, the banking sector has gained much importance as they help contribute 80% of services and 20% of goods in GDP. Today, Jordanian banks are facing several challenges. To deal with challenges and for sustainable banking operations, Jordanian banks have started developing a proactive approach. For this, they need to retain professional employees and to develop their skills. More importantly, most of the banks realized that employees of the bank are the best assets of the firm that they can compete with external as well as internal banks in their sectors [7].

Previous studies confirm that the involvement of the employee in their work is the key factor in improving job performance. Furthermore, the role of human resources in the financial sector is big for moving banks forward and in making the bank to perform its best [66]. Many factors are affecting the management of human resources; one of these factors is Quality of Work Life (QWL) [38, 44]. QWL can be defined as the favorable circumstances of a workplace that endorse employee satisfaction by assuring proper rewards, job security, and growth opportunity [18]. In explaining QWL earlier literature identifies different dimensions such as job security, better reward, and opportunity for growth participation, higher pay, and increased organizational productivity highly discussed. Besides, very prominent research such as [71] proposed eight major dimensions for measuring QWL. For example, adequate and fair reward, safe and healthy working conditions, opportunity to use, develop human capital, the opportunity for continued growth and security, social integration in the workplace, constitutionalism in the work organization, work, and total living space, and social relevance of work life.

The concept of quality of work-life (QWL) indicates that a bank must encourage the process of enhancing the environment of its work through several factors to be provided to the employees such as the feeling of job security, motivations, and opportunities for professional development and career growth [55]. Also, the quality of work-life in a bank is essential for the smooth running and success of its employees. Quality of work-life helps the employees to feel secure and like they are being thought of and cared for by the bank in which they work. For effective development of both individual objectives and the bank's objective, the bank should provide a good overall working environment considering factors like career opportunities participative management working environment. In this process, it can generate a sense of satisfaction for which on his capability to achieve simultaneously both individual and bank's objectives.

Finally, the work environment or climate perception of employees has significant consequences for both individuals and organizations. Climate or atmosphere in the workplace has an impact on employee's motivation, behavior, attitudes, and potential, which, in turn, is predicted to influence organizational productivity [23, 24]. The importance of this study highlights the great competition among commercial banks in Jordan to maintain operating efficiencies. Hence the importance of this study is to try to link the talent management, quality of work-life, work climate, work engagement, and job performance for being influential topics in the Jordanian financial sector.

Because of the importance of job performance and its role in improving the quality of banking service and ultimately improving the quality of services, this paper came to investigate the effects of behavioral factors on job performance, which are engagement, talent management, quality of work-life, and work climate. By reviewing the relevant literature and coming up with a clear model that can contribute to banks' focus on individual employee factors that would improve service quality, which is mainly the focus of bank managers.

1.1 Objectives

1) To examine the effect of Behavioral factors (talent management, quality of work-life, organizational climate) on job performance in Jordanian commercial banks.
2) To examine the effect of employee engagement on job performance in Jordanian commercial banks.
3) To examine the mediating effect of employee engagement between behavioral factors and job performance in Jordanian commercial banks.

Literature Review
A literature review is a body of text that aims to review the critical points of current knowledge including substantive findings as well as theoretical and metrological contributions to the effect of behavioral factors on job performance, and the mediate effect of employee engagement. Commonly, any enterprise relies on a diverse set of factors to maintain the success and the consistency in accomplishments level, also to attain remains improvements [67]. The list of factors compresses but is not limited to the operational factors, the technological factors, the strategic directions, and most vitally the human capital [41]. In any human activity such as industry, agriculture, trade commerce, politics, or government, the source of action and the target of action is the man. The job of man is the mainspring of his life [47]. In this concept, some important factors must be focused on to achieve the highest job performance, especially in light of employee Engagement.

1. **Behavioral Factors**
   [33], referred that Job performance consists of behaviors that can be observed on individuals in their jobs, and are relevant in achieving the objectives of the organization.

   A. Talent Management
   [10] noted that the characteristics of talented people are related to the manifestations of excellence. [11], states that the most important characteristics of talented people are cognitive characteristics and emotional characteristics, studies agree that characteristics of talented people are characterized by cognitive characteristics and the emotional characteristics that distinguish them from others. Also, [13], deduced that the primary role of talent planning is to enhance easy identification of future talent s which are needed at all organizational levels.
   According to [32] reports on the Society for HRM indicates that the talent management model involves an integrated strategy or system that is

designed to improve the processes of recruiting, developing, and retaining people with the required skills and aptitude to meet current and future employee needs, which is essential to employee performance. Talent management is fundamental to any HR department to boost employee performance [20]. As deeply explained in the literature, properly handling talent is considered one of the best options organizations can adopt to enhance performance and beat off competition [79]. According to [59], the concept of talent management has a major impact on the workforce and if properly administered, it can make a difference in how employees perform.

B. Quality of Work-Life

Quality of work-life (QWL) is fundamentally a multidimensional concept that represents the mechanism, which regulates the relationship between the individual and their work [43, 56]. [51], defines QWL as a "process by which an organization responds to the employee's needs by developing mechanisms to allow them to share fully in making the decisions that design their lives at work". Likewise, [73] refer to QWL as "employee satisfaction with a variety of needs through resources, activities, and outcomes stemming from participation in the workplace". Furthermore, "QWL encompasses the physical, technological, psychological, and social dimension of work corresponding to the ideals of a more humane and healthy organization" [27].

The quality of the work-life framework shows a difference between results and chances while proving its worth to public policy [62]. Chances have a place with the space for open strategy prompt activity, while it comes about to constitute the last inspiration of open arrangement. Governments, as well as countless organizations, usually would contemplate the quality of work-life and observe their improvement, but the concluding assessment of open approaches must be considered [68].

Most studies focus on the relationship of QWL with some of the variables such as job satisfaction, organizational commitment, job performance, labor relations, etc. which play a crucial role in determining the overall wellbeing of any industrial organization. However, there is a lack of empirical evidence on the relationship between QWL and employee work engagement. Work engagement is fundamentally a motivational concept that represents the active allocation of personal resources toward the tasks associated with a work role [25].

C. Organizational Climate

Employees who are satisfied with the working environment are both loyal to the organization and capable of providing a better quality of care. The above discussion provides ample evidence to establish the relationship between QWL and performance. But when we put the spotlight on financial organizations, this relationship is of critical importance. [28], investigated the relationship between the QWL and productivity among employees and suggested managers design appropriate strategies for promoting QWL to enhance productivity in the organization.

Organizational climates can be operational zed at any level within the organization (from company-wide to team units), depending on the referent used in the climate measure [30]. The departmental climate is the most common level to operationalize climate [68]; climate at the department level (defined as all employees directly under the same supervisor) has generally been found to have the strongest influence on employee behavior compared to other levels of climate [12, 42, 49].

[64] pointed out organizational climate as the unique styles of an organization, the characteristics composed of leader-subordinate interaction. [46], regarded the organizational climate as a member's direct or indirect perception of the characters of a specific environment. Such perception results would affect the member's attitudes, beliefs, values, and motivation.

2. Job Performance

   Job performance is considered as an important parameter in the hotels' profession, in such a way that in the last pent etic, new innovative ways of calculation and consideration have been invented [17] even for newly registered professionals [63]. After a careful detection of international literature, what can be seen is that job performance is directly and strongly related to stress and burnout [34]. Equally, the crucial role in shaping the professional performance playing both the employee leadership [71] and the procedure of rational decision making [60].

   [26] describes performance as the individual's ability to achieve the objectives of the organization through the optimal use of available resources efficiently and effectively. The definition of Job performance is "a coordinated effort to do tasks include converting inputs to outputs with quality consistent with the skills, abilities, and experience of working, with the help of the supporting factors and the environment, the appropriate action to make this effort precisely, the shortest time and less expensive" [9].

   In the concept of staff Performance, [39] suggested two aspects of employee performance which are task efficiency and contextual success. Task efficiency (or professional job performance) is the actions associated with the management and management of the professional heart of the company. Contextual success (or interpersonal work efficiency) is a feature of one's interpersonal skills awareness that reflects the wider social context in which the technological core would operate.
3. Employee Engagement

   Work engagement is a positive, fulfilling work-related state of mind, and characterized by vigor, dedication, and absorption'. Vigour refers to the willingness to invest effort in one's work, dedication is related to involvement, and absorption is related to concentration and being engrossed in one's work [69].

Employee Engagement and employee-organizational responsibilities are important organizational criteria as companies face globalization and rebound from the global recession. Employee engagement, employee participation, and corporate dedication have become areas of concern amongst many researchers and have gained significant attention from analysts and educators. Therefore, based on the Job demands resource model, the authors surveyed the relationship between work engagement and job resources for employees working in banks [40].

The literature shows that both labor and personal resources are important predictors of engagement; working environments with adequate labor resources foster engagement, especially when the work is highly demanding, and personal resources such as self-esteem, optimism, and self-efficacy are also useful for coping with the everyday demands of working life [16, 36, 81]. Also, [53] conducted a systematic review to synthesize the research about engagement in organizational psychology, business literature.[72], wanted to review engagement in an organization, but given the limited number of publications, she extended it to any working environment. We are now able to overcome this limitation because research on work engagement in employees has greatly increased.

I. Conceptual Framework and Hypotheses Development
Based on the literature, talent management is expected to affect job performance. Quality of work-life is expected to affect job performance. Organizational climate is expected to affect job performance. Besides, employee engagement is expected to mediate the effect of behavioral factors and job performance.

2 Behavioral Factors and Job Performance

The previous studies confirmed the positive effect between behavioral Factors (Talent Management, Quality of Work-Life, and Organizational Climate) and Job Performance. The previous literature found that talent management has a significant positive effect on job performance, quality of work-life has a significant positive effect on job performance, and organizational climate has a significant positive effect on job performance.

In a review of studies investigating organizational climate and employees' performance, the study of [2] found that organizational climates exhibit the clear role clarity dimensions that result in higher satisfaction and performance of employees. [40] explained the characteristics of organizational climate, for instances having a high level of self-governance, giving opportunities for employees, sustaining connections among employees, concerning and demonstrating enthusiasm for employees, perceiving workers' achievements also, holding them in high respect result in more fulfilled employees.

[15], agreed that the organizational climate in the career development of the employees is important for the employee to perform better in work as providing necessary and related training are required. Good communication among the employees and upper management form a good organizational climate to boost up the satisfaction of employees in work [6]. Moreover, organizations with a positive organizational

climate are more productive as employees have higher job satisfaction and are more committed to the organization [1]. Organizational climate serves as an antecedent of knowledge management and therefore results in increased performance measures and outcomes [53]. However, since there are not many studies of job performance in Jordan, thus, the following is hypothesized:

H1: Behavioral Factors have a significant positive effect on job performance in Jordanian commercial banks.

3 Employee Engagement and Job Performance

There are indications that the level of engagement is positively associated with job performance in terms of financial benefits, greater client loyalty, and better adaptation to the working environment [14]. Empirical studies are also available that indicate that engagement is positively related to finance. For example, engaged employees have been shown to suffer less from depression and stress and to have fewer psychosomatic symptoms [78].

The research literature suggests that engagement and performance are related in direct and indirect ways. In terms of its direct relationships, engagement at the individual and group levels is associated with both organizational and employee performance [21, 54]. [21], found that employee engagement was a significant predictor of desirable organizational outcomes such as customer satisfaction, retention, productivity, and profitability. In [54] study of 65 firms in different industries, the top 25% on an engagement index had a greater return on assets, profitability, and more than double the shareholder value compared to the bottom 25%. These studies indicate a direct relationship between engagement and performance.

[61], further asserted that designing performance management processes that foster employee engagement will lead to higher levels of performance, and argued that employee engagement behavior is an antecedent of job performance. [52], explained that the energy and focus inherent in work engagement allow employees to bring their full potential to the job. This energetic focus in turn enhances the quality of their core work responsibilities. This gives them the capacity and the motivation to concentrate exclusively on the task at hand. Other authors and practitioners have similarly discussed how corporate strategies and imperatives that develop employee engagement may be utilized as mechanisms for enhancing organizational productivity and competitive edge [58, 77]. However, since there are not many studies of job performance in Jordan, thus, the following is hypothesized:

H2: Employee Engagement has a significant positive effect on job performance in Jordanian commercial banks.

4 Mediation of Employee Engagement

Researches in the past have examined several elements that can affect job performance. [45] and [5] showed that a crucial element is employees' commitment to their job. There is also a strong connection between being satisfied with their job and their performance [37]. QWL initiatives can greatly help to improve employees' self-esteem and job satisfaction [48], lead workers to provide better services and increase customer satisfaction [35]. Moreover, QWL programs can improve work performance and the quality of life among employees [80]. In a related 47 context, [73] found that QWL was associated with both organizational and individual efficiency. Conversely, a weak level of QWL causes job dissatisfaction, increased absenteeism, demotivation, low morale, rising accident rates, and poor productivity, which therefore causes poor organizational performance [74]. [31], found that QWL is crucial for organizational success and competitive advantage.

[57] identified six elements of working life that lead either to burnout or engagement: "workload", "rewards and recognition", "community and social support", "perceived justice", "choice and control", and "meaningfulness and value of work". These issues are the core constituent variables of QWL; hence, the present study assumes that the improvement of employees' engagement is mediating between talent management, QWL, organizational climate, and job performance. However, since there are not many studies of job performance in Jordan, thus, the following is hypothesized:

H3: employee engagement mediates the effect of behavioral factors and job performance in Jordanian commercial banks

Research Methodology

This paper aims to investigate the mediating effect of employee engagement between behavioral factors and job performance in Jordanian commercial banks, while the quantitative method is the most appropriate to be taken. The reason lies in the fact that the current 56 study focuses on research problems in an objective way rather than subjectively. As stated by [65]; and [76] the quantitative strategy works on objective and measure it through the actions and opinions which helps the researcher to describe the data rather than interpret the data. Additionally, the nature of this study is empirical where researchers emphasize fresh data collection by the research problem. Thus, this study is quantitative using a survey design.

By reviewing the reports of the central bank of Jordan (CBJ), there are 25 operating banks, 15 listed in the Amman Stock Exchange (ASE), led by Arab Bank. Also, the number includes regional institutions (e.g., Kuwait National Bank, National Bank of Abu Dhabi, and Egyptian Arab Land Bank), and Western multinationals banks (e.g., Citibank and Standard Chartered). The banking sector history is dating back to the 1940s when Arab Bank moved to Amman. Banking accounted for 18.82% of the GDP as of mid-2018, this justified according to it the most significant economic sectors in Jordan too.

The managers argue that they would do better if they were empowered to make certain business decisions directly [50]. Therefore, this study focused on all employees.

The size of the sample was calculated with the help of the table proposed by [70]; and [48]. From the table, population with sample sizes are matched with a 95% level of confidence concerning normally distributed data. Therefore, for a population of 17,569 employees, the calculated required size of the sample is 377.

For the methodology section, this paper suggests statistical techniques such as normality tests, multicolinearity test, and EFA using the SPSS. For the measurement model and hypothesis testing using smart PLS.

Even though PLS-SEM is methodologically well-established and frequently applied, editors and reviewers often require researchers to justify their choice of the method [23]. The present study used PLS-SEM for several reasons; In real world scenario and application of complex models, modeling of PLS has better advantage hence its suitable and appropriate compared to other models in the literature [4]. The basic assumption guiding the modelling of PLS including the ability to dynamically develop and measure complicated models, make the model to possessed the superior qualities of evaluating complex and large models [4]. Hence, this study applied a complex model with several variables. This serves as justification for the adoption of PLS technique in this study.

5 Conclusion

This paper covered several phases of research methodologies, for instance; research design, population, and sampling, instrumentation, pre-testing survey instrument, reliability, and validity of survey instrument, data collection procedure and analysis, and technique of data analysis. Samples were identified using a proportionate stratified sampling method. Previously established measurement items were adapted in this study. The hypotheses were developing based on the previous study. For the methodology section, this paper suggests statistical techniques such as normality tests, multicollinearity test, and EFA using the SPSS. For the measurement model and hypothesis testing using smart PLS.

The findings indicate that there are indications that the level of engagement is positively associated with job performance in terms of financial benefits, greater client loyalty, and better adaptation to the working environment. Moreover, organizational climate serves as an antecedent of knowledge management and therefore results in increased performance measures and outcomes, there is also a strong connection between being satisfied at their job and their performance, education was a positive influence on job performance, and a person's self-motivation and efficiency have a positive effect on job performance.

The current study recommends the need to focus on participation among all employees of any sector, and the matter becomes more and more important in the service sectors. To provide the best service and work as one team. Also, the management must focus on showing creative talents in the organizations they belong to. For future research, this paper recommends that all variables affecting the individual factors of employees should be included and applied to a service sector to confirm the results received. Also, to Find a stronger approach to measure the selected variables separately and then testing the model completely and finding differences for future research.

References

1. Agarwal P (2015) The moderating effect of strength of organisational climate on the organisational outcomes. J Indian Acad Appl Psychol 41(1):71
2. Akbaba Ö, Altındağ E (2016) The effects of reengineering, organizational climate and psychological capital on the firm performance. Procedia-Soc Behav Sci 235:320–331
3. Akbarpour M, Li S, Gharan SO (2020) Thickness and information in dynamic matching markets. J Polit Econ 128(3):783–815
4. Akter S, Fosso Wamba S, Dewan S (2017) Why PLS-SEM is suitable for complex modelling? An empirical illustration in big data analytics quality. Prod Plan Control 28(11–12):1011–1021
5. Al-Ahmadi H (2009) Factors affecting performance of hospital nurses in Riyadh Region, Saudi Arabia. Int J Health Care Qual Assur 22(1):40–54
6. Alajmi SA (2016) Organizational climate and its relationship to job satisfaction in Kuwaiti industrial companies. Asian J Manag Sci Econ 3(2):38–47
7. Al-dalahmeh M, Khalaf R, Obeidat B (2018) The effect of employee engagement on organizational performance via the mediating role of job satisfaction: the case of IT employees in the Jordanian banking sector. Mod Appl Sci 12(6):17–43
8. Al-Hawary SIS, Al-Kuwait ZH (2017) Training programs and their effect on the employee's performance at King Hussain Bin Talal development area at AlMafraq Governate in Jordan
9. Al-Hawary S, Banat N (2017) Impact of motivation on job performance of nursing staff in private hospitals in Jordan. Int J Acad Res Account Financ. Manag Sci 7(2):54–63
10. Al-Lozi MS, Almomani RZQ, Al-Hawary SIS (2018) Talent management strategies as a critical success factor for effectiveness of human resources information systems in commercial banks working in Jordan. Glob J Manag Bus Res
11. Al-Zoubi SM, Rahman MSBA (2015) Talented students' satisfaction with the performance of the gifted centers. J Educ Gift Young Sci 4(1):1–20
12. Andreoli N, Lefkowitz J (2009) Individual and organizational antecedents of misconduct in organizations. J Bus Ethics 85(3):309
13. Anlesinya A, Amponsah-Tawiah K (2020) Towards a responsible talent management model. Eur J Train Dev
14. Bailey C, Madden A, Alfes K, Fletcher L, Robinson D, Holmes J, Currie G (2015) Evaluating the evidence on employee engagement and its potential benefits to NHS staff: a narrative synthesis of the literature. Health Serv Deliv Res 3(26):1–424
15. Balkar B (2015) The relationships between organizational climate, innovative behavior and job performance of teachers. Int Online J Educ Sci **7**(2)
16. Barbier M, Hansez I, Chmiel N, Demerouti E (2013) Performance expectations, personal resources, and job resources: how do they predict work engagement? Eur J Work Organ Psychol 22(6):750–762
17. Becton JB, Matthews MC, Hartley DL, Whitaker LD (2012) Using biodata as a predictor of errors, tardiness, policy violations, overall job performance, and turnover among nurses. J Manag Organ 18(5):714

18. Bibi S, Khan A, Qian H, Garavelli AC, Natalicchio A, Capolupo P (2020) Innovative climate, a determinant of competitiveness and business performance in Chinese law firms: the role of firm size and age. Sustainability 12(12):4948
19. Blackman D, Buick F, O'Donnell M (2018) Developing high performance: performance management in the Australian public service
20. Boxall P, Purcell J, Wright P (2008) Scope, analysis, and significance. The Oxford Handbook of Human Resource Management
21. Buckingham M, Coffman C (2014) First, break all the rules: what the world's greatest managers do differently: Simon and Schuster
22. Cascio WF, Montealegre R (2016) How technology is changing work and organizations. Ann Rev Organ Psychol Organ Behav 3:349–375
23. Chin WW (2010) How to write up and report PLS analyses. Handbook of partial least squares, pp 655–690. Springer
24. Chowdhury S (2020) Talent management and its implications on the firm's productivity
25. Christian MS, Garza AS, Slaughter JE (2011) Work engagement: a quantitative review and test of its relations with task and contextual performance. Pers Psychol 64(1):89–136
26. Daft RL (2015) Organization theory and design: cengage learning
27. Daubermann DC, Tonete VLP (2012) Quality of work-life of nurses in primary health care. Acta Paul Enferm 25(2):277–283
28. Dehghan Nayeri N, Salehi T, Ali Asadi Noghabi A (2011) Quality of work life and productivity among Iranian nurses. Contemp Nurse 39(1):106–118
29. Dorsey D, Mueller-Hanson R (2017) Performance management that makes a difference: an evidence-based approach. SHRM science to practice series
30. Ehrhart MG, Raver JL (2014) The effects of organizational climate and culture on productive and counterproductive behavior. The Oxford handbook of organizational climate and culture, pp 153–176
31. Fajemisin T (2002) Quality of work life: a study of employees in West Africa. Afr Bus Rev 13(4):501–517
32. Fegley S (2006) Talent management survey report, society for human resource management (SHRM): Society for human resource management
33. Furnham A (2005) The psychology of behaviour at work: the individual in the organization. Psychology Press, UK
34. Gandi JC, Wai PS, Karick H, Dagona ZK (2011) The role of stress and level of burnout in job performance among nurses. Ment Health Fam Med 8(3):181
35. Griffith J (2001) Do satisfied employees satisfy customers? Support-services staff morale and satisfaction among public school administrators, students, and parents. J Appl Soc Psychol 31(8):1627–1658
36. Grover SL, Teo ST, Pick D, Roche M, Newton CJ (2018) Psychological capital as a personal resource in the JD-R model. Pers Rev 47(4):968–984
37. Gu Z, Siu RCS (2009) Drivers of job satisfaction as related to work performance in Macao casino hotels: an investigation based on employee survey. Int J Contemp Hosp Manag 21(5):561–578
38. Gupta B (2016) Factors affecting the quality of work-life among private bank employees. Pac Bus Rev Int 8(9):1–10
39. Hartini H, Fakhrorazi A, Islam R (2019) The effects of cultural intelligence on task performance and contextual performance: an empirical study on public sector employees in Malaysia. Human Soc Sci Rev 7(1):215–227
40. Harunavamwe M, Nel P, Van Zyl E (2020) The influence of self-leadership strategies, psychological resources, and job embeddedness on work engagement in the banking industry. South Afr J Psychol 50:507–519. https://doi.org/10.1177/0081246320922465
41. Hassan S, Mei TS, Johari HB (2018) Relationship between human capital, it resources and organizational performance: an operational capabilities mediated model. Orient Res J Soc Sci (ORJSS) **3**(1)
42. Hsieh H-H, Wang Y-D (2016) Linking perceived ethical climate to organizational deviance: the cognitive, affective, and attitudinal mechanisms. J Bus Res 69(9):3600–3608

43. Hsu MY, Kernohan G (2006) Dimensions of hospital nurses' quality of working life. J Adv Nurs 54(1):120–131
44. Hyde AM, Gupta B (2018) Factors affecting quality of work life among nationalized bank employees. Int J Manag IT Eng 8(9):285–305
45. Jaramillo F, Mulki JP, Marshall GW (2005) A meta-analysis of the relationship between organizational commitment and salesperson job performance: 25 years of research. J Bus Res 58(6):705–714
46. Kim S, Vandenberghe C (2017) The moderating roles of perceived task interdependence and team size in transformational leadership's relation to team identification: a dimensional analysis. J Bus Psychol 33:1–19
47. Kluckhohn C (2017) Mirror for man: the relation of anthropology to modern life. Routledge, London
48. Krejcie R, Morgan D (1970) Determining sample size for research activities. Educ Psychol Meas 30(3):607–610
49. Kuenzi M, Schminke M (2009) Assembling fragments into a lens: a review, critique, and proposed research agenda for the organizational work climate literature. J Manag 35(3):634–717
50. Kulpa W, Magdon A (2012) Operational risk management in a bank. Intern Audit Risk Manag **28**(4)
51. Lau RS (2000) Quality of work life and performance–an ad hoc investigation of two key elements in the service profit chain model. Int J Serv Ind Manag 11:422–437
52. Leiter MP, Bakker AB (2010) Work engagement: a handbook of essential theory and research. Psychology Press, UK
53. Ling Tan C, Yan Ho M (2015) Organizational creative climate in determining the work innovativeness among the knowledge workers. Glob J Bus Soc Sci Rev 3(2):14–26
54. Macey WH, Schneider B, Barbera KM, Young SA (2011) Employee engagement: tools for analysis, practice, and competitive advantage, vol 31. Wiley
55. Mackay M (2017) Professional development is seen as employment capital. Prof Dev Educ 43(1):140–155
56. Martel J-P, Dupuis G (2006) Quality of work life: theoretical and methodological problems, and presentation of a new model and measuring instrument. Soc Indic Res 77(2):333–368
57. Maslach C, Schaufeli WB, Leiter MP (2001) Job burnout. Annu Rev Psychol 52(1):397–422
58. Masson RC, Royal MA, Agnew TG, Fine S (2008) Leveraging employee engagement: the practical implications. Ind Organ Psychol 1(1):56–59
59. Mellahi K, Collings DG (2010) The barriers to effective global talent management: the example of corporate elites in MNEs. J World Bus 45(2):143–149
60. Mohammed AS, Nassar ME, Ghallab SA, Morsy SM (2013) Nurses managers' decision making styles and it's effect on staff nurses' job performance. J Am Sci 9(12):171–179
61. Mone EM, London M (2018) Employee engagement through effective performance management: a practical guide for managers. Routledge, London
62. Pawirosumarto S, Sarjana PK, Gunawan R (2017) The effect of the work environment, leadership style, and organizational culture on job satisfaction and its implication towards employee performance in Parador Hotels and Resorts, Indonesia. Int J Law Manag
63. Platis C, Reklitis P, Zimeras S (2015) Relation between job satisfaction and job performance in healthcare services. Procedia-Soc Behav Sci 175(1):480–487
64. Qiu H, Haobin Ye B, Hung K, York QY (2015) Exploring antecedents of employee turnover Intention–evidence of China's hotel industry. J China Tour Res 11(1):53–66
65. Rahi S (2017) Research design and methods: a systematic review of research paradigms, sampling issues and instruments development. Int J Econ Manag Sci 6(2):1–5
66. Rahman MA, Qi X, Jinnah MS (2016) Factors affecting the adoption of HRIS by the Bangladeshi banking and financial sector. Cogent Bus Manag 3(1):1262107
67. Reid RD, Sanders NR (2019) Operations management: an integrated approach. Wiley, Hoboken
68. Sari NPR, Bendesa IKG, Antara M (2019) The influence of quality of work life on employees' performance with job satisfaction and work motivation as intervening variables in star-rated hotels in Ubud tourism area of Bali. J Tour Hosp Manag 7(1):74–83

69. Schaufeli WB, Salanova M, González-Romá V, Bakker AB (2002) The measurement of engagement and burnout: a two sample confirmatory factor analytic approach. J Happiness Stud 3(1):71–92
70. Sekaran U, Bougie R (2016) Research methods for business: a skill building approach. Wiley, Hoboken
71. Salanova M, Llorens S, Cifre E, Martínez IM (2012) We need a hero! Toward a validation of the healthy and resilient organization (HERO) model. Group Organ Manag 37(6):785–822
72. Simpson MR (2009) Engagement at work: a review of the literature. Int J Nurs Stud 46(7):1012–1024
73. Singh T, Srivastav SK (2012) QWL and organization efficiency: a proposed framework. J Strateg Hum Resour Manag 1(1):1
74. Stephen A, Dhanapal D (2012) Quality of work life in small scale industrial units: employers and employees perspectives. Eur J Soc Sci. ISSN 1450–2267
75. Supramaniam M, Muniandi T, Ramayah T, Mohamed RKMH, Subramaniam V, Begum M (2020) The relationship between human resource management practices, psychological contract, and innovative behaviours of academicians. Int J Psychosoc Rehabil 24(2):2477–2489
76. Tashakkori A, Creswell JW (2007) Exploring the nature of research questions in mixed methods research. Sage Publications, Los Angeles
77. Wellins RS, Bernthal P, Phelps M (2005) Employee engagement: the key to realizing competitive advantage. Dev Dimens Int 5:1–31
78. Welthagen C, Els C (2012) Depressed, not depressed or unsure: prevalence and the relation to well-being across sectors in South Africa. SA J Ind Psychol 38(1):57–69
79. Wickramaaratchi D, Perera G (2020) The impact of talent management on employee performance: the mediating role of job satisfaction of generation Y management trainees in the selected public banks in Sri Lanka.
80. Xanthopoulou D, Bakker AB, Fischbach A (2013) Work engagement among employees facing emotional demands. J Pers Psychol
81. Zhang L, Zhang X, Xu Y (2017) The sociality of resources: understanding organizational competitive advantage from a social perspective. Asia Pac J Manag 34(3):619–648

The Usage of Social Media for Academic Purposes

Abdulsadek Hassan, Mahmoud Gamal Sayed Abd Elrahman, Enas Mahmoud Hamed Ahmed, Dalia Ibrahim Elmatboly, and Khalid Ibrahem ALhomoud

Abstract The study was undertaken to investigate the effect of WhatsApp on the academic performance of Mass Communication education students. Four research questions were raised to guide the study. The study employed descriptive research design. The population of the study comprised of 275 Mass Communication education students were randomly selected and used as sample for the study. The instrument that was used for the study was a set of structured questionnaires which were validated by the supervisor and one other expert. The results indicated that there is no significant influence of frequency of WhatsApp usage on the academic purposes of this usage, also there is significant influence of the rate of WhatsApp usage on the academic purposes of this usage, and there is significant influence of topics that students communicate through WhatsApp and the academic purposes of this usage and there is no significant influence of patterns of WhatsApp usage on the academic purposes of this usage.

Keyword E- Learning · WhatsApp · Academic performance

A. Hassan (✉)
Ahlia University, Manama, Kingdom of Bahrain

M. G. S. A. Elrahman (✉)
Faculty of Mass Communication, Radio and Television Department, Beni-Suef University, Beni Suef, Egypt

E. M. H. Ahmed (✉)
Faculty of Post Graduate Childhood Studies, Department of Mass Communication and Children Culture, Ain-Shams University, Cairo, Egypt

D. I. Elmatboly (✉)
Mass Communication Faculty of Specific Education, Damietta University, Damietta, Egypt

K. I. ALhomoud (✉)
Faculty of Journalism and Electronic Publishing, Imam Muhammad Bin Saud Islamic University, Riyadh, Saudi Arabia

A. M. A. Musleh Al-Sartawi (ed.), *The Big Data-Driven Digital Economy: Artificial and Computational Intelligence*, Studies in Computational Intelligence 974,
https://doi.org/10.1007/978-3-030-73057-4_28

1 Introduction

WhatsApp is one of the most important social media applications, as it is an alternative messaging application for short text messages, and through it all types of media can be shared, whether image files, audio and video clips, or others, and what most distinguishes the WhatsApp application is its high privacy; As all conversations and files they contain are fully encrypted; No one can see it, and it is worth noting that the WhatsApp application was co-founded by Jan Com and Brian Acton, who were working for Yahoo for a period of about twenty years, and in 2009 they jointly launched the WhatsApp service and provided it to people as a product that can be Using it to communicate with others easily, and despite some obstacles that stumbled upon the application at the beginning of its launch, it is now a place that brings together more than a billion monthly active users who interact with each other, taking advantage of all the features and advantages that the application provides to its users. $ 19 billion in which Facebook seized the WhatsApp application and became its property in 2014 AD, and due to the widespread use of the WhatsApp application among all age groups, some responsible authorities in the educational process decided to use the application to achieve education goals more effectively and successfully. By taking advantage of its popularity and ease of use in the completion of a large number of tasks that improve the performance of students in the classroom.

WhatsApp began in the year 2009, developed by former yahoo managers "Brain Acton and Jan Koum" under the tag line " Simple", this Application facilities communication among individuals for free without any cost and it makes it easier to create groups, send boundless messages, sharing Images, audio messages and video, sharing ideas with other users [1].

WhatsApp has several advantages such as: multimedia, group chat, cross platform engagement (smart phones, tablets), offline messages, no charges involved and pins and uses name, so instructor has also begun to observe the new technologies and explore its impact on student achievement. Consequently, these technologies have a large impact on the academic progress of students [2].

The usage of WhatsApp for academic purposes aims to send instruction to students in the classroom. It provides all the instructions when the instructor and students are separated by distance, time or both. The overwhelming adoption of this app can be linked to the fact that the students have overwhelmingly embraced the use of mobile devices as an integral part of their everyday lives. In recognition of the unprecedented adoption of WhatsApp by many people especially the youths and students, this trend can no longer be neglected, This comes in light of the increased use of the WhatsApp program in the Gulf countries, so the use of WhatsApp in Bahrain reached 581 thousand users out of the total population of one million and 600 users comparing to 320 thousand users of Telegram and 100 thousand users of Messenger [3], and the number of users in Saudi Arabia is 24 million and 700 thousand out of the total population of 34 million and 200 thousand comparing to 12 Million users of Telegram and 7 Million and 650 thousands users to Messenger [4].

There is the need therefore to understand the experience and pattern of use by students in Saudi and Bahraini tertiary institutions and illuminate them on the different effects of this app on their academic performance, regarding that the WhatsApp has become a growing phenomenon in academic use for discussing mutual topic of academic interest.

The aim of this study is to explore the usage of WhatsApp among Mass Communication students for academic purposes in Saudi and Bahraini universities.

2 Literature Review

In qualitative studies, the purpose of literature review is to provide a comprehensive understanding about the research topic, the literature review introduce mainly looks at research related to the usage of WhatsApp among students for academic purposes [5], the researchers display the literature review as follow:

Mohammad Irfan, Sonali Dhimmar [6] found out the influence of WhatsApp on university students, applying to a sample of 105 students, the results showed that WhatsApp is a tool of making communication faster and easier by reinforcing effective flow of information, idea sharing and connecting people easier.

Manpreet Singh Nanda [7] investigated the role of WhatsApp in augmenting learning among the third professional MBBS students, 82 students in total participated in the study, the results showed that WhatsApp can improve learning, especially among slow learners by creating their interest and improving communication among students.

Bilge Çam Aktaşa, Yafes Can [8] conducted a research to find out if using WhatsApp actively in English outside the school has any effect on the students' attitudes, applying to a sample of 20 students. The results proved that the application is effective in the emotions such as happiness, joy, excitement, pride and showed that the students considerably support the use of this application.

Joan Francesc Fondevila-Gascón et al. [9] carried out a study to analyze how the use of instant messaging services impacts the Spanish university context, applying to a sample of 332 students. The results showed that the most of students use instant messaging applications for issues related to university issues -besides their personal lives.

Eucharia Chinwe Igbafe, Chinekpebi Ngozi Anyanwu [10] carried out a study to find out how WhatsApp enhances students' academic performance, applying to a sample of 20 purposively selected students. Findings showed that WhatsApp can cause academic disruption through addiction to non-educational communities, but also, it can enhance academic performance through building and improving students' community of learning.

Levent Cetinkaya [11] found out the effects of WhatsApp use for education, applying to a sample of 30 experimental group students, the results found out that students showed positive attitudes towards WhatsApp usage in their courses and they demanded the same practice in their other courses as well. The results also showed that learning could also happen unconsciously and the messages with images were more efficient for their learning.

Augustine Sandra Eberechukwu, Nwaizugbu Nkeiruka Queendarline [12] found out the effect of WhatsApp when used as a tool to deliver instruction to 400 level trainee teachers who offered computer in education, the results revealed that there was no significant difference existed between the mean values of the two groups at post-test level.

Ruba Fahmi Bataineh et al. [13] conducted a research to examine the potential effect of e-mail- and WhatsApp-based instructional treatment on a sample of students. The findings revealed that students developed positive opinions towards the use of WhatsApp in their courses than e-mailed group.

Hananel Rosenber, Christa S. C. Asterhan [14] this study analyzed the phenomenon of "classroom WhatsApp groups", the methodology combines questionnaires, personal interviews, and focus groups with secondary school students. The findings showed that students are aware of the challenges inherent to the use of WhatsApp for communication with their teachers.

Al-Mothana M. Gasaymeh [15] carried out a study to explore university students' use of WhatsApp and their perceptions regarding its possible integration into their education, applying to a sample of 154 students. the results revealed that the use of WhatsApp for educational purposes was limited. The participants perceived the integration of WhatsApp into their education to be easy, fun and useful.

Alberto Andújar, Maria-Soledad Cruz-Martínez [16] investigated the benefits of WhatsApp to develop oral skills in second-language learners, applying to a sample of 80 Spanish students. The results indicated that WhatsApp offers an environment where learners can ubiquitously negotiate meaning, reflect and evaluate on their own performance through authentic interaction and feedback, constituting a powerful tool for developing second language proficiency.

Zahoor Hussain et al. [17] explained the WhatsApp usage level among the students, their periods choice to use WhatsApp, their times surfing WhatsApp, applying to a sample of 280 students. The results showed that huge segment of the students (125) use WhatsApp in their hostels in terms of placement preferences.

Dhiraj Kumar Malhotra and Sonia Bansal [18] this study aimed to analyze the use of WhatsApp for academic proposes with a special emphasis on its impact on studies, applying to a random sample of 100 postgraduate. The results indicated that More than 90% of the students were using WhatsApp for academic purposes. More UG students were using WhatsApp to chat, share images/videos and so forth for academic purposes as compared to PG students.

Akpan, Kufre P, Ezinne [19] found out the effectiveness of WhatsApp as a collaborative tool for learning among undergraduate students, applying to a sample of 60 students. The findings revealed a significant progress in the level of students taught with WhatsApp application and those taught using the traditional method of teaching.

Sonia Gon, Alka Raweka [20] found out the affectivity of social media like WhatsApp in delivering knowledge to 4th semester MBBS students and to compare the improvement of knowledge gain through e-learning and didactic lecture, applying to a sample of 40 students. The finding revealed that High infiltration of Smartphones has initiated growing use of WhatsApp for groups of teachers and their students to support the learning process by allowing direct access to lots of online resources.

Anshu Bhatt and M. Arshad [21] found out the impact of WhatsApp on youth, applying to a sample of 100 students. Findings show that students are spending more time on this application rather than spending quality time with their family members, many youths are addicted to it and cannot abstain themselves from constantly chatting, replying and sharing of ideas or information.

Md. Golam Rabbani Sarker [22] found out the Impact of WhatsApp on the university level students and the data was collected from 200 students. The results indicated that WhatsApp has a profound negative impact on students and adversely affects their education, behavior and routine lives. It affects their education, behavior and routine lives. It is highly addictive in nature.

It is clear that there are many studies conducted on the usage of WhatsApp for academic purposes, this naturally raise an issue, while the results of some studies agree with one other, others simply disagree with each other on the usage of WhatsApp for academic in purposes in general and teaching and learning particular.

Analyzing literature review regarding the usage of WhatsApp in academic purposes indicates that many characteristics enhancing the learning process such as: encouraging cooperative learning between lecturers and students and they become involved in an activity in lectures and learning any time any place.

From the literature review, there is evidence that WhatsApp performs many tasks through various services it offers to its users such as: voice call, instant message and video uploads.

The previous literature review indicated that WhatsApp seemed to be an effective aid that instructors might use to enhance students' learning. On the other hand, other researchers expressed concerns regarding the impact of WhatsApp on students' flow of information.

3 The Problem Statement

The widespread use of Internet technologies has contributed to changing the way individuals communicate with each other, and the great popularity of social networks has had a great role in meeting various needs, and has become an essential pillar in daily life that cannot be dispensed with, and in light of this development The tremendous technological development was obligatory for the competent authorities in all fields to take this shift as a means through which positive results can be obtained in the education sector in particular, as these developments have become usable and their effects are clear and evident due to the progress they have achieved in terms of communication, motivation, social interaction, educational attainment and others, It provided many possibilities that the educational system was unable to provide. Such as: creating cooperation and interaction between students and teachers, supporting distance learning, exchanging information, and many other capabilities that can be described as the spirit of the educational process [23, 24].

In recent years, technology has become a part of human life, we cannot separate by the technology because all aspects of life connected by technology advances. It is easily to communicate with others by smart phone, besides the ability of getting all information we want [25]. Students have the same option as others. they make their smart phone as inseparable part of themselves. They can use it for many things, play online games, access social networks media, upload and download videos and photos, files, conducting real time conversation or follow the news about events around the world [26].

Consequently, this study is to explore the usage of WhatsApp among Mass Communication students for academic purposes in Gulf tertiary Institutions in Saudi and Bahraini universities and determine the preferences of the students toward the process.

4 Research Hypothesis

- There are significant differences in frequency of WhatsApp usage and the reasons of this usage among Mass Communication students in Saudi and Bahraini universities.
- There are significant differences in the rate of WhatsApp usage and the reasons of this usage among Mass Communication students in Saudi and Bahraini universities.
- There are significant differences in the topics that Mass Communication students communicate through WhatsApp for academic purposes and the reasons of this usage in Saudi and Bahraini universities.
- There are significant differences between the patterns of WhatsApp usage among Mass Communication students and the reasons of this usage in Saudi and Bahraini universities.

5 Population

Students registered for Mass Communication bachelor's degree in Saudi and Bahraini universities were considered as target population of the study during the first semester of the academic year 2018–2019 which started at the beginning of September and finished in mid- January 2019 from 1st year to from 1st year to 4th year.

6 Sample

This study used purposive sampling to select the response of the students. purposive sampling is not only used to select participants, but to select events, incidents, settings and activities to be included for data collection. the purpose was to get views from different sources to avoid biased data. The purposive sampling means that participants are selected because of defining characteristics that makes them the holders of the data in the study [27].

This study is a descriptive survey research that used to explore and describe data collected from purposive sample of 275 Mass Communication students (144 students from Saudi universities and 131 students from Bahraini universities (were purposively surveyed on their usage of WhatsApp services and academic performance. All students were members of at least one joint instructor- student WhatsApp class group.

The researchers distributed a questionnaire to 275 Mass Communication students in Saudi and Bahraini universities in Arabic language Since the native language of the participants is Arabic, the questionnaire was translated into Arabic by the researcher. The following table shows the characteristics of the sample:

7 Testing Hypotheses

Hypothesis 1:
There are significant differences in frequency of WhatsApp usage and the academic purposes of this usage among Mass Communication students in Saudi and Bahraini universities (Table 1)**.**

The study used a Pearson correlation coefficient to test the relationship between frequency of WhatsApp usage and the academic purposes of this usage among Mass Communication students in Saudi universities. The results showed that there is a positive Correlation between the two variables, $R = 0.643$, $N = 88$ and $P = 0.000$. So, we can accept the hypothesis and conclude that there is a significant influence of frequency of WhatsApp usage on the academic purposes of this usage.

Table 1 Correlation between frequency of WhatsApp usage and the academic purposes of this usage among Mass Communication students in Saudi and Bahraini universities

Frequency of WhatsApp usage	The academic Purposes of WhatsApp usage in Saudi universities		The academic Purposes of WhatsApp usage in Bahraini universities	
	Pearson (R)	P Value	Pearson (R)	P Value
	.643	.000	.103	.75
N	88		87	

Correlation is significant at the 0.01 level (2-tailed)

Table 2 Correlation between the rate of WhatsApp usage and the academic purposes of this usage among students in Saudi and Bahraini universities

The rate of WhatsApp usage	The academic purposes of WhatsApp in Saudi universities		The academic purposes of WhatsApp in Bahraini universities	
	Pearson (R)	P Value	Pearson (R)	P Value
	.382**	.014	.274**	.027
N	88		87	

Correlation is significant at the 0.01 level (2-tailed)

The study also used a Pearson correlation coefficient to test the relationship between frequency of WhatsApp usage and the academic purposes of this usage among Mass Communication students in Bahraini universities. The results showed that there is no correlation between the two variables, R = 0.103, N = 87 and P = 0.75. So, we can reject the hypothesis and conclude that there is no significant influence of frequency of WhatsApp usage on the academic purposes of this usage.

Hypothesis 2:
There are significant differences in the rate of WhatsApp usage and the academic purposes **of this usage among Mass Communication students in Saudi and Bahraini universities** (Table 2)**.**

The study used a Pearson correlation coefficient to test the relationship between the rate of WhatsApp usage and the academic purposes of this usage among Mass Communication students in Saudi universities. The results showed that there is a positive Correlation between the two variables, R = 0.382, N = 88 and P = 0.014. So, we can accept the hypothesis and conclude that there is a significant influence of the rate of WhatsApp usage on the academic purposes of this usage.

The study also used a Pearson correlation coefficient to test the relationship between the rate of WhatsApp usage and the academic purposes of this usage among Mass Communication students in Bahraini universities. The results showed that there is a positive Correlation between the two variables, R = 0.274, N = 87 and P = 0.027. So, we can accept the hypothesis and conclude that there is significant influence of the rate of WhatsApp usage on the academic purposes of this usage.

Hypothesis 3
There are significant differences in the topics that Mass Communication students communicate through WhatsApp and the academic purposes of this usage in Saudi and Bahraini universities (Table 3).

The study used a Pearson correlation coefficient to test the relationship between topics that Mass Communication students communicate through WhatsApp and the academic purposes of this usage among students in Saudi universities. The results showed that there is a positive Correlation between the two variables, R = 0.651, N = 88 and P = 0.000. So, we can accept the hypothesis and conclude that there is a

Table 3 Correlation between topics that Mass Communication students communicate through WhatsApp and the academic purposes of this usage in Saudi and Bahraini universities

Topics	The academic purposes of WhatsApp in Saudi universities		The academic purposes of WhatsApp in Bahraini universities	
	Pearson (R)	P Value	Pearson (R)	P Value
	.651**	.000	.365**	.021
N	88		87	

Correlation is significant at the 0.01 level (2 -tailed)

Table 4 Correlation between the patterns of WhatsApp usage among Mass Communication students and the academic purposes of this usage in Saudi and Bahraini universities

The patterns of What's App usage	The academic purposes of WhatsApp in Saudi universities		The academic purposes of WhatsApp in Bahraini universities	
	Pearson (R)	P Value	Pearson (R)	P Value
	.091	.098	.178**	.043
N	88		87	

Correlation is significant at the 0.01 level (2 -tailed)

significant influence of topics that students communicate through WhatsApp and the academic purposes of this usage.

The study also used a Pearson correlation coefficient to test the relationship between the rate of WhatsApp usage and the academic purposes of this usage among Mass Communication students in Bahraini universities. The results showed that there is a positive Correlation between the two variables, $R = 0.365$, $N = 87$ and $P = 0.021$. So, we can accept the hypothesis and conclude that there is significant influence of topics that students communicate through WhatsApp and the academic purposes of this usage.

Hypothesis 4
There are significant differences between the patterns of WhatsApp usage among Mass Communication students and the academic purposes of this usage in Saudi and Bahraini universities (Table 4).

The study used a Pearson correlation coefficient to test the relationship between patterns of WhatsApp usage and the academic purposes of this usage among Mass Communication students in Saudi universities. The results showed that there is no Correlation between the two variables, $R = 0.091$, $N = 88$ and $P = 0.098$. So, we can reject the hypothesis and conclude that there is no significant influence of patterns of WhatsApp usage on the academic purposes of this usage.

The study also used a Pearson correlation coefficient to test the relationship between the patterns of WhatsApp usage and the academic purposes of this usage among Mass Communication students in Bahraini universities. The results showed that there is a positive Correlation between the two variables, $R = 0.178$, $N = 87$ and

P = 0.043. So, we can accept the hypothesis and conclude that there is significant influence of patterns of WhatsApp usage on the academic purposes of this usage.

8 Discussion and Conclusion

Conclusion

The results revealed that the use of WhatsApp has become an inevitable necessity for millions of people around the world. Because it provides entertainment and educational services, and education through its pages and groups has become a common thing among its users, including male and female students, at various educational levels.

The results also revealed that WhatsApp provides many educational services through its various platforms; What contributes to the effective transfer and dissemination of knowledge through its various collections and pages; It allows students to communicate directly and permanently with each other to exchange and transfer ideas and information, and friendly communication between them.

The results also showed the importance of using closed groups on WhatsApp, as a basic tool to enhance the educational process. Where the teacher creates a closed group on "Facebook" that includes all students in the classroom; To publish the content of the scientific material taught to students, to publish discussion topics and to conduct educational dialogues; This helps them to interact and be interested in the content of the scientific material and to help thire countries in developing its financial and nonfinancial systems [28].

The analysis of the data in this study has led to certain findings from which the following useful conclusions were drawn:

The study has revealed that The results indicated that there is no significant influence of frequency of WhatsApp usage on the academic purposes of this usage, also there is significant influence of the rate of WhatsApp usage on the academic purposes of this usage, and there is significant influence of topics that students communicate through WhatsApp and the academic purposes of this usage and there is no significant influence of patterns of WhatsApp usage on the academic purposes of this usage.

Recommendation

Based on the findings and conclusions of this study, the following recommendations are made:

School administrators should advised students during orientation of the dangers of addiction to WhatsApp sites. They should be introduced to sites that can add values to their academic work and research.

Students, especially those willing to record huge academic success should not rely only on class note but guide themselves by using online academic materials relevant to their course together with their class note to read as this will improve learning.

Both governments, communities, school administrators should ensure to help reduce the challenges faced by the students in the implementation of WhatsApp by funding of the facilities necessary in the use of WhatsApp in the university's environment.

Students with phones having internet facility should be encouraged to either use it to supplement their research in the library rather than the usual chatting with friends all the time.

Suggestion for Further Study

Studies could be conducted on the following issues:

Influence of WhatsApp participation on student academic performance across gender lines.

The importance of WhatsApp on the student life in general.

Use of social media and its impact on academic performance of tertiary institution students in Gulf countries.

The use of WhatsApp in education, challenges and opportunities.

The role of WhatsApp on students learning experiences.

References

1. Veena G, Lokesha M (2016) The effect of whatsapp messenger usage among students in Mangalore university: a case study. Int J Library Inf Stud 6(2):122
2. Gon S, Rawekar A (2017) Effectivity of E-learning through WhatsApp as a teaching learning tool. MVP J Med Sci 4(1):19–20
3. 54 million internet users in Bahrain. https://alwatannews.net/article/815542/Bahrain/154-%D9%85%D9%84%D9%8A%D9%88%D9%86-%D9%85%D8%B3%D8%AA%D8%AE%D8%AF%D9%85-%D9%84%D9%84%D8%A5%D9%86%D8%AA%D8%B1%D9%86%D8%AA-%D9%81%D9%8A-%D8%A7%D9%84%D8%A8%D8%AD%D8%B1%D9%8A%D9%86. Accessed 12 Dec 2019
4. Statistics of social media in Saudi Arabia 2018. https://emarketing.sa/%D8%A7%D8%AD%D8%B5%D8%A7%D8%A6%D9%8A%D8%A7%D8%AA-%D9%88%D8%B3%D8%A7%D8%A6%D9%84-%D8%A7%D9%84%D8%AA%D9%88%D8%A7%D8%B5%D9%84-%D8%A7%D9%84%D8%A7%D8%AC%D8%AA%D9%85%D8%A7%D8%B9%D9%8A-%D9%81%D9%8A-%D8%A7/. Accessed 12 Dec 2019
5. Gallardo Echenique EE (2014) An investigation of the social and academic uses of digital technology by university students, Un-Published Ph.D dissertation, Universitat Rovira i Virgili, Department of Pedagogy, p 33
6. Mohammad I, Sonali D (2019) Impact of WhatsApp messenger on the university level students: a psychological study. Int J Res Analyt Rev (IJRAR) 6(1):572–586
7. Singh NM (2019) Role of WhatsApp in improving learning among medical students. Int J Med Sci Public Health 8(2):165–168
8. Çam AB, Yafes C (2019) The effect of "Whatsapp" usage on the attitudes of students toward english self-efficacy and English courses in foreign language education outside the school. Int Electron J 11(3):247–256
9. Francesc F-GJ, Joaquín M-P, Pedro M-B, Marc P-L (2019) Uses of WhatsApp in the Spanish university student. Pros Cons Revista Latina de Comunicación Social 74:308–324
10. Chinwe IE, Ngozi AC (2018) WhatsApp at tertiary education institutions in Nigeria: the dichotomy of academic disruption or academic performance enhancer? J Pan Afr Stud 12(2):179–205

11. Levent C (2018) The impact of Whatsapp use on success in education process. Int Rev Res Open Distrib Learn 18(7):59–74
12. Sandra EA, Nkeiruka QN (2018) WhatsApp utilization and academic performance of computerin education trainee teachers in University of Port-Harcourt. Int J Educ Learn Develop 6(5):15–25
13. Bataineh RF, Baniabdelrahman AA, Khalaf KMB (2018) The effect of e-mail and WhatsApp on Jordanian EFL learners' paraphrasing and summarizing skills. Int J Educ Develop Inf Commun Technol (IJEDICT) 14(3):131–148
14. Hananel R, Asterhan SC (2018) "Whatsapp, Teacher?" - student perspectives on teacher-student WhatsApp interactions in secondary schools. J Inf Technol Educ Res 17:205–226
15. Gasaymeh A-MM (2017) University students' use of WhatsApp and their perceptions regarding its possible integration into their education.Global J Comput Sci Technol G Interdisc 17(1):1–11. Version 1.0
16. Andújar A, Cruz-Martínez M-S (2017) Mobile instant messaging: WhatsApp and its potential to develop oral skills. Media Educ Res J 25(50):43–52
17. Zahoor H, Rameez M, RazaHussain S, Ahmed MN (2017) WhatsApp usage frequency by university students: a case study of Sindh University. Eng Sci Technol Int Res J 1(4):15–20. DEC
18. Kumar MD, Sonia B (2017) Magnetism of WhatsApp among veterinary students. Electron Library 35(6):1259–1267
19. Kufre Akpan P, Ezinne A (2017) Effectiveness of WhatsApp as a collaborative tool for learning among undergraduate students in university of Uyo, Akwa Ibom state. Int J Adv Educ Res 2(5):43–46
20. Sonia G, Alka R (2017) Effectivity of E-Learning through WhatsApp as a teaching learning tool. MVP J Med Sci 4(1):19–25
21. Bhatt A, Arshad M (2016) Impact of WhatsApp on youth: a sociological study. IRA Int J Manage Soc Sci 4(2):376–386
22. Sarker Md, Rabbani G (2015) Impact of WhatsApp messenger on the university level students: a sociological study. Int J Natural Soc Sci 2(4):118–125
23. Al-Sartawi A (2020) Social media disclosure of intellectual capital and firm value. Int J Learn Intell Capital 17(4):312–323
24. Musleh Al-Sartawi AMA (2020) E-learning improves accounting education: case of the higher education sector of Bahrain. In: Themistocleous M, Papadaki M, Kamal MM (eds.) Information systems. EMCIS 2020. Lecture Notes in Business Information Processing, vol 402. Springer, Cham
25. Solomon Tawiah Y, Emmanuel AA, Nondzor H (2014) Usage of WhatsApp and voice calls (Phone call): preference of polytechnic students in Ghana. Sci J Bus Manage 2(4):103–108
26. Hussain Z, Mahesar R, Shah RH, Memon NA (2017) WhatsApp usage frequency by university students: a case study of Sindh University. Op. Cit 16
27. Chinwe IE, Ngozi AC (2018) WhatsApp at tertiary education institutions in Nigeria: the dichotomy of academic disruption or academic performance enhancer, Op. Cit. 188
28. Memdani L (2020) Demonetisation: a move towards cashless economy in India. Int J Electron Bank 2(3):205–211

The Relationship Between Human Resource Practices and Organizational Performance and Their Operation in Light of the Development of Using Big Data Technology

Jasem Taleb Al-Tarawneh, Mohd Saiful Izwaan Saadon, and Aya Naser Maqableh

Abstract This study aims to analyses how Jordanian public universities take advantage of the relationship between HRM practices and organizational performance with their combined dimensions and their operation in light of the development of using big data technology. This study is so important for different sectors to choose appropriate HRM practices which help to organizational performance in the organization. As well as, the organizational performance in the Jordanian universities is low comparing with expected results. However, HRM practices integration formulate strong and influential instruments to put a figure on ways Organizational performance.

The literature review outlined a wide range of relations between HRM practices and it is a positive relationship. The results display gaps in the plane and the implementation of researches. The researchers observed the dominance of positive results, in the studies that seek to evaluate the impact of HRM on performance.

Keywords Organizational performance · HRM practices · Recruitment and selection · Training and development · Rewards systems · Performance evaluation · Big data

1 Introduction

Nowadays institution that follows a successful line in work depends heavily on human capital. In addition to this, human capital is considered a valuable element that may grant benefits to an organization, although human capital alone will not add value to institution t without taking the decisions correctly.

J. T. Al-Tarawneh (✉) · A. N. Maqableh
Faculty of Business, Universiti Malaysia Terengganu, Kuala Nerus 21030, Terengganu, Malaysia

M. S. I. Saadon
Faculty of Business, Economics and Social Development, Universiti Malaysia Terengganu, Kuala Nerus 21030, Terengganu, Malaysia

A. M. A. Musleh Al-Sartawi (ed.), *The Big Data-Driven Digital Economy: Artificial and Computational Intelligence*, Studies in Computational Intelligence 974,
https://doi.org/10.1007/978-3-030-73057-4_29

As well as, all organizations of all levels of business are required to maintain accurate data. Big data is one of the agents of change that challenges organizational leaders. Used effectively, they provide accurate business models and forecasts to better support decision-making in all aspects of the organization.

Moreover, The Human Resources (HR) of an institution offers the possibility of synergy for sustained competitive advantage, particularly, when it is properly deployed and applied. The traditional HR strategy, as well as the formal system for managing people in the institution, is essentially concerned with transactional and administrative support services. To perform successfully, the roles of business partners and change agents under the HR strategy must be highly knowledgeable, multi-skilled, and be able to acquire core competencies like business knowledge, vision strategy, global operating skills, reliability and integrity, internal consulting skills, and many others [3]. Therefore, the human resources management practices affect organizational performance. Whereas, Organizational performance concerning the ideas of effectiveness and efficiency. Because any institution requires to produce the right things by adopting the fewest possible inputs, then just like institution guarantee powerful and strong organizational performance [3].

Commonly, the institution seeks to implement well in different areas: First, they attempt to perform well financially, which seek to accomplish the best return on their investment, further, they must add as much rate as possible in their production operation. Second, they seek to perform well in concepts of the market.in other words, the institution needs to accomplish as much market share as they can and they should be manufactured a product that is in demand and at a price that enables them to compete strongly on the market [32].

Add to that, companies can manage human resources and foster organizational performance, by having imagined of HR practices working alongside other dimensions. According to [27], these are:

- Career planning and HR planning, which includes creating venture teams with a balanced skill-mix, recruiting the right people, and voluntary team assignment.
- Performance Evaluation, which includes the encouragement of risk-taking, demanding innovation, generating or adopting new tasks, peer evaluation, frequent evaluations, and auditing innovation processes.
- Reward systems, which includes freedom to do research, freedom to fail, freedom to form teams, freedom to run businesses, balancing pay and pride, noticeable pay raises dual career tracks, promoting from within, recognition and rewards, and balancing team and individual rewards.
- Training, which includes empowering people and continued education.
- Selection, which includes determining the type of employees to be selected, the skills and motivation of these employees, and the opportunities and incentives that these employees have to design new and better ways of doing their jobs.

On other hand, higher education plays a significant role in various bodies and aspects in the country, such as the national economy and the industrial sector. As well as, the Hashemite Kingdom of Jordan possesses efficient and competent human resource systems that can provide the community with lifelong learning experiences closely related to its current and future needs in response to sustainable economic development and stimulated by preparing educated individuals and skilled workforces. Jordan is witnessing, in general, a qualitative boom in the field of higher education since the early nineties, and the capital Amman in particular, in addition to the major Jordanian cities such as Irbid and Zarqa, contains most of the most important scientific edifices in the Kingdom in most specialties. The private university education sector witnessed a remarkable boom, as Amman was the first in the Middle East.

The students who hold a high school diploma are accepted in public or private universities or colleges. Most universities in Jordan implement the American university model based on the credit hours' system, which gives students the flexibility to choose the number of hours and morning or evening hours. There are ten public universities, most of which are affiliated with universities in the United States and the United Kingdom. Jordanian universities attract a large number of Arab foreign students and non-Arabs every year. Thus, this study aimed to investigate the Relationship between Human Resource Practices and Organizational Performance and their operation in light of the development of using big data technology.

2 Problem Statement

An institution's success and competitiveness, in the long term, count on the existence of qualified managers inside it and how the organization using big data technology in their work. Many benefits are gotten when institutions strongly measure their treatment of Organizational Performance problems, develop and apply Human Resource Practices plans the problem of the study lies in the different role of Human Resource Practices and Organizational Performance and their operation in light of the development of using big data technology that followed HRM department, thus reflecting the level of efficiency and effectiveness of these resources and their contribution to the competitive advantage in different [30]. Hence, it is the main concern of the HRM department to explore all possible ways of creating and sustaining organizational performance and how to use big data technology in human resource management [1, 33].

Add to that effective HRM practices emphasize the organizational performance level of the organization and ward off from being sluggish. They also highlight that an organization's efficiency is proven by a robust package of actions and practices pointed at taking advantage of its strengths. In this regard, HRM provides a better trace of support to the higher management before the time of selecting and hiring suitable personnel, managing, and training them to meet up with the requirements needed to get better of organizational performances and how to use big data technology in human resource management. Moreover, Actual HRM practices are measured as the assurance of unceasing organizational performance. The varying nature of humans makes HRM practices a difficult task. Consequently, it is highly significant for organizations to emphasize on strengthen their HRM practices. Thus, attaining better organizational performance needs successful, effective, and efficient exploitation of organizational competencies and resources to create a competitive advantage. HRM practices so on must be expressed and applied by HRM practices specialists with the help of managers to achieve an efficiency of organizational performance. Organizational performance is an important outcome that the management will use to assess the business procedures and the organization's actions. It is significant because a perfect performance would give them confidence in management and thus secure a sustained and money-making organization future [3].

This study is so important for different sectors to choose appropriate HRM practices that help to organizational performance in the organization and their operation in light of the development of using big data technology. As well as, the organizational performance in the Jordanian universities is low comparing with expected results. However, there is a need to increase research opportunities that can enrich the understanding of human resources practices and their relationship to organizational performance in Jordanian universities through the development of using big data technology [4].

As well, the education sector as such is the single most important constituent of the service sector which is very crucial for the economic development of any country. Moreover, the Jordanian education sector is passing through a very critical phase according to Covid-19 because of its need for the unprecedented pace of automation, communication revolution. All these challenges in the economics, social-cultural, technological, psychological environment have important implications on the HRM practices being followed in the education sector as well as on the performance of the Jordanian universities. All the HRM practices like recruitment, training, and development, compensation and rewards, performance appraisal, promotion and empowerment, sharing of decisions, job security benefits, welfare measures, and separations in the form of lay-offs, retrenchment have come under tremendous pressure. As a result, the performance of the education sector has been affected very seriously.

In universities, HRM practices integration formulate strong and influential instruments to put a figure on ways Organizational performance and their operation in light of the development of using big data technology. Thus, the problem of this research relates to the gap between the expected organizational performance and the actual organizational performance and their operation in light of the development of using big data technology in the administrative departments of the Jordanian government universities. Thus, this study comes to expand the effort in this area and achieve the most distinguished aim in this study, is there a link occurred in the relationship between Human Resource Practices and Organizational Performance and their operation in light of the development of using big data technology at Jordanian public universities were selected as part of the higher education sector in Jordan.

3 Research Questions

Since the success of educational institutions and increase their competitiveness depends on the existence of management excellence, successful and qualified within them, in addition to the ability to provide high quality and efficiency of educational services. Many benefits are gained when educational institutions measure vigorously to address organizational performance problems and apply human resource practice plans by using big data technology. The current research problem lies in the different role of human resources management practices of the Human Resources Management Section, reflecting the level of efficiency and effectiveness of these resources and their contribution to the advantage Competitiveness in the higher education sector. Thus, the main concern of HRM is to maintain organizational performance by using big data technology within educational institutions.

Therefore, this research aims to expand the effort in this field and achieve the most distinctive objective in this research to find out if there is a link between the role in the relationship between Human Resource Practices and Organizational Performance and their operation in light of the development of using big data technology in Jordanian public universities. From this the following main study questions emerge:

The first main question: What is the relationship between Human Resource Practices and Organizational Performance and their operation in light of the development of using big data technology?

4 Research Objectives

The objectives of this study impact to investigate relationship between Human Resource Practices and Organizational Performance and their operation in light of the development of using big data technology at Jordanian public universities.

This study aimed to achieve the following objectives
The main objectives: To investigate the relationship between Human Resource Practices and Organizational Performance and their operation in light of the development of using big data technology at Jordanian public universities.

5 Significance of the Research

The importance of this study comes from the fact that it is based on extrapolation "relationship between Human Resource Practices and Organizational Performance and their operation in light of the development of using big data technology". Therefore, the importance of the study divided into the following:

5.1 *Theoretical Significance*

- The importance of this research considered the importance of the relationship between Human Resource Practices and Organizational Performance and their operation in light of the development of using big data technology. Besides, there are few studies.
- This research will affect the range of studies in this field that can be conducted, which lead to some limitations and complications to the contents. However, this will be considered as a challenge from one side.
- This research contributes definitely by establishing an actual value in the academic research in Human Resource Practices and Organizational Performance and their operation in light of the development of using big data technology field, and encourage other researchers to build up and improve studies, which support this field to achieve the desired goals in the future.

5.2 *Practical Significance*

- The result of the current research expects to provide convincing evidence about the direct relationship between Human Resource Practices and Organizational Performance and their operation in light of the development of using big data technology.
- The results of the study will provide important decision-makers who are interested in informational and analytical studies with important information on the relationship between Human Resource Practices and Organizational Performance and their operation in light of the development of using big data technology and will provide assistance to officials and decision-makers in Jordanian universities to identify the relationship between Human Resource Practices and Organizational Performance and their operation in light of the development of using big data technology, and thus help them in taking appropriate measures to implement these concepts.
- The results of this study will be used in some subsequent studies and research that can address the same topic in different dimensions, and that the results of this study contribute to coming up with recommendations showing the relationship between Human Resource Practices and Organizational Performance and their operation in light of the development of using big data technology in Jordanian universities.

6 Literature Review and Previous Studies

Nowadays, Organizations work in an environment that is characterized by uncertainty, unpredictability, and change, which creates several challenges [10]. The modern working environment contains numerous factors, such as increased globalization, speedy technological change, and the increasing need for qualified employees and improved performance. This forces organizations to try to realize the best use of the resources at their disposal to achieve a competitive advantage. Human resources are considered critical factors that contribute to an organization's success [41].

In addition, the appeared need for organizations to change their processes and innovate, as the market requires, the effective response is necessary in order to meet up with the highly competitive market characterized with continuous changing. As a result, to this, managers have realized that the success of any business enterprise in a complex and changing environment would depend on the effective management of its Human Resources and other departments, and through make valued coordination between them.

It is worth to mention, that recently the role of the human resources department has been transformed over the decades from existence as an administrator to a critical component in the competitive success of the business. The Human Resources role, which was primarily represented favourably in administrative worker, personnel data management and legal/legislation requirements, now includes focusing on tangible properties and financial resources. Strategically using human resources is required to overcome different business challenges. Therefore, effectively managing human resources is of importance to all organizations. Managing the human resources of an organization requires the use of different components that play an important role in helping organizations create and maintain the performance they want, as this will affect the attitudes and behaviours of employees [20].

Additionally, some human resource management components can affect employee commitment and motivation. Such as staffing and selection, training and development, performance evaluation, collaboration, and payment and reward [13].

The most important task for many organizations today is to survive the present competitive market through the impact of privatization, liberalization, globalization, competition.

Thus the organization need to achieve Organizational performance concerning the ideas of effectiveness and efficiency to achieve competitive advantage. Commonly, the institution seeks to implement well in different areas: First, they attempt to perform well financially, which seek to accomplish the best return on their investment, further, they must add as much rate as possible in their production operation. Second, they seek to perform well in concepts of the market.in other words, the institution needs to accomplish as much market share as they can and they should be manufactured a product that is in demand and at a price that enables them to compete strongly on the market [18].

The historical directions of the human resource (HR) job explain the rising value of employees, from being just one of the means of production in the 20th-century industrial economy to existence a key source of sustainable competitive advantage in the 21st century. There has been much heed concentrate on HRM in a different type of organizations. Good HRM can be applied as a magic weapon for organizations and it is tricky for other rivals to imitate. Consequently, it will share in making a competitive advantage for the organization in regards to human capital [19]. Therefore, this part will shed the light on human resource management, strategy and practice.

HRM begin in England in half of the 1800s in the apprenticeship period and turn into more developed on the industrial revolt at the end of 1800s. In the 19th century, Taylor explained that scientific management and the industrial psychology of employees must be noted. In this case, it was planned that employees should achieve the happiness of the employees with the job and its efficiencies. They could also feel the same by drastic changes in technology, the development of organizations, the increase of associations, and government involvement result by the development of personnel departments in the 1920s. In this stage, personnel administrators were called 'welfare secretaries' [19].

Additionally, investigators felt that the concept HRM was not appreciably several from "personnel management", as they are both attentive with the function of earning, planning, and motivating human resources major to organizations. The investigators are realizing the concept HRM and personal management in other several ways [38].

Due to the development of and changes in the world of management, the term 'management' to HRM was changed. Thus, a new term would seem appropriate to take new thought, concepts of HR, human resource management (HRM) outlined as an asset of united strategies with a political and philosophical underpinning. Accordingly, there are four characteristics of HRM: a specific set of faiths and anticipations, a strategic thrust informing decisions about individual management, the central involvement of line managers, and reliance upon a group of levers to characterize the employee relations [7].

Moreover, HRM can be defined as "the strategic and rational approach of an organization's most valued assets, which are the people working there, who individually and cooperatively contribute to the accomplishment of its objectives" [25].

Besides, HRM is generally actuated out of HR systems that bring jointly HR attitudes in an intelligible way. These positions help describe the major values and lead principles that one should take on board when managing the individual.

HRM practices include the familiar methods used when managing individuals. HR agendas give HR strategies, policies, and practices to be performed according to plan. Furthermore, the main aim of HRM is to certify that the organization is arable of realizing success by individuals. Moreover, HRM systems can be the substance of organizational abilities that led organizations to pick up and achievement new opportunities [43].

As well as, HRM depends on 12 agreement goals as a following: managing individuals, supporting HRM policies, encouraging team-working, empowering employees. Also, developing planned prize strategies, reading a strong customer-first philosophy, improving employee involvement, building greater employee commitment, raising line management accountability, improving the extended role of managers as enablers, and supporting the emergence of a closer group of HR policies, step and systems [7].

Employees as human resources are considered one of the most important resources within an organisation and are key to accomplishing a competitive advantage. Also, it has struggled that managing HR is more problematic than managing capital [39].

Organisations adopt HRM practice to amend the actions, attitudes, and awareness of employees in a way that developed its performance and anticipated outcomes. In addition, what should be considered is that the HRM strategy is not fixed; they change from one company to another. Many scholars defined different practice that is associated with HRM. These include recruitment, selection, training and development, motivation, and Performance Evaluation [20].

6.1 Recruitment and Selection

Any organisations to be effective its need to catch the best staff by the important and major activity of recruitment. Therefore, recruitment and selection forms as a main part of the main activities depend on HRM: namely, the acquisition, improvement, and reward of staff. It repeatedly shapes a substantial part of the action of HR managers. However, non-specialists, just like line managers, often (for good reason) take recruitment and selection resolution [44].

Recruitment outlined as the operation of searching for and finding candidates for jobs. The resulting group is then interviewed so that the true people can be chosen. According to theory, the recruitment operation is finished off with the reception of applications; in practice, the outcoaching activity boils down to the examination of applications to remove those who are not capable for the work [11].

Likewise, recruitment goals to acquire the correct number of accomplished workers to run into the human resource requirements of organisations. Thus, numerous staffing methods are used to realize this purpose [13].

Recruitment methods can be either internal or external but most companies tend to emphasise three external methods: advertising, online recruitment, and the use of employment agencies. Each of these has been created to increase the possibility of recruiting gifted employees [7].

The importance of this operation can be seen from the two-point view as a following: (1) the organisation, which depends on the recruitment and selection operation to develop and improve individual performance. (2) The individual, who depends on a good recruitment and selection process. The procedure at large affects safety and trust for individuals within the organisation.

As well as, Selection is a technical procedure that includes a variety of actions to ensure that the applicant fits in the right place. It can be outlined as the procedure of selecting the right individual from a list of job searchers. The chosen candidate must suit and have the capabilities and requirements according to the specific standards needed for the job. Moreover, the importance of the selection procedure is that it is one of the most significant processes in employ policy, and HRM must be organised to dissatisfy applicants looking for the job [26].

The most important object when choosing the individual for the job is the effectiveness. The interviewer must share the individual specification, and then through the interview and testing one can regulate the efficiency of the person therefore determine whether they meet the requirements of the job [7].

6.2 Training and Development

Due to the many challenges workers face in the work environment as underappreciated assets, organisations requirement to participate in the training and development of their human capital to get better their abilities and capacities.

Further, the motivation behind having training and improvement is to rise and inform the abilities, knowledge, and experiences of an organisation's workers [13].

Training and Development: Training and development is an effort to develop the current and future performance of workers by developing their basic skills. Training is the attempt create by the firms to facilitate the learning operation for its worker. The knowledge that is on the show is work-related, especially focused on the skills of workers. Thus, it performs an education action of a particular sort, to develop the performance of the individual in the role they occupy. This is a means of worker improvement for the firms [12].

Moreover, the training and development consist of five steps, which are as follows [12]:

1. Needs Analysis: Classify the skills needed to develop performance and productivity. Then worker can have training in correspondence with their education plane, experiment, skills, and personality, and improve specific objectives for the training operations.
2. Instruction Design: Set goals to be educative out of training, and choose suitably, the process to realize them. The sequence and organisation of content must also be designed.
3. Validation: Confirm that the training will accomplish the preferred aims to ensure efficiency.
4. Implementation: After the implementation of the training, a workshop must be prepared to emphases on presenting the knowledge gained from it.
5. Evaluation and Follow-up: The immediate reaction of the trained workers should be measured. The achievement of the training should be verified in the place of work, counting the changes in worker behaviour and production that have followed the procedure.

On other hand, Competencies are applied in development centres, which assist entrant, reinforcement their understanding of the competencies they needed now and in the future. By utilizing this; they can outline their self-directed learning programmes [7].

In addition, training and development rise workers' efficiency and commitment, which in turn improves an organizational performance [42].

The most significant basic components connected with the concept of training and development are culture, training, development, and knowledge Additionally, worker training and development agendas may be seen as activities destined to raise organizational and employee performance in the short run, while job management can be seen as a way for the employee and the organisation to raise organizational and employee performance and long-run satisfaction together [11].

Training and development consist of many components that similarly affect one another as a following [37]:

1. Trainee: Trainees must appreciate the need for training; then, it will outcome in a deprived response and operation of working out objectives.
2. Trainer: Trainers must be accountable for stating the training material in a way that suits trainees and reaches the training objectives.
3. Training Material: This must be acceptable and distributed by the trainer to the trainees.
4. Training Environment: The environment is a place of training, and must be logistically satisfying in terms of its audio and visual instruments.

6.3 Rewards Systems

Contributions made to an organisation by workers and worker achievements should be recognized and reciprocated in some system of reward.

A reward system outlined as "a package/systems that take in rewards and benefits, just like holiday leave, medical benefits, transport payments, and performance bonuses" [23, p. 104].

Furthermore, a reward system is a process used to develop motivation and job engagement by valuing people following their contribution. Also, encouraging productive, discretionary effort by ensuring that people are positive and interested in their jobs, that they are satisfied to work for the organisation and want to continue working for it, and that they take action to accomplish organisational and individual objectives [7].

On other hand, reward systems typically serve many purposes, just like attracting, retaining, and motivating workers [48].

The components of a reward system (such as providing compensation) should be deferential to the major objectives of the system, which are to draw, keep, and inspire workers. In a reward system contain the following components [47]:

1. Direct financial rewards: Compensation received in the form of wages, salaries, bonuses, and commissions.
2. Indirect financial rewards: All financial rewards not detailed in the direct reward.
3. Nonfinancial rewards: the satisfaction an individual obtains from the job itself or from the psychological and/or physical environment in which the individual works.

Reward system is one of the most important parts in organization development and success. In order to maximize the rally performance of the organization, it is the requirement for the worker to understand what motivates the workers and how to raise their job satisfaction. Furthermore, the reward system designed according to the employee is an efficient method to raise work motivation. The sustainable type of reward is developed following the organization reward philosophy, strategies [2].

6.4 Performance Evaluation

Performance Evaluation is an old matter deepened on developing the assessment style, of try to objectify the performance quantification, of decrees measurement errors and even restructuring the philosophy at the estimate systems. Performance is set via a specific organizational behaviour to which the worker is recompense rewarded by salary. Additionally, staff Evaluation systems were based just on a job analysis and were only indiscriminately linked to the goals or profitability of the organization [24].

Performance Evaluation (PE) is an official system of check and evaluation of worker or team task performance [23]. Moreover, employee performance Evaluations are held out by individuals, if there is an understanding by the worker that the operation has not been lead with most topicality and they are not pleased with the Evaluation system this may outcome in complaints and petitions which in turn, may outcome in damage of management's valuable time to address these matter [8].

PE is the assessment of an individual's performance in a methodical way. An improvement tool adopts for the overall development of the worker and the organization. On another hand, the performance is measured with different factors as job knowledge, quality and quantity of output, initiative, leadership abilities, supervision, dependability, co-operation, judgment, versatility and health. Assessment should be confined to the past as well as potential performance also. Further, PA focused on behaviours as a side of assessment because behaviours do affect job outcomes. Performance concerned with the employee and their performance, typically line managers, to look at their performance and development, also to support they need in their role. It is adopted to assess recent performance and focus on future goals, opportunities and resources requirement [17].

This is the systematic evaluation of the performance of a worker by the superior and is an instrument for discovering, examining and classifying the dissimilarities between members of staff concerning job standards. Furthermore, HR professionals are usually accountable for improving performance Evaluation systems. Many of these offer a basis for pay, promotion, and disciplinary action. The performance Evaluation info is vital for worker development since knowledge of feedback is essential to motivate and guide performance developments [12].

On other hand, in performance Evaluations, capabilities are adopted to ensure that performance Evaluations do not simply focus on results, but also reflect the behavioural characteristics of how the work is carried out. Thus, Performance Evaluations are adopted to report personal improvement, development plans, and other learning and development creativities [7].

6.5 Organizational Performance and Big Data

Experts define big data as any set of data that is larger than the ability to process it using traditional database tools from capturing, sharing, transferring, storing, managing and analyzing within an acceptable period of time for that data; From the point of view of service providers, they are the tools and processes that organizations need to deal with a large amount of data for the purpose of analysis. The two parties agreed that it is enormous data that cannot be processed by traditional methods in light of these aforementioned restrictions. Because of the time, effort, and large cost that big data needs to analyze and process, technologists have had to rely on Artificial Intelligence systems that have the ability to learn, conclude and react to situations not programmed into the machine using complex algorithms to work on, in addition to using cloud computing techniques to complete their work [45].

On the other hand, big data has high importance, as it provides a highly competitive advantage for companies if they can benefit from it and process it because it provides a deeper understanding of its customers and their requirements, and this helps to make appropriate and appropriate decisions within the company in a more effective way, based on the information extracted from customer databases and thus increase Efficiency, profit and loss reduction [46].

Organizations have a substantial role in daily life. Thus effective organizations perform a key ingredient for developing nations. Therefore, several economists consider institutions similar to an engine in determining the social, economic and political progress. Precisely for this reason. Thus, organizational performance is one of the most important variables in the management studies and arguably the most imperative factor of the organizational performance. Though the concept of organizational performance is so common in academic literature, its definition is tricky because of its several moral. Accordingly, there is not a Specific definition of organizational performance. In the '50s, the organizational performance was outlined as the range to which organizations, showed as a social system achieve their objectives. Performance evaluation over this time was focused on work, people and organizational structure. Further, in the 60s and 70s, organizations have started to explained new paths to estimate their performance so organization look to performance as the capability to achieve its environment for accessing and using the limited resources. In the 80s and 90s were marked by the investigation that the similarity of organizational objectives is more complex than at first beheld. Managers start to understand that an organization is effective if it achieves its goals by using a less number of resources. Thus, organizational theories that followed confirming the thought of an organization that realize its performance objectives based on the constraints imposed by the limited resources [14].

The concept of performance is one of the most common and used concepts in the field of economics and management of institutions. It has received wide attention by researchers and thinkers, especially in economics, and performance has a broad concept and its contents are renewed with the renewal and change of each component of the institution of different types. Interest and analysis in research and administrative studies.

According to Girard et al. [15] Performance means achievement, implementation, results, and therefore performance is intended to perform business based on a goal or set of objectives taking into account the consequences of such actions.

In terms of management, performance can reflect three general concepts of work, work, and success in performance. Performance is expressed in a rather complex set of indicators, and the term success requires the satisfaction of the internal and external parties of the institution. External recognition of the market [35].

Moreover, performance is defined as: "the result of any activity that includes knowing what to do, who performs, how performance is performed and how performance evaluates?" [29].

According to Bianchi [9], Organizational performance can be defined as the output of the effort and behaviour of all individuals, employees, and institutions in all departments and departments, which determines the extent to which the organization can achieve the outputs and objectives of its business through excellence in its performance. Organizational performance points out to how well an organization meets its financial objectives and market criteria.

As well as, the organizational performance is the integrated system of the organization's work in light of its interaction with the elements of its internal and external environment. It includes: the performance of individuals in their organizational unit, the performance of organizational units within the general policies of the organization, the performance of the organization in the context of the economic, social and cultural environment, It is linked to the efficiency and effectiveness of organizations in achieving their goals and to achieve the aspirations of the visions and goals and values are essential, and hence increased attention to human resources management and improve the level of performance of employees because the success of any organization linked to the level of performance of its members and this will improve the institutional performance of the Organization and carry out its duties to the fullest extent possible [6].

It can be noted that organizational performance is a reflection of the organization's ability and ability to achieve its objectives through the use of available resources efficiently and effectively.

Organizational performance defined as a continuous holistic activity that reflects the organization's ability to exploit its capabilities according to certain criteria and criteria, based on its long-term objectives. If the performance results differ from the criteria, the management corrects the procedures to address the imbalance between the expected results and actual results [21].

Also, organizational performance defined as the performance and implementation of the functions and tasks assigned to each component and each element of the institution individually or collectively leading to the completion of the mission of the institution or organization and the achievement of its objectives [16].

Therefore, organizational performance can be seen as a reflection of how the organization uses its human and material resources in such a way that it can achieve its objectives and thus is a function of organized indicators and characteristics as well as the expertise of the organization's leaders.

Further, Oluoch [31] investigates the relationship between human resources management practices and the performance of the college of humanities and social sciences of the University of Nairobi, Kenya. The result showed that human resource management practices have a positive relationship with organizational performance.

From the above-mentioned literature and after a deep review of previous studies human resource practices and organizational performance, it can be noted that various studies have explored and researched the impact of human resource practices on many aspects of working life, such as productivity, efficiency, competitive advantage, and organizational performance, whether for employees or managers or the entire organization In different sectors and countries.

In a study, Usrof and Elmorsey [40] study aimed to look at (HRM) practices as the right mix in the implementation of (TQM) effectiveness. Such practices. The result represented that recruitment and selection process, teamwork and employee empowerment, training and development, performance Evaluations and compensation have the potential to explore and generate new knowledge for the organizations. The outcomes also recognized the advantages from practices during TQM implementation and they believe this approach could give them a chance to achieve their goals. However, several organizations do not adopt HRM and TQM practices because of the lack of identified reasons and less serious outcomes manifested in the effectiveness of such practices.

Furthermore, Saifalislam et al. [36] study examined the impact of HRM practices as well as the factors that affect recruitment and selection and training and development on organizational performance of the Jordanian Public University in the Kingdom of Jordan. The result showed that recruitment, selection, training, and development significantly correlated with the organizational performance of the Jordanian Public University.

Furthermore, Lambooij et al. [22] aimed to examine the relationship between HRM and organizational performance can be explained by the effect of the internal and strategic fit of HRM on the cooperative behaviours of employees. The result showed that cooperation with co-workers is negatively concerning to turn over and positively related to sick leave.

Sacchetti et al. [34] study aimed to deals with the mediating role of immaterial satisfaction between substantive HR features and organizational performance. The acquired outcomes offer that HRM practices influence immaterial satisfaction and, satisfaction positively effects on organizational performance. Therefore, the effect of several HRM practices is not the same. On other words, employee involvement and workload pressure have a positive effect on organization performance; but task autonomy does not affect organizational performance.

It has become clear that organization performance has essential implications for workers and organizations thus the study aimed to examines HR practices and the impact of incentives on manufacturing companies in the Malaysian context. In addition, the study aimed to explain the role of incentives as a moderator on organizational performance and HR practices. The outcomes show that HR practices training and information technology have a direct impact on organizational performance. The outcomes found that incentive is positively related to organizational performance but did not moderate the relation between both HR practices and organizational performance [28].

Sabiu et al. [33] the study aimed to investigate the influence of human resource management (HRM) practices, (recruitment and selection) and organizational performance (OP) through mediation role of ethical climates (ECs) in Nigerian educational agencies. The results revealed strong support for the mediating role of ECs on the relationship between HRM practice (recruitment and selection) and OP.

Alshammari [5] aimed to investigate the mediating role of knowledge management for organizational performance (OP), and human resource management (HRM) practices and the moderating role of organizational learning (OL) on the relationship between organizational performance and knowledge management capability (KMC). The findings of this study look at a significant effect of HRM practices on knowledge management capabilities, organizational culture, organizational performance, and organizational learning in Saudi Arabia during 2019.

Pradhan [32] aimed to focuses on assessing human resource practices in the manufacturing and service sector institutions in Nepal. This study also attempts to find out the relationship between the adoptions of such practices in their performance. The results suggested that there is a negative association between HRM practices and employee turnover whereas there is a positive association between HRM practices and productivity. Moreover, the result suggested that there is a negative association between HRM practices and employee turnover whereas there is a positive association between HRM practices and productivity.

Garcia-Arroyo and Osca [49] study aimed to conduct a review of the literature to systematize the academic inputs so far and to explain what it means for HRM that the data is big. the Results represented that that big data means a new method and methodology to head up data on employees and many opportunities for HRM, however, it's essential challenges at a technological, methodological and ethical plane.

7 Methodology

An inductive research method was followed in the completion of this research paper, by reviewing previous literature on the topic, in addition to reviewing recent documents, articles and seminars in this field.

The inductive approach is a delicate process aimed at collecting data and observing the phenomena associated with it to link them to a group of general macro-relations, and also the inductive approach is a research method that the researcher uses in generalizing his study to the general studies related to the topic he is researching, i.e. linking the study that He worked to implement them as part of a whole, and the inductive approach relies on the use of a set of conclusions based on observations, estimates, and experiences related to the previous literature review.

The researcher is interested in reading all the information about the research part before issuing a general decision about it, but it is considered one of the types of induction that is slow to apply, or far from the process. Because the researcher needs complete care while observing the components of the inductive approach, which takes a long time in order to know all the components, but it contributes to providing accurate and correct results.

8 Research Contribution

This study plays a significant role, contributing to knowledge by presenting original results from the relationship between Human Resource Practices and Organizational Performance and their operation in light of the development of using big data technology at Jordanian public universities.

As mentioned before, this research will enjoy uniqueness and importance in the field of studies that explore the relationship between Human Resource Practices and Organizational Performance and their operation in light of the development of using big data technology in Jordanian public universities. Also, the result of this research contributes to assisting specialists and managers in this field in regards to the methods of important decision-makers who are interested in informational and analytical studies with important information on the relationship between Human Resource Practices and Organizational Performance and their operation in light of the development of using big data technology and will provide assistance to officials and decision-makers in Jordanian universities to identify the relationship between Human Resource Practices and Organizational Performance and their operation in light of the development of using big data technology and thus help them In taking appropriate measures to implement these concepts. Besides that, the results of this research may draw the attention of managers to the importance of the relationship between Human Resource Practices and Organizational Performance and their operation in light of the development of using big data technology, which could reflect positively on public universities in Jordan.

Moreover, the results of this research affect the range of studies in this field that can be conducted, which leads to some limitations and difficulties to the contents. Nevertheless, this will be considered as a challenge from one side. From the other side, it contributes positively by establishing a real value in the academic research in this field, which will encourage other researchers to build up and develop studies that support this field to achieve the desired goals in the future.

9 Conclusion

In our time, we are witnessing a huge explosion in data. The analysis and processing of this data mainly increases the understanding and absorption of customer requirements, thus increasing efficiency and productivity and reducing losses for companies. However, there are many challenges and obstacles that hinder the use or expansion of big data.

As well as, the importance of human resource management comes from critical factors that contribute to an organization's success. In addition to realizing a good relationship and mutual understanding between management and workers by provided that different big data about the activities of employees in the organization, which contribute to evaluating performance; the human resource information systems reveal any changes in the internal and external environment and allow the administration to ready for the changes. It also realizes integration and coordination between the different activities of human resources management and other departments by using big data technology.

In addition, the importance of human resource management comes from the appeared need for organizations to change their processes and innovate, as the market requires, an effective response is necessary in order to meet up with the highly competitive market characterized by continuous changing. As a result, of this, managers have realized that the success of any business enterprise in a complex and changing environment would depend on the effective management of its Human Resources and other departments, and through make valued coordination between them in light of the development of using big data technology.

Additionally, some human resource management components can affect employee commitment and motivation. Such as staffing and selection, training and development, performance evaluation, collaboration, and payment and reward. The most important task for many organizations today is to survive the present competitive market through the impact of privatization, liberalization, globalization, competition. Thus the organization needs to achieve Organizational performance concerning the ideas of effectiveness and efficiency to achieve competitive advantage by using big data technology.

The previous study has suggested connections between HR practices and organizational performance. The present research aims were to analyse how Jordanian public universities take advantage of the relationship between human resource practices and organizational performance in light of the development of using big data technology with their combined dimensions. Its relation is in the recognition of the importance of individuals in the improvement of resources and abilities and in the making of possible competitive advantages in light of the development of using big data technology As well, some studies have developed a conceptual framework for people management practices (such as recruitment, training, job design, engagement, rewards, and evaluation) that create value for influencing the creation, transfer, and integration of knowledge that forms the basis of a company's competencies. To maintain continuous improvement and development of organizational performance, organizations need to emphasize learning at the level of individuals, teams, and organizations. Human resource practices contribute to the organization's learning and facilitate the development of organizational knowledge in light of the development of using big data technology.

t is believed that higher education institutions need a system of committed human resources practices to improve individual knowledge, skills and ability, and to stimulate knowledge sharing within projects and the institution, learned from the external environment to facilitate access to organizational knowledge and learning. Therefore, this paper proposes a positive relationship between the Relationship between Human Resource Practices and Organizational Performance and their operation in light of the development of using big data technology in higher education institutions in general and Jordanian government universities in particular. Thus, according to resource-based theories, the influence of organizational learning on the Relationship between Human Resource Practices and Organizational Performance and their operation in light of the development of using big data technology is assumed.

To understand the complex situation between human resource practices, organizational learning, and organizational performance in light of the development of using big data technology in higher education institutions in general and Jordanian government universities in particular, inductive research must be investigated. As a determination, the study highlights the time structure and the beginning examination, done only from the abstracts. The literature review outlined a wide range of relations between HRM practices and make go measure the relationship between these structures. The results also display gaps in the plane and the implementation of researches.

With consideration to performance measures, there is a sustainable lack in the measurement of variables, in the integration of more than one dimension of performance in light of the development of using big data technology, in the comparative analysis between Jordanian public universities, and the temporary analysis of performance.

With esteem to HRM, the verified the concomitant use of variables related to HR practices and outcomes, which is not recommended. The researcher also specified the employ of a wide range of variables, a point also noted in the international studies (and that has been the base of criticism since it forces determination on knowledge accumulation.

In the analysis of the studies that seek to evaluate the impact of Human Resource Practices and Organizational Performance and their operation in light of the development of using big data technology, the researcher observed the dominance of positive results, in line with evidence found in international studies. Moreover, considering the dispersion of variables and the methodological matter, the researcher could not draw more overall conclusions about the Jordanian public universities' reality, suggesting a fruitful field for future research.

In light of the study's results and findings, the researcher concludes that it is essential to conduct more research that aims towards filling some of the gaps in this area. In particular, it is important to focus on the relationship between Human

Resource Practices and Organizational Performance and their operation in light of the development of using big data technology and loyalty towards the university.

Moreover, further research is necessary to achieve more diversified samples, thus reaching better generalizability. Further studies are needed to shed light and focus on the impact of the Relationship between Human Resource Practices and Organizational Performance and their operation in light of the development of using big data technology, to potentially uncover more thorough results.

References

1. Abu-Mahfouz SS (2019) TQM practices and organizational performance in the manufacturing sector in Jordan mediating role of HRM practices and innovation
2. Adam TC, Epel ES (2007) Stress, eating and the reward system. Physiol Behav 91(4):449–458
3. Alkhazali Z, Aldabbagh I, Abu-Rumman A (2019) TQM potential moderating role to the relationship between HRM practices, Km strategies and organizational performance: the case of Jordanian banks. Acad Strateg Manag J 18(3):1–16
4. Al-Nsour M (2012) Relationship between incentives and organizational performance for employees in the Jordanian Universities. Int J Bus Manag 7(1):78
5. Alshammari AA (2020) The impact of human resource management practices, organizational learning, organizational culture and knowledge management capabilities on organizational performance in Saudi organizations: a conceptual framework. Revista Argentina de Clínica Psicológica 29(4):714
6. Argyris C (2017) Integrating the individual and the organization. Routledge, London
7. Armstrong M (2010) Handbook of human resource management and practice. 11th edn. Kogan Page, British Library Cataloguing, London
8. Awunyo-Vitor D, Hagan MA, Appiah P (2014) An assessment of performance appraisal system in savings and loans companies in Ghana: evidence from Kumasi metropolis. J Fac Econ Adm Sci 4(2):191–206
9. Bianchi M (2019) Beyond the structural modelling for the analysis of organizational performances in the resilience management. Economia Aziendale Online 10(1):63–74
10. Bimpitsos C, Petridou E (2012) A transdisciplinary approach to training: preliminary research findings based on a case analysis. Eur J Train Dev 36(9):911–929
11. Edwards J (2011) A process view of knowledge management: it is not what you do; it's the way that you do it. Electron J Knowl Manag 9(4):297–306
12. El-Ghalayini Y (2013) Measuring the readiness of managers to effectively implement a new human resources (HR) process - an empirical study on the international committee of the red cross (ICRC). Department of Business Administration. Faculty of Commerce. Islamic University, Gaza
13. Fong C, Ooi K, Tan B, Lee V, Chong A (2011) HRM practices and knowledge sharing: an empirical study. Int J Manpow 32(5/6):704–723
14. Gavrea C, Ilies L, Stegerean R (2011) Determinants of organizational performance: the case of Romania. Manag Mark 6(2):285–300
15. Girard J, Snow R, Cavataio G, Lambert C (2007) The influence of ammonia to NOx ratio on SCR performance (No. 2007–01–1581). SAE Technical Paper
16. Hamann PM, Schiemann F, Bellora L, Guenther TW (2013) Exploring the dimensions of organizational performance: a construct validity study. Organ Res Methods 16(1):67–87
17. Hassan S (2016) Impact of HRM practices on employee's performance. Int J Acad Res Account Financ Manag Sci 6(1):15–22
18. Inyang BJ (2010) Strategic human resource management (SHRM): a paradigm shift for achieving sustained competitive advantage in organisation. Int Bull Bus Adm 7:23–36

19. Ivancevich J (2007) Human resource management. McGraw-Hill/Irwin, New York
20. Juhdi N, Pa'wan F, Hansaram RM, Othman NA (2011) HR practices, organizational commitment and turnover intention: a study on employees in Klang Valley, Malaysia. Recent Res Appl Econ 16:30–36
21. Katsikeas CS, Morgan NA, Leonidou LC, Hult GTM (2016) Assessing performance outcomes in marketing. J Mark 80(2):1–20
22. Lambooij M, Sanders K, Koster F, Zwiers M (2006) Human resource practices and organisational performance: can the HRM-performance linkage be explained by the cooperative behaviours of employees? Manag Revue 17:223–240
23. Lim LJW, Ling F (2012) Human resource practices of contractors that lead to job satisfaction of professional staff. Eng Constr Archit Manag 19(1):101–118
24. Lobontiu G, Birle V, Petrovan A (2015) Human resource performance appraisal: ranking the importance of derived information. In: Managing intellectual capital and innovation for sustainable and inclusive society: managing intellectual capital and innovation; proceedings of the MakeLearn and TIIM joint international conference 2015, pp 1105–1111. ToKnowPress
25. Marchington M, Wilkinson A (2002) People management and development, 2nd edn. CIPD, London
26. Mavondo F, Chimhanzi J, Stewart J (2004) Learning orientation and market orientation relationship with innovation, human resource practices and performance. Eur J Mark 39(11):1235–1263
27. Minbaeva DB (2005) HRM practices and MNC knowledge transfer. Pers Rev 34(1):125–144
28. Mohamad AA, Lo MC, La MK (2009) Human resource practices and organizational performance. Incentives as moderator. J Acad Res Econ **1**(2)
29. Mori N, Golesorkhi S, Randøy T, Hermes N (2015) Board composition and outreach performance of microfinance institutions: evidence from East Africa. Strateg. Chang. 24(1):99–113
30. Obeidat D, Yousef B, Tawalbeh HF, Masa'deh RE (2018) The relationship between human resource management (HRM) practices, total quality management (TQM) practices and competitive advantages. Total Quality Management (TQM) Practices and Competitive Advantages. Mod Appl Sci 12(11) (2018)
31. Oluoch JO (2013) Influence of best human resource management practices on organizational performance: a case of college of humanities and social sciences university of Nairobi, Kenya. Unpublished Thesis of Master of Arts in Project Planning and Management of the University of Nairobi
32. Pradhan GM (2019) Impact of human resource management practices on organizational performance (a case of Nepal). Batuk 5(2):14–31
33. Sabiu MS, Ringim KJ, Mei TS, Joarder MHR (2019) Relationship between human resource management practices, ethical climates and organizational performance, the missing link. PSU Research Review
34. Sacchetti S, Tortia EC, Lopez Arceiz FJ (2016) Human resource management practices and organizational performance. The mediator role of immaterial satisfaction in Italian social cooperatives
35. Safa S (2006) Study the influence strategist call choices on the performance of companies implicated in one public offer of purchase and exchange. Arab J Econ Sci Adm. University Rohe el kods, Elkaslik, Lebanon, n2
36. Saifalislam KM, Osman A, AlQudah MK (2014) Human resource management practices: influence of recruitment and selection, and training and development on the organizational performance of the Jordanian Public University. Organization 3:1–08873
37. Sakarnah B (2009) Managerial training. Dar AL Wa'al for Publishing
38. Schuller T (2000) Social and human capital; the search for appropriate techno methodology. Policy Stud 21(1):25–35
39. Tiwari P, Saxena K (2012) Human resource management practices: a comprehensive review. Pak Bus Rev 9(2):669–705
40. Usrof HJ, Elmorsey RM (2016) Relationship between HRM and TQM and its influence on organizational sustainability. Int J Acad Res Account Financ Manag Sci 6(2):21–33

41. Vanhala S, Stavrou E (2013) Human resource management practices and the HRM—performance link in public and private sector organizations in three Western societal clusters. Baltic J Manag 8(4):416–437
42. Vlachos IP (2009) The effects of human resource practices on firm growth. Int J Bus Sci Appl Manag 4(2):18–34
43. Welbourne T (2008) Editor's-in-chief note: technology, HRM, and "Me." Hum Resour Manag 47(3):421
44. Wheelen L, Hunger J (2013) Strategic management and business policy: toward global sustainability. Pearson, New Jersey
45. Wang Y, Hajli N (2017) Exploring the path to big data analytics success in healthcare. J Bus Res 70:287–299
46. Wang Y, Kung L, Byrd TA (2018) Big data analytics: understanding its capabilities and potential benefits for healthcare organizations. Technol Forecast Soc Chang 126:3–13
47. Yap JE, Bove LL, Beverland MB (2009) Exploring the effects of different reward programs on in-role and extra-role performance of retail sales associates. Qual Mark Res: Int J 12(3):279–294
48. Zhou Y, Zhang Y, Montoro-Sánchez A (2011) Utilitarianism or romanticism: the effect of rewards on employees' innovative behavior. Int J Manpow 32(1):81–98
49. Garcia-Arroyo J, Osca A (2019) Big data contributions to human resource management: a systematic review. Int J Hum Resour Manag 1–26

The Effect of Information Systems for Human Resources on the Capability of Individual Innovation in Jordanian Companies: A Conceptual Review

Aya Naser Magableh and Jasem Taleb Al-Tarawneh

Abstract In recent decades or so, the use of big data has become very common among organizations). The reasons for such an interest have been well proven: data storage costs have dropped (in all formats) significantly and data generation technology (such as sensors and wearables) has at the same time become inexpensive. At the same time, techniques for manipulating and managing data stored by organizations are now incorporated in standard software to allow practitioners to rapidly gain insights and use them to enhance organizational performance. The ability of a company to recognize, integrate, turn, and apply useful external information is individual innovation capacity. For business growth, it is regarded an essential. In the formation and evaluation of the Individual Innovation Capability of a company, new technological systems play a vital role. This study provides an evaluation of the capacity of individual innovation in the literature of information systems. HRIS researchers have used the aim to establish individual innovation in different and sometimes conflicting directions. Controversy relates to how to conceptualize individual capacity for innovation, the acceptable degree of study, and how it can be assessed. In reviewing this framework, our goal is to reduce such uncertainty by enhancing the awareness and appreciation of individual innovation and directing its successful use within HRIS study. This study describes the origins in the larger strategic literature of the Individual Innovation Capability Framework and gives specific attention to its conceptualization, conclusions, and connection to knowledge management. Regarding this, this study discusses whether individual innovation capacity in HRIS study has been conceptualized, assessed, and used. This study further explores how the capacity for individual innovation integrates into separate HRIS concepts and promotes comprehension of different HRIS concepts. Based on this, this provide a framework in which HRIS researchers can more effectively utilize the valuable factors capacity for innovation while analyzing the influence of HRIS on individual capacity for innovation in companies.

Keywords HRIS · Individual innovation capability · Big data

A. N. Magableh (✉) · J. T. Al-Tarawneh
Faculty of Business, Universiti Malaysia Terengganu, 21030 Kuala Nerus, Terengganu, Malaysia

A. M. A. Musleh Al-Sartawi (ed.), *The Big Data-Driven Digital Economy: Artificial and Computational Intelligence*, Studies in Computational Intelligence 974,
https://doi.org/10.1007/978-3-030-73057-4_30

1 Introduction

A company's capability for innovation is related to the internal efforts of human, technical and organizational capital, coupled with the ability to engage with the external world in order to seek the resources, expertise and skills to be integrated into the enterprise in order to develop products and procedures that investors interpret and appreciate [24]. Developing innovation capabilities is not a simple task from a management perspective, as it involves a decision-making process that guides the company's innovation activities and generates an innovation and creativity between employees and inside the company benefit of the entire [68]. Innovation capability conceptually means a team of committed workers to develop systems, goods and services that could be appreciated by consumers and suppliers, adding confirmation to the company and long-term competitive edge constraints [2]. In the internal environment, the capacity to innovate could be studied, understanding how capacities are produced, designed, formed and controlled, as well as in the external environment, aimed at provider relationships, building links between socialization and trust, and exchanging knowledge to generate innovations. It has become generally accepted that within organizations, internet technology has a critical role [72]. The value of innovation and technical development has improved with the addition in internet and compact device use. Because of its benefits, generalized HRIS cannot be avoided all over the world. Companies are involved in integrating advanced in HRIS for their development and boosting competition to benefit from it. Innovation capabilities have been increasingly recognized because many companies are starting to make their decision-making initiatives part of innovation. In a specific company, the degree of innovation represents the willingness of its managers to implement value-added ideas that influence a company, its production capacities and efficiency [28]. In order to support the innovation investment in terms of the company's competitiveness, the result of innovation must be expressed in the company's results. In order to thrive and stay competitive, companies should also have an information technology capable of making available accurate and clear information on all the processes of the company. It is data that now directs decision-making because it produces knowledge within the organization and, thusly, it is critical for the consistency of the operations of the company. Companies find it difficult to handle a vast volume of knowledge coming from their various partners with traditional (non-integrated) information technology [44]. This has prompted companies around the world to concentrate on implementing information systems that can incorporate data on all the operations of the company. Similar technologies provide a forum of expertise that enables users to exchange information from various industries, such as industry, operations and human resources. Such technology is capable of promoting decision-making and giving the company competitive advantages. Likewise, Jordanian SMEs are gradually moving into new technologies, such as Enterprise Resource Planning [8]. Even so, such optimization techniques help companies to utilize information that is more accurate and predictable. Since awareness and interaction with companies are much more about competition, human capital is now the critical competency. In the growth and management of the company, [59] sees human capital as such a significant aspect. The workforce has been a profitable asset and not a hassle that the company has to suffer, as per [36]. [4] described human capital as the expertise, abilities, innovation and health of the individual within a strategic concept. Correspondingly, [82] and [44] interested in enhancing, training, job experience and individual aspects such as attitude, individual differences, learning ability, intentions to human capital. Human capital is a vital consideration for the overall success of company and, in general, creativity. Evidently, the greater the human ability, the greater the ability of employees to discover. This, in essence, increases employee innovation capacity. Companies should pay special attention to the human resources management role in order to preserve and grow this valuable human capital (remunerations, training, and recruitment).This role is supported by the implementation of information systems for human resources. The literature classifies the spectrum of technological innovations related to human resources into different classes: virtual human resources, interactive human resource groups and information systems for human resources [30]. Upgrading the HR role made it possible to expand the various services it provides, to share tacit information around the company and to give a very significant influence on the success to the directly affect. Exceedingly, global companies use information systems as technology advances and innovation improves. Administrators primarily use information systems for management. In all companies, there are many electronic technologies that are essential for improving and coordinating hiring and training. Modern major advancements in implementations for the HRIS have demonstrated their significance and their possible implications on the efficiency of a company [18]. There is increasing concern regarding the effects on organizational performance and effectiveness of HRIS implementations. HRISs relate to the Organizational structure being developed and evolved. The threefold effect of digitalization on the HR role is the emphasis of [23]. Their results are confirmed by [64]. They observed that the effect of HRIS is strategic and contextual on the Organizational structure. The strategic influence can be illustrated in its opportunities to boost the efficacy of the activities of human resources; to simplify the common duties of the HR function; and, ultimately, to improve the efficiency of HR employees. It contributes to decreasing the time complexity of client needs for the transactional effect; and enhancing the effectiveness of satisfaction of HR workers and their recognition by company employees. [36] claim that employees would be prepared to operate with a high degree of accuracy, quality and efficiency regarding HRIS use. HRIS is vital to the HR function's effectiveness. [48] noted that responsiveness, quick availability of information, improved serve as guidance, enhanced monitoring and evaluation are the advantages of HRIS. [49] found out that the strategic and financial management of human resources is also influenced by HRIS. The role of information systems in inclusive growth was demonstrated by [16], as the use of paper is reduced significantly. [64] suggests, in a specific manner, that HRIS actually strengthens organizational change. Obviously, at a high level in the organization, the information created by the system will be used to facilitate better decision. Information decreases the rate of uncertainty, minimizes the distance between predictions and behavior, and creates information that serves as the basis for decision making. This illustrates why, in order to improve organizational skills and guarantee their longevity, companies are now investing in human resources. In reality, in a knowledge-based economy, human resources efficiency alters organizational success criteria. Companies should dispose of the most modern technical equipment and systems to increase the efficiency of employees. The significant factor of HRIS in enhancing individual quality and reliability has been revealed by several studies. For all companies, HRIS has become the main factor [78]. Given the comparatively high failure rate of ERP, before and after its introduction, businesses are called upon to involve their workers. Therefore, the use of HRIS must be followed by affective employee involvement to ensure a greater chance of improving innovation capability. Studying the influence of this particular instrument on the attitudes of users demonstrates the many advantages that can be produced by a company [74]. On individual users, HRIS has numerous effects. It has the ability to improve individual imagination, in addition to improving their skills. Inventiveness is therefore the initial prerequisite in the development of innovation. To summarize, numerous studies [54] have emphasized the role of the information system for human resources. They identified a range of advantages, such as quick access to information on human resources, decreased time to complete duties, enhanced preparation systems [83]. Very little study, however, has contributed significantly to HRIS' enhancing the competitiveness of individual capacity for innovation. Our interest lies in the study of the conceptual analysis of the relationship among HRIS and the innovation capacity of individuals in this paper. It is therefore evident that much more conceptual and analytical work would be required to explore the effect of general capacity f or innovation on the success aspects of the company. This research adds to the literature gap mentioned above, because one of its key goals is to provide additional proof of the effect of a HRIS on company's overall innovation capacity.

2 Conceptual Literature

2.1 Individual Innovation Capability

Innovation capability is characterized as the capacity to consistently change information and thoughts into new items, cycles and frameworks to assist the firm and its partners [63]. An innovation can be another item or administration, another creation cycle innovation, another structure or managerial framework, or another arrangement or program relating to authoritative individuals. Managerial innovations are identified with hierarchical structure and regulatory cycles, though specialized innovations are associated with items, administrations, and creation measure innovation. As per [57], item innovations are new items or administrations acquainted with address market issues, while measure innovations are new components brought into an association's creation or administration activities (for example input materials, task details, work-and data flow instruments, and hardware used to deliver an item or render an administration. Innovation ability is affected by outside and inner components that are for the most part informative variables of firms' innovation cycle and additionally the result of the cycle. The current study recognized not many literature reviews on innovation ability [20, 31, 39, 57, 77]. These surveys have an alternate attention on innovation capacity. For instance, [10] literature review zeroed in on inhibitors of troublesome innovation ability. He directed broad survey to distinguish inhibitors of troublesome innovation capacities. The principle inhibitors distinguished incorporate powerlessness to forget out of date mental models, a fruitful prevailing plan, a risk averse corporate atmosphere, innovation measure bungle, absence of sufficient finish skills, and the failure to create compulsory inside and outer framework. [29] literature survey zeroed in on cycle innovation ability; explicitly, they analyzed the degree to which wanted innovation measure results is appeared in the assembling area. Their review proposed technique, coordinated effort, and culture as the principle measurements of innovation capacity. Conversely, [31] gave a literature review on extremist item innovation capacity; they built up a model of revolutionary item innovation achievement. Their survey distinguished senior initiative, hierarchical culture, authoritative design, extremist innovation item improvement cycle, and item dispatch system as the primary innovation capacity measurements. Zeroing in on firms in agricultural nations, [84] literature review zeroed in on innovation capacity of firms from creating/arising economies (newbie firms). Their survey shows that newbie firms' innovation capacity relies upon their capacity to take part in conscious endeavors to develop, use, and oversee distinctive learning systems inside their limits and as a team with purchasers, providers, makers, clients, colleges, R&D initiates, and counseling firms. All the above articles assessed so far utilized account or conventional approaches in their reviews. Hence, the uniqueness of this survey is the utilization of a deliberate literature review way to deal with produce the innovation capacity measurements, which was then used to build up the reasonable structure Second, the survey gave operationalization of the measurements to help future observational examination required for hypothesis improvement in the region of hierarchical abilities.

3 Innovation Capability Dimensions

The accompanying segment presents a combination of key topical territories (innovation capacity measurements) distinguished in this survey. In the majority of the study, the innovation capacity measurements are expressly expressed, for instance, vision and procedure, tackling ability base, authoritative insight, creativity and thought the board, hierarchical structure and framework, culture and atmosphere, and the executives of innovation were referenced by [50]. Innovation strategy, creativity, organizational culture, leadership, organizational learning, knowledge management and linkages, execution, frameworks and choice standards, authoritative setting and learning frameworks and choice guidelines, hierarchical setting and learning were referenced by [53].

4 Innovation Strategy

Innovation system decides how much a firm prepares accessible assets to accomplish hierarchical objectives notwithstanding unsure promoting conditions. An innovation procedure encourages association's capacity to distinguish outer chances and match those open doors with interior abilities to investigate new business sectors and convey inventive items [75]. [8] considers uncovered that innovation procedure and organization had critical effect on innovation ability. Additionally, the accentuation on innovation procedure in innovation ability was examined by [77]. The creators contend that supervisors can create ingenuity by invigorating inventiveness, planning, actualizing and observing suitable innovation procedure. [55] explored innovation ability working in the Swedish guard industry to react to troublesome, non-specialized changes to its current circumstance. Their proof shows that the Swedish protection industry innovation capacity exertion requires away from changes of creating innovation ability. An analysis by [4] gives experimental proof with respect to the part of vital ingenuity in innovation capacity. Their relapse investigation shows that essential imaginativeness had the second most elevated coefficient estimation of 0.79. This implies that essential ingenuity has possible effect on current and future imaginative capacity of associations. [69] found that definition of danger strategy, setting of needs, and asset designation added to innovation abilities of the assembling firms examined. Procedure decides the arrangement of the current assets, frameworks, and cycles that organizations need to embrace to meet market vulnerability. A firm without innovation procedure probably won't be on the pathway toward innovation.

5 Idea Management

An association's capacity to change over thoughts into better than ever items, administrations, or methods of getting things done has been perceived as a significant supporter of innovation ability building. Thought the board cycle empowers a central association's connection with clients, providers, workers, and other colleagues to create and execute inventive items or administration thoughts. The aftereffects of an examination by [61] show that coordinated thought the board assisted the organizations with get-together countless thoughts and commitments from providers, clients, and rivals in their innovation exercises. Additionally, exact discoveries from [17] show that the UK supply chains in oil and gas, semiconductor, IT gear, synthetic compounds, and aviation areas all look to draw in members as dynamic trendsetters via contributing thoughts and encounters along their worth chain. In like manner, observational discoveries from [78] propose that SPEC, a main eyeglasses producer situated in China, utilized thought the executive's frameworks to gather and examine various types of information, including existing clients' inclinations and attributes, for example, recordings and photographs of accessible eyeglasses items. The thought the executives frameworks empowered SPEC's various units and divisions to synchronize their exercises in another item innovation and to upgrade the assembling cycle along its worth chain, consequently, captivating clients, providers, and different partners in thought age just as producing thoughts from base up inside an organization tends to encourage innovation ability.

6 Creativity

Creativity has been recognized as a significant measurement that takes care of into innovation. Creativity can contrast between a gathering, an association, or a culture, and it can likewise change after some time since it is setting explicit [37]. Subsequently, [28] proposes that imagination can be assessed at the degree of individual, association, industry, calling, or more extensive. Accordingly, in this audit, inventiveness is assessed based on the capacity of a central firm to establish the empowering climate to upgrade imaginative conduct of representatives. By and large, associations are needed to upgrade innovation by guaranteeing a climate that upholds creativity and thought age. [64] distinguished imagination, inspiration, initiative, correspondence channels, thought creation and evaluation, and new methods as a portion of the vital components of innovation ability that impact business execution. Their outcomes demonstrate that singular inventiveness emphatically impacts an organizations' innovation ability. They further contend that people's innovation ability fills in as a reason for an organizations' general innovation capacity. Innovation rotates around human action, subsequently upgrading skill of workers according to imagination, cooperation, learning, administration, network abilities; and business could be basic for age of innovations effectively. [13] observational proof proposes that directors can uphold innovation capacity by invigorating inventiveness, experimentation, and receptiveness to groundbreaking thoughts inside the firm. Creativity is discovered to be reliant on administration's capacity to establish openings and an empowering climate and to give the required assets. Observational proof from [59] shows that imagination improves authoritative creativity. Along these lines, associations can encourage innovation by establishing and keeping a climate that upholds thought age and creativity. Such empowering conditions incorporate the arrangement of assets and openings. Hierarchical consolation and arrangement of sound workplace can animate creativity and innovation.

7 Collaboration

The surviving writing has focused on the significance of cooperation in innovation producing exercises [81]. [57] recognized outside coordinated effort and inside cooperation as a portion of the key factors that add to firms' innovation ability building. Cooperation causes firms to divide data and information between associating parties. The estimation of joint effort with respect to capacity is appeared in the exact work and her decision is upheld by [39]. In examining Volvo Cars fabricating organization's innovation capacity, it was taken note of that all together for Volvo vehicle to accomplish its vision 2020, Volvo vehicle teamed up with new outside accomplices, including colleges, to acquire information in the field of purchaser conduct and energy use, a specific information discipline that Volvo Cars' needed inside. The continuous cooperation with outer gatherings added to a difference in context and the structure of new organizations, which thus encouraged information an innovation required for innovation exercises in Volvo cars. [48] study shows that the biotechnology firms in Australia regularly bring out R&D through coordinated effort with research organizations, colleges, providers, and clients during the time spent structure their innovation capacity. Likewise, experimental proof by [46] shows that little cutting-edge firms' coordinated effort with colleges and examination organizations prompted the testing of novel thoughts and development of specialized information and capacities that the little innovative firms needed house. The significance of joint effort is additionally exhibited observationally by [73]. The innovations researched how coordinated effort encouraged innovation capacity in the organizations' store network. Their outcome demonstrated that coordination's firms occupied with joint arranging, shared cycle, and information with their organization individuals. The synergistic activities assisted the central firms with setting out on both gradual and revolutionary innovations. Hence, when firms team up remotely, through joint arranging, shared cycle and information, and shared data, the organizations can encounter high expansion in ingenuity.

8 Leadership

The job of administration in supporting and invigorating innovation by establishing generally speaking helpful climate has been set up in the literature [38, 45, 47]. Observational proof from [22] shows that an essential command from an organization's administration to a gathering of laborers who were liable for making extremist innovation prompted a positive outcome. An examination by [58] shows that authority animates innovation execution through innovation the executives. Their examination demonstrated that overseeing individuals, inventiveness, and thought age through solid organization's administration impacted innovation exercises. This worth shows a positive connection among initiative and innovation. An examination by [66] exhibits that solid initiative and job displaying drove supported innovation capacity among assembling firms researched. Along these lines, the capacity of hierarchical administration to set up a structure of organizing representatives, empowering worker work investment, producing thoughts without dread, and inspiring the remainder of the organization can essentially add to innovation ability building.

9 Knowledge Management

Studies by [67] explore fundamental the significance of information in innovation ability. [3] contended that an association's capacity to abuse its information and the neglected capability of the innovation advances odds of innovation, endurance, and innovations. Building innovation ability includes securing information, aptitudes, and different components of capacity from outside sources and those that are interior to the firm. [59] explored feasible learning for creating innovation abilities at Volvo Cars. She found that creating abilities is identified with change, however creating capacities requires insider information that is significant for comprehension of innovation ability building measure. An examination by [26] shows that the worth chain of certain organizations turned out to be important for co-working frameworks and shared information frameworks. That helped innovation of normalized estimating frameworks and improved recuperation rates. This improved the store network effectiveness among sawmill and furniture producers. Exploration by [1] recommends that assembling organizations' capacity to oversee information the executive's cycle altogether, improved their innovation ability. Their investigation shows that information the executives decidedly impacted the assembling firms' innovation capacity. Moreover, an examination by [53] zeroed in on the part of implicit information in assembling and administration firm's innovation ability building exertion. The outcomes uncovered that unsaid information move decidedly affected the organizations' innovation capacity. Exploration by [40] shows that the coordination's director imparted information to clients and providers, to the degree that a portion of the organizations were even ready to get to providers data set, which prompted an improvement of innovation ability among coordination's firms explored. Factual examination by [32] recommends that innovation includes a wide scope of information sharing cycle that encourages the usage of cycles, thoughts, and items. They likewise found that laborers capacity to share information essentially impacts the organizations innovation ability. [79] inspected information the executives zeroing in on authoritative capacities. Their outcome demonstrates that the association's way of life, structure, and innovation along with viable procedures of information securing, change, application, and assurance are significant for hierarchical execution. Information the executive's frameworks of firms create, store, and offer information and data that can uphold authoritative innovation exercises.

10 Organizational Culture

[7, 11, 17, 78] have all perceived organizational culture as a main thrust for innovation capacity. An association that advances strengthening of workers, resistance, successful correspondence between central firms and huge accomplices, and inspirational mentality towards accomplishing hierarchical objectives can be believed to advance an innovation culture. Innovation culture can be depicted as a company's disposition towards investigating and executing thoughts that encourage the association's imaginative reasoning and exercises [27]. Exploration by [9] exhibited that innovation culture was positive in all the organizations researched. This is on the grounds that the way of life of the organizations allowed people who were not exactly effective trying to be innovative to be given a subsequent chance. This demeanor urges workers to be imaginative. An examination by [71] uncovers that every one of the assembling organizations explored had set up an innovation situated culture that molded practices and designated the assets important to accomplish deliberate innovation and positive business results. An investigation by [16] inspected authoritative culture and strengthening of innovation ability of SMEs. The outcomes from their exploration demonstrate that hierarchical culture decidedly impacts innovation ability through the intercession of worker strengthening. Accordingly, advancing a culture of worker strengthening, open correspondence, uphold for change, and representative danger taking activities can fundamentally impact the capacity to develop.

11 Organizational Learning

Learning has been featured as quite possibly the main elements of innovation capacity [49]. [45] portrayed learning as an extension among working and developing. In this way, learning exercises ought to be available in the way of life of an association to allow the use of inner and outer mastery expected to advance innovation. The impact of learning on innovation ability has been broadly recorded in innovation the executive's literature [69, 71]. For instance, proof from [18] shows that Volvo's Vision 2020 undertaking 'getting the hang of by doing' or 'active learning' included experimentation with moderately little innovations (e.g., vehicle subsystems). The bits of knowledge created through the learning cycle were shared all through the Volvo vehicle producing organization. The outcomes uncovered that imaginative ability and innovation exercises all in all were affected by learning by doing. Likewise, an exploration [33] uncovered that learning encourages the usage of cycles, thoughts, and items among the wide range of US enterprises. Accordingly, uphold for gathering of work encounters into schedules, interfirm trade of encounters and data, just as inclusion of clients and providers in learning exercises, can add to innovation ability building.

12 Human Resource Information System (HRIS)

Recent innovation in technology have made it conceivable to make some genuine memories, data based and self-serving intuitive workplace [62]. The improvements went significantly further, and brought about 800 000 petabyte information flowing the world over today. These improvements show that data is presently a piece of association capital with ideas like large information, distributed computing and web of things, and characterize the elements of data economy which is intensely talked about in the new occasions. The data which is anticipated to arrive at 35 zettabytes in 2020 features its essential significance and force as a likely future capital venture and most important resource of firms. Associations attempt to shape their data the board works around this system. With this point of view, they think about clients, workers and hierarchical structure as a type of capital [51]. Information the board is characterized as three essential information activities: securing, stockpiling and move [60]. Fundamentally, information the executives is a coordinated and methodical methodology related with deciding, overseeing and sharing all information resources including work's normal information and aptitudes with the end goal of arriving at association's exceptionally extraordinary objectives. The objective of information the board is to deal with the information which is essential and valuable for the association. Hierarchical proficiency starts to increment when the correct information arrives at the perfect spot at the perfect time [70]. The computational conditions where gained information is ordered, coordinated, systemized, shared and incorporated by needs are classified "information frameworks". Information frameworks are utilized in associations for information the executives, and they are tended to regarding three frameworks, in particular Operational Process Systems. Management Information Systems and Expert Systems. MIS has subsystems, for example, Decision Support Systems and Human Resources Information System. MIS which changes the crude information from the inside and outside of the association to arranged and organized valuable information plays a more compelling and significant part in choice help measure by passing on such to DDS subsystem for an answer for convoluted administration issues and more powerful and proficient dynamic [21, 34]. Accordingly, MIS is characterized as a coordinated framework supporting a successful dynamic cycle or giving the information needed to management of authoritative tasks. In another definition, the executive's information frameworks can be portrayed as PC based information frameworks which can join information from different sources to give the information needed to administrative choice [80]. As of late, MIS robotization has been progressively utilized. These programmed frameworks have yielded huge positive change constantly in administrative dynamic of the associations [65] HRIS, as a subsystem of MIS, is an information framework approach coordinated for estimating and surveying current HR activities as to creation (results) in the post-mechanical age. HRM's getting progressively more mind boggling and information serious in its utilization in execution, all things considered, and in decision-making and anticipating such activities has carried HRIS to the front, and this framework has helped HR administrators and experts to take more fast, precise and compelling inputs and choices [41].

HRIS mainly keeps the information related with the employees. System is made out of three coherent subsystems: HR arranging and enlistment subsystem which incorporates work necessities and current work information, and wages and advantages the board subsystem which incorporates representative's compensation history, finance practice, wellbeing and protection information, execution installments and deals commissions, and preparing subsystem which tracks vocation improvement with the consideration of unique abilities and capabilities notwithstanding the information on workers' presentation examination, aptitudes and gifts [76]. HRIS is a cycle permitting a business to accumulate, store, protect, refresh and dissect the information, and plan related reports it needs regarding its own HR, faculty exercises and hierarchical offices and their highlights [35]. With expanded utilization of HRIS, the work of Human Resources experts has gotten simpler, and along these lines they had chance to save additional time on in-house counseling exercises. With a further methodology, it is contended that HR the executive's experts enhance the associations while simultaneously reinforcing their impact and status in the association [15]. With the goal for HRIS to be fruitful, it should give administrators and clients required and adequate information in different stages, for example, continuing, arranging, controlling and dealing with the information related with HR (Buzkan 2016). Such information provided by Human Resources Information System should bear explicit highlights, for example, updatability, coming from right source through right strategies, culmination and precision, and being examined and quickly submitted to the clients through right techniques appropriately for the reason. Where no update is settled on, the administrative choices to be taken won't be sound [64]. [39] accept that when HR capacities are stacked into PCs inside a HRIS, more quick dynamic is accomplished being developed, arranging and the executives of HR because of a lot simpler putting away, refreshing, grouping and dissecting of the information. One of the significant advantages offered by HRIS is its commitment to the productivity of control capacities. Besides, it carries vital advantages to the information sharing, production of a hierarchical memory, sped up dynamic and foundation of a culture of shared characteristic. Moreover, some different commitments yielded by the utilization of HRIS by organizations in their HRM tasks can be considered saving fixed costs constantly of looking through a work candidate, and danger the board, monetary arranging, position control (staff prerequisite), participation reports and their investigation, and faculty arranging, innovation arranging, mishap revealing and avoidance, and unsafe material divulgence reports [75].

HRIS applications are the quantity of HR-related administrations accessible on an association's HRIS conveyance channels. Scientists have made different proposals with respect to how HRIS applications may uphold hierarchical cycles. For instance, [56] have revealed a few applications for HRISs, for example, enlistment and choice; turnover following/work examination; finance, advantages and pay (the board); preparing and improvement; execution evaluation; interior and outside correspondence and progression HR arranging. Notwithstanding information stockpiling, the HRIS likewise permits certain errands to be finished more effectively than they would be by hand, while decreasing the measure of paper that HR divisions should store. Past investigations have indicated that HRIS applications are all the more generally utilized in little and medium-sized organizations for regulatory purposes, for example, worker record-keeping and finance, while in huge organizations, a HRIS is frequently utilized for key purposes [6]. Sizeable examination has additionally been distributed with respect to business related results of HRIS execution. Moreover, quantities of workers (size) influence HRIS utilization. For example, [19] saw that various gatherings of clients utilized HRISs in an unexpected way. HR experts utilized HRISs in a few jobs including: administrative revealing and consistence; remuneration investigation; finance; annuity commitments; benefit sharing organization; aptitude advancement and ability stock; and advantages organization. Supervisors in utilitarian territories utilized the HRIS to meet their particular information needs, while the last gathering, the workers, utilized a HRIS for regulatory other options. Hence, the capacities and utilizations of HRISs change with end-clients and kind of associations. [42] reports that HRISs created in-house or in exceptionally altered adaptations can fabricate the dynamic capacities needed by an organization and guarantee a company's seriousness. Bamel et al. (2014) fortify this view, expressing that legitimate execution of a HRIS enlarges authoritative adequacy through the acknowledgment of human asset and hierarchical system. [14] perceive the connections between HRIS usage, representative occupation fulfillment and staff maintenance. With an end goal to expand the extent of HRISs, [60] report that the HRIS assumes a basic part in choice, application and utilization of human asset measurements and investigation, and adds to key progression of associations. The phase of financial advancement of a nation likewise has some effect on the use and utilizations of HRISs, since in creating economies, HRISs fundamentally satisfy regulatory jobs.

13 Big Data and HRIS

Big data is a label used to describe vast quantities of data created by wearables, sensors and social media platforms. Big data may be of various sizes, such as structured or unstructured data, but the most common type of big data is unstructured (that means data with an undefined underlying data system) [25]. Big data are typically set to reflect number, speed, and variety, using the 3 Vs [5]. The amount of information generated by different sources, including social media, business transactions and the Internet of Things, refers to volume. The second V is the speed of the generation of data, while the third V reflects the variety of formats. Kwon and [73], namely variability and complexity, have been described as essential in other Big Data dimension [52].Variability refers to the frequency of data (i.e. it can be irregular, hourly or real-time), while complexity refers to the fact that due to diverse data schemes that underlies the processing of data, the multiplicity of data sources makes it difficult to deal with them. Many companies store complex data using databases. Databases are typically cloud-based and very useful for reporting and data visualisation. The stored information contains information about your staff, holidays, working hours, rotation, time sheets, etc. The idea of a 'data lake' has become very common in the sense of broad data management. Data lakes permit companies to store multiple data types at a low cost because they do not need to process the data in a specified data model. For perspective rather than analyses, data lakes are very useful. Data lakes augment data warehouses in terms of data storage, typically using an already defined data model, which can facilitate reporting and advanced analysis (usually requiring a data scheme).

14 Theoretical Foundation

The individual innovation capacity of a company depends fundamentally on its individuals' specific experience and behavior patterns, which search the world, carry awareness into the company, and gain information to products and processes [20]. Restricted awareness, however, restricts people's ability to obtain, process, recreate, and send information without failure [19]. Behavioral approach thus puts a limit on the erosive capacity of an individual, which in turn reduced the individual creativity capacity of the company. However, HRIS can affect individuals' communication and computation skills [16], thus extending the limits of rationality and, in turn, the constraints of dynamic capabilities. Even though HRIS may extend the limited rationality and acquisition capacity of a company, investing in HRIS alone is sometimes inadequate to sustain inherent benefits. [24, 61]. As a consequence, studies are starting to examine how other organizational capabilities (e.g., existing learning) are enhanced by efficiencies extracted from HRIS capabilities and complementary capabilities [7, 20, 28, 83]. Interaction is also used to characterize experiences that contribute to positive results [45, 51, 67, 79]. Cohesion is characterized as the value increase that involves the interaction of adequate human capacities. The importance of an organizational capacity will increase in the existence of other organizational learning capacities, according to the theory of comparative advantage [54] In particular, when the returns to a capacity differ in the levels of returns to the other capacities, complementarity exists. Capabilities are different, but they are interdependent as well. With this view in mind, the next section explores how value synergies between complementarities between HRIS improve the individual innovation capacity of a company.

15 Empirical Literature

15.1 HRIS and Individual Innovation Capability

Innovation is viewed as the foundation of the success and endurance of organizations [43, 55, 69, 79, 82] in this instructive time. Actually, it gives the foundations to the organization to build its presentation [83]. Assets hypothesis specifies that the person in an organization speaks to a holy asset. In accordance with this line of thinking, we uphold the significant job of representatives in the production of innovation [81]. Innovation and innovativeness indicated by employees are the initial phase in the innovation cycle [42, 47, 64, 77, 81]. This readiness and capacity of people to enhance guarantee the exchange of innovation within the association [80]. To better under-stand the idea of innovation limit, most importantly it is essential to characterize it. For [47], the capacity to improve speaks to "a passionate and a psychological supportive of comparative with imagination". As indicated by [21], creativity or innovation of an individual could be characterized as "Novel and useful thoughts, cycles or items offered by a representative, as judged by important others". In an overall way, innovation of an individual or innovation as a rule should meet an all-around decided need and uncommonly ought to be helpful. In a similar viewpoint, [27] guarantee that innovation is a cycle that comprises of passing from uniqueness of thoughts to assembly of arrangements. The literature examined singular innovation limit under different perspectives. A few investigations called attention to that inventiveness speaks to an important factor for innovation [19, 43, 48, 80]. To this end, associations should encourage their representatives to be more creative by putting at their removal suitable systems. In the same vein, [12] advocate that organizations ought to know about the basic requirements of the permanent improvement of individual innovation limit. They add that innovation limit improvement requires including managers throughout the learning cycle, and setting up sufficient policies and changes like the use of accessible resources. Previous examines featured elements that are probably going to support individual inventiveness. Among those variables, a few examinations demanded that transformational administration with forefront work has direct and significant sway on representatives' innovativeness. In addition, natural inspiration additionally ends up being a factor with a critical impact on imagination. [82] highlighted the immediate and positive effect of the measurement 'Opening up to encounter' on individual innovation. As for [76], singular innovation results from creativity and self-initiative. In reality, the creator specifies that creativity should be reinforced without anyone else administration which causes the individual to create and support their innovative thoughts. HRIS is the focus of our investigation. Specifically, we are intrigued then with regards to the influence of this instrument on innovation at an individual level. Numerous studies revealed the idea of the connection among HRIS and individual innovation limit. [47] showed that the use of HRIS instruments and applications improves and supports operational productivity. On the one hand, according to these creators, these apparatuses take into account sharing and transferring information. Then again, and as a result, this knowledge helps people, in the firm, to produce thoughts that are innovative and possibly valuable simultaneously. Similarly, [15] recognized the immediate and huge impact of ERP usage on the limit of representatives to advance in telecommunication firms in Jordan. In an extremely ongoing investigation, [60] guarantee that information and advanced technologies are prone to improve and advance innovation of people and the association itself. As per them, these advanced technologies guarantee and give the fundamental conditions to the employees to amazing their positions and particularly produce inventive thoughts. This study features the essential part of various apparatuses, information sup-port and applications in the formation of new remarkable ideas and the molding of these plans to be really actualized into an innovative item. [71] express that instantaneous admittance to new information furnishes the

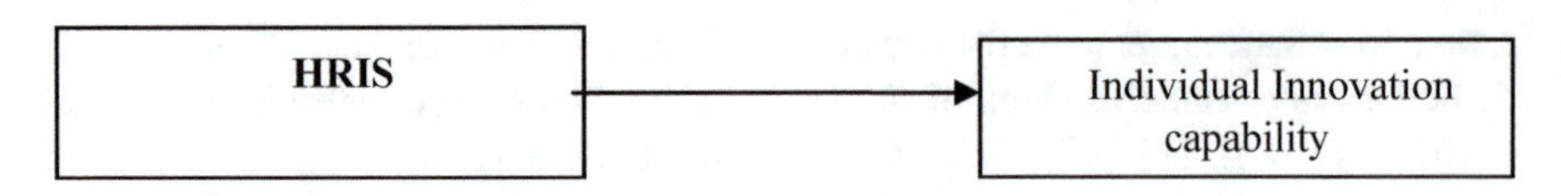

Fig. 1 Proposed research framework (developed by authors)

worker with new approaches and thoughts. In addition, PC applications promote individual innovativeness.

16 A Framework for Examine the influence of HRIS on Individual Innovation Capability

The capacity for individual innovation is a powerful concept that can be used to analyses a variety of HRIS phenomena. We are, however, quite interested in developing the interaction among HRIS and individual capacity for innovation than in encouraging the integration of individual capacity for innovation into HRIS function. With our analysis as a basis, we propose a framework in which researchers may discuss how the valuation efficiencies among HRIS have a direct influence on employee capacity for innovation. Hence, the proposed framework was depicted as follows (Fig. 1):

17 Conclusion

In this study, the conceptual review demonstrates that HRIS is involved in many significant flows of individual capacity for innovation. Individual innovation capacity, nonetheless, has been conceptualized and evaluated in various ways, leaving its wider function in the theoretical framework of HRIS uncertain and underutilized. In addition, HRIS tends to be absent from any of the research on individual capacity for innovation.In establishing a suitable framework in which HRIS researchers can examine how HRIS affects individual capacity for innovation, we discussed these limitations. This study is anticipation this framework will offers an inducement for future HRIS studies, individual capacity for innovation and organizational efficiency.

18 Research Methodology

The study will employing a quantitative approach with a descriptive research style. Furthermore, 384 employees of Jordanian companies are the sample sizes of this study. To avoid response bias, the sample will be increased to 400 [60]. As a result, the research will be conducted in Jordan and the data will be provided through a self-administered questionnaire in which respondents were asked to complete the survey on their own. The purpose of this study is to reduce costs by removing from the respondent devices and objects, including computer software (Abdulla et al. 2014). The researcher will performs a data mining technique to ensure sufficient data representation before actual data analysis is conducted. Furthermore, using SmartPLS 3.0, the collected data will be analysed to check the fitness of the model or structure proposed by the analysis and to test the research hypotheses proposed. Partial Least Square Structiral Equation Model (PLS-SEM) will be used for the study of the impact of HRIS on human creativity capabilities.

References

1. Abbas J (2020) Impact of total quality management on corporate sustainability through the mediating effect of knowledge management. J Clean Prod 244:118806
2. Abongo B, Mutinda R, Otieno G (2019) Innovation capabilities and process design for business model transformation in kenyan insurance companies: a service dominant logic paradigm. J Inf Technol 3(1):15–45
3. Abubakar AM, Elrehail H, Alatailat MA, Elçi A (2019) Knowledge management, decision-making style and organizational performance. J Innov Knowl 4(2):104–114
4. Ahn JM (2020) The hierarchical relationships between CEO characteristics, innovation strategy and firm performance in open innovation. Int J Entrep Innov Manag 24(1):31–52
5. Akter S, Wamba SF, Gunasekaran A, Dubey R, Childe SJ (2016) How to improve firm performance using big data analytics capability and business strategy alignment? Int J Prod Econ 182:113–131
6. Alam MGR, Masum AKM, Beh LS, Hong CS (2016) Critical factors influencing decision to adopt human resource information system (HRIS) in hospitals. PLoS ONE 11(8):e0160366
7. Al-Dmour RH, Masa'deh RE, Obeidat BY (2017) Factors influencing the adoption and implementation of HRIS applications: are they similar? Int J Bus Innov Res 14(2):139–167
8. Ali M, Miller L (2017) ERP system implementation in large enterprises – a systematic literature review. J Enterp Inf Manag 30(4):666–692
9. Ameen AM, Ahmed MF, Abd Hafez MA (2018) The impact of management accounting and how it can be implemented into the organizational culture. Dutch J Finan Manag 2(1):02
10. Andersson M, Moen O, Brett PO (2020) The organizational climate for psychological safety: associations with SMEs' innovation capabilities and innovation performance. J Eng Tech Manag 55:101554
11. Arditi D, Nayak S, Damci A (2017) Effect of organizational culture on delay in construction. Int J Project Manag 35(2):136–147
12. Argote L, Lee S, Park J (2020) Organizational learning processes and outcomes: major findings and future research directions. Manag Sci
13. Aulawi H (2018) Improving innovation capability trough creativity and knowledge sharing behavior. In: IOP conference series: materials science and engineering, vol 434, p 012242
14. Bamel N, Bamel UK, Sahay V, Thite M (2014) Usage, benefits and barriers of human resource information system in universities. VINE J Inf Knowl Manag Syst
15. Baswardono W, Cahyana R, Rahayu S, Nashrulloh MR (2019) Design of human resource information system for micro small and medium enterprises. In: Journal of physics: conference series, vol 1402, no 6. IOP Publishing, p 066056
16. Bayanova AR, Vodenko KV, Sizova ZM, Chistyakov AA, Prokopyev AI, Vasbieva DG (2019) A philosophical view of organizational culture policy in contemporary universities. Eur J Sci Theol 15(3):121–131

17. Begum H, Bhuiyan F, Alam AF, Awang AH, Masud MM, Akhtar R (2020) Cost reduction and productivity improvement through HRIS. Int J Innov Sustain Dev 14(2):185–198
18. Bondarouk T, Parry E, Furtmueller E (2017) Electronic HRM: four decades of research on adoption and consequences. Int J Hum Resour Manag 28(1):98–131
19. Buzkan H (2016) The role of human resource information system (HRIS) in organizations: a review of literature. Acad J Interdisc Stud 5(1):133
20. Castela BM, Ferreira FA, Ferreira JJ, Marques CS (2018) Assessing the innovation capability of small-and medium-sized enterprises using a non-parametric and integrative approach. Manag Decis
21. Chandru S, Jothimani V, Rajalakshmi G (2020) A study on the effectiveness of human resource information system (HRIs) on bureau veritas consumer product services (India) Pvt. Ltd. NOLEGEIN-J Hum Resour Manag Dev 1–9
22. Daniëls E, Hondeghem A, Dochy F (2019) A review on leadership and leadership development in educational settings. Educ Res Rev 27:110–125
23. De Bellis E, Johar GV (2020) Autonomous shopping systems: identifying and overcoming barriers to consumer adoption. J Retail 96(1):74–87
24. de Mattos CA, Kissimoto KO, Laurindo FJB (2018) The role of information technology for building virtual environments to integrate crowdsourcing mechanisms into the open innovation process. Technol Forecast Soc Chang 129:143–153
25. Dedic N, Stanier C (2016) Towards differentiating business intelligence, big data, data analytics and knowledge discovery. In: International conference on enterprise resource planning systems. Springer, Berlin, pp 114–122
26. Di Vaio A, Palladino R, Pezzi A, Kalisz DE (2021) The role of digital innovation in knowledge management systems: a systematic literature review. J Bus Res 123:220–231
27. Felipe CM, Roldán JL, Leal-Rodríguez AL (2017) Impact of organizational culture values on organizational agility. Sustainability 9(12):2354
28. Fernando Y, Jabbour CJC, Wah WX (2019) Pursuing green growth in technology firms through the connections between environmental innovation and sustainable business performance: does service capability matter? Resour Conserv Recycl 141:8–20
29. Ferreira J, Coelho A, Moutinho L (2020) Dynamic capabilities, creativity and innovation capability and their impact on competitive advantage and firm performance: the moderating role of entrepreneurial orientation. Technovation 92:102061
30. Florkowski GW (2018) HR technology systems: an evidence-based approach to construct measurement. In: Research in personnel and human resources management. Emerald Publishing Limited
31. Ganguly A, Talukdar A, Chatterjee D (2019) Evaluating the role of social capital, tacit knowledge sharing, knowledge quality and reciprocity in determining innovation capability of an organization. J Knowl Manag
32. Garcia-Perez A, Ghio A, Occhipinti Z, Verona R (2020) Knowledge management and intellectual capital in knowledge-based organisations: a review and theoretical perspectives. J Knowl Manag 24(7):1719–1754
33. Gomes G, Wojahn RM (2017) Organizational learning capability, innovation and performance: study in small and medium-sized enterprises (SMES). Revista de Administração (São Paulo) 52(2):163–175
34. Haeruddin MIM (2017) Should I stay or should I go? Human resource information system implementation in Indonesian public organizations. Eur Res Stud 20(3A):989
35. Hosain MS, Arefin AHMM, Hossin MA (2020) The role of human resource information system on operational efficiency: evidence from MNCs operating in Bangladesh. Asian J Econ Bus Acc 29–47
36. Hosseini S, Kees A, Manderscheid J, Röglinger M, Rosemann M (2017) What does it take to implement open innovation? Towards an integrated capability framework. Bus Process Manag J
37. Hughes DJ, Lee A, Tian AW, Newman A, Legood A (2018) Leadership, creativity, and innovation: a critical review and practical recommendations. Leadersh Quart 29(5):549–569

38. Hui L, Phouvong S, Phong LB (2018) Transformational leadership facilitates innovation capability: the mediating roles of interpersonal trust. Int J Bus Adm 9(3):1–9
39. Iddris F (2016) Measurement of innovation capability in supply chain: an exploratory study. Int J Innov Sci 8(4):331–349
40. Jabeen F, Al Dari T (2020) A framework for integrating knowledge management benefits in the UAE organisations. Knowl Manag Res Pract 1–15
41. Jayabalan N, Makhbul ZKM, Selvanathan M, Subramaniam M, Nair S, Perumal I (2020) HRIS contributions and impact on strategic employee engagement and participation in private education industry. Int J Manag (IJM) 11(10)
42. Karikari AF, Boateng PA, Ocansey EO (2015) The role of human resource information system in the process of manpower activities. Am J Ind Bus Manag 5(06):424
43. Kim H (2020) Analysis of how tesla creating core innovation capability. Int J Bus Manag 15(6)
44. Kwon O, Sim JM (2013) Effects of data set features on the performances of classification algorithms. Expert Syst Appl 40:1847–1857
45. Le PB, Lei H (2019) Determinants of innovation capability: the roles of transformational leadership, knowledge sharing and perceived organizational support. J Knowl Manag
46. Lee R (2020) An analysis of structural relationship between technological innovation capability, collaboration and new product development performance in small & mid-sized venture companies. Asia-Pacific J Bus Ventur Entrep 15(1):185–195
47. Lei H, Leaungkhamma L, Le PB (2020) How transformational leadership facilitates innovation capability: the mediating role of employees' psychological capital. Leadership Organization Development J
48. Liao SH, Hu DC, Shih YS (2018) Supply chain collaboration and innovation capability: the moderated mediating role of quality management. Total Qual Manag Bus Excel 1–19
49. Liu CH (2017) Creating competitive advantage: Linking perspectives of organization learning, innovation behavior and intellectual capital. Int J Hosp Manag 66:13–23
50. Lubis FR, Hanum F (2020) Organizational culture. In: 2nd Yogyakarta international conference on educational management/administration and pedagogy (YICEMAP 2019). Atlantis Press, pp 88–91
51. Masum AKM, Beh LS, Azad MAK, Hoque K (2018) Intelligent human resource information system (i-HRIS): a holistic decision support framework for HR excellence. Int Arab J Inf Technol 15(1):121–130
52. McAfee A, Brynjolfsson E (2012) Big data: the management revolution. Harvard Bus Rev 90:60–68
53. Migdadi MM, Zaid MKA, Yousif M, Almestarihi RD, Al-Hyari K (2017) An empirical examination of knowledge management processes and market orientation, innovation capability, and organisational performance: insights from Jordan. J Inf Knowl Manag 16(01):1750002
54. Moussa NB, El Arbi R (2020) The impact of human resources information systems on individual innovation capability in Tunisian companies: the moderating role of affective commitment. Eur Res Manag Bus Econ 26(1):18–25
55. Müller JM, Buliga O, Voigt KI (2020) The role of absorptive capacity and innovation strategy in the design of industry 4.0 business models-a comparison between SMEs and large enterprises. Eur Manag J
56. Nagendra A, Deshpande M (2014) Human resource information systems (HRIS) in HR planning and development in mid to large sized organizations. Procedia-Soc Behav Sci 133:61–67
57. Najafi-Tavani S, Najafi-Tavani Z, Naudé P, Oghazi P, Zeynaloo E (2018) How collaborative innovation networks affect new product performance: product innovation capability, process innovation capability, and absorptive capacity. Ind Mark Manag 73:193–205
58. Oc B (2018) Contextual leadership: a systematic review of how contextual factors shape leadership and its outcomes. Leadersh Quart 29(1):218–235
59. Oktari RS, Munadi K, Idroes R (2020) Knowledge management practices in disaster management: systematic review. Int J Disaster Risk Reduct 101881

60. Qadir A, Agrawal S (2017) Human resource information system (HRIS): re-engineering the traditional human resource management for leveraging strategic human resource management. MIS Rev 22:41
61. Quandt CO, Silva HDFN, Ferraresi AA, Frega JR (2019) Idea management and innovation programs: practices of large companies in the south region of Brazil. Int J Bus Innov Res 18(2):187–207
62. Rahman MA, Islam M, Qi X (2017) Barriers in adopting human resource information system (HRIS): an empirical study on selected Bangladeshi garments factories. Int Bus Res 10(6)
63. Rajapathirana RJ, Hui Y (2018) Relationship between innovation capability, innovation type, and firm performance. J Innov Knowl 3(1):44–55
64. Reckwitz A (2018) The invention of creativity: modern society and the culture of the new. Wiley, Hoboken
65. Remawati D, Harsadi P, Nugroho RD (2020) PenerapanSistemPenunjang Keputusan MenggunakanAlgoritma Naive Bayes Pada konsep Human Resource Information System (HRIS) (Studikasus: PenerusanKontrakKerjaKaryawan di PT. XYZ). JurnalIlmiah SINUS 18(1):63–74
66. Rudolph CW, Rauvola RS, Zacher H (2018) Leadership and generations at work: a critical review. Leadersh Quart 29(1):44–57
67. Santoro G, Vrontis D, Thrassou A, Dezi L (2018) The Internet of things: building a knowledge management system for open innovation and knowledge management capacity. Technol Forecast Soc Chang 136:347–354
68. Sartori R, Costantini A, Ceschi A, Tommasi F (2018) How do you manage change in organizations? Training, development, innovation, and their relationships. Front Psychol 9:313
69. Schulz M (2017) Organizational learning. In: The Blackwell companion to organizations, pp 415–441
70. Shahreki J, Ganesan J, Raman K, Chin ALL, Chin TS (2019) The effect of human resource information system application on employee satisfaction and turnover intention. Entrep Sustain Issues 7(2):1462
71. Shahzad F, Xiu G, Shahbaz M (2017) Organizational culture and innovation performance in Pakistan's software industry. Technol Soc 51:66–73
72. Silva JJD, Cirani CBS (2020) The capability of organizational innovation: systematic review of literature and research proposals. Gestão Produção 27(4)
73. Singhry HB (2015) Effect of supply chain technology, supply chain collaboration and innovation capability on supply chain performance of manufacturing companies. J Bus Stud Quart 7(2):258
74. Soliman MSM, Noorliza K (2020) Explaining the competitive advantage of enterprise resource planning adoption: insights Egyptian higher education institutions. J Inf Technol Manag 12(4):1–21
75. Song W, Yu H (2018) Green innovation strategy and green innovation: the roles of green creativity and green organizational identity. Corp Soc Responsib Environ Manag 25(2):135–150
76. Srivastava S, Bajaj B, Dev S (2020) Human resource information system adoption and implementation factors: a theoretical analysis. Int J Hum Capital Inf Technol Prof (IJHCITP) 11(4):80–98
77. Sulistyo H, Ayuni S (2020) Competitive advantages of SMEs: the roles of innovation capability, entrepreneurial orientation, and social capital. Contaduría y administracíón 65(1):10
78. Taruwona GA (2017) The application of information and communications technology, (ICT) for the purposes of human resource management in the Ministry of Education: an exploratory study. Doctoral dissertation, University of Namibia
79. Ukko J, Saunila M, Parjanen S, Rantala T, Salminen J, Pekkola S, Mäkimattila M (2016) Effectiveness of innovation capability development methods. Innovation 18(4):513–535
80. Wandhe P (2020) A Role of Effectiveness of Human Resource Information System (HRIS) in 21st Century. Available at SSRN 3718247
81. Wang F, Zhao J, Chi M, Li Y (2017) Collaborative innovation capability in IT-enabled inter-firm collaboration. Ind Manag Data Syst

82. Wibawa JC, Izza M, Sulaeman A (2018) HRIS (human resources information system) design for small for micro, small and medium enterprises. In: IOP conference series: materials science and engineering, vol 407, no 1. IOP Publishing, p 012134
83. Wibowo YC, Christiani N (2020) The effect of affective organizational commitment towards innovation capability and its impact to job performance in family business. J Entrep Entrepreneurship 9(2):99–110
84. Yeh CC, Ku EC (2019) Process innovation capability and subsequent collaborative team performance in travel planning: a knowledge exchange platform perspective. Curr Issues Tour 22(1):107–126

Exploring a Broadband Marketing Strategy to Build Customers' Relationship Management: Buzz Marketing Perspective. A Case Study of Virgin Broadband in Stafford-Shire University

Hyder Kamran, Mudassar Mahmood, and Sherine Badawi

Abstract UK Broadband Industry has become highly competitive of late with the advent of numerous broadband service providers. Different service providers follow different marketing strategies to woo customers. Virgin has acknowledged the growing importance of these young consumers and reaped in huge benefits by targeting them by better marketing and formulating strategy that is based on the needs and wants of university students. Although lot of studies have been done on marketing strategies for young customers, there is still a gap in the understanding of consumer behaviour dynamics of a university student and his choice of broadband service provider. The researchers strive to analyse the marketing strategy of Virgin vis-à-vis University student's needs and requirements. The research reviewed current literature, questionnaire surveys and interviews with University students, Virgin Sales staff and Sales managers. To reinforce the findings, an online survey has been undertaken to add rigor to the argument. The findings show that young customers like university students need to be treated differently rather than as normal residential customers. A young customer also brings life time value and increased business with word-of-mouth marketing. An excellent marketing strategy can be a source of competitive advantage for Virgin in a market that has become highly competitive of late.

Keywords Marketing strategy · Buzz marketing · CRM

H. Kamran · M. Mahmood
College of Business, University of Buraimi, Al Buraimi, Oman

S. Badawi (✉)
College of Business and Finance, Ahlia University, Manama, Bahrain
e-mail: sbadawi@ahlia.edu.bh

A. M. A. Musleh Al-Sartawi (ed.), *The Big Data-Driven Digital Economy: Artificial and Computational Intelligence*, Studies in Computational Intelligence 974,
https://doi.org/10.1007/978-3-030-73057-4_31

1 Introduction

According to UK National Statistics Office, virtually all adults aged 16–44 years in the UK were recent internet users (99%) in 2019. The major players in the broadband industry in UK are BT, Virgin and Sky and with the entry of different players like Vodaphone, Tiscali, O2, Pipex, Orange and Carphone Warehouse, UK broadband industry has become highly competitive in the last couple of years. At the backdrop of such highly competitive market, it has become very important for firms to understand the dynamics of consumer behaviour and more important, to formulate a successful marketing strategy that shall bring competitive advantage to the firm in this cut-throat competitive arena. A marketing strategy should be centred around the key concept that customer satisfaction is the main goal [1]. In UK, telecom service providers like BT, Virgin, Orange, Pipex, O2 and others follow different marketing strategy to attract customers. One of the groups that have heavily fuelled the growth of internet usage is university and college students all over UK. The total number of students starting university in 2017–18 was 353,960 (UCAS 2018 as cited on BBC 2018). Interestingly, not many service providers have been targeting this group as most of companies have wider target groups when it comes to marketing strategy. One of the potent effects of targeting this university student group is 'buzz marketing' or 'word of mouth marketing' where one happy user recommends the service provider to his/her peers in university and thus it brings more and more customers to the company. It is to be noticed that in this age group, peer recommendation plays crucial role in decision making when it comes to purchasing broadband. Virgin is one of the companies that have taken this age group into consideration and it has a focused marketing strategy targeting this age group in universities and thus attracting lot of young customers that respond to its successful marketing strategy.

Staffordshire University is one of the most thriving universities of UK and a popular educational institution in midlands of UK. The university currently has 14,910 students registered at its campuses located at Stoke-on-Trent, Stafford and Shrewsbury. There are also a huge number of international students. Since Staffordshire University is located very close to Stoke-on-Trent city centre, there is wide range of broadband service providers like BT, Virgin, Sky, O2 and others available to the students at the university. As soon as university term starts, there is a mad rush for student accommodation in and around Staffordshire University and along with it comes the requirement of a broadband connection for the students at their respective on and off campus accommodation. Virgin has been successful in reading the pulse of students and every year Virgin set up its stalls in the campus and in city centre to attract this vast number of students that add to their increasing customer base. The successful marketing strategy of Virgin to bring more and more students under its banner and reaping enormously from the 'buzz marketing' that it generates.

The research critically analyzes the marketing strategy of broadband providers, especially Virgin and how do they cater unique needs of students in university. The marketing strategies of companies aimed to attract such students. The research is carried out at Staffordshire University and since Virgin company is a popular choice among the university students, the research has kept the focus of study on Virgin's marketing strategy for university students. The research aims to fulfill a general overview of broadband providers and their marketing strategies in UK, explore the unique needs and requirements of broadband usage of university students, critically analyze marketing strategy of Virgin in context of university students, and identify the importance of unique marketing strategy of Virgin and how it helps create brand awareness by buzz marketing as well as contributing to the areas of one-to-one marketing, and CRM.

2 Literature Review

2.1 *Broadband Service Providers in UK and Their Marketing Strategies*

There are now nearly 28 million broadband connections in the UK, recently-released figures from Ofcom show. The broadband providers have followed different marketing strategies to promote their broadband products. More or less, all of them promote broadband as a complete package tied up with other products, popularly called 'bundle' package. BT, for instance, promotes broadband as a 'BT Total Broadband', complete package along with its phone line (www.bt.com). Sky, on the other hand, promotes broadband alongside its TV package that remains its strong point in market as there are more than 8 million Sky TV subscribers in UK (www.sky.com). In a nutshell, all service providers despite having different target groups aim to get bigger share in UK broadband market by deploying different marketing strategies, some even offering now free laptops with broadband connection (www.pcworld.co.uk). Due to such extensive marketing strategies to reach out to maximum customers, broadband industry is thriving and broadband users are growing fast.

As it can be seen that UK broadband market is growing steadily, broadband service providers need to keep assessing their marketing strategies so that more and more customers join their respective companies. In the context of UK economy, Virgin had formulated an effective marketing strategy, not only for wider target groups but also for university students that form one of the most active broadband user groups. The marketing strategy of Virgin fits well in the philosophy of Kotler and Keller [9] who stated that marketing involves satisfying customers' needs and wants.

2.2 *Buzz Marketing*

One instant benefit for companies of targeting such young customers in university is 'word of mouth marketing'. Word-of-mouth (WOM) is communication about products and services between people who are perceived to be independent of the company providing the product or service, in a medium perceived to be independent of the company. These communications can be conversations, or just one-way testimonials. Such phenomenon can be seen and established between students in campus where they share with their peers the valued feedback of products they are using. "Word-of-mouth is thousands of times more powerful than conventional marketing" [13]. According to Yang [18], word-of-mouth marketing is more credible than other marketing techniques because only 14% of people believe what they see, read or hear in advertising. Surprisingly, 90% of the folks will believe their family, friends, or colleagues who endorse a service or product because they know they do not have a vested interest in it.

Buzz is all the word of mouth about a brand. It's the aggregate of all person-to-person communication about a particular product, service, or company at any point in time. Companies, like Virgin, that can generate and sustain positive buzz, can enhance the success of their products without the often-large expenses of traditional marketing strategies. In the middle lies the realm of invisible networks of interpersonal connections among consumers. As the world is waking up to this potent tool of marketing, most of the companies, including Virgin, are trying to reap the benefit of this phenomenon. The emphasis here is on the active role of the marketer in stimulating people to talk about his or her product.

2.3 Customer Relationship Management of University Students

Customer relationship management, is imperative to understand its three components; customers, relationships and their management. Barnes (2014) argues that if we were to truly understand the word *relationship* in our modern approach to business building, and in the context of long-term customer loyalty, then it is critical that enterprises understand what relationships are all about. The students' needs to feel that company is understanding their unique needs and status and thus reaching out to them with flexible offer, rather than treating them same like every other customer. It is this 'special status' that students are searching for unconsciously while deciding broadband service providers for their university term. It is important to be highlighted that in this age group (17–25 years), peer recommendation plays an important part in decision making and satisfied customers provide psychological assurance to his or her peers about the product [9].

For Virgin, a satisfied customer may just not take a higher priced broadband package, but he or she may also be interested in other related services like Virgin's television package or even Virgin mobile phone service. Szymanski and Henard's [16] metaanalysis show that satisfaction has a positive impact on the self reported customer loyalty. Virgin has put in extra efforts to render such satisfaction to its young customers in university. The sales staff at the stalls set up in university campus show lot of interest in students' individual needs and requirements. Studies have linked the performance of frontline employees, like Virgin Sales Staff at university stalls, with such factors as customer satisfaction, customer loyalty, organization performance, and strategy emergence [2]. A happy and satisfied customer that Virgin finds in a young student carries a high probability of brining life time value as a customer to Virgin. As many studies [5, 12] have shown that a satisfied customer remains loyal for long period of time and adds to brand equity, Virgin is doing the right thing by catering the needs of students, delivering excellent service and satisfaction and thus increasing its chance of getting a long time loyalty from students even when they leave university and move on in their careers.

3 Data Collection Methodology

The research analyses the marketing strategy of Virgin for university students and consumer behaviour of such students from the prism of marketing management. The research therefore covers the emotional component as well as the physical component of the study. The primary aim of this research is to critically analyze the marketing strategy of Virgin for university students and how it creates buzz marketing. In order to fulfill this objective, the researchers attempt to answer the following questions: *Question 1—Is a successful marketing strategy understands individual needs of customers? Question 2——Is Buzz marketing a potent marketing tool for university students? Question 3—Is a satisfied young customer brings life time value of customer to the company?*

3.1 Research Design

The foremost action while performing research is to evaluate the research strategies. Since the testing of research questions was done with university students and sales staff of Virgin, both quantitative and qualitative techniques have been used during the data collection. Qualitative researches typically rely on four methods for gathering information: (1) participation in the setting, (2) direct observation, (3) in depth interviews, and (4) analysis of documents and materials [10]. This research has tried to put balance of both positivism and interpretivism in the approach. Qualitative research allowed for obtaining more in-depth information about why students choose Virgin as their service providers and why they recommended to their peers. Virgin Sales staff has their own view-points, perceptions plus their own competencies while handling students so it is evident to see differences in the view of respondents. Thus, researcher used qualitative methods to solicit information from the respondents on the all possible aspects of a successful marketing strategy and its importance from CRM point of view to get larger picture of things. The researchers believe that the interviews that have been conducted provide both validity and reliability. The interviewees were well informed about different choices of broadband service providers, were full time students at University and intend to stay in university for 2–4 years. Virgin Sales team have been involved with the sales campaign for university students for couple of years and had good grasp of the existing marketing and CRM practices and challenges both. This satisfies the requirement for validity.

In order to collect the information related to research topic, it was imperative to choose the right medium of data collection. The researchers availed the opportunity of creating an online questionnaire, whereas one-to-one interviews were at Stoke-on-Trent campus of Staffordshire University. The goal of the interviews was to obtain in depth information about how the students perceive different broadband providers, how do they choose and recommend Virgin to their peers as well.

3.2 Data Analysis

The researchers took grounded approach for data analysis where interpretations of the findings were carried out. A total of 52 questionnaires were given to students and virgin sales staff. Respondents of this research include students of both under and post graduate levels and sales team of Virgin at Stoke-on-Trent. A total of 50 questionnaires were returned and used for analysis, represented response rate of 96.15%. There were interviews and discussions with the Virgin Sales managers in order to get a better perspective on the issue.

4 Research Findings and Discussion

The research explores the results of the survey questions and the interviews with the university students, Virgin sales advisors and managers in accordance with the methodology set out in the previous section. Aimed to understand the marketing strategy of Virgin in particular reference to Staffordshire University students and to answer the research proposed questions.

Question 1 – Is a successful marketing strategy understands individual needs of customers?

The findings that pertain to this question were extracted from the results of the questionnaire survey. Findings from the survey questionnaire showed, a successful marketing strategy puts customer's perspective first and view things from customer point-of-view. Since needs and wants of a customer is dynamic in nature and keeps changing with the time, a company's marketing strategy should be able to meet customer's expectation and it can be achieved by learning more and more about the customer with each interaction. The findings suggest that Virgin has considered the unique needs and requirements of university students rather than treating them at par with normal house-hold customer. The findings also reflected that once student feels cared and valued for and realizes a 'special status' (as discussed by John Czepiel [6]) in the customer relationship, he is more than happy to recommend Virgin to his peers around in the campus. Thus, this hypothesis is tested positive and the success of a marketing strategy depends upon its ability to understand subtle needs and requirements of customers as seen in Virgin's successful strategy of marketing broadband in Staffordshire University.

Question 2 – Is Buzz marketing a potent marketing tool for university students?

This question was strongly agreed by most of the students as well as Virgin sales staff that Word-of-mouth and Buzz marketing has become a potent tool of marketing Virgin broadband in the university campus. It was also evident in the findings that with so much to choose from the market and advertising clutter all around university campus, students are more receptive to WOM and thus forming positive perception about Virgin broadband after sharing enthusiastic experience of their peers who have been using Virgin broadband already.

The ability of WOM to operate within a consumer network (like student community in Staffordshire University) appears to be influenced by the tie strength, or the intensity of the social relationship between consumers [4, 7]. Considering these previous studies, researches and the findings of the university students, this question provided agreeableness. WOM and Buzz marketing are accepted to be the potent tools of marketing for university students. However, testing ROI is still an issue with successful WOM campaigns in the industry.

Question 3 – Is a satisfied young customer brings life time value of customer to the company?

The findings revealed that many students, who are happy and satisfied by Virgin broadband, showed inclination not only to use additional services from Virgin like TV and mobiles but also expressed desire to stay with Virgin even after completing university term. Since needs and wants of a customer is dynamic in nature and keeps changing with the time, a company's marketing strategy should be able to meet customer's expectation. Since many students also agreed that Virgin has been futuristic in terms of bringing cutting edge technologies through its high speed (50 MB) broadband over its optical fiber cable network, they showed inclination of staying with Virgin rather than switching over providers who they think are lagging behind. As evident from the Relationship Life Cycle concept [2], the intensity of relationship is directly proportional to the duration of customer relationship and if the changing needs and wants of customer are not met with the passage of time, customer may enter the imperilment and dissolution phase where customer begins to mull over the idea of not using company's products and may consider switching over to competitors. But with Virgin, since they are already the leaders in providing high speed broadband and integrated web services, the changes of young customers staying loyal with Virgin are high. It can be safely said that a satisfied young customer does bring the life-time value to the company like Virgin but having said that, the author feels that this can be the area where competitors can catch up by upgrading their technologies and imitating services thus nullifying the competitive advantage.

5 Conclusion and Recommendations

Word of mouth (WOM) is becoming increasingly recognized as an important form of promotion, particularly within professional services environments, where credence qualities play a critical role in consumers' choices. The research paper strived to explore the successful marketing strategy of Virgin in Staffordshire University campus and how it triggers WOM in the campus. In an environment in which there has been a reduction in consumer trust of both organizations and advertising, as well as a decrease in television advertising, word of mouth (WOM) offers a way to obtain a significant competitive advantage [18]. However, there has been very little research into this important topic, especially in context to student communities like university campus. Further, little past research has focused on what happens when WOM is received [17] or on the conditions in which WOM will be most effective in enhancing a receiver's perceptions or actions. The important role WOM plays has been long recognized by diffusion of innovation researchers and has been acknowledged as the most important communication source between consumers [3].

The author made an attempt to test effectiveness of Virgin broadband marketing strategy in Staffordshire University and how it created buzz in the campus, thus attracting more and more customers to Virgin. It also analysed how Virgin treat students and considers their unique requirements and doesn't put them at par with other residential customers. While the potential power of WOM as a form of promotion is generally accepted, it is important to understand that the generation of positive WOM is not sufficient for it to be an effective source of communication; the recipient also needs to react positively to the WOM. The author made an attempt to test this further at Staffordshire University campus and explored how students reacted to the WOM generated by a successful Virgin broadband marketing strategy targeting university students.

Virgin Acknowledges University Student's Unique Customer Requirements - Virgin has been successful in forging a relationship with university students by understanding their unique needs and requirements. In Staffordshire University most of the students tend to stay for a time period of one to three years. Most of the Master's degrees are of one year and so are many programmes. Only bachelor degrees are of longer period. In this period, most of the students tend to move houses within on and off campus accommodation. They need flexibility in terms of broadband contract length, home-move and even changing packages and ownership of contracts. Since most of students share accommodation, changing ownership of contract is a regular feature. Also students don't want to pay connection fee or even installing a telephone landline to get broadband, as most of them have their own mobile phones. In some cases, the standard connection charge of £122.33 for a landline is a put off for students. So any company who has to target the students has to consider all these factors in formulating its marketing strategy and customer relationship management. So, while comparing the marketing strategies of other broadband service providers, first round goes to Virgin as it has definitely created a competitive advantage by treating university students as separate entity rather than putting them at par with residential customers which most of companies do. As evident in the survey, most of the students preferred Virgin since it is flexible in its terms and conditions and accommodate students' requirements in its scheme of things. The fact that Virgin can provide broadband without a hefty connection charge of £122.50 for landline, move accommodations and ownership of contract without much hassle and also award incentive for recommending a friend makes it a popular choice among Staffordshire University students. It is evident in findings that one-to-one relationship marketing is the need of the hour, especially when UK broad-band industry is growing highly competitive. The most successful company will not be the one with the most customers, but the one that has the most knowledge about its individual customer's needs. Another myth that has been exploded from the research is that in this age of one-to-one marketing, market segmentation is no longer meaningful. In fact, customer relationship strategy begins with segmentation and ends with the individual customer, for example, students are classified as residential customers but have individual needs that should be addressed in formulating successful marketing strategy.

Virgin Generates Positive WOM - One major benefit of having such large army of young, satisfied and happy virgin customers among student community is the WOM that it triggers in the campus. Staffordshire University student community is highly interactive and good word about broadband service spreads very quickly. Since in this young age-group, peers play a crucial role in decision making process (Kotler and Keller, 2006), most of students trust their friend's recommendation and take Virgin broad-band as a result. In the survey and interviews with university students, WOM outcomes were investigated from a receiver's point of view. Overall, positive messages led to a sense of enthusiasm, confidence and optimism in the students and they felt more inclined to use Virgin broadband as a result. Positive WOM also led to an improved opinion of Virgin. Even those students who were not using Virgin broadband accepted the growing popularity of Virgin in the campus and said they might switch over to Virgin in future after seeing so many happy users in their friend circle. In some cases however, few students were somewhat doubtful of the credibility of the WOM. Some students mentioned that one of reasons why a friend would recommend Virgin broadband is that they get £50 incentive and they suspected that it can be primary motive rather than enthusiasm of sharing a good service with peers. It is worth mentioning that the author observed that, WOM was more effective when there was a close relationship and a good rapport between a sender and a receiver that was based on trust and mutual respect (for example, students residing in same accommodation or classmates trusted WOM from each other more than from other students). These characteristics are similar to the factors associated with a sender's opinion leadership role [11] and are likely to impact on the effectiveness of WOM communication.

Overall, it is evident from the findings of survey and interviews that Virgin broadband did create a WOM and also benefits from the buzz that it created in Staffordshire University campus. By formulating a successful marketing strategy that considers the unique requirements of the students, Virgin has been able to relate with needs and wants of university students and thus attracting them as customers. This bunch of young happy customers then in turn created a positive image of Virgin in university campus and by the influence of WOM more and more students chose Virgin broadband thus created a 'ripple' effect. Virgin has reaped enormous benefits from this phenomenon and has taken a clear lead over its competitors like BT, Sky, O2 and others in terms of serving student community. From here, the survey brought out three statements which has long term implications for success of marketing strategy of Virgin.

1. *A successful marketing strategy understands individual needs of customers.*
2. *Buzz marketing is a potent marketing tool for university students.*
3. *A satisfied young customer brings life time value of customer to the company.* Virgin should carry on the good work and by understanding the needs and wants of customers on individual basis, it will definitely become leaders in broadband market in.

UK.

Recommendations

It is evident that Virgin has already done an excellent job by giving 'special status' to university students and accommodating their requirements in their scheme of things and marketing strategy. It is recommended that Virgin carry on with this marketing strategy and to keep an eye on latest trends and fads as needs and wants are dynamic in nature, especially of young students who are more influenced than others by latest technologies and its applications in day-to-day life. Virgin should also consider hiring youth icons and young celebrities as their brand ambassadors with whom young students can relate with.

Virgin has successfully created a positive WOM in university campus but it was noticed that WOM is not homogeneous in its impact as people vary in how they react to WOM. Hence, an understanding of the factors that enhance WOM effectiveness would help marketers like Virgin determine how to better harness WOM as a promotional and relational tool.

References

1. Baker M (2014) Marketing strategy and management, 5th edn. Palgrave, London
2. Bruhn M (2003) Relationship marketing management of customer relationship. Prentice-Hall, Harlow
3. Derbaix and Vanhamme (2003) Marketing and Consumer Behavior: Concepts, Methodologies, Tools and Applications. Business Science Reference. IGI Global
4. Gerpott TJ, Rams W, Schindler A (2001) Customer retention, loyalty and satisfaction in German mobile cellular telecommunication market. Telecommun Policy 25(4):249–269
5. Hansemark Ove C (2004) Customer satisfaction and retention: the experience of individual employees. J Mark 14:40–57
6. Czepiel J (1990) Service encounters and service relationships: implications for research. J Bus Res 20:13–21

7. Kim HW, Kim YG (2001) Rationalizing the customer service process Bus Process Manag J 7(2)
8. Kotler P, Keller K (2006) Marketing Management, 12th edn. Prentice Hall, London
9. Kotler P, Keller K (2016) Marketing Management: 15th ed., Pearson
10. Marshall C, Rossman GB (1998) Designing qualitative research. Sage, Thousand Oaks
11. Mazzarol et al. (2007) New prespectives in Marketing by word of mouth. Emerald
12. Reinartz, WJ, Kumar V (2002) The mismanagement of customer loyalty. Harvard Bus Rev 80, 86–94
13. Silverman D (2011) Interpreting Qualitative Data, 4th edn. Sage Publications, London
14. Fournier S (1998) Consumers and their brands: developing relationships theory in consumer research. J Consum Res. 343–373
15. Szymanski and Henard's (2008) Citizen Relationship. A Study of CRM in Government. Peter Lang.
16. Walters D, Lancaster G (1999) Using internet as a channel for commerce. Manag Decis. 37(10):800–817
17. Winer RS (2001) A framework for customer relationship management. California Manag Rev 43:89–108
18. Yang FX (2017) Effects of restaurant satisfaction and knowledge sharing motivation on WOM intentions: the moderating role of technology acceptance factors. J. Hosp. Tour.

E-procurement Significantly Affects Supply Chain Performance

Ali M. Albinkhalil and Razzaque Anjum

Abstract The advent of the internet has revolutionized the procurement's sourcing processes in firms among multiple sectors and industries. It has stimulated cost-effective ways besides various innovative applications that enterprises could leverage from it. In this paper, we are going to spot the light on the main factors which are impacting the supply chain performance. These factors are E-procurement, Supplier relationship and supplier integration. The literature review will discuss the benefit of automating procurement processes and how it will add value to the organization, followed by the need for strengthening the relationship between the organization and their key suppliers, especially for complex and bottleneck products. Finally, supplier integration to which extent it will assist the organization to be adaptive to the unforeseen situation in the external environment. This paper will examine the impact of the E-procurement on the performance of the supply chain and discusses the main benefits and challenges that the organization may face during its implementation. Moreover, to demonstrate that supplier relationship contributes to influencing supply chain performance in a positive way. Furthermore, to analyze the perception of business users.

Keywords Electronic procurement · Supply chain management · Supplier relationship · Supplier Integration · Supply chain performance

1 Introduction

Business is all about the commercial transaction. These transactions result in the essential revenue that businesses need to survive in the market. Procuring of services and product is a vital mission of any organization; it can directly influence an organization's profitability [1]. Purchasing the right goods and services at the right quality

---

A. M. Albinkhalil
Bahrain National Gas Company, Sakhir, Bahrain

A. Razzaque (✉)
Ahlia University, Manama, Bahrain

A. M. A. Musleh Al-Sartawi (ed.), *The Big Data-Driven Digital Economy: Artificial and Computational Intelligence*, Studies in Computational Intelligence 974,
https://doi.org/10.1007/978-3-030-73057-4_32

on the right time with the right price and deliver it to the right place are the main target of procurement department [2] The evolution of the Internet has revolutionized the sourcing and procurement processes in organizations in every sector and has stimulated cost-effective ways besides various innovative applications that enterprises could take advantage of. Nowadays, many businesses application is using information technology (IT) as a way of procurement for B2B, B2C, which is also known as E-procurement represents the flow of information among internet base networks, at all stakeholders in the "supply chain," whether within the enterprise, among businesses or between (B2C). Simply, supply chain operation technologies can be described as the use of ICT to enhance the output, operational routine, and logistics activities in the supply chain channel. The productivity involves improvement of normal working processes responsible for management of the capacity and inventory.

E-procurement can be described as the automation of the firm's procurement functions using web-based applications. E-procurement refers to the procure of materials and services to use in the organization [3, 4]. Procurement considers as the largest expense goods in an organization's cost structure [5]. Many firms use e-procurement for obtaining contracts to attain advantages such as increase the efficiency of their employees and save costs by getting quicker and cheaper goods and services. In addition, it minimizes corruption and improves transparency in procurement services through the employees and manager in the firms. In earlier days, procurement transactions were processed manually. Thus, it takes a long time to accomplish a single task [6], as the buyer drafts request for quotations (RFQ), prepare purchase requests, obtain management endorsement, and issue purchase orders in manual ways. While commercial transactions and communications between parties have occurred manually in earlier days, these transactions' mechanisms have evolved rapidly, especially with the advent of the internet, computers, and other communication technologies. Procurement process have had re-engineered and restructured with the advent of the internet; thus, business around the world tend to implement "e-procurement" to improve the management of their supply chain and to streamline the procurement functions [7, 8].Information technology (IT) has an important role in evolving supply chain management (SCM). Hence, an ideal system that helps to aggregate of scattered information and sharing information through applications in secure and flexible manner is highly desirable [9, 10].

In 1970, the available technologies used for supply chain were finite such as customized forms, innovative warehouse management techniques, and straightforward information processing procedures. In the 1980s, the most significant technological innovation has arrived, such as electronic equipment able to read the bar code such as, optical scanner, which is customized for each item. Enters the bar code for each item into the store server database and establishes the current cost of that product. In the 1990s, improved inventory cost-saving, and customer service attained through cooperation among supply chain partners. For instance, invest in supply chain technologies, for example, EDI "Electronic data interchange", "E-commerce", "Vendor Managed Inventory" VMI. In the 2000s, cooperation among supply chain partners increased by adapting the "Enterprise Planning system" (ERP). Which in turn lead to additional developments in the organization such as "warehouse management system" WMS, "Transportation management system" TMS, "Customer relationship management" CRM, and "Vendor Relationship Management" VRM; these systems result in improving the accuracy of forecasting on consumer-demand and has changed the way of interaction between supply chain partners. E-procurement has been attracted more attention from industries and individuals. In Nowadays global business scenario, the use of recent technology in organizations has gained immense importance. As a result, organizations are obliged to restructure their business process from traditional manners to electronic platforms [3].

2 Literature Review

2.1 E-procurement

The procurement function is the "process of acquiring goods, works, and services, covering both acquisitions from third parties and in-house providers" [11], It has an influential role in attaining cost competitiveness. In most manufacturing firms, the cost of materials varies between 40–70% of the cost of goods sold. The cost of holding and handling material could be substantial [12, 13]; they have stated that the estimation price of preparing a single purchase order could vary between 50$ to 200$. Furthermore, the transaction cost could be higher if the products procured through the tendering process. These extra costs could appear in inviting suppliers to participate in the specific tender, evaluating suppliers' technical capabilities, preparing items or contract specifications, technical and commercial evaluation of the received offers and selecting the winner bidder, and other related activities could incur very high costs.

In earlier days, procurement transactions processed manually, resulting in slow transaction accomplishment and even slow handling of the whole procurement processes. However, nowadays, firms become more conscious of the way they react to different influences and pressure in the procurement department. Earlier, before the launch of "e-procurement", Usually, strategic procurement interact with routine administrative work, for example, turning purchase requisition to purchase order, guarantee the proper distribution of the received invoices for each transaction. Simultaneously, the strategic sides are usually ignored in the process, where the buying organization has a slight impact over suppliers' selection and required items. Multiple and frequent changes in business advancements and technological shifts have increased the level of competition in the industry. Hence, it has increased the need for innovation as an inevitable condition for survival and growth in an intensively competitive environment. These advancements are realized in cost improvements, process improvements, new markets, customer satisfaction, and technological competence [14, 15] Thus, the use of web-based technologies in procurement aims to boost procurement processes and to be more efficiently, by allowing buyers to focus on strategic tasks [16]. E-procurement is business-to-business "B2B" purchasing process through electronic platforms. It has been used to facilitate, manage, and report on the enterprises purchasing functions. E-Procurement system assists managers in the decision-making stage by providing all related information up to date and organized. Nowadays, many firms worldwide strive for e-procurement implementation to digitize their operations, which will allow the organization to improve their efficiency and effectiveness [17, 18].

2.2 Definition of E-Procurement

E-procurement defines as a comprehensive procedure in which organizations utilize information technology system to establish a contract to procure items or services or to purchase products or services for instant payment "Purchasing" [6]. E-procurement used to streamline the process between organizations in procurement transactions [17] Likewise, [19] described e-procurement as "the use of information technology to facilitate business-to-business purchase transactions for materials and services." From both diffusions we realize that: E-procurement is not just online purchasing process system. However, it considers as a communication tool between organization and it suppliers. Boer et al., 2002 defined that e-procurement is "the application of e-commerce in procurement, It involves the use of various forms of Information Technology (IT), such as electronic mail (e-mail), Electronic Data Interchange (EDI), and electronic marketplace (e-marketplace), to automate and streamline the procurement process in business organizations." Simultaneously, the [20] describes e-procurement system as "I have creating private web-based that automate communications, transactions, and collaboration between supply chain partners".

2.3 E-Procurement Functions

E-procurement involves activities such as "Enterprise Resources Planning" ERP; "E-maintenance, Repair, and Operations" E-MRO; E-tendering, E-reverse auction; E-marketplaces; E-sourcing, and E-informing [21]. E-procurement consists of e-catalogue, e-dossiers, e-submission, e-signature, e-auction, and e-notice [21]. [22] emphasized that e-tendering, e-evaluation, e-negotiation, e-informing e-design, and e-sourcing are the prime functions through which e-procurement participates in supply chain performance. The automation of these purchasing processes by leveraging information technologies such as EDI, Internet, and ERP will help the organization's linking their material management system with suppliers. E-Procurement system consists of six essential processes: e-design, e-sourcing, e-negotiation, e-evaluation, e-tendering, and e-sourcing [16].

- E-design: refer to "setting of purchasing requirements on an online procurement system" [23]. E-design enables vendors for early involvement to develop product specifications and come up with new ideas by utilizing their expertise.
- E-Informing refers to "gathering and distributing of purchasing information from both internal and external parties using Internet technologies" [14, 15].
- E-negotiation refers to "entering into a negotiation between business partners regarding a specific subject by using the internet" [24]. Thus, the primary purpose of e-negotiation is to make a significant cost saving for the organization in their purchasing of goods and services through the internet.
- E-evaluation refers to "the stage when there is extensive information about vendors performance already collected for further evaluations and transactions via the Internet" [23]. [24] indicated that organizations that have implemented e-procurement tools should evaluate and evolve their purchasing process to attain their desire goals and benefits.
- E-tendering is "the process of sending requests for information and prices to suppliers and receiving the response through internet-based technology [6]. E-tendering, according to [5], refers to "the electronic integration and management of all supply chain activities inclusive of the purchase requests, authorization, ordering delivery, and payment between a supplier and the buyer".
- E-sourcing refers to the process of searching for new capable vendors using internet technologies to minimize search costs [25]. It merely uses the internet to support the process of sourcing, involving expenditure analysis, aggregation of demand, precise requirement specification, sourcing of supplier, negotiations, reverse auctions, evaluation, and analysis of offers, and contract management.

2.4 E-Procurement Features

The application of e-procurement functions has several advantages to the organization, such as fast accomplishment of orders, cost reduction, reduce purchasing cycle time, better budgetary control, avoid of administrative errors. IT maximizes the productivity of buyers and lowering the process through consortium and consolidation of orders by utilizing buyer purchasing power, and better information management [6].

E-procurement increase geographical outreach [26]; it allows building a new relationship with new vendors and provides better control over supplier relationships [6]. The benefits of e-procurement have been proved in several research [27], according to these research, e-procurement allows organizations to operational decentralized procurement processes and functions and centralize strategic processes due to the higher transparency offered by the e-procurement system. For example, in Africa, they require eliminating corruption; enhance the responsibility for procurement activity in public sector, and openness enforced the e-procurement system's adaption.

Most African countries decided on legal reform to facilitate the implementation of the procurement framework and public procurement. In Tanzania, for example, the government supports implementing an e-procurement system to enable proper communication, invoicing, valuation, allotment, oversight, openness in the advertisement, and ensure that all purchasing activities done by public sector is integrated and processed through internet [17]. A research conducted by [28] revealed that e-procurement contributed to minimizing total costs expenses in the ministry of economy and finance to 18.6 million in 2016 compared with the year 2015 as the total expenses were 24.4 million. Organizations who have implemented e-procurement have attained significant cost saving by 8–15% [6], resulting, improved coordination and collaboration within the organization, saving on transactional costs and increase competitive sourcing [20].

2.5 *Drawback of E-procurement*

Few authors have mentioned the obstacles of the application of an e-procurement system. According to [6, 29] employees are unwilling to change employees, lack adequate knowledge of the technology, lack reliance on the technology. Whereas [30] indicated that the system's high initial investment cost, less flexibility, and the inability to access organizations that do not have internet services, lack of confidentiality, and information insecurity in an electronic transaction. Members of the management team face challenges on the proper way to convince top management to invest in technology; they must show that these technologies are in line with the strategic objectives of the enterprise [30]. Macro environment factors such as market, government, industry, and ICT change are out of control of the companies. Nevertheless, these obstacles can be reduced and even eliminated through proper planning and research. The technology challenges to suppliers involve realization and commitment to specialist software and start-up costs required by the suppliers that are usually over their financial capabilities, especially SMEs or they will not interest to implement such a high investment system. Supporting and proper infrastructures such as sufficient broadband coverage are the main reason for the widespread e-procurement system. For instance, in developing countries, there is inadequate infrastructure and coverage of the internet [31].

Inadequacies in government policies and legislation are area to be highlighted in the system. The government tendering procedures obliged the buying companies interested in participating in the tender to purchase the printed tender documents in person from the physical offices. This manner prevents using the e-tendering system and presents a massive relapse for the government trying to initiate the electronic government system. The use of different standards of e-procurement system leads to prevent the users from communicating with each other due to that difference in systems [32]. The most common reasons indicated by firms regarding implementing e-procurement are that they do not intend to adopt e-procurement due to their high costs. Some firms believe that their size is too small to benefit e-procurement—resistance of the management to support the idea of implementing e-procurement in their organization [33].

3 Supplier Relationship

One of the primary functions of procurement; they are responsible for searching and evaluating new vendors, supporting strategic objectives of the organization by improving sourcing portfolio, strengthening the long-term relationship with key partner, selecting the suitable ICT to streamline procurement activities and related processes. Thus, it will create a fit environment for effective cross-functional teams [35]. Due to the intense demand for complex products, buyers and sellers increase their collaboration, resulting in reliable, trustworthy, and partner key vendors. In this scenario, the vendors must be flexible and can take the risk to co-develop customized items [14, 15]. [36] described SRM as a "process involved in managing preferred suppliers and finding new ones whilst reducing costs, making procurement predictable and repeatable, pooling buyer experience and extracting the benefits of supplier partnerships."

According to [37], supplier relationship management is of five types: (1) market buying: which is considered as a short term contracts which is at arm's length, (2) ongoing relationship: that is a medium term contract, (3) partnership: considered as a long-term contracts, (4) strategic alliance: which is meant to also be a long-term contract, and (5) backwards integration: which is an ownership of the supplier. [38], through their study, revealed that different procurement's technologies have a positive relationship with procurement functions such as EDI, e-marketplace, e-catalogue, e-Auctions, and supplier relationship. It also indicates that the buyer–supplier relationship does not affect by e-procurement technologies, whereas it contributes indirectly to sustainable and lean operation manners.

3.1 Supply Chain Performance

Successful supply chain performance relies on the robust commitment and mutual trust between supply chain members. Effective supply chain planning represents trust and shares information between the partners as it considers as a primary requirement to attain successful supply chain management [39]. [40] Identified supply chain performance as "on-time delivery, reduced lead time, and responsiveness, and cost reduction, conformance to specifications, process improvements, and time-to-markets as constituents of supply chain performance." Cook and Hagey (2003) debate that organizations that do not measure their supply chain performances are in the dark about their supply chain efficiency as they estimate their target inventory should be. These organizations are probably bear a high price for their goods and services; thus, they become less competitive in the market.

The relationships between integration and SME performance have been extensively studied by [41] These findings confirm that integration can be transformed into competitive capabilities, thus contributing to positive supply chain performance. A study conducted by [8] showed that e-procurement and supply chain integration is responsible for enhancing supply chain performance. E-procurement is considered a vital resource, based on the "Resource-Based View" (RBV) theory; RBV theory is an organization. Framework used to define the strategic resources an organization can take advantage of to attain competitive advantage sustainability. This study's findings will be presented in the form of five propositions that have been based on the argument portrayed in section two and three. The five propositions, also depicted in Fig. 1 of this study's model, are:

P1: The uses of e-procurement system can significantly strengthening the buyer-seller relationship especially in partnership relationship which is important for high involvement procurement object in the evolvement phase.
P2: The high level of supplier integration is positively related to technology and systems implementation perspective of an e-procurement initiative.
P3: The implementation of e-procurement initiative in any organization positively associated with the high level of support by top management.
P4: The high level of performance of an organization and management implementation of e-procurement initiative positively linked with the satisfaction of users and suppliers.
P5: The high level of utilization of the performance measures is positively related with the implementation of e-procurement system.

4 Research Methodology and Descriptive Statistics

4.1 Survey Instrument

The research methodological approach is quantitative research via a method of identifying the role of the independent on the dependent variable/s, to confirm the validity of the 3 hypotheses of this study. The survey instrument appoints a 79-item online series of questions based on five-point-Likert-scale- from 1 "Strongly Disagree" to 5 being " Strongly Agree" and was developed through Google Forms.

The survey includes three parts: (1) a cover letter demonstrating the purpose of this study's research. Furthermore, this paper proved the participants that the study achieved ethical approval from Ahlia University, ethical review board: allowing it to gather data, and proving that this gathered data be considered confidential. (2) Five demographic items (Table 2) designed to emphasize participant's works, department, gender, age, and educational level. The survey questions (Table 1) which pertained to the independent and the dependent variables of this study and it consist of seven elements (E-procurement, partnership relationship, and supply chain performance).

Table 1 Descriptive statistics

Variables	Model's variable definition				
	N	Max	Min	Mean	SD
Work At	255	2	1	1.36	0.480
Department	255	4	1	2.30	1.193
Gender	255	1	0	0.73	0.447
Age	255	4	1	2.80	0.905
Educational level	255	4	1	1.64	0.998
E-procurement (EP)	255	5	1	3.61	0.959
Supplier Relation (SR)	255	5	1	3.71	0.904
Supply Chain Performance (SCP)	255	5	1	3.63	1.023

Table 2 Demographics characteristics of respondents

Sample characteristics	Frequency	Percent (%)
Works At		
Private Sector	164	64.3
Public Sector	91	35.7
Total	255	100
Department		
Management	93	36.5
Professional	53	20.8
Administration	48	18.8
Operations	61	23.9
Total	255	100.0
Gender		
Female	70	27.5
Male	185	72.5
Total	255	100.0
Age		
18–24 years	16	6.3
25–31 years	86	33.7
32–38 years	85	33.3
39 and above	68	26.7
Total	255	100.0
Education Level		
Bachelor's degree	157	61.6
Master's degree	65	25.5
Doctor of Philosophy	1	0.4
Other	32	12.5
Total	255	100.0

4.2 Model Hypotheses

This study assesses three hypotheses:

E-procurement and Supply Chain Performance. E-procurement is anticipated to be positively related to supply chain performance. E-procurement system facilitates the flow of information between an organization and its suppliers; therefore, the quality of information can be maintained [41]. E-procurement streamlines the communication between the firm and its vendors by utilizing the internet-based system [41]. Also the e-procurement system could minimize transaction expenses among supply chain members. Therefore, *H1is* expected:

H1: *E-procurement has significant impact on supply chain performance.*

Supplier Relationship and Supply Chain Performance. The impact of supplier relationship on supply chain performance is anticipated to be positive. Depending on the mutual and ongoing beneficial supplier relationship, this relationship could help the organization introduce a successful product/service rapidly than its rivals [42, 43]. Firms that involve their strategic in the early stage in the product design process could further minimize the cost and time of improving and launching a new product or service [33]. Furthermore, supply chain performance could be enhanced by supplier integration. Process automation is a significant factor in enhancing process efficiency [30]. It could be anticipated that the product's customer service and quality could be improved by supplier integration [6]. Therefore, *H2* is expected:

H2: *Supplier relationship has significant impact on supply chain performance.*

E-procurement and Supplier Relationship. The relationship between e-procurement and supplier relationship is anticipated to be positively related. E-procurement could solve the problem in the process between the organization and its suppliers in the supply chain [27]. Through this streamline, the relationship between organization and supplier will be built on a high level of trust; therefore, their relationship will strengthen in the long-run. E-procurement is positively related to supplier integration. The e-procurement system can facilitate communication between the buying organization and its vendors [17]. Previous research emphasized that cooperation can be improved by information technology [19, 40]. Therefore, it can be concluded that the collaboration between organization and suppliers can be improved through the e-procurement system. Hence, it could be anticipated the e-procurement system could advance that supplier integration. Therefore, *H3* is expected:

H3: *E-procurement has significant impact on supplier relationship.*

4.3 Sample and Data Collection

Survey validity and reliability were re-assessed through 255 voluntary participants. The link of the online survey was shared to the buyers and all employees who are interacted with procurement departments in private and public sector in Bahrain. These groups will be appropriate to gather the required information from various organizations which can help in assessing the buyer's satisfaction for their current purchasing system and testing the impact of the E-procurement. Validity of the data was proved after data was manually entered in SPSS. Then, the entered data was screened to determine missing values and outliners. As a result, 255 valid responses, of 555,307 the total employees working in private and public sector in Bahrain as reported by LMRA (2019), enough to generalize over the entire target population of public and private sector working in Bahrain. A sample of 255 is over the minimal requirement given a 5.5% margin of error, 90% confidence level, and 50% response distribution of the online sample size calculator (Raosoft, no data). Table 1 describes the information of the descriptive statistics, such as the value of standard deviation (SD), maximum (max), minimum (min), and mean. Table 2 demonstrate that 72.5% of the demographics of the 255 participants were males (n = 185) while 27.5% were females (n = 70), Most of them 33.7% (n = 68) were 25–31 years of age with the majority 61.6% holding bachelor's degrees (n = 157). Most of them 64.3% working in private sector (n = 164) and 36.5% working in management department (n = 93).

5 Data Analysis

To analyze the data, the mean score, T-test, frequency counts, correlation, and percentage are be implemented, especially to determine the focal variables of this study. SPSS statical tools "Statical package for social sciences'" is applied to analyze the collected data. This tool was selected for its efficient data management, superior output organization, and vast range of options. This research examined the composite reliability for which items relating to the constructs EP, SR, and SCP should indicate Cronbach's α that should be surpass 0.6 [19]. Table 3 shows that all Cronbach's α values surpasses 0.6; hence are acceptable, and since their ranged was from 0.779 to 0.941 this indicates that all variables items are reliable. Consequently, no items need to be eliminated. The validity is the extent to which the instrument that was chosen showed the reality of the constructs that were being measured [11]. This Instrument discriminant validity was performed to examine the correlation analysis, Table 4, among variables; where the p-value that examines the level of significance should fall below 0.05 to express a correlated relation among the.

Table 3 Cronbach's α value of variables

Variables	Cronbach's α	N of Items
EP	0.962	15
SR	0.969	16
SCP	0.975	12

Table 4 Pearson correlation statistics with p-value = 0, N = 255

Variables	EP	SCP	SR
EP	1.000	0.709**	0.809**
SCP	0.709**	1.000	0.806**
SR	0.809**	0.806**	1.000

NOTE **. Correlation is significant at the 0.01 level (2-tailed).

Table 5 Summary of the regression analysis

Model	Hypotheses	Relationship	F	t	R^2	β
M1	H1	EP → SCP	243.186 Sig. 0.000	2.682	65.9%	0.168
	H2	SR → SCP		10.713		0.670
M2	H3	EP → SR	477.557 Sig. 0.000	21.853	65.4%	0.0809

The variables as depicted in the Table 4 most of the relation e.g., SR and EP: r = 0.809, $p < 0.01$. The Correlation lower than 0.7 are acceptable. Multiple regression analysis (Table 5) was performed to assess the three hypotheses, to examine convergent validity; summarize hypotheses H1, H2 being positively significant: all supported by the analysis of this study. According to table 5 the e-procurement positively and significantly influence the supply chain performance (β = 0.168 and $p < 0.01$) model 1 (M1). Accordingly, hypotheses 1 (H1) is supported. Interestingly, the supplier relationship is positively and significantly impact the performance of supply chain (β = 0.670 and $p < 0.01$). Accordingly, hypotheses 2 (H2) is also supported and significantly influence the supply chain performance due to the overweight than e-procurement. More particularly, empirical testing shows that supplier relationship positively significant influence the use of e-procurement (β = 0.809 and $p < 0.01$). This indicates that the empirical results of this research support H3 and interpret that the suppliers prefer using e-procurement system over the traditional procurement.

6 Findings. Implications, Limitations and Implications for Future Research

The research's findings show that the use of e-procurement and supplier relationships positively and significantly influences the supply chain performance and contributes to enhancing the process. Though the extent of the impact of e-procurement and supplier relationships, supplier relationship has a more substantial impact than e-procurement on the supply chain performance. Strengthening the relationship between the buying organization and its suppliers will positively affect the supply chain performance.

The study also revealed that the suppliers prefer using e-procurement systems than traditional procurement due to their immense advantages, such as facilitating communication between the supply chain partners, sharing information, and cost-saving. Process automation and developments considered as the top encouragement for organizations to implement procurement system. There is a clear need for companies to enhance and improve their business function, operational efficiency, and supply chain performance [6, 44]. Such findings motivate the firms to implement the e-procurement system. It will provide several advantages and solutions for the whole firm and enhance their supply chain performance.

6.1 Recommendations

Based on the study results, the implementation of e-procurement system will add value and bring immense advantages to all sectors and industries. Since the e-procurement system strongly relies on ICT, the organization need to build a suitable infrastructure based on recent technologies that will enable the procurement department to do procurement function effectively and efficiently. Additionally, managers in future can leverage this study and design policies based on this model. Also, the organization should move toward a collaborative partnership relationship with its supplier rather than an arms-length relationship to ensure the sustainability of their processes.

6.2 Theoretical, Practical Implication

One of the top challenges to e-procurement implementation in any organization are the high investment costs, made worse by existing tight budgetary measures [33, 45]. The costs to implement e-procurement could be quite significant, and the returns of investment could take more than three years. This will be considered as a burden on an organization since every organization seeks for fast return to their investment. However, the proposed model of this study constitutes a couple of theoretical implications, as well as practical implications.

A vast number of empirical studies regarding the implementation of e-procurement been accomplished using questionnaires. Whatever, there is a relative absence of deep research of e-procurement implementation using qualitative methods. For instance: [3, 6, 46] conducted their studies regarding the impact of e-procurement on supplier relationship. This is where this study shows off its importance as a significant contributor to the wide body of global knowledge in the e-procurement area. Future research can take advantage of the viability of this study's model and empirically examine it cross sectionally within various industrial sectors, as well as examine it longitudinally so to eliminate any biasness caused by cross-sectional studies.

A representative sampling design should be used in the future to be able to make claims for generalizable results. Managers should first empirically examine the viable model and after assuring empirical fitness this model should be reflected into policies and strategies which will aid the respective paganizations to streamline their procurement strategies. Managers should particularly consider this study's model since it will aid their e-procurement to improve, and better management, customer relations as well as the performance of the supply chain (holistically speaking).

6.3 Conclusion and Limitation

This research conducted to recognize how e-procurement participates to supply chain performance. E-procurement is a web-based process involving of four main aspects: e-sourcing, e-evaluation, e-negotiation, and e-design [40, 47, 48] Based on the systemic nature of e-procurement, supplier relationship proposed as an independent variable which reflects the influence of e-procurement on supply chain performance. The empirical study stated that supplier relationship has a significant effect on supply chain performance. Therefore, it could be concluded that e-procurement system can streamline the activities coordination and flow of information between supply chain members.

This study spots light on the main factors that affect the performance of the supply chain. These factors are e-procurement, supplier relationship, and supplier integration. This study provided the evidence through the literature review that these factors have a significant influence on the supply chain performance in a positive way. E-procurement contributes to eliminate paperwork, and significantly reduce the costs, improve transparency, minimize the purchasing cycle, and streamline the procurement process. E-procurement improves the effectiveness and efficiency of procurement functions, its assist the organization to geographical outreach to build a new collaborative relationship with multiple suppliers worldwide.

Supplier relationship management contributes to strengthening the relationship with key suppliers to ensure a sustainable flow of materials and services. There are two main types of the supplier relationship, relationship with the OEM; this relationship mainly focuses on building a long-term relationship with a key supplier to ensure sustainable flow for complicated items. The other type is multiple sourcing where the organizations endeavor to attain the best price available at the market. Furthermore, supplier relationship management provides better control over supplier relationships. Finally, the literature revealed the fundamental role of supplier integration in supporting SCP in different forms of enterprises. Supplier integration helps the organization to collaborate with its vendors to manage their inter-organizational processes that lead to customer demand fulfillment [48, 49]

Supplier integration contributes to minimizing the risk of receiving wrong items or late in delivering the required material by working closely with the supplier and proper sharing and communicating the related information to its vendors to ensure adequately performing of their obligation and to avoid such issues. The analysis of the data showed that the effective adoption of e-procurement specifically e-design allows the buying organization integrate with their vendors, consequently, it will contribute to cutting the supply chain costs and enhancing overall performance.

6.4 Future Studies

This research conducted at ministries and private organization level; future research could focus on the potential relationship based on a personnel level. For instance, the way that the organizations' staff behavior concerning technology usage impact the relationship between e-procurement and supply chain performance is deserve exploration for cross-level analysis [40].

References

1. Sharif AM, Alshawi S, Kamal MM, Eldabi T, Mazhar A (2014) Exploring the role of supplier relationship management for sustainable operations: an OR perspective. J Oper Res Society 65(6):963–978
2. Razzaque A, Karolak M (2011) Building a knowledge management system for the e-health knowledge society. J Econ Dev Manage IT Fin Market 2(2):23–40
3. Faheem M, Siddiqui DA (2019) The impact of E-procurement practices on supply chain performance: a Case of B2B Procurement in Pakistani Industry. SSRN 3510616
4. Gunasekaran A, McGaughey RE, Ngai EW, Rai BK (2009) E-Procurement adoption in the Southcoast SMEs. Int J Prod Econ 122(1):161–175
5. Autry CW, Golicic SL (2010) Evaluating buyer–supplier relationship–performance spirals: a longitudinal study. J Oper Manage 28(2):87–100
6. Choy KL, Lee WB, Lo V (2003) Design of an intelligent supplier relationship management system: a hybrid case based neural network approach. Expert Syst Appl 24(2):225–237
7. Nicoletti B (2013) Lean Six Sigma and digitize procurement. Int J Lean Six Sigma 4(2):184–203
8. Pattanayak D, Punyatoya P (2019) Effect of supply chain technology internalization and e-procurement on supply chain performance. Bus Process Manage J 26(6):1425–1442
9. Al-Sartawi A (2020) Information technology governance and cybersecurity at the board level. Int J Crit Infrastruct 16(2):150–161
10. Al-Sartawi A (2019) Assessing the relationship between information transparency through social media disclosure and firm value. Manage Account Rev 18(2):1–20
11. Madzimure J (2020) Enhancing supplier integration through e-design and e-negotiation in small and medium enterprises. South Afr J Entrepr Small Bus Manage 12(1):8
12. Madzimure J (2020) Examining the influence of supplier integration on supply chain performance in South African small and medium enterprises. South Afr J Entrepr Small Bus Manage 12(1):7
13. Gupta M, Sikarwar TS (2020) Modelling credit risk management and bank's profitability. Int J Electron Bank 2(2):170–183
14. Gupta N (2019) Influence of demographic variables on synchronisation between customer satisfaction and retail banking channels for customers' of public sector banks of India. Int J Electron Bank 1(3):206–219
15. Razzaque A, Eldabi T, Chen W (2020) Quality decisions from physicians' shared knowledge in virtual communities. Knowl Manage Res Pract 1–13. doi:https://doi.org/10.1080/14778238.2020.1788428

16. Puschmann T, Alt R (2005) Successful use of e-procurement in supply chains. Supply Chain Manage 10(2):122–133
17. Musleh Al-Sartawi AMA (2020) E-learning improves accounting education: case of the higher education sector of Bahrain. In: Themistocleous M, Papadaki M, Kamal MM (eds) Information systems, EMCIS 2020. Lecture Notes in Business Information Processing, vol 402. Springer, Cham
18. Marra M, Ho W, Edwards JS (2012) Supply chain knowledge management: a literature review. Expert Syst Appl 39(5):6103–6110
19. Matano F, Musau E, Nyaboga YB (2020) Effects of e-procurement implementation practices on procurement of goods, works and services in the national youth service, Nairobi City County. Int Acad J Procure Supply Chain Manage 3(2):63–82
20. Afolabi A, Ibem E, Aduwo E, Tunji-Olayeni P (2020) Digitizing the grey areas in the Nigerian public procurement system using e-Procurement technologies. Int J Constr Manage 1–10
21. Li X, Olorunniwo F, Fan C, Jolayemi J (2016) Improving performance in supplier relationship management with lower-tier supplier visibility and management. Ann Manage Sci 5(1):19
22. Wieteska G (2016) Building resilient relationships with suppliers in the B2B market. Management 20(2):307–321
23. Mohd Nawi MN, Roslan S, Salleh NA, Zulhumadi F, Harun AN (2016) The benefits and challenges of E-procurement implementation: a case study of Malaysian Company. Int J Econ Fin Issues 6(S7):329–332
24. Razzaque A, Jalal-Karim A (2010) Conceptual healthcare knowledge management model for adaptability and interoperability of EHR. In: European, mediterranean and middle eastern conference on information systems (EMCIS 2010), Abu-Dhabi, UAE
25. Moon MJ (2005) E-procurement management in state governments: dffusion of e-procurement practices and its determinants. J Public Procure 5(1):54–72
26. Vaidya K, Sajeev ASM, Callender G (2006) Critical factors that influence e-procurement implementation success in the public sector. J Public Procure 6(1/2):70–99
27. Anekal P (2018) Supplier relational integration under conditions of product complexity. Am J Manage 18(1)
28. Bhagwat R, Sharma MK (2007) Performance measurement of supply chain management: a balanced scorecard approach. Comput Ind Eng 53(1):43–62
29. Razzaque A, Jalal-Karim A (2010) The influence of knowledge management on EHR to improve the quality of health care services. In: European, mediterranean and middle eastern conference on information systems (EMCIS 2010), Abu-Dhabi, UAE
30. Croom S, Brandon-Jones A (2007) Impact of e-procurement: experiences from implementation in the UK public sector. J Purchas Supply Manage 13(4):294–303
31. Abdulrasool FE, Turnbull SI (2020) Exploring security, risk, and compliance driven IT governance model for universities: applied research based on the COBIT framework. Int J Electron Bank 2(3):237–265
32. Pishevar S, Morris DE, Freewebs Corp (2006) Collective procurement management system, U.S. Patent 7,124,107
33. William SP, Hardy (2007) E-procurement: current issues and future challenges. In: ECIS 2007 Proceedings, vol 133
34. Seetharaman A, Patwa N, Wai SLK, Shamir A (2020) The impact of E-procurement systems in the biomedical industry. J Account. Bus Manage (JABM) 27(1):66–85
35. Pujawan IN, Goyal SK (2005) Electronic procurement and manufacturing strategic objectives. Int J Logist Syst Manage 1(2–3):227–243
36. Herrmann JW, Hodgson B (2001) SRM: leveraging the supply base for competitive advantage. In: Proceedings of the SMTA international conference, vol. 1, pp 1–10
37. Chen YJ, Deng M (2015) Information sharing in a manufacturer–supplier relationship: suppliers' incentive and production efficiency. Prod Oper Manage 24(4):619–633
38. Doherty NF, McConnell D, Ellis-Chadwick F (2013) Institutional responses to electronic procurement in the public sector. Int J Public Sector Manage 26(6): 495–515

39. Kwon IWG, Suh T (2004) Factors affecting the level of trust and commitment in supply chain relationships. J Supply Chain Manage 40(1):4–14
40. Razzaque A, Eldabi T, Jalal-Karim A (2013) Physician virtual community and medical decision-making: mediating role of knowledge sharing. J Enterpr Inf Manage 26(5):500–515
41. Alhakimi W, Esmail J (2019) The factors influencing the adoption of internet banking in Yemen. Int J Electron Bank 2(2):97–117
42. Liker JK, Choi TY (2004) Building deep supplier relationships. Harvard Business Rev 82(12):104–113
43. Razzaque A (2020) M-Learning Improves Knowledge Sharing Over E-Learning platforms to build higher education students' Social Capital. SAGE Open 10(2):1–9
44. Shang W, Ha AY, Tong S (2016) Information sharing in a supply chain with a common retailer. Manage Sci 62(1):245–263
45. Al-Sartawi A (2020) Does it pay to be socially responsible? Empirical evidence from the GCC countries. Int J Law Manage 62(5):381–394
46. Madzimure J, Mafini C, Dhurup M (2020) E-procurement, supplier integration and supply chain performance in small and medium enterprises in South Africa. South Afr J Business Manage 51(1):1–12
47. Lambert DM, Schwieterman MA (2012) Supplier relationship management as a macro business process. Supply Chain Manage 17(3):337–352
48. Carter TC (2003) Supplier relationship management: models, considerations and implications for DOD. Industrial collage of the armed forces University Fort McNair Washington, DC, pp 20319–25062
49. Jain M, Abidi N, Bandyopadhayay A (2018) E-procurement espousal and assessment framework: a case-based study of Indian automobile companies. Int J Technol Manage Sustain Dev 17(1):87–109

Review of Effects of the Virtual Reality in Education

Zakaria Alrababah and Samer Shorman

Abstract Virtual Reality (VR) have been shown to be effective instruments for teaching students, particularly in some disciplines such as Science, Technology, Engineering and Mathematics (STEM) fields that need more learners' engagement as an immersive, whole-body interactive simulation with environment components for experiments, this will affect students' interactions and conceptual understanding. In this article, we focus on the effects of the using of VR on learning Biology and Medicine courses. The paper summarizes the outcomes of some related articles; it presents a comparison between some methods that are used to support the educational process. Furthermore, it discusses some challenges of the use of VR in different fields of education.

Keywords Virtual reality · Interactive learning · Biology · VR in education

1 Introduction

Virtual Reality is a computer-based technology that facilitates creation of a simulated environment. VR functions by involving the individuals in an overall 3D experience of the subject, as opposed to the traditional user interfaces that only provides a 1D or 2D experience. It is believed that instead of just viewing a screen, the users find VR much friendlier owing to its provision of being immersed and to interact with 3D worlds [24]. Virtual reality is one of the modern technologies that can enhance students' interactions, engagements, and achievements if it is used effectively in classroom. Virtual reality has captured the world's imaginations. It seems to be the natural next step for the evolution of education. Virtual reality is used widely in computer gaming, medicine, collaboration in work places, recruiting and training,

Z. Alrababah
Naseem International School, A'ali, Kingdom of Bahrain

S. Shorman (✉)
Department of Computer Science, Applied Science University, Eker, Kingdom of Bahrain
e-mail: samer.shorman@asu.edu.bh

A. M. A. Musleh Al-Sartawi (ed.), *The Big Data-Driven Digital Economy: Artificial and Computational Intelligence*, Studies in Computational Intelligence 974,
https://doi.org/10.1007/978-3-030-73057-4_33

creating ideas and forecasting trends, training medical students, and many others. In classroom, VR technology it is relatively a new technology. Even if VR technology is used actively at classrooms in some countries, but it still needs more educational researches to know it effects on students' interactions, engagements, and achievements. The advent of Virtual Reality has been of particular interest to many critics and scholars. Not only has it been hailed as a significant means of ushering a change but it has also been seen responsible in promoting interactions with virtual people, and allowing them to get a true sense of adventures in a simulated world or time. Several studies have been devoted to how VR can be intrinsic in understanding the roots of the problem that are plaguing education [9].

2 Research Significant

Virtual Reality has been a new entrant in the educational field and has undoubtedly ushered in a revolution in the learning outcomes. Several researches have been conducted over the years that indicated the implications of VR enhancing the learning outcomes. Three-dimensional based virtual representation allows to the students to visualize and interact with the subject thus allowing them to better understand the study topic. Besides that, it also provides a real time experience and can be aid in comprehending the abstract concepts that might be difficult to perceive. Although, implications of VR in biology, have been discussed in few studies, a detailed insight is not yet available. Biology, is the field of science wherein experimentation holds the key to effective learning. One has to gain perspective regarding many viewpoints which might be hard to follow through the traditional means of education. The major reason for incorporation of VR in the domain of biology is to attain a level of reality through a virtual environment. The current study intents to fill the research gap by conducting a study to evaluate the effects of virtual reality on interaction, engagement and achievement for biology lessons. The objective of the work is to gain an understanding of the current situation of education that is followed in biology lessons with primary focus on the type of technology that is used in studying the dynamic subjects like biology.

Figure 1 shows the learning technologies that are included On-the job, hands-on or experiential learning, Mentoring or coaching, Instructor-led or classroom training, Formal curriculum or courses, Videos, Online simulations games, Virtual instructor-led training, eLearning courses, Podcast, Augmented reality, Virtual reality, Social collaboration or learning application, Mobile device notifications or text messages. These statistics present most of learning technologies effective on-the-job experience. The statistic shows fluctuation between learning technologies starting from the highest percentage 90%, which is on-the-job experience, and then to lowest, which is Mobile device notifications or text messages with 20%. Relative to virtual reality got 29% that have seen this technology is effective as learning methods. This percentage is good for this time and still this virtual reality technology not wide used as other methods.

Fig. 1 Learning technologies are effective in 2018, by type of technology [22]

3 Virtual Reality Literature Review

In the following section will discuss the articles related to Virtual Reality and its application in educational process and followed by Table 1 that represent summary for this discussion.

As well in [12] which titled "Learning with desktop virtual reality: Low spatial ability learners are more positively affected" attempted to identify the correlation between desktop virtual reality (VR) and learning environment. It is believed that VR-based learning environment renders a positive impact on learning efficiency. To investigate the same a quasi-experimental design comprising of pre-test and post-test was adopted. The study involved 431 high school students recruited from four randomly selected schools. The respondents again were randomly divided into two groups namely- experimental or control groups. Measurement of the learning outcome was performed cognitively through academic performance. Results revealed a much better performance for students who were using desktop virtual reality. Further, they exhibited greater performance achievement levels as opposed to the ones who were not exposed to the VR learning environment. In context of performance achievement, significant interaction effect was noted between learning mode and spatial ability. Extending the analysis to explore potential impact of spatial ability, results reported a significant difference in the performance for low spatial ability learners compared to the high spatial learners in the experimental and control groups. In concluding remarks, the authors thus opined that high potential of desktop virtual reality instructional intervention facilitates to reduce extraneous cognitive load. Such a reduction in the cognitive load allows the students to learn more effectively and perform better.

Table 1 Summary of virtual reality literature review

Authors	Study	Outcomes
1. Lee and Wong (2014) [12]	Identify the correlation between desktop virtual reality (vr) and learning environment	The study involved 431 high school students recruited from four randomly selected schools
2. Brinson (2015) [3]	Experimental groups based on the learning outcome achievement when traditional lab (TL) such as hands-on and non-traditional lab (NTL)	89%), out of a total of 56 studies to demonstrate a positive impact for ntl on student learning outcomes as opposed to Tradition Learning
3. Huang, liaw and lai (2013) [7]	Evaluating the acceptance for virtual reality (VR) platforms among the learners is of significant importance owing to its implications in ensuring its complete utilization for greatest effect	Results of the survey analysis revealed 'immersion' and 'imagination' as the two primary features of the VR that positively impacts
4. Makransky et al. 2019) [15]	Desktop based virtual reality (VR) science lab simulation	
5. Lindgren et al. (2016) [13]	The learners to be embedded in a realistic representation of planetary astronomy	Higher levels of learning toward simulation and learning environment compared to traditional instruments such as mouse and keyboard controls
6. Allcoat and Muhlenen (2018) [2]	Assess the multifaceted effects of VR. Three learning conditions, namely, traditional (textbook style), VR and video	Performance for the participants in the VR condition was also much better compared to the participant of both traditional and the video conditions
7. Radianti et al. (2020) [19]	Evaluated the outcomes of immersive VR in the higher educational context	Immersive VR lacks provision of being applied regularly in actual teaching
8. Hadiprayitn, Muhlis, and Kusmiyati (2019) [6]	Learning biology for senior high schools	
9. Parong and Mayer (2018) [18]	Impact of VR on learning outcomes of students taking biology lessons	No difference noted in the interest, engagement, and motivation of the students for both the groups
10. Shim et al. (2003) [20]	Virtual reality technology (VRT) in biology education	The VRT will be highly promising in influencing the students biology learning compared to the other multimedia programs

(continued)

Table 1 (continued)

Authors	Study	Outcomes
11. Muhamad et al. (2010) [17]	Virtual Laboratory For Biology (VLab-Bio) and its scope in the near future	Virtual laboratory depending on the complexity of the topic would indeed facilitate the students in their biology courses
12. Garcia-Bonete, Jensen, and Katona (2018) [4]	Virtual reality and augmented reality (VR and AR) in teaching of structural biology	VR and AR to be instrumental in extending the visualization the student's toolbox to interaction with content related to structural biology. However, some critical topic in biology could limit the utilization of VR and AR
13. Another study by Keaveney et al. (2016) [11]	Implications of advanced 3D imaging, modeling, and printing techniques in the biological sciences sector	Indicative of the significant potential for advanced 3D imaging, modeling, and printing techniques in zoological sciences
14. Wang et al. (2016) [20]	3D VR simulation system its application in training the students for dental crown preparation	Facilitates the students to experience realistic clinical situations and also creates provision for greater practice
15. Alfalah et al. (2019) [1]	3D representation of heart structure was projected in the study utilizing a VR system	Model rectification to clearly fathom the anatomical relations of different parts of the heart. To make the experience more realistic,
16. Lv et al. (2017) [14]	VR in learning outcomes of geography lessons	The study indicated a positive impact of the software on the learning outcomes
17. Makransky, Terkildsen, and Mayer (2017) [16]	Virtual reality to a science lab simulation	No positive impact of VR on the learning outcomes of the students of science background
18. Kaminska et al. (2019) [10]	Virtual reality and its applications in education: survey	

As well in [3] which titled - "Learning outcome achievement in non-traditional (virtual and remote) versus traditional (hands-on) laboratories: A review of the empirical research" made an attempted to critically assess the reports related with direct comparison of the experimental groups. It based on the learning outcome achievement when traditional lab (TL), such as hands-on and non-traditional lab (NTL) comprising of virtual and remote are implemented for the learning purpose. Results of the study reported majority of the reviewed studies (n = 50, 89%), out of a total of

56 studies to demonstrate a positive impact for NTL on student learning outcomes as opposed to TL. The results hold true for different learning outcome categories such as "knowledge and understanding, inquiry skills, practical skills, perception, analytical skills, and social and scientific communication". However, the outcome category largely dictates the extent of the learning achievement. The difference in the learning outcomes for TL and NTL could stem from its contrasting areas of focus. Greater emphasis was on content knowledge and understanding for the studies that reported higher achievement in NTL. The research instrument in such studies has been quizzes and exams that facilitated the assessment of degree of achievement. On the contrary, qualitative studies were the mainstay for studies supporting higher achievement in TL seemed. Research instruments such as surveys were implemented to collect the qualitative data related to student and/or instructor perception. Till date still there has been a considerable amount of debate among the science educators regarding the efficiency of NTL versus TL for instructional purpose of the laboratory (i.e. learning outcome preference). The study thus indicated the need for greater number of studies to assess the potential of NTL such as virtual reality in enhancing learning outcomes.

In [7] which titled - "Exploring learner acceptance of the use of virtual reality in medical education: a case study of desktop and projection-based display systems" evaluated the acceptance of advanced technologies such as virtual reality that finds application in modern day education. Evaluating the acceptance for virtual reality (VR) platforms among the learners is of significant importance owing to its implications in ensuring its complete utilization for greatest effect. The different VR based learning environments that are taken into consideration while evaluating the acceptance are 'human-patient simulators', 'immersive virtual reality Cave Automatic Virtual Environment systems', and 'video conferencing'. In this regard, the authors herein described how high performance real-time interactive software (VR4MAX) has been utilized to create a prototype 3D VR learning system. A questionnaire-based survey involving 167 university students was conducted to perceive their opinion on VR applications and its applicability in learning. Results of the survey and its consequent analysis revealed 'immersion' and 'imagination' as the two primary features of the VR-mediated course contents that positively impacts and predicts, perceived usefulness and perceived ease of use respectively. The factors are highly significant owing to its contribution towards setting the behavioral intention of learners that encourages utilizing VR learning systems. In conclusion, it can be thus said that the study unraveled the relationship between the VR features and how it affects the behavioral intention of the learners to use the same.

The multiple intentions that drive the researchers to study implications of VR is not just limited to its learning implications but also towards its potential in enabling companies and educational institutions to expand their reach, reduce the overhead cost, and elevate understating of the students. In the recent times, virtual laboratories have made students more accessible to hands on training thus preparing them for becoming future scientists. Such virtual setting helps overcome the boundaries of classroom setting thus making way for greater student opportunities in reduced amounts of time and money. The popularity of VR based learning environment has been raising and the same is evident from the initiatives of greater number of new higher education institutions to outsource lab activities in a virtual environment [15]. Realising the increasing trend of utilizing VR in the learning environment,

In addition, the researchers in [15] compared the outcomes in terms of learning and motivation when desktop-based virtual reality (VR) science lab simulation is used under two conditions, the first one being on the internet at home while the second, under the supervision class teacher. In this regard a between-subjects experimental design was adopted with 112 university biology students. The participants were made to learn on the topic of microbiology at either in home or in class from the same virtual laboratory simulation. Results revealed no significant difference in the scores for post-test learning outcome scores. Similarly, self-report measures of intrinsic motivation or self-efficacy also exhibited no distinct variation. The results thus have been indicative of the efficiency of virtual simulations in engaging students in learning activities even outside of the classroom environment [15]. Virtual Reality Technology (VRT) thus poses as the most suitable 3D technology that has immense potential in simplifying the complex subjects for the learners and reducing the need the labs and real-life experiments demanding subjects such as nature science, biology, chemistry, and physics [26]

In [13], which titled - "Enhancing learning and engagement through embodied interaction within a mixed reality simulation" attempted to compare the students based on their learning outcomes and attitude towards learning. The learning conditions used for the comparison purpose are immersive, whole-body interactive simulation, and desktop version of the simulation. To fulfill the study objectives a room-sized simulation called Meteor, it was designed and developed. MEteor is a laser-based motion tracking system that remains integrated with both floor- and wall-projections. This facilitated the learners to be embedded in a realistic representation of planetary astronomy. The authors believed that when the students get an opportunity to enact the ideas it allows them to connect with the subject matter related graphs and other visualizations. Further, it is presumed that with embodied engagement the students would experience heightened feelings of agency and efficacy, which in turn would result in increased positive attitudes towards science and perceived value of the simulation experience. In actual terms, the study results revealed that simulations involving whole-body activity significantly impacts learning gains, engagement levels, and positivity in attitude towards science. Embodied interaction facilitated by a dynamic system with complex components and relationships has been experimentally proven to render both cognitive and motivational benefits. Students who engaged with their whole bodies in enacting of the predictions exhibited higher levels of learning and positive attitudes toward both simulation experience and learning environment compared to the ones using the identical simulation that made use of traditional instruments such as mouse and keyboard controls.

The researchers Allcoat and Muhlenen in [2] in their study titled - "Learning in virtual reality: Effects on performance, emotion and engagement" identified virtual reality (VR) as one of the most prominent technology of the modern era that hold immense potential in revolutionizing the learning experience in the educational fields. To root for the same the authors attempted to assess the multifaceted effects of VR.

Three learning conditions, namely, traditional (textbook style), VR and video (a passive control) were utilized and out of the total 99 participants each one was assigned to any one of the three available options. Making all the learning materials to use the same text and 3D model under all conditions, participants belonging to the traditional and VR conditions as opposed to the ones in video condition exhibited improved overall performance measured in terms of learning comprising of knowledge acquisition and understanding. Performance for the participants in the VR condition was also much better compared to the participant of both traditional and the video conditions. Analysis of the results that depicted the emotional status of the participants revealed VR condition to increase positive emotions and consequently decrease negative emotions, prior and post the learning phase. Higher engagement for the participants under VR conditions as opposed to the traditional and the video conditions has been evident from the evaluation of the Web-based learning tools. Thus, when compared to traditional and video learning methods, VR undoubtedly stands out to be a displayed most efficient tool in enhancing the overall learning experience [8]. Thus reported virtual reality to be "an immersive, hands-on tool for learning". VR in its unique manner allows solving most of the educational challenges of the modern day.

VR owing to its promising contributions in the educational field has made it a popular research topic among the modern-day researchers. However, there has been limited amounts of systematic work that evaluated the outcomes of immersive VR in the higher educational context. Much of this research gap has been filled in by [19] that is a recent study titled - "A systematic review of immersive virtual reality applications for higher education: Design elements, lessons learned, and research agenda". Herein, the authors attempted to explore how virtual reality finds its application in different scenarios thus benefiting the students to enhance their learning experience. To fulfill the study objectives systematic mapping existing articles that assessed the application of VR in higher education were reviewed and the design elements of such studies were identified. With the aim to achieve a foundation for successful VR-based learning, the study emphasized on three key points- "the current domain structure in terms of the learning contents, the VR design elements, and the learning theories". Mapping between application domains and learning contents and between design elements and learning contents revealed several gaps when considering VR to be implemented in the higher education sphere. This has been evident from the ignorance towards learning theories when VR application was considered in the development process that assisted and guided toward learning outcomes. Another, limiting factor could be the primary focus on the usability of the VR apps when evaluating educational VR applications instead of learning outcomes. Immersive VR on the other hand lacks provision of being applied regularly in actual teaching owing to the fact that much of the studies related to the immersive VR have been at experimental and developmental stages. Despite the challenges, VR yet poses to be a promising sphere.

As per a study contemplated by [10] the results of analysis of their study on "Problems in learning biology for senior high schools in Lombok Island", it was highlighted that the abstract concept was hard to conceive by students of biology in these particular regions. All of these abilities require memorizing power and cognitive skills in order to fully fathom it. The major predicament arises due to the complexity of the subject also because the materials and resources were not available. The teaching method is monotonous, the collaboration quotient is minimal and the students exclaim that their academic environment is boring as opposed to exciting and supportive. To fathom the perception of the students and the teachers a survey was conducted with 568 students and 24 biology teachers of the senior high schools located all around the Lombok. The data obtained from the questionnaire-based survey were then subjected to qualitative and quantitative statistical analysis. Results revealed that for the difficult topics the perceived difficulty has been as follows- bacteria and viruses (18.64%), endocrine system (10.63%), cell structure (8.81%), genetics (8.41%), and nervous system (8.28%). The problems mainly remained associated with the scientific name, the topic complexity, and the students learning habits [6].

In [18] the study titled - "Learning Science in Immersive Virtual Reality" focused particularly on the impact of VR on learning outcomes of students taking biology lessons. The objective of the study has been to compare the effectiveness in terms of instructions provided when immersive virtual reality (VR) is used as opposed to desktop slideshow for teaching scientific knowledge. Further, the study also attempted to examine whether the efficacy of VR lesson is enhanced with addition of generative learning strategy. To fulfill the study objectives two experiments were conducted. The first one dealt with college students who were made to take a biology lesson related to the working of the human body using either immersive VR or a self-directed PowerPoint slideshow on a desktop computer. Two theories that govern the outcomes of the experiment are the interest theory, and the cognitive theory of multi-media learning. The former beliefs that the students were exposed to immersive VR learning environment would report more positive ratings of interest and motivation compared to the non-VR students and would also score higher on a post-test based on the teachings of the lesson. The latter or the cognitive theory of multimedia learning, on the other hand predicts the students learning on a well-designed slideshow to score higher on a posttest, but with lower levels of interest and motivation compared to the VR based learning students. Results of the study complied with the cognitive theory reporting high scores on the posttest for the students who viewed the slideshow as opposed to the VR group. But the VR group students exhibited greater motivation, interest, and engagement ratings. The second experiment was set to investigate VR in totality. VR lessons were now divided into two parts- one that was segment and the other being a continuous one. The students for the segmented VR lesson attempted the test after each segment while the continuous VR students like experiment 1 took the test at the end of the lesson. Results revealed students of segmented VR lessons to perform significantly better compared to the continuous VR group. There was no difference noted in the interest, engagement, and motivation of the students for both the groups. Findings of both the experiments thus reinstate the belief in the current trend of utilizing VR for education purposes. Further, the results also provide support for the cognitive theory of multimedia learning, also demonstrating how generative learning strategies could enhance immersive VR environments.

A study that directly cross-linked VR with biology lessons was by [20] titled - "Application of virtual reality technology in biology education". The study utilized three-dimensional virtual reality technology (VRT) learning programs to evaluate the programs' efficiency in facilitating education for middle school students. A 3D Webmaster software - a Windows based application that facilitates creation of interactive 3D web pages, was used for authoring virtual reality biology simulations (VRBS) programs. The topic 'structure and function of the eye' was covered by the simulation. Results of the study revealed VBRS to be effective in teaching the students the structure and function of eye [25]. VRBS paved way for greater understanding of the by utilizing visualization in reality and immersion to explain the abstract science concepts and events. Thus, in concluding remarks the authors proposed the VRT to be highly promising in influencing the student's biology learning. Owing to the provision for active participation and immersion in learning activities, VRBS poses as the most preferred when compared to the other multimedia programs.

In [17] conducted a preliminary investigation in order to ascertain the efficiency of Virtual Laboratory for Biology (VLab-Bio) and its scope in the near future. To fulfill the study objectives 72 students comprising of 38 girls and 34 boys of age varying 16–17 years were recruited from two secondary schools. The respondents also comprised of 10 teachers who taught biology in the selected schools. Based on the data collected from the interviews conducted with the respondents it was revealed that despite the need for the much-required modernization in the educational sector, teachings were still limited to use of traditional power point-based practice. Building of a virtual laboratory depending on the complexity of the topic would indeed facilitate the students in their biology courses. However, setting up such virtual laboratories could be both time consuming and economically intensive. In order to overcome the challenges measures are required to be taken to enhance the adaptability for virtual laboratories when used as an educational tool.

In addition in [4] which titled "A practical guide to developing virtual and augmented reality exercises for teaching structural biology". They are attempt to explore the implications of virtual and augmented reality in teaching of structural biology, primarily raised three questions which were based on the preparation and dissemination of the VR/AR based study materials and development of the user-friendly tools and the perception of the users towards such VR/AR based tools. In response to the questions, the authors concluded VR and AR to be instrumental in extending the visualization the student's toolbox to interaction with content related to structural biology. Such interaction in turn leads to immersion of the students in the subject further enhancing their understanding of the topic. However, some critical topic in biology could limit the utilization of VR and AR. In this regard, the authors opined that despite the challenging nature, this emerging field of utilizing VR and AR in biology lessons holds promise, but its implementation calls for greater number of studies that would provide a clear vision of the study arena.

Another study in [11] implications of advanced 3D imaging, modelling, and printing techniques in the biological sciences sector. In zoological sciences a significant part of biology, there exists, a clear link between increasing rarity and access to specimens is considered. With people being least aware of such rarity, introduction of fully digital specimens could pose as an ideal solution to the existing concern.

Digital specimens could be made utilizing 3D geometry, visual textures, mechanical properties and specimen functionality. Such specimens when combined with VR, AR, or mixed reality (MR) could easily satisfy the educational and research needs without causing any harm to the ecosystem sustainability. The results of the study also have been indicative of the significant potential for advanced 3D imaging, modelling, and printing techniques in zoological sciences. Some of the most notable benefits being low cost associated with the production of the exact replicas of the original specimens. Digital specimens also broaden the research perspective which is now not being limited by availability of the specimens.

4 Virtual Reality in Medical Education

In Medical education, VR holds significant importance much of which stems from the great opportunities it provides in the medical field. VR by creating real-life scenarios allows the physicians, nurses and students to better understand the circumstances because of which they become capable and confident of responding to any such situation with enhanced quality of medical skills. VR thus paves the way for implementation of the concept of learn-by-doing. Although, implications of VR in medical science have been a naïve one, its positive implications in medical education have been confirmed by many clinical researchers and medical practitioners [5].

In [20] have reported another, interesting application of VR in dentistry. The study utilized "Simodont - a 3D VR simulation system" to find its application in training the students for dental crown preparation. Added advantage of the simulator to differentiate between dental students and prosthodontics residents based on their time-and skill makes it suitable for implementation as a teaching tool. Such modern-day technology thus facilitates the students to experience realistic clinical situations and creates provision for greater practice to become an expertise in that sector.

A recent study by [1] conducted "a comparative study between a virtual reality heart anatomy system and traditional medical teaching modalities". A real-time 3D representation of heart structure was projected in the study utilizing a VR system with provision of an interactive environment. Such real-life representation facilitated free manipulation, and model rectification to clearly fathom the anatomical relations of different parts of the heart. To make the experience more realistic, different shades of flesh colors were used in the model. Further, hands on experience at correct positioning of the heart was also provided by such 3D technology-based VR models.

Application of VR in the education field is varied and is not just limited to biology or medicine. The study titled - "Virtual reality geographical interactive scene semantics research for immersive geography learning" by [14] commented on the scope of VR in learning outcomes of geography lessons. Herein the authors utilized a virtual reality based immersive glasses technology to ascertain whether it could provide an immersive environment of geographic structure. Results of the study indicated a positive impact of the software on the learning outcomes also indicated its immense potential to be used for learning purposes in the near future.

In [16], which titled - "Adding immersive virtual reality to a science lab simulation causes more presence but less learning" however contradicted the positive impact of VR on learning outcomes. Indeed, Virtual reality (VR) has been a path breaking revolution that ushered in a paradigm shift in education and training, but empirical evidences that support its educational value are rare. In this regard the authors, attempted to fill the gap by determining the consequences when virtual learning simulations incorporates immersive VR. Thereafter, investigations were made to check whether the principles of multimedia learning could be generalized to immersive VR. Direct measurement of cognitive processing was executed using electroencephalogram (EEG). To fulfill the study objectives, 52 university students were recruited and made to participate in a 2×2 experimental cross-panel design. The 2×2 experimental cross-panel design comprised provision for desktop display (PC) or a head-mounted display (VR) wherein students learned science simulation. Results of the study revealed students to be more present in the VR condition ($d = 1.30$); but exhibited less learning ($d = 0.80$), and increased cognitive load based on the EEG measure ($d = 0.59$). Results thus have been indicative of no positive impact of VR on the learning outcomes of the students of science background. The authors opined that when VR is utilized for learning science it might distract the learner and create an overload on the student thus interfering with the learning outcomes resulting in poor learning outcome test performance.

However, a recent survey-based study by [10], titled as - "Virtual Reality and Its Applications in Education: Survey" overruled the belief that VR has no impact on learning outcomes. It is believed that VR would be instrumental in providing a solution to much of the problems that students face when they intent to fathom the meaning of a complex topic. It makes teaching process more attractive to the students thus capable of attracting the maximum attention of the students towards the study lessons. Realising the implications of VR in making education much more student friendly, a large number of educational centers around the world now have taken up initiatives to introduce modern technology-based tools that would ensure fulfillment of the diverse student needs. Over viewing VR in education, the authors opined that with the dynamic changes in the education system over the centuries, it is imperative for the scholars, educators and teachers to realize and embrace the change. It is only when the teachers make themselves adaptable to the change, they can make the students adapt to the same. The 21st century being the era of digitization, the education for the new generation even at the initiatory stages has been online. Digital world thus poses to be equally significant and immersive, if not less when compared to the real one. VR thus holds immense potential maximizing the efficiency and engagement of students of Generation Z is considered.

5 Virtual Reality Challenges

However, VR does have some challenges that could intervene with its efficiency. Traditional classes provide the students with opportunities to clarify their doubts through asking questions and conducting discussion schedules. However, a virtual reality headset being based on specific software, there remains no opportunity for such instant interactions as such application-based learning involves the students to follow a set of rules that have been pre-set. Discrepancies among the students may also stem owing to the fact that some students find themselves easy to adapt to the change but other may require help to familiarize with the technology to make its introduction a successful one. Teacher–student interaction holds a critical position in education, which is only achievable in the traditional mode of teaching. The human teacher serves as a natural filter and moderator validating the relevance of the obtained data. Lastly, complete reliance on digital education solutions could interfere with the necessary balance between teaching hard and soft skills [10]. Nevertheless, implications of VR could not be neglected in the modern-day education and greater number of studies is essential to explore the scope of its application in the near future so that learning outcomes can be enhanced to its highest potential.

Virtual Education (VE) has been the need of the hour. As well, greater understanding of the same is required apart from just visualizing it from a technological perspective. The application of both Virtual and Mixed Reality (VMR) technology in the fields of teaching and learning will certainly be part of the future of education across all disciplines, and the formulation of evidence-based guidelines for the creation of VMR educational material is urgently needed.

6 Conclusion

This study investigates the effects of Virtual Reality on education process and how to be implemented in education fields such as (STEM). Teachers face many challenges to introduce some unfamiliar subjects in medicine and biology education; in addition, students have many issues related to understanding these subjects. This study, answers the questions of what are the systems that used and its effects on the students. Our findings that apply of Virtual Reality in education is promising results and there is the opportunity of the employment of Virtual Reality application to develop the educational processes in biology and medicine. Likewise, this technology required a special skills, tools and software that able to create the proper VR systems that are compatible with the educational curricula. Future wok will focus on building and creating models to implement the VR technology in education. These models should be derived from the modern educational theories, especially the theory of the 21st century, which depends on employing technology in the educational process and introduce many skills for students such as, Information and Communication Technology (ICT) literacy, media and internet literacy, data interpretation and analysis, computer programming.

References

1. Alfalah SF, Falah JF, Alfalah T, Elfalah M, Muhaidat N, Falah O (2019) A comparative study between a virtual reality heart anatomy system and traditional medical teaching modalities. Virtual Real 23:229–234
2. Allocat D, von Muhlenen A (2018) Learning in virtual reality: effects on performance, emotion and engagement. Res Learn Technol 26:1–13
3. Brinson JR (2015) Learning outcome achievement in non-traditional (virtual and remote) versus traditional (hands-on) laboratories: a review of the empirical research. Comput Educ 87:218–237. https://doi.org/10.1016/j.compedu.2015.07.003
4. Garcia-Bonete M-J, Jensen M, Katona G (2018) A practical guide to developing virtual and augmented reality exercises for teaching structural biology. Biochem Mol Biol Educ 47(1):16–24. https://doi.org/10.1002/bmb.21188
5. Gorski F, Bu´n P, Wichniarek R, Zawadzki P, Hamrol A (2017) Effective design of educational virtual realityapplications for medicine using knowledge-engineering techniques. Eurasia J Math Sci Technol Educ 13:395–416
6. Hadiprayitno G, Muhlis G, Kusmiyati G (2019) Problems in learning biology for senior high schools in Lombok Island. J Phys Conf Ser 1241:012054. https://doi.org/10.1088/1742-6596/1241/1/012054
7. Huang H-M, Liaw S-S, Lai C-M (2013) Exploring learner acceptance of the use of virtual reality in medical education: a case study of desktop and projection-based display systems. Interact Learn Environ 24(1):3–19. https://doi.org/10.1080/10494820.2013.817436
8. Hu-Au E, Lee J (2018) Virtual reality in education: a tool for learning in the experience age. Int J Innov Educ 4(4):215–226
9. Jain LC, Peng S-L, Alhadidi B, Pal S (Eds) (2020) Intelligent computing paradigm and cutting-edge technologies. Learning and Analytics in Intelligent Systems. https://doi.org/10.1007/978-3-030-38501-9
10. Kamińska D, Sapiński T, Wiak S, Tikk T, Haamer R, Avots E, Anbarjafari G (2019) Virtual reality and its applications in education: survey. Information 10(10):318. https://doi.org/10.3390/info10100318
11. Keaveney S, Keogh C, Gutierrez-Heredia L, Reynaud EG (2016) Applications for advanced 3D imaging, modelling, and printing techniques for the biological sciences. In: 2016 22nd International Conference on Virtual System & Multimedia (VSMM). https://doi.org/10.1109/vsmm.2016.7863157
12. Lee EA-L, Wong KW (2014) Learning with desktop virtual reality: low spatial ability learners are more positively affected. Comput Educ 79:49–58. https://doi.org/10.1016/j.compedu.2014.07.010
13. Lindgren R, Tscholl M, Wang S, Johnson E (2016) Enhancing learning and engagement through embodied interaction within a mixed reality simulation. Comput Educ 95:174–187. https://doi.org/10.1016/j.compedu.2016.01.001
14. Lv Z, Li X, Li W (2017) Virtual reality geographical interactive scene semantics research for immersive geography learning. Neurocomputing 254:71–78. https://doi.org/10.1016/j.neucom.2016.07.078
15. Makransky G, Mayer RE, Veitch N, Hood M, Christensen KB, Gadegaard H (2019) Equivalence of using a desktop virtual reality science simulation at home and in class. Plos One 14(4):e0214944. https://doi.org/10.1371/journal.pone.0214944

16. Makransky G, Terkildsen TS, Mayer RE (2017) Adding immersive virtual reality to a science lab simulation causes more presence but less learning. Learning and Instruction. 10.1016/j.learninstruc.2017.12.007
17. Muhamad M, Zaman HB, Ahmad A (2010) Virtual laboratory for learning biology – a preliminary investigation. Int J Soc Behav Educ Econ Bus Ind Eng 4(11):2179–2182
18. Parong J, Mayer R (2018) Learning science in immersive virtual reality. J Educ Psychol 110(6):785. http://dx.doi.org/10.1037/edu0000241
19. Radianti J, Majchrzak TA, Fromm J, Wohlgenannt I (2020) A systematic review of immersive virtual reality applications for higher education: design elements, lessons learned, and research agenda. Comput Educ 147:103778. https://doi.org/10.1016/j.compedu.2019.103778
20. Shim K-C, Park J-S, Kim H-S, Kim J-H, Park Y-C, Ryu H-I (2003) Application of virtual reality technology in biology education. J Biol Educ 37(2):71–74. https://doi.org/10.1080/00219266.2003.9655854
21. Wang F, Liu Y, Tian M, Zhang Y, Zhang S, Chen J (2016) Application of a 3D haptic virtual reality simulation system for dental crown preparation training. In: Proceedings of the 2016 8th International Conference on Information Technology in Medicine and Education (ITME). Fuzhou, China, 23–25 December 2016, pp 424–427
22. https://www.statista.com/statistics/885952/effectiveness-of-learning-technologies-worldwide-by-type/
23. Alqahtani AS, Daghestani LF, Ibrahim LF (2017) Environments and system types of virtual reality technology in stem: a survey. International Journal of Advanced Computer Science and Applications (IJACSA), vol 8, no 6
24. Bardi J (2019) What is Virtual Reality? [Definition and Examples]. https://www.marxentlabs.com/what-is-virtual-reality/
25. Shorman SM, Al-Shoqran M (2019) Analytical study to review of arabic language learning using internet websites. International Journal of Computer Science & Information Technology (IJCSIT), vol 11
26. Le Thi P, Do Thuy L (2019) Applying virtual reality technology to biology education: the experience of vietnam. In: International Conference on Information, Communication and Computing Technology. Springer, Cham, pp 455–462

The Relationship of Electronic Human Resource Management Systems on Artificial Intelligence and Digital Economy in the Organization

Jasem Taleb AL-Tarawneh and Aya Nasar Maqableh

Abstract The study aimed to investigate the relationship of Electronic human resource management systems on artificial intelligence and digital economy in Organizations. An inductive research method was followed in the completion of this research paper, by reviewing previous literature on the topic, in addition to reviewing recent documents, articles, and seminars in this field. The main conclusion of this study is that even although the application of AI in E-HRM is relatively new, it is an increasing area HR field. To do that, the implications for the digital economy in organizations are important to study. Moreover, the study represents the There are major trends that are emerging due to the quick changes in HR technology, containing privatization, as there is a real opportunity to move far from the way HR programs were offered in the past. It is considered to understand the importance of digital transformation the digital technologies play a gradually strong role in both the employed lives of workers and electronic human resource management that is to be affected in many ways.

Keywords Electronic human resource management systems · Artificial intelligence · Digital economy · Organization

1 Introduction

The human resource management is the effective engine for the development of economic and commercial institutions, as it is the sum of the activities that seek to bring in, employ, develop, and maintain the human element in these institutions, and it pertains to employment movements, training, planning and evaluation. As well, human resources management recorded a quantum leap with the beginning of the current century and moved from managing personnel affairs to developing human capital, as human resources became an unrivalled wealth that achieves dynamism for

J. T. AL-Tarawneh (✉) · A. N. Maqableh
Faculty of Business, Universiti Malaysia Terengganu, 21030 Kuala Nerus, Terengganu, Malaysia

A. M. A. Musleh Al-Sartawi (ed.), *The Big Data-Driven Digital Economy: Artificial and Computational Intelligence*, Studies in Computational Intelligence 974,
https://doi.org/10.1007/978-3-030-73057-4_34

the institution and elevates it in the future as it is one of the competitive advantages that need to be valued and invested.

On this basis, the technological worker has a large role in these changes, which has resulted in wide shifts in the responsibilities of the human resources manager, who has become an expert in administrative matters, familiar with labor laws, can psychoanalysis, is a planner and has a strategic vision, motivates employees as well as marketing to the image of the institution abroad.

Further, several major changes and developments have emerged in the regulatory environment as a result of the development of new production, information and communication technologies; Therefore, it is necessary to provide sufficient, current and appropriate information at any time, to make decisions effectively, and this can only be achieved using computers and electronic means provided by information technology according to logic theory and developments in artificial intelligence [11].

Add to that, Artificial intelligence is representing an actual development in business management and will have a deep impact on the way employees work, particularly in the human resources and employment departments. Artificial intelligence (AI) technologies have an effect on the management of human resources in a different way. For example, plan training and development for all employees from background operations, based on major data or data analytics concerning employment practices in real-time. Artificial intelligence points out the technology applied to do a task that seeks some level of intelligence to accomplish [15].

The importance of artificial intelligence is due to its ability to document and preserve human experiences by transferring them to smart machines, and to use natural human language in dealing with machines, which allows the ability to deal with them by all groups of society, in addition to carrying out dangerous and arduous work that is complex and requires high concentration and a burning mind All the time and a high capacity for endurance, as the artificial intelligence has independence, accuracy and objectivity without bias or emotion, which contributes to deciding with a great degree of accuracy and health, and artificial intelligence helps in many fields such as economics, administration, science, health, interactive education, and legal advice, etc. [12].

As well, many researchers emphasize that artificial intelligence and cognitive technology have the promising opportunities that employees and institutions can achieve as a result of employing this technology, as it can play a vital role in various institutional processes, as some reports confirm that 25% of future employment methods that will be adopted On artificial intelligence technology, and 30% of the major technology-oriented companies spend nearly five million dollars on artificial intelligence technology and knowledge technology, and accordingly, employing these technologies in the organization will contribute to many benefits, especially in the decision-making process and thus improving customer service [9].

There are major trends that are emerging due to the rapid changes in HR technology, including privatization, as there is a real opportunity to move away from the way HR programs were presented in the past. As it was one large size that fits all, now using artificial intelligence technology and the digital economy can create specialized environments for each employee so that they have a personal experience that encourages them to use the facility platforms and provides them with data as they do in their consumer life, which means that the system identifies them and helps them improve their experience the work. Among them is also lifelong learning, as with the revolution in development in automation and the use of technology, full and continuous support for lifelong learning will be necessary for the success of human resources in their work in the future. It's not about supporting employees' human resources when it comes to outsourcing new tasks or hiring them in new positions, but it is also about ensuring that employees are constantly thinking about how technological changes and changes in the business world will affect their jobs, and the skills they must acquire to continue their success in the future. The establishments must build an infrastructure that supports employees in their efforts to learn new skills to develop their capabilities in line with changes in the field of work [4, 14].

On other hand, it is significant to understand the importance of digital transformation the digital technologies play a progressively powerful role in both the employed lives of workers and human resource management (**HRM**) that is to be affected in numerous ways. Further, digital transformation strategies can best be comprehended in a business-centric viewpoint these strategies have as their emphasis on the transformation of products, processes, and all organizational features as an outcome of new technology. Digital transformation strategies transport about changes to and have insinuations for business models as a whole. Likewise, human resource management is also business-centric and takes about change by adding value to organizations as a strategic partner, administrative expert, and employee champion. Digitalization is achieving organizations and human capital there has been slight research about how digitalization is knowledgeable by HR managers in practice and how it can affect the intellectual capital of the organization through developing the human capital strategies [3, 10]. Thus, this study aimed to investigate the relationship of Electronic human resource management systems on artificial intelligence and digital economy in the organization.

Problem Statement and Study Questions

The business world is witnessing rapid changes, with which human resources departments find themselves before a new scene. The World Economic Forum's Future Jobs Report 2018 predicted that 75 million jobs will disappear by 2022, and 133 million new jobs will be created, thanks to robotics and artificial intelligence [7].

On other hand, according to the rising developments in the use of modern technology and computer applications, the usage of electronic human resources management became an obligation and a requirement in human resources management in all sectors to attain positive results in the direction of improving the performance of workers in organizations and organizations and raise their efficiency. Add to that, EHRM as a new managerial approach essentially requires high human abilities, also need a change in management approaches and organizational assemblies and the growth of electronic infrastructure, in order to allow organizations from side to side which raising the level of their facilities and efficiency of their employees.

Moreover, the mission of artificial intelligence depends on achieving cooperation between man and machine to harness the potential of technology to enhance human capabilities, and human resources departments may be the source of this mixing between machine capabilities and human capabilities. Based on the above, this research paper came to answer the following main question:

The Main Question: What is the relationship of Electronic human resource management systems on artificial intelligence and digital economy in the organization?

From this main question the following sub-questions emerged:

- What is the relationship of Electronic human resource management systems on artificial intelligence in the organization?
- What is the relationship of Electronic human resource management systems on the digital economy in the organization?

Importance of the Study

The importance of this study comes from the fact that it is based on extrapolation "The relationship of Electronic human resource management systems on artificial intelligence and digital economy". Therefore, the importance of the study divided into the following:

Theoretical Significance

- The importance of this research considered the importance of the relationship of Electronic human resource management systems on artificial intelligence and the digital economy. Besides, there are few studies.
- This research will affect the range of studies in this field that can be conducted, which leads to some limitations and complications to the contents. However, this will be considered as a challenge from one side.
- This research contributes definitely by establishing an actual value in the academic research in Electronic human resource management systems, artificial intelligence and digital economy in Organizations field, and encourage other researchers to build up and improve studies, which support this field to achieve the desired goals in the future.

Practical Significance

- The result of the current research expects to provide convincing evidence about the relationship of Electronic human resource management systems on artificial intelligence and digital economy in Organizations.
- The results of the study will provide important decision-makers who are interested in informational and analytical studies with important information about Electronic human resource management systems, artificial intelligence and digital economy in Organizations.
- The results of this study will be used in some subsequent studies and research that can address the same topic in different dimensions, and that the results of this study contribute to coming up with recommendations showing the Electronic human resource management systems, artificial intelligence and digital economy in Organizations.

Objectives of the Study The objectives of this study to investigate the relationship of Electronic human resource management systems on artificial intelligence and digital economy in Organizations. From these main objectives the following sub-questions emerged:

- Investigate the relationship of Electronic human resource management systems on artificial intelligence in the organization?
- Investigate the relationship of Electronic human resource management systems on the digital economy in the organization?

Literature Review The recent and rapid developments in the digital stage have prompted companies in the private and public sectors alike to fundamentally change the way they work [22]. as these companies themselves are required to innovate new ways of thinking about providing their services, which forces them to change the way their business models are designed. In turn, it directly affects the functions of the human resources departments in these companies and their role in defining new methodologies for managing these resources.

No doubt planning the future capabilities of the workforce in the digital economy involves complex challenges and tasks for human resources departments, just as the ability to integrate the right employees into a dynamic corporate environment and help existing employees and leaders to acquire new digital competencies so that they can drive the transformation process. Therefore, human resources management must do the following to overcome the revolutionary transformation in the coming years [1, 2, 20]:

- **Ensure that the new developments are noticed**: Since the day when the employees of the customer service departments are replaced by robots is soon, the director of the human resources department should establish certainty among his colleagues that change is coming, and enable them to form a clear vision of their future. It should also sense the difference that will occur in managing human resources, in addition to developing an implementable plan that serves as a transformation strategy
- **Ensuring the participation of every one**: The director of the Human Resources Department's writing of what he aspires to allows him to achieve his goals, and the delivery of the digital transformation message to all his colleagues contributes to spreading harmony and harmony among them and makes them try to achieve the same goal. Therefore, the human resource manager should instil a culture in which everyone feels enthusiastic, positive, harmonious, and strives to achieve the same goals. It should also involve everyone in the organization.
- **Ensuring that employees are taught digital skills**: One of the responsibilities of the HR department manager is to educate himself, update and renew his colleagues' knowledge and skills, and recruit new experts in the digital field.

As well, the modern period considered the period of transformation from an asset economy to an information economy or a knowledge-based economy and one of its most prominent and important features, but its most important requirements are the growing employment of computing and communication means in various walks of life and the increasing dependence on information technology in performance, service and production, which led and leads to the expansion of programs Education and training to raise the level of productivity of working individuals, develop the communication network that contributes to diversifying materials and means, and thus raising the level of material productivity [6].

On other hand, the concept of artificial intelligence science has gained great importance in recent years due to its many applications in many basic and vital fields, such as computing, human resource systems, machine translation, and others. Teamwork through the participation of computer scientists, mathematics, psychology, linguistics, philosophy and logic. Artificial intelligence science aims to understand the nature of human intelligence by creating computer programs capable of simulating intelligent human behaviour and expressing the ability of a computer program to solve a problem or make a decision in a situation. Artificial intelligence science is also concerned with the cognitive processes resorted to. People can perform various tasks [8].

Accordingly, AI is an order that appraisals PCs re-enact certain points of opinion and wise practices, (for example, learning process, thinking, thinking, arranging, and so on.). The Massachusetts Institute of Technology (MIT) teacher called attention to that Artificial Intelligence mainly distillates on how to apply PCs to light Artificial intelligence work. The impression of Artificial Intelligence has first developed abroad in 1956 and started to be considered. McCarthy initially cited the term Artificial intelligence to make a hypothetical point of reference for the progression of Artificial Intelligence. During the 1990s, Kasparov fought with the "dark blue" PC and "dark blue" gained, which is an important accomplishment in the progression of AI. In 2006, Varian Hinton planned the idea of deep learning and Artificial Intelligence has arrived at a phase of the fast turn of actions. Moreover, AI hiding place many orders, for example, arithmetic, the board, PC, phonetics, and so on, and has a solid breadth. By the incessant progression of innovation, artificial intelligence has made onward bounds in picture acknowledgement, dissertation salutation and different fields. Although AI is still in the incessant advancement stage, many organizations have unspoken the development potentials and treatment estimation of AI [1].

Add to that, AI innovation can similarly assistance organizations of all sizes to achieve their development purposes. AI has a place with software engineering, yet it isn't only software engineering. It frequently includes human science, brain science, arithmetic, and so forth, and level proficient hypothetical information and capabilities in clear application fields, just as human experience gathering in related fields [21].

On other hand, starting here view, AI has wide extensiveness and multi-layered nature in the theoretical information level and won't have a place with a specific division of knowledge. Human asset the board mainly makeup and contracts with agents' practices over spacing out, association, control and unlike approaches and techniques, with a conclusive objective of helping out endeavours to attain most extreme advantages. The centre physical of conventional human asset the managers depends on the improvement of big business advantages and the management of workforce in the undertaking while supervising the improvement requirements of abilities as people and the joining between the singular turn of events and undertaking advancement. In the period of the internet, the human asset the board of present-day endeavours should give more consideration to the advancement needs of gifts, from concentrating on the executives to concentrating on individuals [5].

Human asset the executives is of extraordinary vital criticalness to the improvement of ventures. It is an important piece of big business active asset the managers and is additionally an important reason for activities to keep up the sound turn of events. So, logical techniques must be comprised to improve the ability and level of the board and development of the solid and economic upgrading of ventures. In the time of the information economy, HR is the source of information advancement and improvement. To recover the accentuation on illustrative preparing, activities need to put more cost in human asset the managers, to improve the nature of workers, and concurrently appreciate the maintainability of typical and venture development [16].

As well as, to discuss the relationship between Electronic HRMS and artificial intelligence; it is unlike HR technology, intelligent "employee systems" can help manage the career, teams of employees, and the future of any organization. Human resource management technology enhanced with artificial intelligence can provide organizations with ways to make the work system less complex and smarter, and these methods can be summarized as follows [19, 21]:

- **Knowledge**: When the database is constantly updated, managers have a complete picture of the skills and experience they have. Then they can find the right person to perform a specific task in a few seconds. Using predictive analytics tools, managers can create a chart that outlines the skills and people they will need for the next year and the year after. This is an easy and fast way to organize your workforce.
- **Support:** Whether the employee wants to know their remaining leave balance or other routine inquiries, catboats answer traditional employee questions instantly. This frees up the HR analyst to tackle more complex issues. The Human Resources Department can also closely follow the morale of the employees by performing a regular, almost instantaneous analysis for them.
- **Development:** Employees are guided directly by an intelligent system that knows the skills they need to excel in their current and future roles. They also receive proactive alerts that provide them with recommendations that help them keep up with the necessary training and know the extent of their compliance with the rules and regulations. Managers also receive customized reports and recommendations based on employees' tasks and actions.
- **Recruitment:** Instead of manually searching for candidates, the institution's future system can find individuals with the required skills and communicate with them automatically, and through artificial intelligence techniques, this system can answer any question that candidates may ask before the personal interview.
- **Focus on value**: When AI tools take over the repetitive manual tasks, employees can focus on the tasks that add value to the company, and those that need their skills and experience to accomplish. This will enable HR professionals to devote more time and resources to taking care of other employees personally.

On other hand, among the main areas in which human resource technology is affected by artificial intelligence, techniques include [13, 17]:

- **Recruitment:** Many establishments use some form of artificial intelligence in their basic system for the recruitment process, whether in the selection, evaluation and recruitment process or by using catboats or others.
- **Employee experience**: For example, using catboats, employees can get instant answers to their questions, get information about tasks they have to complete at work, information about the company's pension policy, book a vacation or a day off, or even book trips. The good about technology like this is that it can be integrated with work messaging apps, where employees can get answers in the same way they text their co-workers.
- **Learning and development**: Here, you can begin to track some of the interactions that employees make in modern learning platforms, to be able to understand the types of information they are trying to obtain and how to obtain it. This enables the company to deduce the skills that an employee has and the skills that he aspires to grow, and in return, a personal experience can be provided to the employee, as this technology changes the way of learning in the workplace. As well as using other technologies such as virtual reality techniques, it is possible to add more excitement to the learning process, as some facilities use virtual reality to train workers by simulating activities while working in dangerous conditions, for example.
- **Employee Analytics**: Artificial intelligence mechanisms work to understand all text-based data in facilities, either through opinion polls, or through communication messages on the intranet, or wherever written discussions take place within the facility, then focus is placed on how to understand the features In that data to demonstrate new insights and take action to improve employee engagement.

Here, some previous work in this field will be reviewed, [1] study aimed to draw an understanding of the phenomenon of using artificial intelligence (AI) in human resources, especially in the Kingdom of Bahrain. The study revealed that in the Kingdom of Bahrain with the application of its vision (2030 vision), the public sector will have a large opportunity to hold up with the digital transformation. This has driven a change in the installation of the workforce within business organizations. Besides, this allows for the major integration of the feminist element. It is urged that the implementation of modern artificial intelligence (AI) is a fundamental method for organizations that work in an inconsistent environment.

Moreover, [17] study aimed to explain the impact of emerging technologies on work and what is the role of HR concerning these changes. The result of the evidence review proposed that emerging technologies just like AI, robotics, VR, and AR, digital technologies, wearables, and blockchain have the potential to affect work and employees significantly. The point and speed of this impact rely to a large extent on improvements in the technologies themselves and the willingness of organizations to adopt them. Add to that, the role of the HR function may become even more important as both the effort advantages and risks of emerging technologies for employees improve.

Furthermore, [18] study represent that digital economy and continuous introduction of innovations, as well as decrease of enduring costs of organizations to management by the mechanisation of numerous business, processes the automation of HR management develops particularly applicable as the organization personnel is its key capital. The up to date level of development of information and communication technologies, availability of the internet and various technical means allow to automate the collection, storage and processing of personal data on potential or current employees, deliver business estimation, provide 24-h admission to various educational resources. Therefore, the range of suggested software and services is wide. The study also analysed the key functionality of combined software keys for HR management as well as potentials of services mechanising individual workers' management procedures for an example video interview, estimation, electronic training.

On other hand, [11] study aimed to discover the exchanging role of human resource management in an era of digital transformation. They represent that whilst digitalization influences day-to-day HR practices and procedures mainly with the use of human resources information systems there is less stress on the role of HR in causal to the strategy of digitalization.

2 Methodology

An inductive research method was followed in the completion of this research paper, by reviewing previous literature on the topic, in addition to reviewing recent documents, articles, and seminars in this field.

The inductive approach is a delicate process aimed at collecting data and observing the phenomena associated with it to link them to a group of general macro-relations, and also the inductive approach is a research method that the researcher uses in generalizing his study to the general studies related to the topic he is researching, i.e. linking the study that He worked to implement them as part of a whole, and the inductive approach relies on the use of a set of conclusions based on observations, estimates, and experiences related to the previous literature review.

The researcher is interested in reading all the information about the research part before issuing a general decision about it, but it is considered one of the types of induction that is slow to apply, or far from the process. Because the researcher needs complete care while observing the components of the inductive approach, which takes a long time to know all the components, but it contributes to providing accurate and correct results.

3 Conclusion

The digital transformation of human resources is an important topic for many human resources specialists because of the ability and potential of technology to change the image of human resources operations that all companies and institutions know in their traditional form. Today; Human resource digital transformation is simply the process of changing human resource operational processes and functions into a technology-based and data management mechanism. On the one hand, the digital transformation process relates to the transformation of human resources processes and on the other hand the transformation of work teams and the way of working. The transformation process includes the organization or the company as a whole and not just a process of transformation in human resource management only. The digital transformation of human resources is the transformation of the processes and functions of human resource management, using data to direct all areas of human resources such as payroll, benefits, performance management, learning and development, rewards and recognition, and recruitment.

The starting point of the paper was to set the current body of knowledge concerning the relationship of Electronic human resource management systems on artificial intelligence and the digital economy in the organization. Indeed, one can dispute that without a grasp of electronic human resources and their likely consequences for organizations it will not be possible to improve a new design of the electronic human resource. Thus, this paper reviewed extant research on the relationship of Electronic human resource management systems on artificial intelligence and the digital economy in the organization.

The main conclusion of this study is that even although the application of AI in E-HRM is relatively new, it is an increasing area HR field. In order to do that, the implications for the digital economy in organizations are important to study.

There are major trends that are emerging due to the quick changes in HR technology, containing privatization, as there is a real opportunity to move far from the way HR programs were offered in the past. It is considered to understand the importance of digital transformation the digital technologies play a gradually strong role in both the employed lives of workers and electronic human resource management that is to be affected in many ways. Further, digital transformation strategies can best be understood in a business-centric viewpoint these strategies have as their assurance on the transformation of products, processes, and all organizational features as an outcome of new technology. Digital transformation strategies transport about changes to and have insinuations for business models as a whole. Also, electronic human resource management is business-centric and takes about change by adding value to organizations as a strategic partner, administrative expert, and employee champion. Digitalization is achieving organizations and human capital there has been slight research about how digitalization is knowledgeable by HR managers in practice.

The digital economy affects Electronic human resource management activities and changes the role of HR managers from static to dynamic and strategic. To achieve successfully, their new role HR managers want to act firstly as strategic positioners (requirement to inform the business context and the grand external factors), as change agents, and as technology supporter (need to inform how to arrive, resolve, assess and share information and how to adopt new technologies of information). Moreover, human resource managers themselves need to fain digital skills and raise their digital economy.

Additionally, Human resource management technology enhanced with artificial intelligence can provide organizations with ways to make the work system less complex and smarter, and these methods can be summarized as follows:

- Knowledge: When the database is constantly updated, managers have a complete picture of the skills and experience they have.
- Support: This frees up the HR analyst to tackle more complex issues. The Human Resources Department can also closely follow the morale of the employees by performing a regular, almost instantaneous analysis for them.
- Development: Employees are guided directly by an intelligent system that knows the skills they need to excel in their current and future roles.
- Recruitment: through artificial intelligence techniques, this system can answer any question that candidates may ask before the personal interview.
- Focus on value: When AI tools take over the repetitive manual tasks, employees can focus on the tasks that add value to the company, and those that need their skills and experience to accomplish.

References

1. Abdeldayem MM, Aldulaimi SH (2020). Trends and opportunities of artificial intelligence in human resource management: aspirations for public sector in Bahrain. Int J Sci Technol Res 9(01)
2. Abdurakhmanova G, Shayusupova N, Irmatova A, Rustamov D (2020) The role of the digital economy in the development of the human capital market. Архив научных исследований (25)
3. Al-Sartawi A (2020) Social media disclosure of intellectual capital and firm value. Int J Learn Intell Capital 17(4):312–323
4. Armstrong M, Taylor S (2020). Armstrong's handbook of human resource management practice
5. Barr A, Feigenbaum EA (Eds.) (2014). The handbook of artificial intelligence, vol. 2. Butterworth-Heinemann
6. Britton DB, McGonegal S (2007) The digital economy factbook. The progress & freedom foundation, Washington, DC
7. Carson C, Robison J, Reinsch W, Caporal J, Chatzky A (2019) Training the next revolution in American manufacturing. Center for Strategic & International Studies
8. Cath C, Wachter S, Mittelstadt B, Taddeo M, Floridi L (2018) Artificial Intelligence and the 'good society': the US, EU, and UK approach. Sci Eng Ethics 24(2):505–528
9. Davenport T, Guha A, Grewal D, Bressgott T (2019) How artificial intelligence will change the future of marketing. J Acad Mark Sci. https://doi.org/10.1007/s11747-019-00696-0
10. Fenech R, Baguant P, Ivanov D (2019) The changing role of human resource management in an era of digital transformation. J Manag Inf Dec Sci 22(2):1–10
11. Giles, KM (2019) How artificial intelligence and machine learning will change the future of financial auditing: an analysis of the university of tennessee's accounting graduate curriculum. Chancellor's Honors Program Projects
12. González García C, Núñez Valdéz ER, García Díaz V, Pelayo García-Bustelo BC, Cueva Lovelle JM (2019). A review of artificial intelligence in the internet of things. Int J Interact Multi Artif Intell

13. Johansson J, Herranen S (2019). The application of artificial intelligence (AI) in human resource management: Current state of AI and its impact on the traditional recruitment process
14. Kshetri N (2021). Evolving uses of artificial intelligence in human resource management in emerging economies in the global South: some preliminary evidence. Manag Res Rev
15. Mahamud T, Suttikan M (2020) Modern artificial intelligence in human resource management in an organization. RMUTT Glob Bus Econ Rev 15(1):75–89
16. Naseem M, Akhund R, Arshad H, Ibrahim MT (2020) Exploring the potential of artificial intelligence and machine learning to combat COVID-19 and existing opportunities for LMIC: a scoping review. J Primary Care Comm Health 11:2150132720963634
17. Parry E, Battista V (2019) The impact of emerging technologies on work: a review of the evidence and implications for the human resource function. Emerald Open Res 1(5):5
18. Pulyaeva V, Kharitonova E, Kharitonova N, Shchepinin V (2019). Practical aspects of HR management in digital economy. In: IOP Conference Series: Materials Science and Engineering, vol. 497, no. 1, p 012085). IOP Publishing
19. Rogiers P, Viaene S, Leysen J (2020) The digital future of internal staffing: a vision for transformational electronic human resource management. Intell Syst Acc Financ Manag 27(4):182–196
20. Vladimirovna VY, Yurievna CE, Dmitrievich LE (2019) Problems of reproduction of human resources towards the formation of the digital economy. J Appl Eng Sci 17(4):514–517
21. Yabanci O (2019) From human resource management to intelligent human resource management: a conceptual perspective. Hum Intell Syst Integr 1(2):101–109
22. Memdani L (2020) Demonetisation: a move towards cashless economy in India. Int J Electr Bank 2(3):205–211

Printed in the United States
by Baker & Taylor Publisher Services